The Urban Book Series

Igor Vojnovic, Department of Geography, Michigan State University, East Lansing, MI, USA

Claudia van der Laag, Oslo, Norway

Qunshan Zhao, School of Social and Political Sciences, University of Glasgow, Glasgow, UK

The Urban Book Series is a resource for urban studies and geography research worldwide. It provides a unique and innovative resource for the latest developments in the field, nurturing a comprehensive and encompassing publication venue for urban studies, urban geography, planning and regional development.

The series publishes peer-reviewed volumes related to urbanization, sustainability, urban environments, sustainable urbanism, governance, globalization, urban and sustainable development, spatial and area studies, urban management, transport systems, urban infrastructure, urban dynamics, green cities and urban landscapes. It also invites research which documents urbanization processes and urban dynamics on a national, regional and local level, welcoming case studies, as well as comparative and applied research.

The series will appeal to urbanists, geographers, planners, engineers, architects, policy makers, and to all of those interested in a wide-ranging overview of contemporary urban studies and innovations in the field. It accepts monographs, edited volumes and textbooks.

Indexed by Scopus.

Avi Friedman

Fundamentals of Innovative Sustainable Homes Design and Construction

Springer

Avi Friedman
School of Architecture
McGill University
Montreal, QC, Canada

ISSN 2365-757X ISSN 2365-7588 (electronic)
The Urban Book Series
ISBN 978-3-031-35367-3 ISBN 978-3-031-35368-0 (eBook)
https://doi.org/10.1007/978-3-031-35368-0

This Springer imprint is published by the registered company Springer Nature Switzerland AG
The registered company address is: Gewerbestrasse 11, 6330 Cham, Switzerland

Preface

Climate change, the adaptation and depletion of natural resources, housing affordability, and sociodemographic transformations are some of the challenges that communities are facing in the twenty-first century. In addition, the effects of the recent COVID-19 pandemic have exacerbated the above-mentioned trials, while simultaneously creating many of their own and highlighting gaps in our social systems that have long been causing problems.

Current modes of dwelling design need to respond to a world which faces climate emergency, depleting natural resources, an aging population, dense and rapidly growing urban environments, expanding virtual communication networks, and health-conscious inhabitants. *Sustainability* is a term that best captures these new sought-after directions. The thrust of sustainable thinking in its most rudimentary form is that one needs to consider the future consequences of present actions. We can no longer assume, for example, that our natural resources will last forever. The years ahead will require designers and builders of homes and communities to rethink old concepts and retool past practices. Two key areas are likely to guide future design: responding to social and lifestyle changes and considering environmental pressures.

Societal transformations affected the relationships between residents and their dwellings. Economic realities forced the need to consider smaller dwellings, and progressive building as a means of reducing initial cost and fostering better pre- and post-occupancy adaptability between residents and their space needs. In addition, a rapidly aging population in most countries has led to a quest for housing solutions that accommodate seniors. The need to reduce the home's physical and carbon footprint, densify communities, and lower their construction costs makes concepts such narrow and small area dwellings highly relevant.

Lifestyle changes renewed the age-old tradition of working at home. The need to minimize construction time heightens interest in prefabrication and Plug and Play homes—those that are constructed in a plant and then transported to a site for rapid assembly.

This book explores how these issues manifest themselves in a dwelling's design and construction—in the many forms they come in and provide the reader with innovative design solutions. It aims to inform the reader on the elements of building dwellings as sustainable systems. In the context of development, a sustainable system is one in which all the four pillars of sustainability: social, cultural, economic, and environmental function both dependent and independently of one another.

Chapter 1 introduces these pillars in further detail, along with the origin, evolution, and significance of sustainability. Chapter 2 validates the importance of sustainable high density residential communities and homes. It provides insight into their conception and ability to achieve carbon neutrality, specifics on high-density dwelling forms such as typology, footprint, foundations, and orientation among others are covered along with softer effects such as those on community form.

Chapter 3 addresses the widening gap between household income and house price becoming a significant barrier to homeownership and the subsequent need for affordable dwellings. One of the main solutions proposed is downsizing dwellings while maintaining if not improving the quality and standards of the physical build. Other cost reduction strategies on micro and macro scales are discussed with an emphasis on cost savings through urban planning, volumetric arrangement, lot sizing, housing shape, and material selection.

While the above strategies appear as tall orders for designers, builders, and homebuyers. Chapter 4 finds creative opportunities to address the comfortability of small interiors. The conception of such homes needs to be approached through a broad design process. As such the specifics of this chapter include zoning, access, circulation, and spatial configuration. Finally, flexible design, space-making strategies, and finishings will be explored.

Moving to the exterior of the home, Chapter 5 sees the customization of outer facades as a way for homeowners to find individuality among increasing urbanization rates while using energy efficient materials. Specifically investigated are homogeneity and diversity, flexible exterior design according to layout, and sustainable energy efficient design principles of facades. A menu of choices for façade elements such as materials, energy efficient windows, daylight, and ventilation is also proposed and discussed.

Chapter 6 approaches sustainable housing from the construction side as opposed to the aesthetic. The negative effects of conventional building practices on the environment coupled with an urgent need for affordable housing has encouraged a prioritization of resource efficiency and conservation. For this and many reasons to be discussed in this chapter, off-site fabrication is becoming an increasingly viable construction method. Readers will understand the fundamentals of the main types of prefabrication construction: mass customization, time saving interior construction methods, and design for growth and adaptability.

Due to conventional utility systems within homes and buildings being uncoordinated and difficult to access, Chapter 7 proposes a single passage for all utilities that uses an open web floor joist system. This passage will allow for greater ease in system upgrades and repairs, resulting in more efficient and therefor sustainable use of time and materials.

Chapter 8 hones in on the feasibility of increasing environmental sustainability in the construction sector. Various green and healthy building materials, those that are both in use and in trial are introduced and discussed to provide the reader with a comprehensive understanding of available sustainable building materials.

A selection of renewable energy sources, Net-Zero homes, and energy performance monitoring devices among other energy efficient systems are introduced in Chapter 9. These initiatives are inspired by the rising global energy demands in the residential sector that have accompanied increasing urbanization and technology uptake. Current energy habits and sources are unsustainable, hence the need for this coverage of energy efficient dwellings.

Home automation covered in Chapter 10 also contributes to increased energy usage within the home. The presence of smart household appliances is on the rise, a cause of the internet of things (IoT), and widespread uptake of smartphones and other internet connected devices. These new appliances create opportunity to improve the efficiency and quality of domestic life while contributing to the home's sustainable performance.

Narrowing in, kitchen development has been greatly influenced by current trends of sustainable action. Chapter 11 discusses the important functional and social roles kitchens now play in the home. Readers will learn about the key design concepts of past, present, and future kitchens such as layout, modularity, adaptability, building materials, growing food, recycling, and composting.

With talks in earlier chapters of the practicalities of downsizing homes, Chapter 12 necessarily covers sustainable storage and furnishing for dwellings of all sizes. It highlights the importance of multifunctional furnishing techniques with a single piece of furniture having multiple purposes. This in turn addresses the unsustainable accumulation of "stuff" many homeowners with excess space have developed.

Chapter 13 addresses the growth of elderly populations, the problems faced by the aging themselves, and sequential changes that need to be made to physical dwellings for accommodations. The ability to age safely, comfortably, and independently is prioritized in this section with adaptions of the kitchen, bathroom, bedroom, and outdoor environments discussed. This chapter goes further than physical design by proposing different social structures such as multigenerational living, aging in place, and retirement communities.

Lastly, Chapter 14 covers working from home (WFH) the uptake of which has recently been accelerated due to isolation during the COVID-19 pandemic and the current information age. It offers a sense of freedom and flexibility of time management to the worker all while eliminating the cost, stress, and loss of time associated with commuting. WFH of course doesn't come without its barriers which will also be explored.

As environmental issues become increasingly pressing, designing sustainable homes will require more innovative solutions. This book aims to provide the reader with a comprehensive understanding of how to build more sustainable dwellings, be it from the ground up, repurposing, or renovating. As will be mentioned countless times throughout this book: sustainable communities are inclusive communities that provide access to all ages, stages, and abilities.

Montreal, Canada Avi Friedman

Acknowledgments

Studying housing and developing prototypes was a subject of my work for many years. It included collaboration with and contribution by numerous colleagues, assistants, and students who directly and indirectly inspired my work. My apology if I have mistakenly omitted the name of someone who contributed to this book. I will do my best to correct such omissions in future editions.

This book could not have been written without contribution to the background research, compiling information, and the writing by a team of highly dedicated assistants. It included my outstanding former students Lucy Anderson, Synthia Lagurre—Hemelear, Rosslyn Sinclair, and Jiahui (Cindy) Duan. Their dedication, hard work, talent, and punctuality is most appreciated.

Copyediting was done by Simone Tharp. I truly appreciate Simone's work, dedication, insistence on clarity, and accuracy.

Special thanks are extended to Charles Grégoire, Elif Kurkcu, Elisa Costa, Jeff Jerome, Ehsan Daneshyar, Zhong Cai, Diana Nigmatullina, Josie White, Emma Greer, Juan Mesa, Jing Han (Jay), David Auerbach, Maria Teleman, Masa Noguchi, David Cameron, Danielle Kastner, Sorel Friedman, Isabella Rubial, Rainier Silva, JJ Zhao, Mahya Sabour, Thomas King, Kurin Qianhui Wang, Michelle Coté, Xiong Wu Fa, Lucy Zhang, Amelie Lessard, Nyd Garavito-Bruhn, Na Zhang, and Jing Yan Liu who drew the illustrations and Juan Osorio for slide scanning. Their talent and insistence on achieving excellence is truly appreciated and admired.

My appreciations go to my design team members, who are listed in the Projects' Teams page for their utmost dedication.

To Juliana Pitanguy, Publishing Editor Geography and Sustainability Research at Springer many thanks for the guidance. To Corine van der Giessen and Sanjievkumar Mathiyazhagan Production Administrator thank you for managing the production process.

Finally, my heartfelt thanks and appreciation to my wife Sorel Friedman, Ph.D., and children Paloma and Ben for their love and support.

Contents

1 Homes for Changing Times 1
1.1 Adjusting to New Realities 1
1.1.1 The Environmental Challenge 3
1.1.2 Sociodemographic Transformations 5
1.1.3 A Shifting Global Economy 7
1.1.4 Changing Lifestyle 7
1.1.5 Technological Innovation 8
1.1.6 Design Appreciation 11
1.2 Origins and Evolution of Sustainability 12
1.2.1 The Four Dimensions of Sustainable Practices 13
1.2.2 The Social Dimension 14
1.2.3 Cultural Dimensions 15
1.2.4 Economic Dimensions 15
1.2.5 Environmental Dimensions 16
1.3 The Guiding Principles of Sustainable Systems 17
1.3.1 The Path of Least Negative Impact 19
1.3.2 The Self-Sustaining Process 19
1.3.3 Supporting Relationships 19
1.3.4 Circular Approach 20
1.3.5 Circular Economy, Adaptive Reuse, and Design-for-Disassembly 21
1.4 Skaftkarr: Green Place for the People and the Planet 22
1.5 Sustainable Community in Komoka, Ontario, Canada 24
1.6 Final Thoughts 27
References 28

2 Denser Living 31
2.1 Urban Planning Aspects for High-Density Environments 31
2.1.1 The Form of Communities 32
2.1.2 Planning for Active Mobility 32
2.1.3 Green Open Spaces 36

2.2 Higher Density Dwellings' Forms 38
2.2.1 Typology and Footprint 39
2.2.2 Attachment 41
2.2.3 Foundations 43
2.3 Denser Dwellings' Design 43
2.3.1 Dimensions and Orientation 44
2.3.2 Access and Egress 45
2.3.3 The Roof 46
2.3.4 Mixed Use Residences 48
2.4 Weaving-In a Denser Building in Melbourne, Australia 50
2.5 A Sustainable Apartment Building 53
2.6 Final Thoughts 56
References 58

3 Quality Affordable Dwellings 61
3.1 The Housing Affordability Challenge 61
3.2 Urban Planning Strategies for Affordability 63
3.2.1 Lot Configuration 63
3.2.2 Streets and Parking 65
3.2.3 Open Outdoor Space 66
3.2.4 Infill Housing 72
3.3 Architectural Strategies for Cost Reduction 72
3.3.1 Floor Staking 73
3.3.2 Plan Simplification 73
3.3.3 Joining Units 74
3.3.4 Ground Relation 75
3.3.5 Chosen Materials 75
3.3.6 Modular Design, Dimensioning, and Efficient Framing Practices 76
3.4 An Affordable Community in Vijfhuisen, the Netherlands 76
3.5 The Grow Home 78
3.6 Final Thoughts 79
References 85

4 Comfortable Small Interiors 87
4.1 The Need for Smaller Dwellings 87
4.2 Macro Design Aspects of Small Spaces 88
4.2.1 Zoning Interior Space 89
4.2.2 Access and Circulation 91
4.2.3 Spatial Configurations 94
4.3 Micro Design Aspects of Small Spaces 97
4.3.1 Accommodating Flexibility 98
4.3.2 Space-Making Strategies 99
4.3.3 Light, Colors, and Finishings 102
4.4 Small Comfortable Interior in Zeeburgereiland, the Netherlands 106

4.5 The Next Home ... 107
4.6 Final Thoughts ... 111
References ... 115

5 Attractive and Energy Efficient Facades ... 117
5.1 A Need for Diversity and Energy Efficiency in Building Envelopes ... 117
5.2 Principles of Designing Attractive Façades ... 118
5.2.1 Logic of Façade Design ... 119
5.2.2 Principles of Designing for Choice ... 120
5.2.3 Menu of Design Choices ... 124
5.3 Energy Efficient Envelopes ... 124
5.3.1 Design Principles for Energy Efficiency ... 126
5.3.2 Choice of Materials ... 129
5.4 Energy Efficient Windows ... 132
5.4.1 Window Composition and Heat Transfer ... 133
5.4.2 Daylight and Natural Ventilation ... 135
5.4.3 Energy-Saving Features ... 137
5.5 Diversity of Facades in Västra Hamnen, Sweden ... 138
5.6 Affordable Prefab Home ... 141
5.7 Final Thoughts ... 149
References ... 150

6 Innovative Construction Practices ... 153
6.1 A Need for Resource Conservation and Affordability ... 153
6.2 Fundamentals of Prefabricated Homebuilding ... 154
6.2.1 Main Types of Prefabrication Methods ... 155
6.2.2 Other Prefabrication Typologies ... 156
6.3 Selecting Fabrication Methods and Building Communities ... 160
6.3.1 Mass Customization and Digital Fabrication of Housing ... 162
6.4 A Mass Customized Prefabricated Community in Ijburg, the Netherlands ... 163
6.5 Dry Interior Construction ... 165
6.6 Design for Growth and Adaptability ... 166
6.7 The Pod Home ... 167
6.8 Final Thoughts ... 170
References ... 174

7 Utilities Systems for Sustainability ... 177
7.1 Evolution and Challenge of Utilities' Insulations ... 177
7.2 Plumbing and Electrical Systems ... 180
7.3 Water-Saving, Harvesting, and Recycling Systems ... 185
7.4 Installing Heating, Ventilation, and Air Conditioning (HVAC) Systems ... 189
7.5 A Home with Well-Integrated Utilities ... 191

7.6 The Internet of Things (IoT) and Wi-Fi 192
7.7 Domus Ex Machina 193
7.8 Final Thoughts 196
References 200

8 Green and Healthy Materials 203
8.1 A Circular Approach to the Use of Materials 203
8.1.1 Systems Thinking 204
8.1.2 Life Cycle Assessment 205
8.1.3 Material Passports, Separation of Cycles, and Recycling 207
8.2 Green Building Products 208
8.2.1 Insulated Concrete Forms (ICFs) 208
8.2.2 Structural Insulated Panels (SIPs) 209
8.2.3 Engineered wood and Laminated Timber 211
8.3 Products Made from Recycled Materials 212
8.4 Innovative Sustainable Materials 215
8.5 Materials for a Healthy Indoor 218
8.5.1 Indoor Air Quality (IAQ) 219
8.6 Using Green Materials in Copenhagen, Denmark 221
8.7 The Green Grow Home 222
8.8 Final Thoughts 224
References 225

9 Energy Efficient Dwellings 229
9.1 A Need for Energy Efficient Dwellings 229
9.2 Selecting an Efficient Heating System 232
9.2.1 Commonly Available Heating Systems 232
9.3 District Heating 237
9.3.1 A Sustainable Community with District Heating in Stockholm, Sweden 238
9.4 Design and Construction Principles for Energy Efficiency Dwellings 241
9.4.1 Building Size and Shape 241
9.4.2 Thermal Insolation 243
9.4.3 Ground Cover and Passive Solar Design 245
9.5 Net-Zero Homes 247
9.6 Renewable Energy Sources 249
9.7 Landscaping for Energy Efficiency 251
9.8 Energy Efficient Home 254
9.9 Final Thoughts 256
References 258

10 Home Automation 261
10.1 Digital Advancement and Sustainability 261
10.2 Defining Smart Homes 264
10.3 Technology and Automation in the Residential Environment 266
10.3.1 Energy Conservation 266
10.3.2 Water Conservation 268
10.3.3 Smart Mobility 271
10.3.4 Innovations in Telehealth 272
10.3.5 Smart Home Security 273
10.3.6 Lifestyle and Technology 274
10.4 A Connected Home in Puglia, Italy 275
10.5 The Smart Home 276
10.6 Final Thoughts 278
References 279

11 Cooking and Dining at Home 281
11.1 The Evolution of Cooking and Dining Spaces 281
11.2 The Kitchen as a Social Hub 284
11.3 The Sustainable Kitchen 286
11.3.1 Modular Customizable Components 286
11.3.2 Sustainable Building Materials 289
11.3.3 Natural Light and Energy Efficient Appliances 292
11.3.4 Growing Food in the Kitchen 293
11.3.5 Composting, Recycling, and Reusing 294
11.3.6 Place-Making for Social Interactions 296
11.3.7 Accommodating People with Special Needs 297
11.4 A Family Kitchen in a Tokyo Apartment 299
11.5 A Home with a Sustainable Kitchen 300
11.6 Final Thoughts 302
References 305

12 Storing Stuff and Furnishing a Home 307
12.1 A Consumer-Oriented Society 307
12.2 Household Demographics and Storage Needs 308
12.3 Organizational Space Practices 309
12.4 The Tiny Home Movement 310
12.5 Sustainable Storing Practices 312
12.6 Residual Spaces for Added Storage 313
12.6.1 Wall-Mounted Seating and Bedding 313
12.6.2 Storing Bicycles 315
12.6.3 Floor Cabinets 316
12.6.4 Staircase Storage 317
12.7 A Home with Creative Storage 318
12.8 Multifunctional Furnishing 319
12.8.1 Bedroom Furniture 319

12.8.2 Fold-Down and Convertible Furniture ... 319
12.8.3 Space Dividers ... 320
12.8.4 Modules for Micro Units ... 321
12.9 The Max Storage Home ... 322
12.10 Final Thoughts ... 326
References ... 327

13 Getting Old at Home ... 329
13.1 The Growing Aging Population ... 329
13.2 Aging in Place ... 331
13.3 The Kitchen ... 334
13.4 The Bedroom and Living Room ... 336
13.5 The Bathroom ... 336
13.6 The Home's Exterior ... 340
13.7 The Potential Benefits of Digital Technologies ... 341
13.8 Multigenerational Living Arrangements ... 341
13.9 A Home for Aging in Place ... 343
13.10 An Adaptable Home ... 343
13.11 Final Thoughts ... 351
References ... 352

14 Working from Home and in Common ... 355
14.1 History and Evolution of Live-Work Arrangements ... 355
14.2 Advantages and Challenges of Working from Home ... 358
14.3 Types of Remote Work and Needed Accommodations ... 360
14.4 Selecting the Location of the Workspace ... 363
14.4.1 Eco-Consciousness in Workspace's Design ... 364
14.5 Designing a Live-Work Home ... 365
14.6 Working in Common ... 366
14.7 A Live–Work Home ... 367
14.8 Final Thoughts ... 368
References ... 370

Illustrations Credits ... 373

Project Teams ... 375

Bibliography ... 381

Chapter 1
Homes for Changing Times

Abstract Climate change, the adaptation and depletion of natural resources, housing affordability, and socio demographic transformations are some of the challenges that lead a call for new housing design and construction paradigms. These challenges are listed, investigated, and discussed in this chapter. Sustainability and its roots are also defined as basis to understanding the emergence and growth of "green thinking" over the last decades. Social, cultural, economic, and environmental dimensions are introduced as pillars of sustainability to act as a segway into the principles of sustainable systems. Finally, the author argues that the aforementioned challenges are leverage points for innovative ideas and practices in the contemporary housing market of various cities and nations.

Keywords Affordable housing · Climate adaptation · Four pillars · Generational inclusivity · Lifestyle changes · Sustainable design · Sustainable systems

1.1 Adjusting to New Realities

On March 11, 2020, the declaration of a global COVID-19 pandemic by the World Health Organization (WHO) caught many nations by surprise. The pandemic has changed the ways in which humans interact, work, and deliver health care to name a few transformations. Technology has become increasingly present in daily life as more citizens worked and studied from home. Although this may seem grim, there is an important silver lining that will last for years to come. The slowing of the world's pace has allowed nations to re-envision how they conduct affairs. New realities have given leaders time to evaluate how cities are to be planned—reconceptualizing housing, address the climate emergency, as well as the way society views human health and wellbeing. Working from home has left office spaces vacant and led to an increasing interest in converting them to housing (Cooke 2021). Civic leaders and planners acknowledge that cities need to become more sustainable, decongested, and cleaner to attract people back to live, work, and visit city centers.

A. Friedman, *Fundamentals of Innovative Sustainable Homes Design and Construction*, The Urban Book Series, https://doi.org/10.1007/978-3-031-35368-0_1

Planners and developers hailing from Barcelona, New York, Milan, Paris, and Stuttgart among other cities are following through with conversion of unused commercial spaces, turning streets and parking lots into urban parklands, and creating natural landscapes on vertical surfaces and roof tops (Fig. 1.1). Paris, for example, future host to the 2024 Olympics is creating comprehensive plans in preparation for the events led by Mayor Anne Hidalgo and architect Philippe Chiambaretta. The intent is to create an extraordinary park at the heart of Paris' which stretches from Place de la Concorde through Champs Élysées to Champ de Mars beneath the Eiffel Tower (Cooke 2021). Such an example supports the thought that a new consensus is brewing among the community to renew and redesign cities and housing for changing times. The dimensions discussed below will provide a better understanding of these new realities and how they might affect the future of housing design in communities both large and small.

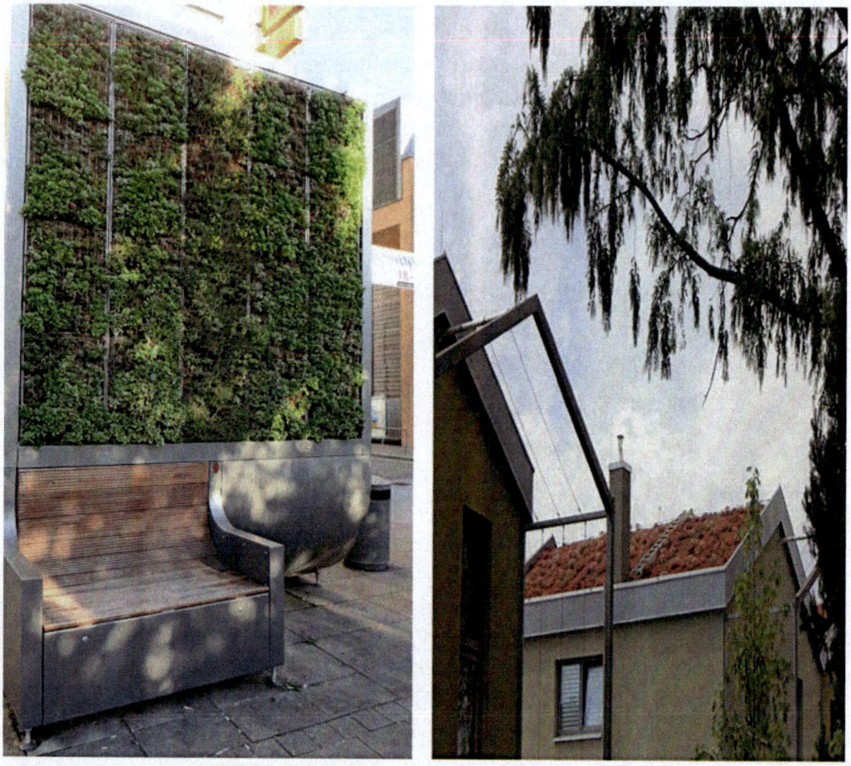

Fig. 1.1 Designers and developers in cities such as Stuttgart, Germany (left) and Vienna, Austria (right) are greening vertical surfaces and roof tops

1.1.1 The Environmental Challenge

Global rise in environmental concern has accelerated the introduction of sustainable town planning and architectural design ideologies. Events such as Australia's 2020 immense bushfires and same year flooding in British Columbia, Canada, and throughout Europe have created an urgent sense to act upon climate change. In the Global South and urban coastal areas around the world, designing homes that are adaptable to climate emergencies is increasingly essential. The consequences of not doing so can be seen in the case of 2005 Hurricane Katrina in New Orleans, US (Fig. 1.2). To mitigate the effects of climate change on communities, much attention needs to be dedicated to improving the resiliency of homes to last extreme weather events. Examples include designing flood-resistant houses with raised foundations and materials such as preservative-treated wood framing and fiberglass faced drywall that can survive water damage (Ward and Wilson 2009). Other adaptation-oriented approaches include designing landscapes resilient to drought and water shortages, wildfires, and power interruptions.

In recent years, with the climate emergency becoming a central preoccupation of governments, urban sustainability subsequent innovation has been included in the 2030 Agenda of United Nations (UN) sustainable development. The new agenda pushes beyond the implementation of the eleven Sustainable Development Goals

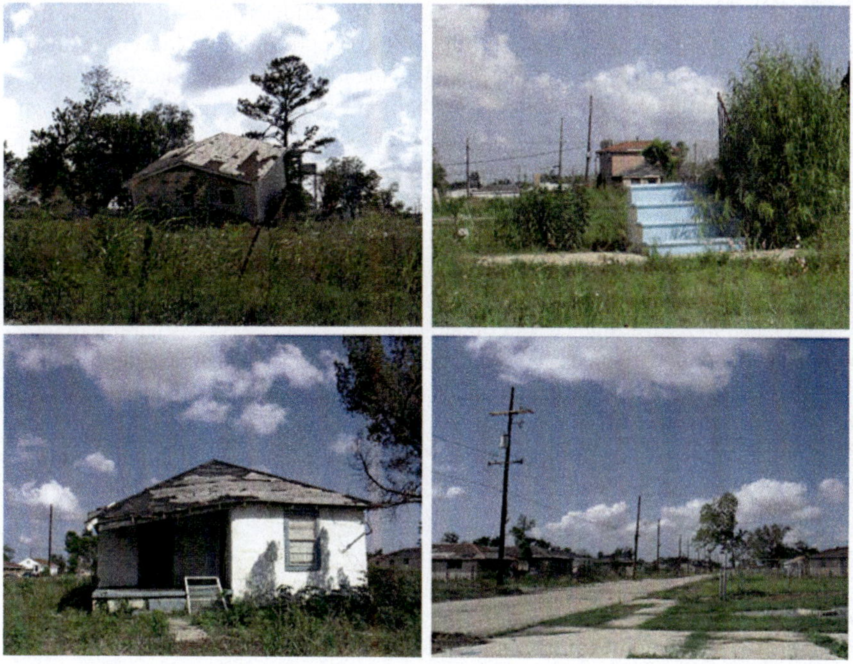

Fig. 1.2 Destruction left in the aftermath of the 2005 Hurricane Katrina, in New Orleans, US

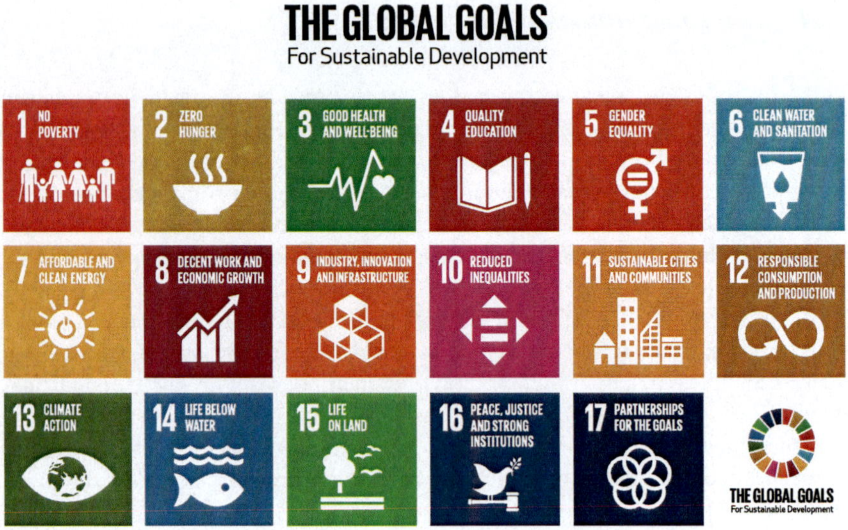

Fig. 1.3 The UN's eleven Sustainable Development Goals

(SDGs) by identifying processes and issues such as urban legislation and policies, local finance frameworks, and spatial planning (UNH 2020) (Fig. 1.3). The UN Habitat's 2020 report on sustainable urbanization recognizes that cities lie at the forefront to climate change solutions, offering various opportunities to develop climate-oriented strategies through adequate management practices, architectural design, and urban planning. It has become clear that players in these fields: architects and planners can play a crucial role in adding lasting value to community design.

With the politicization of the climate crisis, governments are also implementing new regulatory policies that will require designers to accommodate and apply these legislative changes to housing design and cities as a whole. COP26, the 2021 climate summit hosted in Glasgow, UK also put forth global agreements for countries to further reduce carbon emissions, protect communities and habitats from natural disasters, and mobilize finance for climate action (UNCC 2021). As part of an international agreement to reduce carbon and greenhouse emissions, many countries have also pledged to stop the production and selling of gas-powered vehicles as early as the year 2025 due to worrying projection models about global emissions (Fig. 1.4). In response to the ban of fusel fuel powered vehicles, homes, and public buildings will have to be equipped with charging spots for electric vehicles. In addition, cities and neighborhoods will have to make these stations customary elements of the urban streetscape.

As a result of changing policies, many countries are providing subsidies to builders and consumers for the construction of energy efficient homes. Examples include the 2 billion GBP green homes grant put forth by the UK in 2020. This grant would allow 600,000 homeowners in England to receive up to 10,000 GBP each to reduce the carbon footprint of their homes through the installation of heat pumps, insulation,

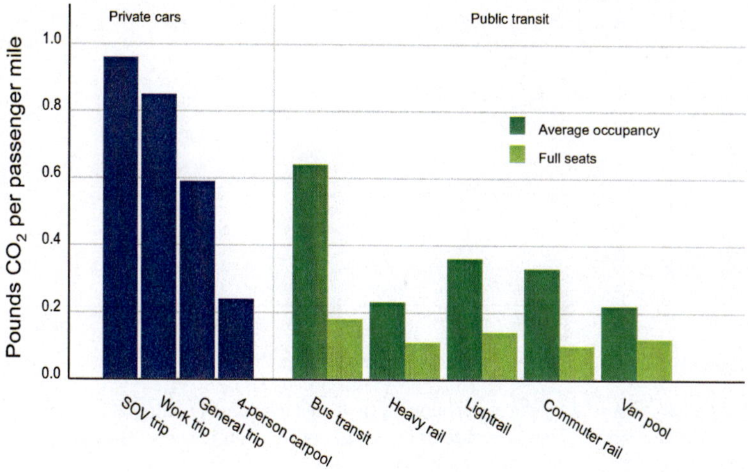

Fig. 1.4 Tailpipe emissions levels of common transport modes per passenger mile traveled

and draft proofing (IISD 2020). The proliferation of policies that are concerned with environmental sustainability of urban spaces and homes is bound to provide planners and architects with far reaching innovative design opportunities.

1.1.2 Sociodemographic Transformations

In recent decades, there has been a general recognition in the Global South, that the aging population is growing as birth rates decline. For example, Canadian homes house an average family size of 2.5 people (Statistica 2021a) (Fig. 1.5). As the Canadian population continues to age, architects and planners must pay greater attention to socioeconomic inclusivity and active engagement of seniors. In addition, the homebuilding industry must come to realize that due to greater demographic diversity, the existing housing stock is not adequately designed to accommodate certain demographic cohorts such as single-person and single-parent households (Fig. 1.6).

By analyzing such trends, dwellings can be designed and adapted to fit the needs of families without oversizing living spaces that lead to overconsumption and eventually to urban sprawl. These market trends offer opportunity to design housing with small footprints that is not only efficient and affordable but also sustainable. Fortunately, new tools and data are being provided to help planners and architects determine population's needs more efficiently. In Canada, the Social Inclusion Index (SII) is a new assessment tool which will be useful in determining the housing needs of Canadians. The index measures the impacts of various housing policies on Canadians' wellbeing and identifies neighborhoods and communities where the sense of household inclusion can be increased through the development of affordable housing.

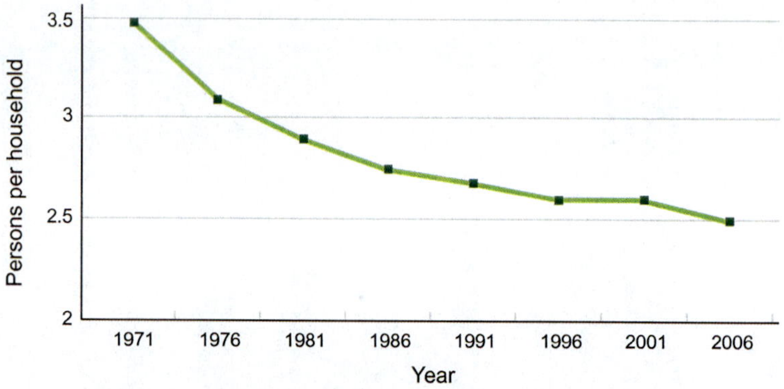

Fig. 1.5 Recent statistics in developed countries show that the aging population is growing as birth rates decline. For example, Canadian homes house an average family size of only 2.5 people

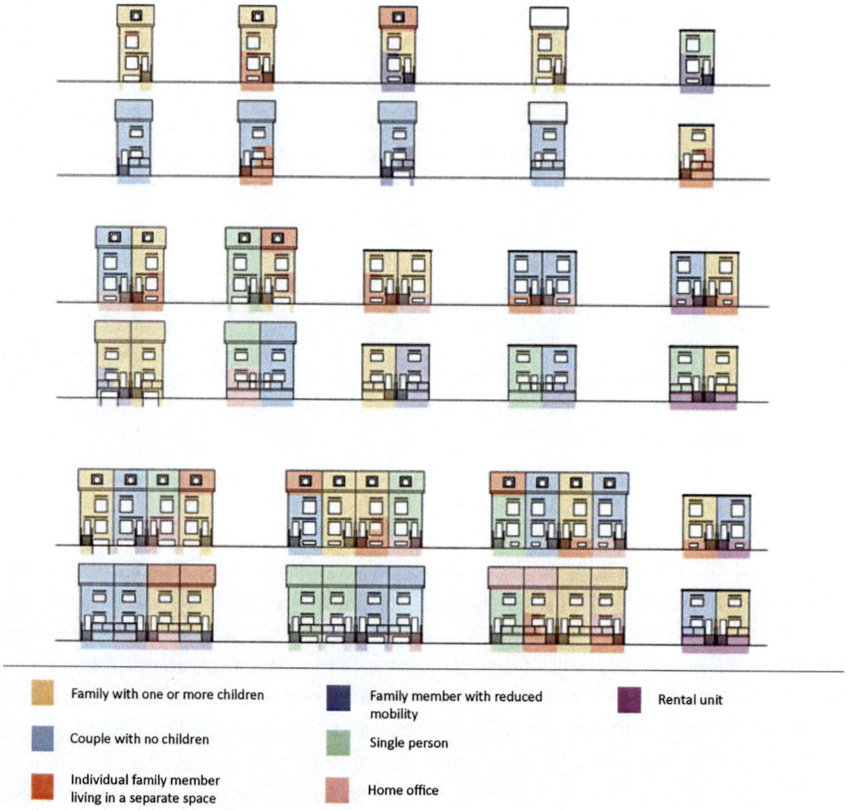

Fig. 1.6 Due to greater demographic diversity the existing housing stock needs to be adequately designed to accommodate certain demographic cohorts such as singles and single parent

The index is calculated by a weighted average of survey responses to five dimensions such as dwelling and neighborhood satisfaction, feeling part of the community, economic hardship, and sense of safety (CMHC 2021). By utilizing available data and indexes like the SII, new housing and design initiatives can address social and economic needs that favor greater urban generational inclusivity.

1.1.3 A Shifting Global Economy

In a time of global economic uncertainty, a shift in housing consumption patterns has been observed. Notably, housing prices are becoming increasingly unaffordable in many countries. As a result, the existing imbalance of the demand and supply of single-family homes has been aggravated and resulted in a significant price increase (Khan et al. 2021). For example, in 2020, Hong Kong was ranked the most expensive city for housing as its average dwelling cost over $1.25 million US (Statistica 2021b). This economic reality places a disproportionate burden on first-time home buyers as they struggle to gain a foothold in the housing market in many cities. For those who can afford a home, mortgage prices account for a large portion of a household income. In some cities, like Vancouver, Canada, mortgage payments for a single-family home amounted for a whopping 85.4% of family income in 2021 (Statistica 2021c).

This data clearly demonstrates an urgent need for affordable housing that can both adapt to the consumer market's changing needs and be financially sustainable. Although much policy and planning are an essential part of the creation of affordable housing, architects play an integral role in providing solutions to the housing crisis through low-cost design strategies. These involve dwellings with small footprint, simplifying facades while creating variation through colors and materials, designing unit layouts and dimensions for construction efficiency and adaptability as well as repurposing existing buildings to reduce costs (Hoyt and Schuetz 2020).

1.1.4 Changing Lifestyle

With the increasing digitization, people are having new means to access information that effect their lifestyles and to some measure the use of their homes (Fig. 1.7). The internet provided people with greater opportunities to connect, learn, shop, and conduct business in new ways. An acceleration in the development of the digital world has given rise to changes from the way people eat to physical activity and mental health. These changes give rise to many hopeful prospects. Whether it is adjusting work-life balance, time spent with loved ones, and/or the time dedicated to increasing physical and mental wellbeing, lifestyle changes are happening on many levels. Some of these changes call for a reevaluation of residential design and spaces within a dwelling unit that will be outlined throughout this book's chapters.

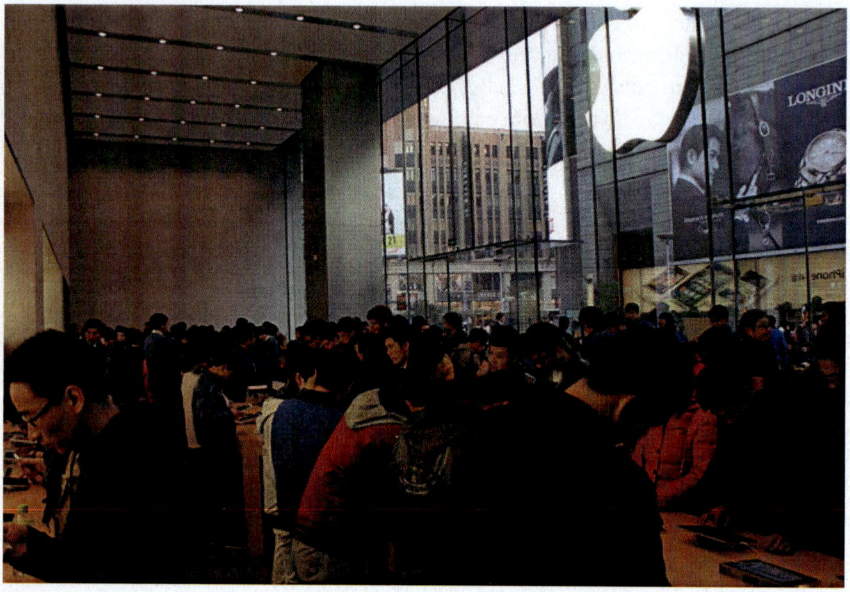

Fig. 1.7 People examine digital gadgets in a Shanghai's Apple store

Along with strong policy recommendations and planning for active living, architects and planners can help achieve activity-driven goals by setting up higher-density communities close to parks, play areas, green space, and pedestrian-friendly locations (Fig. 1.8). Importantly, designing homes that allow for greater residential density can be one among many solutions as research has shown that urban dwellers are more active than suburban ones. High- to medium-density communities need to be encouraged as surrounding commerce will eventually be attracted to these housing units as desirable places for doing business. As a result, more people will be willing to walk to nearby commerce using active modes of transportation such as biking and walking. Not only will encouraging physical activity be beneficial for strengthening the social fabrics but will also provide solutions to reducing pressure on the healthcare system.

1.1.5 Technological Innovation

In recent decades, much effort has been dedicated to developing and testing new state-of-the-art technologies and tools to assist architects, planners, and policymakers to plan more sustainable environments. United Nations Habitat (UNH) for example is developing technologies and practices that will serve as learning tools for planning and community design. Keen to stay technologically up to date, UNH is currently working and experimenting with blockchain technology policy recommendations,

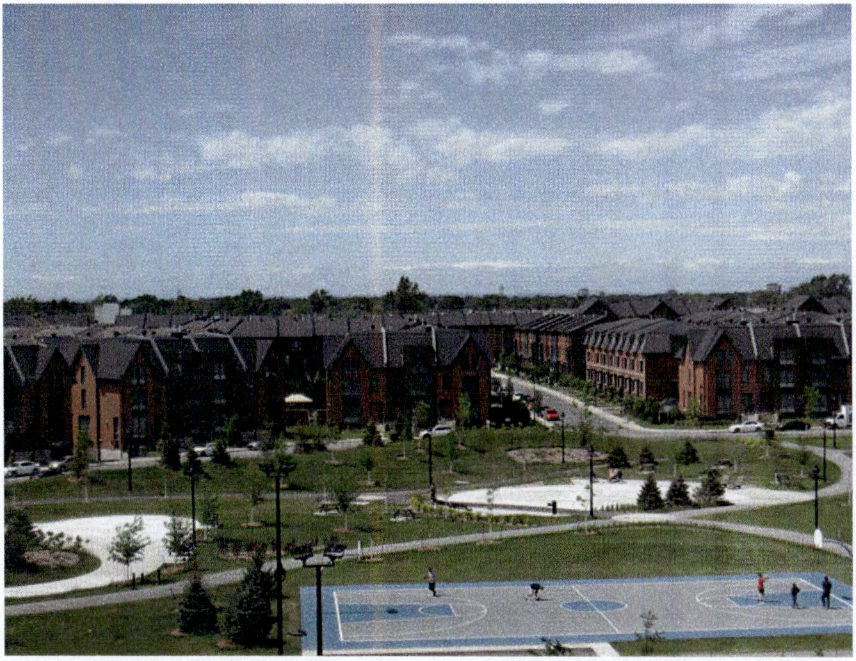

Fig. 1.8 Architects and planners can achieve human activity-driven design goals by setting up higher density communities close to play areas like this one near Montreal, Canada

collecting big data, and using spatial integration models to facilitate decision-making and design practices (UNHI 2021).

Among the many innovative technologies being developed, some recent ones are becoming readily available for architects and builders. Tools like robotic prefabricated production and 3D printers make it possible to create affordable and resistant building components which reduces dependency on natural or expensive resources (Fig. 1.9). This technology allows for construction materials to be customized on demand without generating excess waste as a printer can accurately produce pieces according to a blueprint. The most common raw materials used to print 3D objects for building purposes include concrete, sand, fiber, and geopolymers. Other biodegradable materials that are being experimented with for a large-scale use are straw and soil.

Recent advances have made it possible to simplify production of plumbing components as well as electrical fixtures. Homes have even been made from this printing technology for as little as $4,000 US (Ohio University 2020). This offers opportunity to integrated to integrate these building components in affordable housing production. Although 3D printing also comes with many challenges and may not be suitable for all building projects, its potential is promising. For example, Apis Cor a 3D printing company created the biggest printed building in the world located in Dubai in just two weeks (Apis Cor, n.d.).

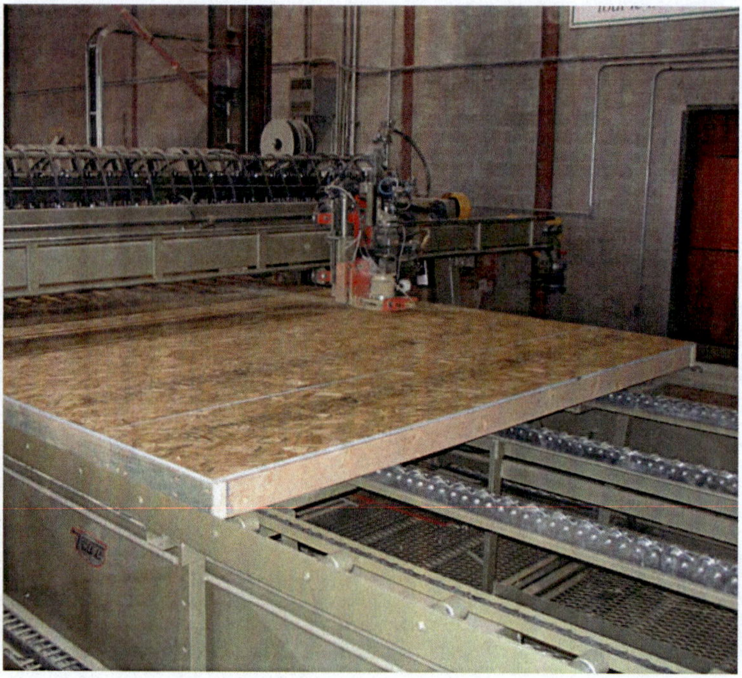

Fig. 1.9 An automated production of prefabricated wood-frame panels for homes

Another available 3D modeling technology, building information modelling (BIM) is an affordance of big data to architects and urban planners. While other 3D models depict much of the physical infrastructure, BIM goes one step further as to model the supporting infrastructure and function of buildings. The vast number of sensors implanted or tacked onto buildings enables the quantification and observation of water, energy, waste, noise, usage, and any other building activity to be synthesized into a single complex model (Douay 2018). This information is valuable to all stakeholders involved in a project with BIM capabilities, architecture, construction, operation, and management. While BIM refers to individual buildings, urban planners become involved when numerous BIM equipped buildings are connected resulting in city information modelling (CIM) (Douay 2018). Utilizing one's creativity in combination with innovative technologies is an essential part of putting forth sustainable solutions to the issues produced by many of today's urban spaces.

The emergence of smart cities is also opening new opportunities for architects to affect the interaction and link between a single home and the city. A smart city is one that utilizes communication, information technologies, and other means to improve quality of life, the efficiency of urban services, operations, and competitiveness while ensuring the needs of future generations with respect to social, economic, and cultural dimensions. Technologically smart city components include smartphone apps, information screens in public spaces, and intelligent public-facing websites and operation centers equipped with critical information and feedback mechanisms that produce big data (UNH 2020).

1.1.6 Design Appreciation

Using a combination of creativity and technology, architects are now able to create more appealing and efficient designs which the buying public can benefit from (Fig. 1.10). Both the interiors and exteriors of homes are now better looking as a result of these innovations. The buying public seems to be drawn toward appreciating designs that are meant to coexist in harmony with the natural world. To put it simply, people are paying attention to, care for, and increasingly appreciate well-thought-out sustainable design. Interior design and architectural trends will have to adapt to changing consumer preferences to satisfy buyers. Importantly, understanding and being in-tune to socioeconomic and as a result what consumers want, provides a better picture of how architects can use social change as a leverage point to drive demand toward more efficient and sustainable future.

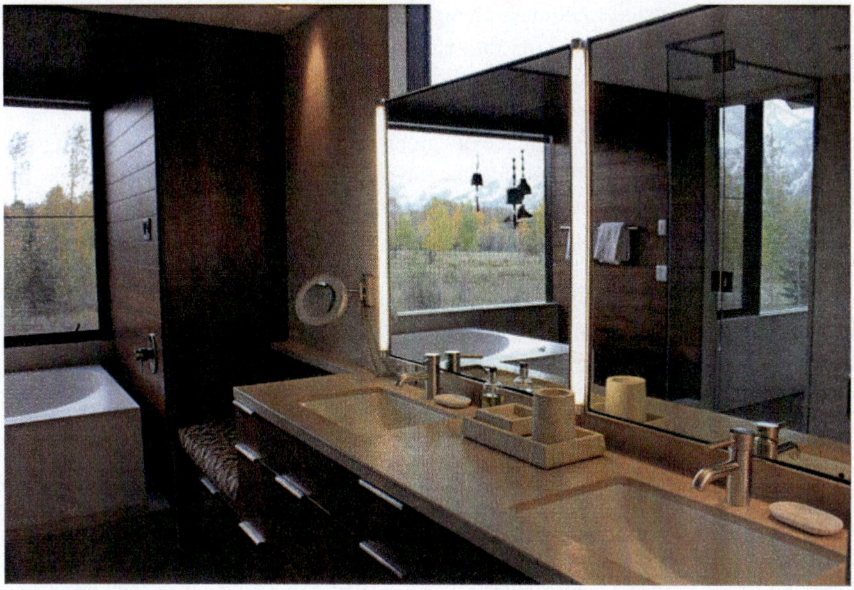

Fig. 1.10 Using a combination of creativity and technology, architects are now able to create more visually appealing designs

1.2 Origins and Evolution of Sustainability

The term *sustainability* is noted in many of today's conversations and professional literature making it arguably trivial. Nonetheless, its importance resonates throughout all aspects of the natural and built environment. Although many interpretations of the term have been put forward, a working definition of sustainability can be understood as a holistic approach that considers ecological, social, and economic dimensions, recognizing that all must be regarded if society is to find a lasting prosperity (University of Alberta Office of Sustainability 2021).

The contemporary concept of sustainability is dated to the 1970s. The movement as a whole is rooted in social equity, environmentalism, and international cooperation. In 1983, former Norwegian prime minister Gro Harlem Brundtland was appointed to head the World Commissions on Environment and Development. After four years, the Brundtland Commission released the *Our Common Future* report which famously defined sustainable development as "development that meets the needs of the present without compromising the ability of future generations to meet their own needs" (University of Alberta Office of Sustainability 2021). By officially consolidating the importance of sustainability, the Commission was able to begin unifying the environment with the social and economic priorities of the world's development agenda. Throughout the 1990s, much work was put into creating policies and conventions to address issues related to climate change and human development (Fig. 1.11). In 1997, the Kyoto Protocol took its first step toward combating climate change and in the year 2000, the Millennium Development Goals were put forth by the United Nations (University of Alberta Office of Sustainability 2021).

With this also came the emergence of "green thinking" which involves the conscientious incorporation of ecology into the environment and society. The process by which the environment is seen as an integral part of the economy was referred to as *green growth.* According to the Organization for Economic Co-operation and Development's (OECD) 2019–20 report, green growth involves fostering economic growth and development, while ensuring that natural assets continue to provide the resources and environmental services on which peoples' wellbeing relies. To do this, investment and innovation must catalyze to underpin sustained growth and give rise to new economic opportunities for all in inclusive ways.

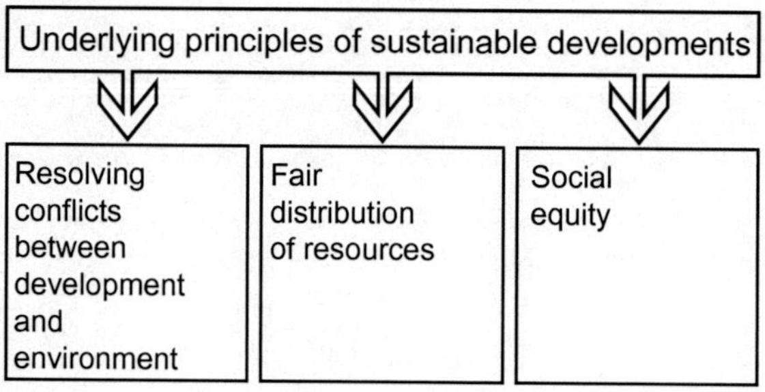

Fig. 1.11 Throughout the 1990s, much work was done by the UN on creating policies to address issues related to climate change and human development

Green thinking also spans beyond the economic scope into the built environment. It shows architects considering the impacts of design on the environment as well as how the environment can be better integrated into their design. Today, there continues to be an increase in the number of available green building certifications to improve sustainability efforts. The world's most widely used green building rating system is the North American Leadership in Energy and Environmental Design (LEED) which is available for virtually all types of buildings, communities, and home projects (Canada Green Building Council, n.d.). The LEED certification encourages cost effectiveness, improved indoor environmental quality, and reduced energy use, pollution, and carbon emissions. Studies have shown that certified green residential buildings (including those LEED certified) have an estimated 4% lower vacancy rate than non-green properties and that lease-up rates typically range 20% above average (USGBC, n.d.). As such, sustainability is not merely a word used to *greenwash* topics of conversation. It is increasingly becoming part of the social, economic, and developmental priorities to improve quality of life in a durable fashion as stated by the Brundtland Commission.

1.2.1 The Four Dimensions of Sustainable Practices

Since sustainability is regarded as a holistic paradigm encompassing both humans and nature, understanding its founding principles are essential to providing innovative solutions for future architectural design. It is more important than ever that designers move away from viewing and conceptualizing the built environment as a *product*. Instances of urban sprawl and low-income neighborhoods with lack of services and transportation are among many of the issues architects and planners are now called upon to rectify. Rather, the built environment should be thought of as a *process* involving interrelated social, cultural, economic, and environmental dimensions. These dimensions are often referred to as the *four pillars of sustainability* (Fig. 1.12). Notably, the four pillars paradigm is not universal as some models also encompass other dimensions like political or institutional dimensions. However, no matter how many other dimensions are added to the four pillars, the idea remains the same. Complex relationships, both independent and dependent, happen consistently between the dimensions. These relationships must work together in equilibrium to create a sustainable system that sometimes requires trade-offs. By applying this understanding into the built environment, a higher quality of life can be achieved for all people no matter their socioeconomic background or their geographic location. Details of each dimension are outlined below to provide a better picture about their importance when designing sustainable dwellings.

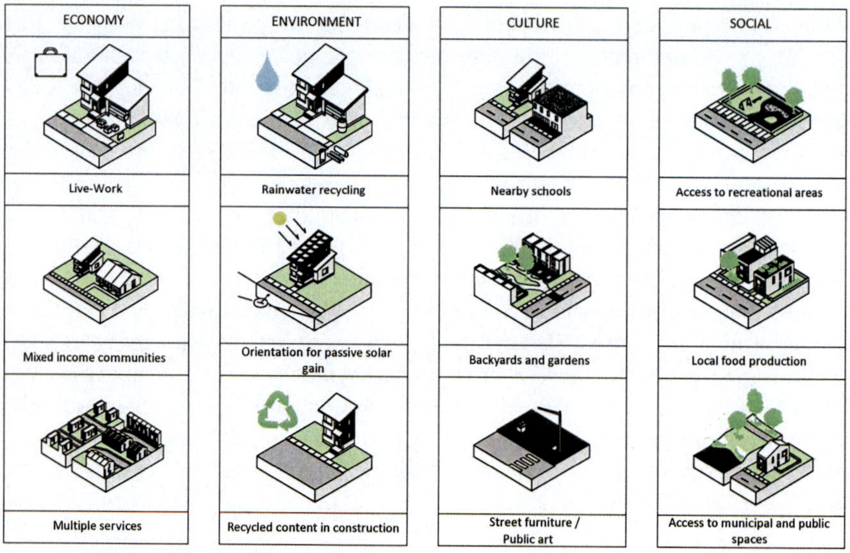

Fig. 1.12 Detailed features of sustainable development arranged according to the four pillars of sustainability

1.2.2 The Social Dimension

The social dimension is about personal security—the needs people must meet to sustain themselves, their families, and their communities. Aspects like the public health care system and governance such as installation of fitness equipment in parks ensure that citizens are fit, engaged, and healthy (Fig. 1.13). Ensuring public safety and respect by weeding out discrimination is also an integral part of this pillar (University of Alberta Office of Sustainability 2021). The social dimension is additionally about fostering human interaction and growth through effective forms of social cohesion supported by the mechanisms mentioned above. A strong social fabric is the basis for fostering cultural and economic development. Designing homes that can help achieve such objectives is possible and many places around the world are currently successful at it. Designing dense housing units with places for outdoor social gathering and gardens where residents can interact and grow food increases their wellbeing and physical activity and can be a vehicle to achieve social sustainability.

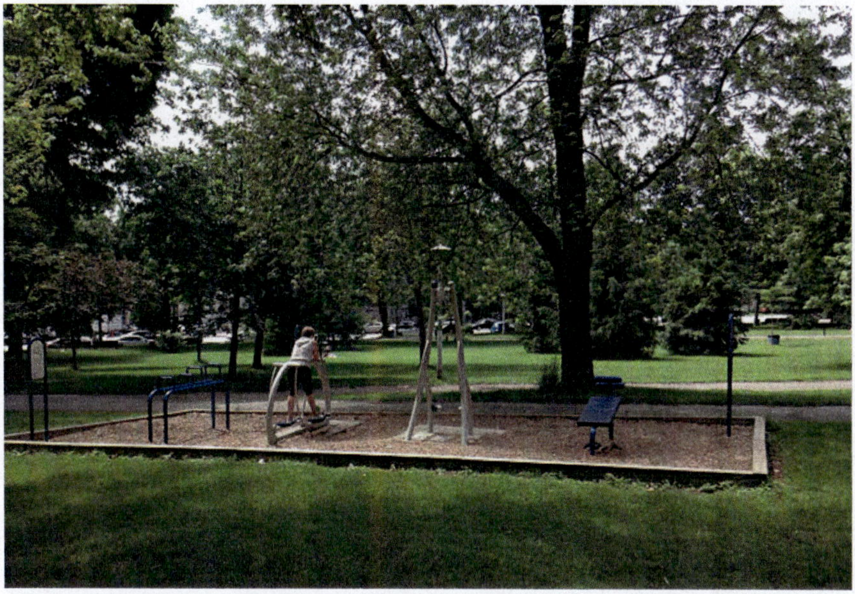

Fig. 1.13 Proactive public health care systems and initiatives such as installation of fitness equipment in parks ensures citizens' wellbeing

1.2.3 Cultural Dimensions

Culture has a codependent relationship with social integrity. It is an important aspect of determining the uniqueness of people and places by creating a sense of belonging through shared artifacts, food, music, clothing, and beliefs. Cultures and traditions are the foundations of society and a product of human history that are also important to highlight and preserve in an architectural context (Fig. 1.14). Cultural sustainability can be achieved by combining architectural creativity and social integrity; preserving and/or repurposing old buildings; and expressing elements of vernacular culture through new designs are two effective and relevant examples.

1.2.4 Economic Dimensions

Economic sustainability entails that a nation's or a municipality's financial systems are intact, and their activities are accessible to everyone. Under the economic dimensions, communities and people should be able to have access to resources required to meet their needs and maintain their financial independence as well (University of Alberta Office of Sustainability 2021). This pillar stimulates societal growth when combined with the latter dimension. The importance of economics when considering sustainable development lies in their ability to allow individuals to realize a full

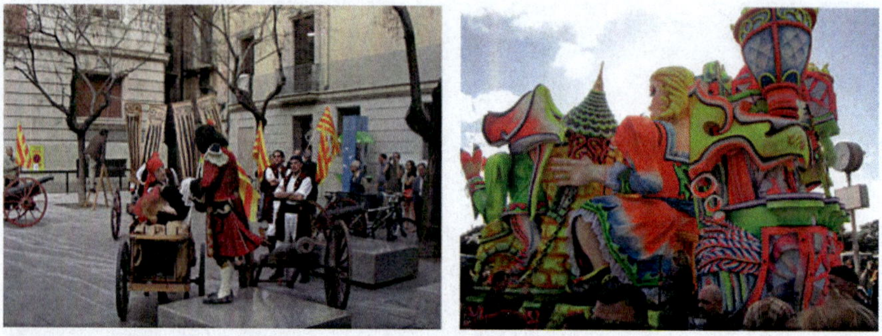

Fig. 1.14 Cultures and traditions like the ones seen in Barcelona, Spain (left) and Valletta, Malta (right), are fundamental to society and a product of history that is important to preserve

potential and become productive citizens. In terms of sustainable architectural development, providing building solutions that counteract inflationary prices of the market is essential as growing income disparities effect families. Along with the examples mentioned above, designing dwellings for maximum energy efficiency and reduced water consumption can decrease the cost of home ownership over the long term. Some methods to achieve this include the installation of energy efficient windows, heating and cooling systems, and water preserving faucets. Lower monthly expenses are possible all while contributing to reducing one's environmental footprint.

1.2.5 Environmental Dimensions

In these dimensions, the earth's environmental systems should ideally be kept in equilibrium while ecological integrity is maintained (University of Alberta Office of Sustainability 2021). In terms of consumption, resources used should be able to replenish themselves naturally which can be done by moving away from reliance on non-renewable resources. For true sustainable development to occur, working alongside nature is critical to ensuring its viability for all living organisms. Just like the social dimensions, ecology is an important determinant in the balance between the four pillars. Inhabitants are increasingly feeling connected to the natural environment as they become subject to the pressures and consequences of climate change. As a result, green growth and green thinking that were explained above are being incorporated as solutions to harmonizing the human to nature relationship. Building dwellings in accordance with green thinking includes using high-quality biodegradable or recycled materials. Essentially, treating the four dimensions as a single organism is key to producing architectural designs that will nourish societies and the landscapes around them in a sustainable manner.

1.3 The Guiding Principles of Sustainable Systems

Before investigating approaches to achieving sustainable systems, it is first important to reiterate what it means in the context of development. A sustainable system is one in which all of the four pillars of sustainability function both dependent and independently of one another to provide social, economic, and cultural growth and benefits to societies while remaining in *coexistence* with the natural environment. This coexistence implies mitigating the human footprint on the natural environment through good design practices.

An example to better illustrate this is the prototype construction of Stefano Boeri's award-winning *Vertical Forest* which aims to become a new format for architectural biodiversity. Two tall residential towers stand in the Porta Nuova area of Milan, Italy, and house 800 trees, 5,000 shrubs, and 15,000 ground plants and perennials which provide an equivalent of 30,000 m^2 (322,917 square feet) of vegetation in a concentrated urban surface of 3,000 m^2 (32,292 square feet). The towers feature large balconies for each unit which can accommodate the trees and vegetation while providing an outdoor space for its residents. The vegetation produces oxygen, absorbs CO_2 and microparticles as well as regulates levels of humidity. In addition, over 1,600 specimens of butterflies and birds have been recorded visiting the towers (Cooke 2021). Greening requirements exist in Singapore where plant material commonly covers the facades of public buildings (Fig. 1.15). To achieve such a sustainable system, many actors are responsible for the processes including governments and financial institutions, design firms, contractors, and real estate developers along with product manufacturers. The experience of the end user as well as the benefits provided to the environment are the ultimate values generated by integrating sustainable systems.

Now that the sustainable systems are better understood, we can begin to break down the main approaches to achieving them. These approaches consider the delicate

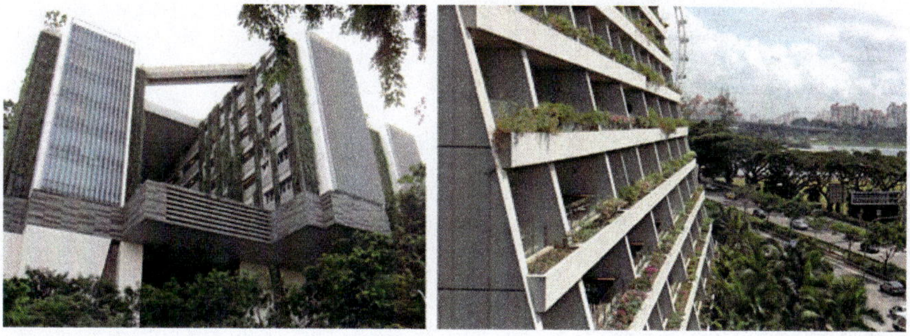

Fig. 1.15 Greening requirement exists in Singapore where plant material commonly covers building facades

balance of the four pillars of sustainability to accomplish rural and urban development. These include the path of least negative impact, the self-sustaining process, supporting relationships, lifecycle approach, and renewable resources that will be explained below (Fig. 1.16).

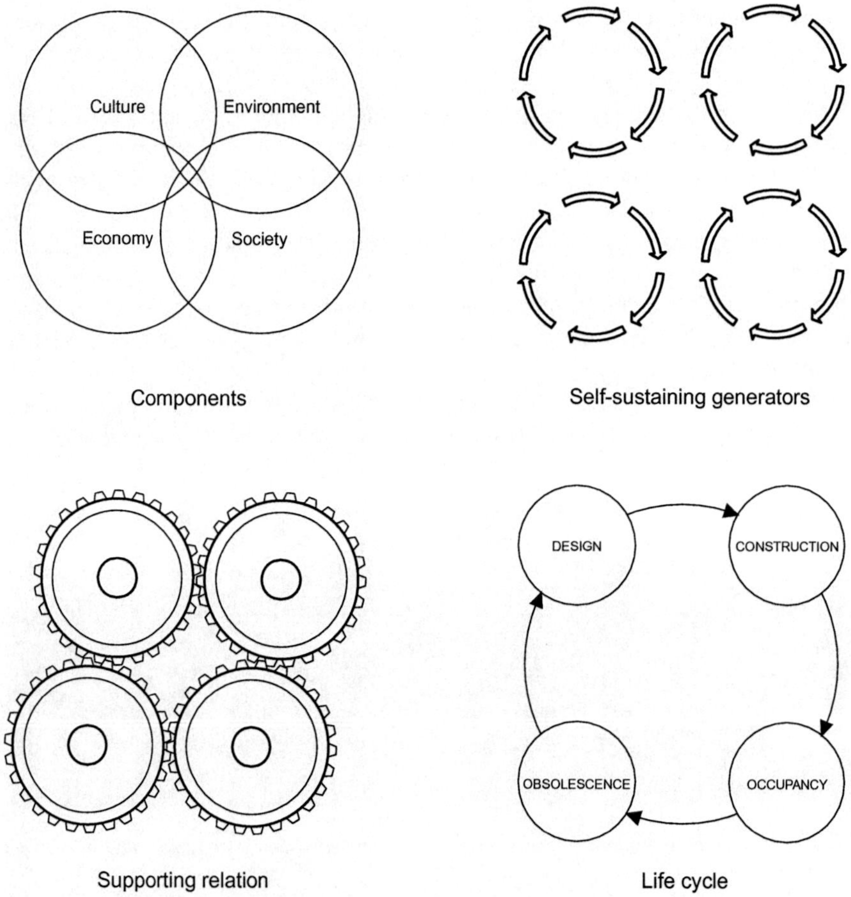

Fig. 1.16 The guiding principles of sustainable systems include the path of least negative impact, the self-sustaining process, supporting relationships, and lifecycle approach

1.3.1 The Path of Least Negative Impact

The *path of least negative impact* argues that decision-makers need to implement processes that will have the smallest negative footprint on social, cultural, economic, and environmental dimensions when conceptualizing a project. To do so, assessments of both the short and long-term effects of a particular project and its disruptive potential need to be integrated in the endeavor's overall process before it begins to physically take place. This will allow the project to be designed to have minimal adverse effects, on any of the four dimensions. From an environmental standpoint, ways to reduce one's footprint can include designing functional components that are easy to disassemble and recycle once the product or building becomes obsolete (Illinois University Library 2021). Oftentimes when a project is very complex, design-for-disassembly becomes an afterthought ultimately requiring the demolition or disposal after its useful life ended. By being mindful of a building's life cycle it is possible to mitigate potential negative impacts.

1.3.2 The Self-Sustaining Process

This approach involves a system of feedback loops where resources and activities are maintained together. The *self-sustaining approach* is meant to produce outcomes that will strengthen one another through inflows and outflows of resources. The goal of this principal is to provide enough energy to guide the perpetual existence of a project and even contribute to creating additional ones. One of the easiest ways to illustrate the self-sustaining approach is to think about personal outdoor water collection containers people tend to place outside of residential homes. No matter how complex or simple these containers may be, the principle remains relatively identical. By collecting rainwater, homeowners can use it to water plants and gardens without having to use additional water resources for maintenance. Through this action, water is returned into the natural environment fostering the growth of fruits and vegetables which will grow to be consumable produce. By growing one's own produce, perhaps less trips can be made to the local market to obtain fresh ingredients since they can be grown at home when the season permits. Therefore, this process is said to be self-sustaining as it provides many overall benefits that support one another in a feedback process.

1.3.3 Supporting Relationships

Symbiotic and *supporting relationships* can enforce the balance within projects that make them viable in varying socioeconomic conditions and the environment. Relationships are argued to support one another when attributes of a particular component can reinforce or support the growth of another. Relationships can exist between actors such as municipal governments and design firms or specific processes. For example, recent studies on the effects of Barcelona's greening project on health, wellbeing, and pollution reduction have estimated that its implementation could increase life expectancy gains by almost 200 days due to the reductions of harmful urban pollutants. The plan involves further greening up superblocks which already have in-house parks and gardens as well as constructing new parkland projects around many Spanish neighborhoods (Cooke 2021). Design can strengthen the quality of life and provide ecological biodiversity in an urban environment therefore demonstrating a symbiotic relationship between social and environmental processes. Here, the built environment acts as the main mechanism for allowing these relationships to take place.

1.3.4 Circular Approach

A circular approach acknowledges the process of change and evolution of the built environment. Known also as *cradle-to-cradle,* this approach uses flexible strategies to achieve planning and design objectives. Being flexible reduces unexpected shocks in the overall process by being able to adapt to an array of potential challenges be them regulatory, environmental, or structural. This approach argues that each step on the path to a final product is unique and involves varying capacities requiring a range of human, capital, and material resources. These steps produce varying sets of outcomes that are in constant evolution. These changes in the built environment ultimately affect the four sustainable dimensions that one must consider throughout the entire process from design to construction and thereafter.

Renewable sources of materials and energy are also an important component of these principles. As society adopts climate-focused strategies for the built environment, reducing dependence on harvesting materials that produce emissions or waste finite resources is key. According to the OECD's 2019 annual report, projections for the global demand of water used for irrigation, domestic, livestock, manufacturing, and electricity purposes are set to increase exponentially by 2050.

Of course, water is only one example of how trends show the increasing demand for natural resources. As some parts of the world already struggle with acute water crisis's, worsening depletion would further devastate social and economic implications for development. Part of the solution to this problem will be to design homes and other infrastructures that promote water conservation to prevent shortages. The overarching theme among all these statistics and projections involves designing for climate change adaptation as a future method. Part of the process will involve widespread integration of renewable resources and energy to sustain development.

1.3.5 Circular Economy, Adaptive Reuse, and Design-for-Disassembly

Circular economy (CE) is a regenerative model that aims at reducing waste and emissions through maintenance, long-lasting design, repair, and reuse. This model has important implications in the housing sector that struggles with waste management. In a linear economy, emphasis is on the quick flow of material to its point of sale and use, creating enormous waste. Loop economy and performance economy are two ways to conceptualize circular economy's framing on stock value, which shifts from a strong emphasis on flows to a maintenance of stock value for a long period of time.

The solution to creating housing that fulfills societal needs, as well as reduces waste and cost, and circular economy is adaptive also known as adaptive reuse housing. Adaptability relies on rethinking current building systems, which includes construction methods. In a completely circular building logic where homes would function as material banks. Components could be easily arranged to create a customized and particular home, or completely taken apart to be used for a new project. This could become a reality by rethinking construction, and designing for disassembly, in which dwellings can be easily broken down to their components, which allows them to be reconfigured or reused at any point in time. In order for this to be achievable, certain construction practices need to be implemented. Through prefabrication and digitization, homes can become entirely adaptable. These methods can be incorporated into any housing design, and yield benefits in terms of environmental, financial, and social issues.

While prefabrication and dry construction represent effective tools in achieving adaptable construction, a holistic approach needs to be implemented in the design process to transform the building industry. Design-for-Disassembly (DfD) is the ultimate solution for a sustainable approach to construction it aims to make any given product easy to disassemble into all of its individual components. In the case of housing, disassembly refers to the disconnection of individual components, including the wall cladding, non-structural wall panels, flooring, kitchens, and internal finishes. Design-for-disassembly is involved throughout the entirety of the design process as it informs decisions and material choices and affects how components are joined and layered. The outcome of this implementation is construction that is accessible, reversible, and robust. Ultimately, the goal of design-for-disassembly is to eliminate waste with closed loops, by creating enduring, adaptable, and flexible buildings. Much like most prefabricated approaches, this approach offers financial and environmental incentives. During the building's use, operation and maintenance are optimized through this method. If a product is made easier to assemble, then it is simpler and, therefore, less costly to produce. Easily disassembled materials are likely less contaminated, which allows them to conserve their characteristics when recycled. Subsequently, this reduces its environmental footprint, as the product is easier to produce, repair, maintain, and upcycle.

1.4 Skaftkarr: Green Place for the People and the Planet

The community of Skaftkarr, in the Finnish city Porvoo, boasts impressive low carbon-emission and healthy lifestyle features. The design firm Sitra also put various planning techniques into place in an effort to minimize energy consumption of its 6,000 residents. Since its first buildings were constructed, the community has inspired the design of similar neighborhoods across Europe (Danish Architecture Center 2014).

Planners of Skaftkarr recognized a direct correlation between the activity levels of its residents and their carbon emissions; residents with healthier, had more active lifestyles, and on average generated less carbon. This is because these individuals walked or cycled to complete their daily routines. This active transport replaced car usage which of course makes up a large portion of the world's carbon emissions. Showing that the fewer cars that are needed within a community, the more sustainable it is (Fig. 1.17).

Skaftkarr has been designed to promote walking and cycling over driving. A large, high-speed bike lane has been installed in the area, and connects residents to the city center. This lane offers cyclists hassle-free commutes and leisure rides with minimal interruptions. It also connects with several bordering services—making it the most efficient and oftentimes fastest means of travel. These optimal bike lanes serve as strong encouragement for residents to bike rather than drive as cyclists experience fewer disturbances in traffic flow. To further enhance the biking experience around Skaftkarr, there are even plans to cover the high-speed lanes with solar panels. This would serve as a barrier against harsh sun and wind conditions while providing the community with an easy source of renewable energy.

The same principles behind the high-speed bike lanes drove the designs of pedestrian pathways in Skaftkarr. Efforts were made to minimize interferences in each path's flow, therefore, maximizing pedestrian convenience. For example, the system has been planned so that pedestrians need not walk across busy streets, in front of the high-speed bike lanes, or any other barrier that would cause an obstacle in the path. As a result, these routes are extremely safe, and decrease the overall commute time of those traveling by foot. There are abundant paths throughout the community. As with bike lanes, pedestrians can reach the city center and many services on Skaftkarr's periphery exclusively by using the walking path.

Green space also plays an important role in the active lifestyles of Skaftkarr's residents. Large parks surround the community and small play areas are designated within, offering an inviting environment for children to play and adults to socialize. The provision of green space encourages residents to leave the home—and further, to walk, play sports, and partake in leisurely activities in a comfortable environment. The green space of one park in the community cuts through the heart of it, connecting with another park on the opposite end of the neighborhood. This maximizes the accessibility of green space. Residents are always close to a park, whether they are on the edge, or in the community's hub.

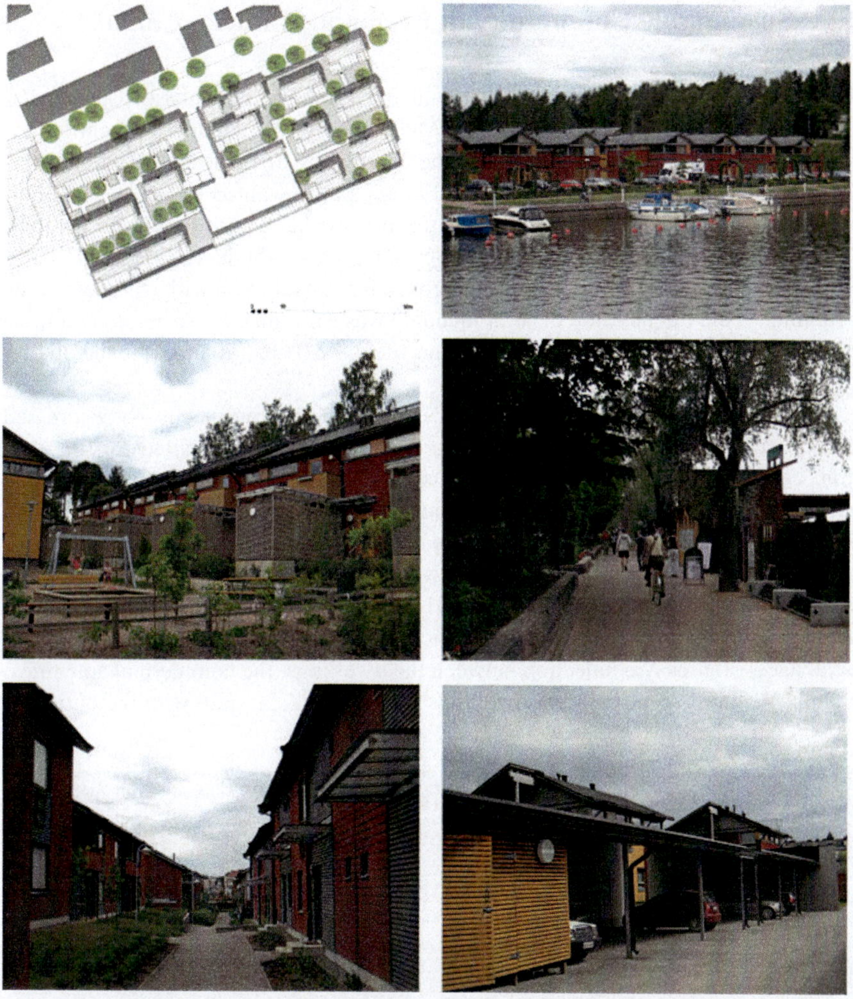

Fig. 1.17 The community of Skaftkarr, in the Finnish city Porvoo, is a pedestrian-oriented community that boasts impressive low carbon-emission and healthy lifestyle features

The planners of Skaftkarr have proven the relationship between active lifestyles, and carbon emissions. As a result, they have encouraged modes of transport that minimize car usage, while keeping residents healthy, safe, and energized. The social sustainability of the community relies on the healthy lifestyles of its inhabitants, as much as it does its meticulously planned infrastructure.

1.5 Sustainable Community in Komoka, Ontario, Canada

The Municipality of Middlesex Centre in the Province of Ontario, Canada has witnessed rapid growth since the 1990s. Most notably was the development of new low-density subdivisions ever closer to the nearby city of London, Ontario. As this development trend is set to continue, the municipality wanted to have a neighborhood built around the new Wellness Centre in Komoka, one of its former hamlets. As a result, the author was invited to propose a plan for the land (Friedman 2014). While the specific objectives were gathered from a consultation with members of the community and council, the general mandate was for a mixed-use residential town center based on sustainable principles. The site is 22.63 ha (56 acres) made up of 5 parcels of a flat terrain (Fig. 1.18).

In setting design objectives, Komoka's current and future needs—as expressed during various public consultations—were examined and accounted for. In addition, Leadership in Energy and Environmental Design (LEED) criteria and suitable sustainable planning principles were considered. Key aspects within the planning and environmental design strategies were to reduce automobile dependency and design a walkable community. This was to be achieved by developing an extensive network of pedestrian and bike paths, considering and including the site's natural features, having medium-density, mixed-use, and diverse housing prototypes to accommodate people of all ages, offer easy connection between the dwellings, the commercial amenities, and the Wellness Centre, orient dwellings for passive solar gain with the option of

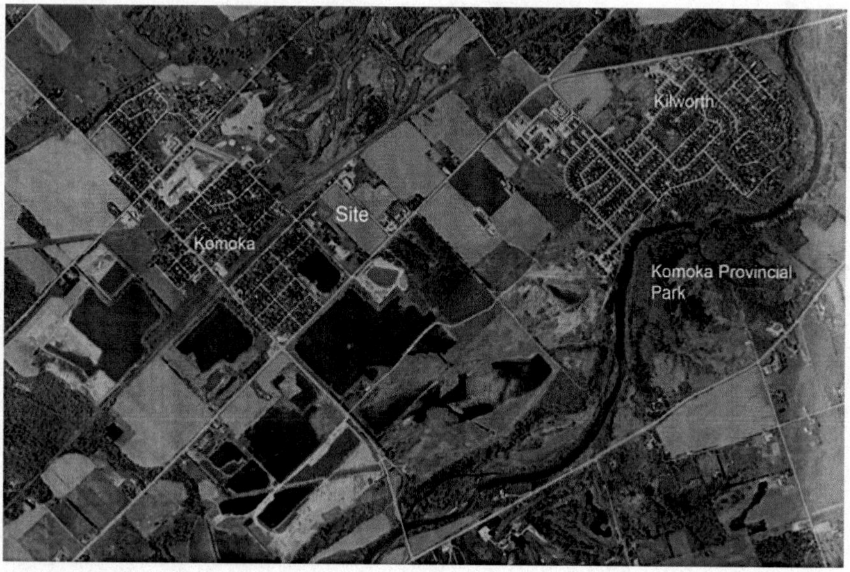

Fig. 1.18 The Komoka community site is 22.63 ha (56 acres) made up of 5 parcels of a flat terrain between the neighborhoods of Komoka and Kilworth

adding-on photovoltaic (PV) panels, include trees and native plants and landscaped xeriscaping practice as well as allow areas for community gardens (Figs. 1.19 and 1.20).

As for the residences, it was decided that most of the housing will be medium-density single and multifamily townhouses affordable for first-time buyers. Apartment buildings were located west of the Wellness Center to provide affordable accommodations for seniors. A small percentage of the land toward north of the site was dedicated to single-family detached housing as well. The objective was to have a medium-density community of between 9.8 and 22.2 units per hectare (4–9 units per acre) for a diverse population.

In the designs, all medium- and high-density housing has shared parking either behind or to the side of each residential block. This minimizes the amount of land

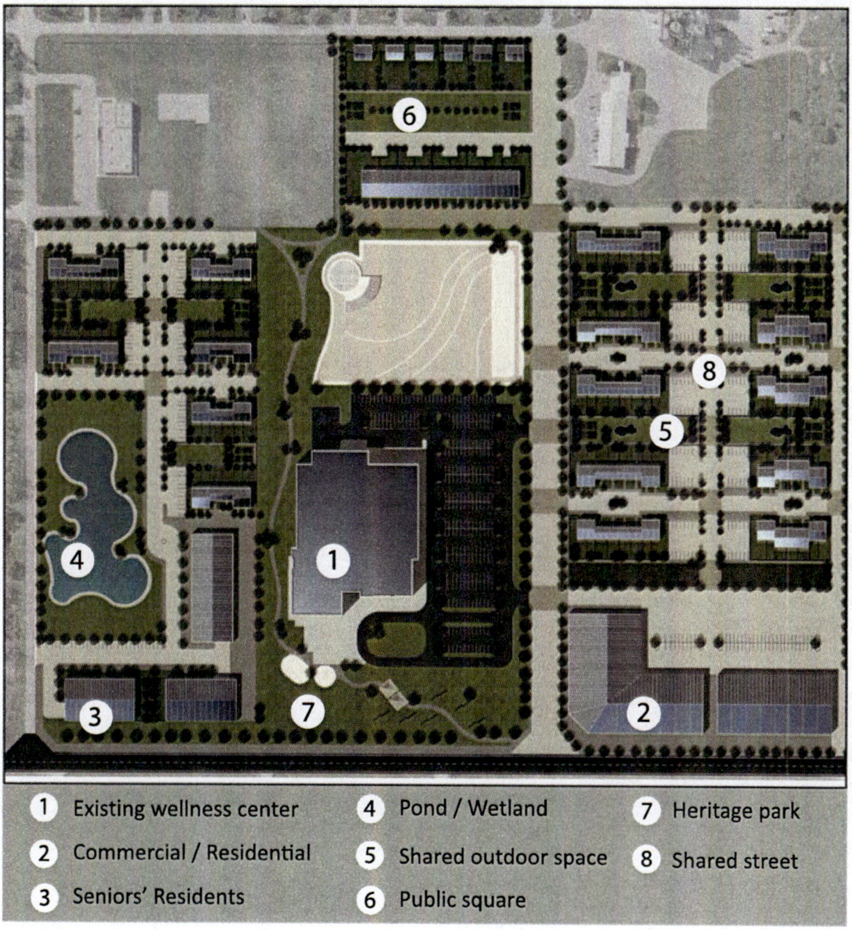

Fig. 1.19 Medium-density housing—some mixed-use—was designed to accommodate people of all ages

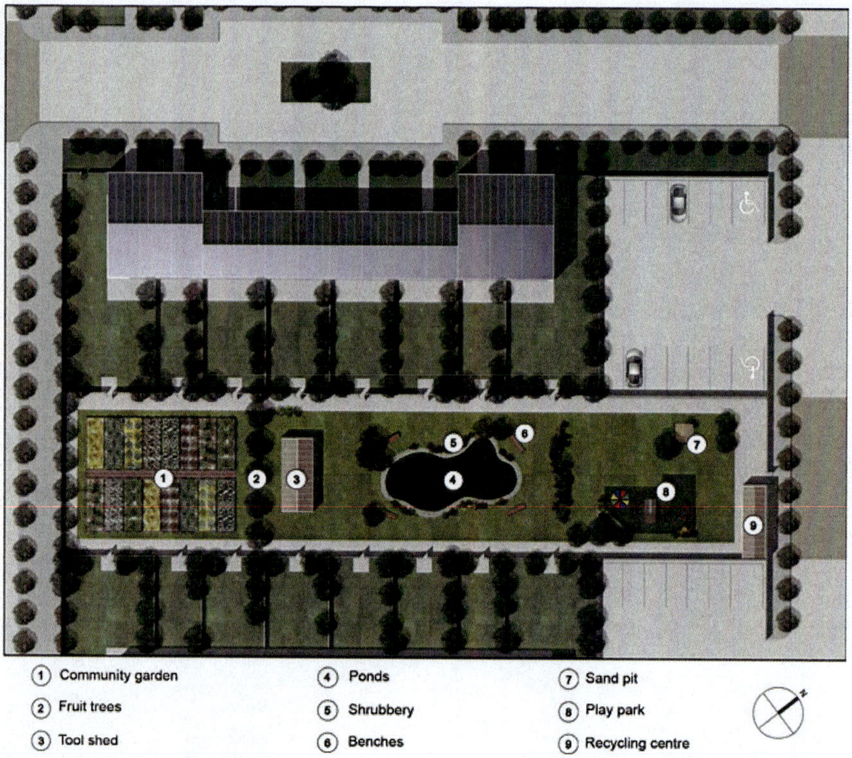

Fig. 1.20 Common green space between housing rows where landscaping provides areas for recreation and urban agriculture

dedicated to cars and improves the usefulness and attractiveness of the street. All streets and lanes have been designed to be shared surfaces that prioritize the pedestrian and encourage cycling. These measures will include a change of surface texture at the entrance to the town, planter boxes, sidewalks projections, and the inclusion of bike lanes.

Architectural guidelines have been prepared to offer design principals of massing, exterior façade, windows, doors, streetscape, and parking. They were developed while considering the various housing types listed in the master plan. In addition, a separate set of environmental recommendations have been introduced to address orientation, nature, public health, urban agriculture, landscaping, and composting and recycling (Fig. 1.21).

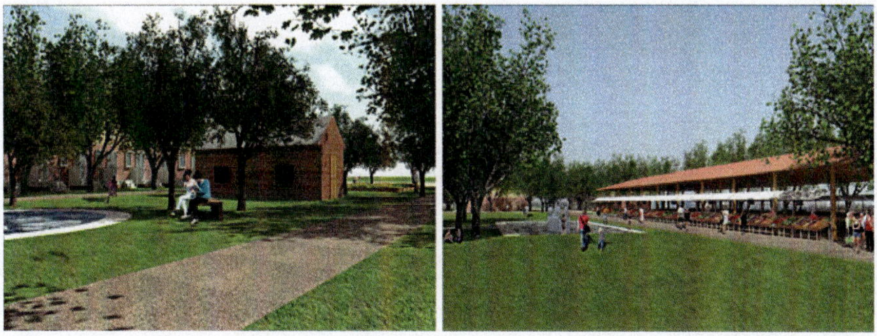

Fig. 1.21 Close-up of the common area (left) and the market square (right) in Komoka

1.6 Final Thoughts

Today's social complexities may seem overwhelming as constant changes are occurring in all spheres of the human and natural environment. These challenges, however, present an opportunity like no other to create positive changes to the built environment. The destabilization of the current system provides a point of leverage for innovation in housing. According to Meadows (1999), leverage points are places within a complex system to include a corporation, an economy, a living body, a city, and an ecosystem where a small shift in one aspect can produce big changes in others. Therefore, what is proposed in this book is a new way of envisioning the future of housing design to reach a much-needed equilibrium among the social, cultural, and economic dimensions of communities. Enlightenment about today's challenges and examples of possible design strategies can give architects and any others interested, insight into good design practices that can be adopted to alleviate some of the pressures we face as a globe.

Questions for a Follow-Up Discussion

1. What are the key social challenges that are likely to affect residential design in the coming decade?
2. What are the key design features of homes and communities that would make them sustainable?
3. What are the four dimensions of sustainable practices and the guiding principles of sustainable systems to effect residential design?

References

Apis Cor (n.d.) The world's largest 3D printed building. https://www.apis-cor.com/dubai-project. Accessed 23 Sept 2022

Canada Green Building Council (n.d.) LEED: the international mark of excellence. https://www.cagbc.org/our-work/certification/leed/. Accessed 28 Sept 2022

Canada Mortgage and Housing Corporation (CMHC) (2021) Understanding canadian households self-reported sense of social inclusion: The Social Inclusion Index (SII). Research Insight June 2021. https://assets.cmhc-schl.gc.ca/sites/cmhc/professional/housing-markets-data-and-research/housing-research/surveys/canadian-housing-survey/2021/canadian-households-self-reported-sense-social-inclusion-index-69774-en.pdf?rev=f51ebc5b-ef8e-4f66-b988-1e9809841a3a. Accessed 23 Sept 2022

Cooke P (2021) Future shift for 'big things': from starchitecture via agritecture to parkitecture. J Open Innov Technol Mark Complexity 7(236):119. https://doi.org/10.3390/joitmc7040236.Accessed23Sept2022

Danish Architecture Centre (2014) Porvoo: energy efficient residential area. Last updated January 21, 2014. https://www.slideshare.net/SitraEnergia/porvoo-skaftkrrsyksy2011eng. Accessed 29 Sept 2022

Douay N (2018) Urban planning in the digital age, vol 6, Chapter 1. Wiley. https://www.wiley.com/en-us/Urban+Planning+in+the+Digital+Age-p-9781119539476. Accessed 27 Sept 2022

Friedman A (2014) Planning small and mid-sized towns; designing and retrofitting for sustainability. Routledge, London, UK

Hoyt H, Schuetz J (2020) Thoughtful design can create high-quality affordable multifamily housing. Joint Center for Housing Studies of Harvard University. https://www.jchs.harvard.edu/blog/affordable-housing-doesnt-have-to-look-cheap-inside-or-out. Accessed 23 Sept 2022

Illinois University Library (2021) Sustainable product design: sustainable design principles. https://guides.library.illinois.edu/c.php?g=347670&p=2344606. Accessed 23 Sept 2022

International Institute for Sustainable Development (IISD) (2020) Green homes grant: £2 billion for energy efficiency in UK. https://www.iisd.org/sustainable-recovery/news/green-homes-grant-2-billion-for-energy-efficiency-in-uk/. Accessed 23 Sept 2022

Khan M, Bilyk O, Ackman M (2021) Update on housing market imbalances and household indebtedness. Bank of Canada. https://www.bankofcanada.ca/2021/04/staff-analytical-note-2021-4/. Accessed 23 Sept 2022

Meadows D (1999) Leverage points places to intervene in a system. Sustainability Institute. http://www.donellameadows.org/wp-content/userfiles/Leverage_Points.pdf. Accessed 23 Sept 2022

Ohio University (2020) 3D printed buildings guide: required materials, tips, and resources. https://onlinemasters.ohio.edu/blog/3d-printed-buildings-guide/. Accessed 23 Sept 2022

Organisation for Economic Co-operation and Development (OECD) (2019) OECD work on green growth. https://issuu.com/oecd.publishing/docs/gg_brochure_2019_web. Accessed 23 Sept 2022

Statistica (2021a) Average number of people per family in Canada from 2000 to 2019. https://www.statista.com/statistics/478948/average-family-size-in-canada/. Accessed 23 Sept 2022

Statistica (2021b) Affordability of single-family detached homes in Canada 2nd quarter 2021, by market. https://www.statista.com/statistics/720848/affordability-of-single-family-detached-homes-by-market-canada/. Accessed 23 Sept 2022

Statistica (2021c) Global housing market—statistics & facts. https://www.statista.com/topics/5466/global-housing-market/#dossierKeyfigures. Accessed 23 Sept 2022

United Nations Climate Change (UNCC) (2021) "COP26 explained". https://ukcop26.org/uk-presidency/what-is-a-cop/. Accessed 28 Sept 2022

United Nations Habitat (UNH) (2020) World cities report 2020 the value of sustainable urbanization. https://unhabitat.org/World%20Cities%20Report%202020. Accessed 23 Sept 2022.

United Nations Habitat Innovation (UNHI) (2021) Current projects, initiatives and partnerships. https://unhabitat.org/sites/default/files/2021/01/innovation_unit_brochure_final_1.pdf. Accessed 23 Sept 2022

University of Alberta Office of Sustainability (2021) What is sustainability? https://www.mcgill.ca/sustainability/files/sustainability/what-is-sustainability.pdf. Accessed 23 Sept 2022

U.S. Green Building Council (USGBC) (n.d.) Value of LEED. https://www.usgbc.org/leed/Why-leed. Accessed 23 Sept 2022

Ward A, Wilson A (2009) Design for adaptation: living in a climate-changing world. Building Green. https://www.buildinggreen.com/feature/design-adaptation-living-climate-changing-world. Accessed 23 Sept 2022

Chapter 2
Denser Living

Abstract Design and construction of future dwellings pose several challenges. They must be functional, comfortable, energy efficient, and address lifestyle trends. To be sustainable, communities also need to be dense and have a low carbon footprint. This seems to be a tall order for designers, builders, and homebuyers. Yet, it offers opportunity to be creative and address issues that in the past were considered marginal. This chapter provides an insight into the conception of sustainable high density residential communities and homes. Planning aspects include mobility, open spaces, and identity. On a smaller scale, the chapter addresses issues related to the dwelling's form.

Keywords Attachment · Density · Energy · Footprint · Ground relation · Liveability · Sustainability · Volumetric arrangement

2.1 Urban Planning Aspects for High-Density Environments

In the last three decades, the number of world's megacities has tripled forcing a need to plan for a high-density sustainable future (Wong et al. 2016). Sustainable residential design is a multi-faceted topic which requires careful planning and architectural conception to be successful. High-density communities and dwellings are often viewed as the most sustainable model for efficient land and energy use. The favorable fundamentals of such housing include longevity, flexibility, and walkability (Bunbury 2015). Regarding energy consumption, Europe for example, an estimated one-third of the energy use is related to the housing sector, and another third goes into transport (Marckmann et al. 2012). Additionally, Fuller and Crawford (2011) found that transportation energy use per capita can be reduced by up to 96 percent in inner city high-rise apartments compared with traditional suburban detached homes. Higher density neighborhoods justify the introduction of public transit thereby limiting dependence on private cars and reducing greenhouse gas emissions (Balling 2016; Steiner et al. 2013).

A. Friedman, *Fundamentals of Innovative Sustainable Homes Design and Construction*, The Urban Book Series, https://doi.org/10.1007/978-3-031-35368-0_2

With high-density dwellings come unique design challenges and considerations since designers must explore different typologies, forms, and planning arrangements. This section discusses key elements of high-density environments including form, identity, mobility, and open spaces.

2.1.1 *The Form of Communities*

The form of densely built environments is important to their liveability. A concentric model streamlines access to central services and allows the neighborhood to grow naturally around a central district. Lee et al. (2018) identify that neighborhood characteristics associated with walkable access to amenities are mixed-use and have multifamily dwellings (Fig. 2.1). Therefore, the success of high-density neighborhoods depends among others on a diversity of the built form. Additionally, Wood (2015) found that positive characteristics identified in denser communities were decreases in car dependencies and increases in social interaction, patronage of local business, and neighborhood vitality.

Cohesive organization of buildings is key to establishing character within the urban fabric. High-density towers should be placed strategically in higher traffic areas while medium and lower density developments can establish a more human scale on secondary streets (Bharne 2011) (Fig. 2.2). In high-density urban environments, maintaining a sense of privacy for residents is important. A perceived lack of privacy can result in social withdrawal leading to poor mental health (Lindsay et al. 2010). Additionally, Lindsay et al. (2010), state that noise pollution which may result from increased density can lead to poor relationships between neighbors. Knowing this, designers should pay careful attention to maximize a sense of privacy by considering window placement and sound barriers.

Social cohesion, which includes fostering a sense of belonging, community and encouraging citizen participation, is an important factor in the character of a neighborhood. In general, social interaction is stronger in places with high-quality spatial organization, creatively built environments, and embedded public amenities (Raman 2010). Furthermore, Raman (2010), found that resident's number of social contacts declined in more isolated high-density environments, such as high-rise buildings or dead-end streets. Therefore, access to communal space is important to fostering social cohesion and engagement.

2.1.2 *Planning for Active Mobility*

Transit oriented development (TOD) is key to minimizing greenhouse gas emissions (Fig. 2.3). It requires diminishing reliance on cars by locating dwellings close to public transit hubs, jobs, and community services (Trubka et al. 2010). Supportive of this is Hammoud et al. (2018) who propose that future sustainable cities ought to

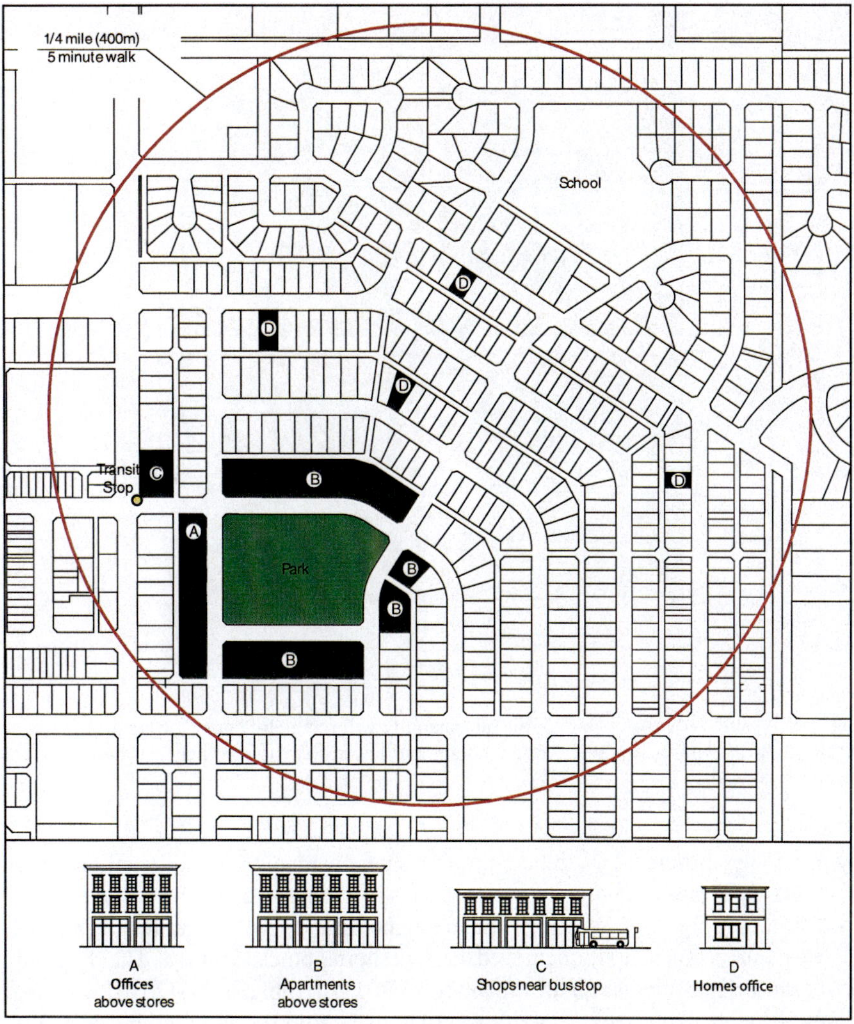

Fig. 2.1 Neighborhoods associated with walkability are commonly mixed-use and include multi-family dwellings

follow a multi-nodal polycentric model planned around transit hubs, with multiple central business districts. As noted above, organization in concentric layouts reduces the travel distance between the center and outer edges of a city. Important amenities such as transit stations, community services, and grocery stores should be located strategically within a reasonable distance from residential areas.

According to Lee et al. (2018), walkable neighborhoods are increasingly high in demand. The two primary factors affecting the walkability of a neighborhood are the quality of the walking environment and the availability of amenities within walking distance (Lee et al. 2018). Walking environment can first be made safer for pedestrians

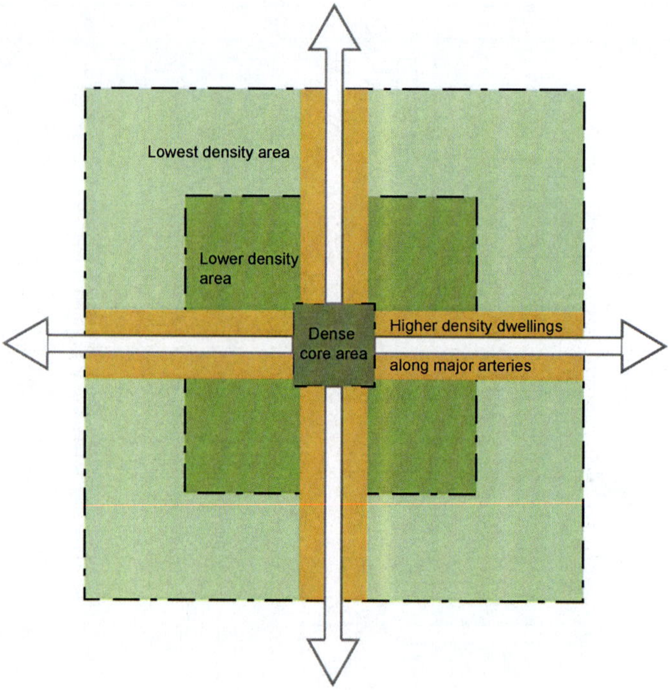

Fig. 2.2 A preferred urban planning for sustainability will see high-density towers placed strategically in areas with busy traffic while medium and lower density dwellings can be located on secondary streets

by improving motorists' awareness through safety campaigns. Additional measures to improve pedestrian safety include reduced speed limits, road bumps, and textured road surfaces (Fig. 2.4). Walking time in neighborhoods can be diminished through increased street connectivity, achieved through shorter block lengths and land parcels, fewer cul-de-sacs, and straighter roads (Oka 2011). Trubka et al. (2010) state that people are more likely to use active modes of transportation such as walking or cycling the closer they are located to their desired destinations. Therefore, the integration of commercial amenities within residential neighborhoods is crucial to improving walkability. Cycling should also be prioritized as a means of active transportation. Adequately marked paths, signage, and traffic calming devices on roads are integral in facilitating bicycle usage. Additional bike parking and implementation of bike sharing programs can also make cycling more accessible.

Public transit is key in facilitating local active mobility and reducing automobile use. Dense communities are necessary to the implementation of good quality public transit, as it is only considered a viable investment when there is sufficient ridership to sustain their services (Fig. 2.5). Transit systems consisting of buses, subways, and trains should provide a network of connections within and around urban areas and be

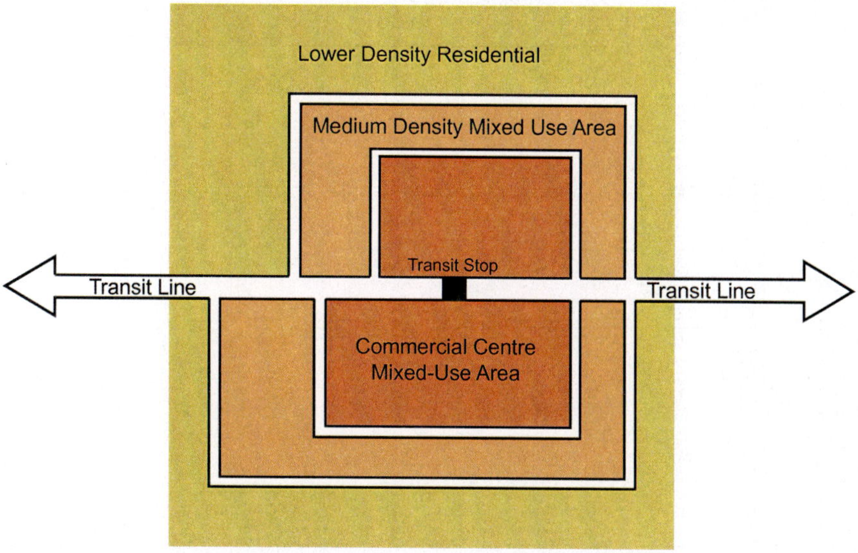

Fig. 2.3 Transit oriented development (TOD) minimizes greenhouse gas emissions due to the use of public transportation by their occupants

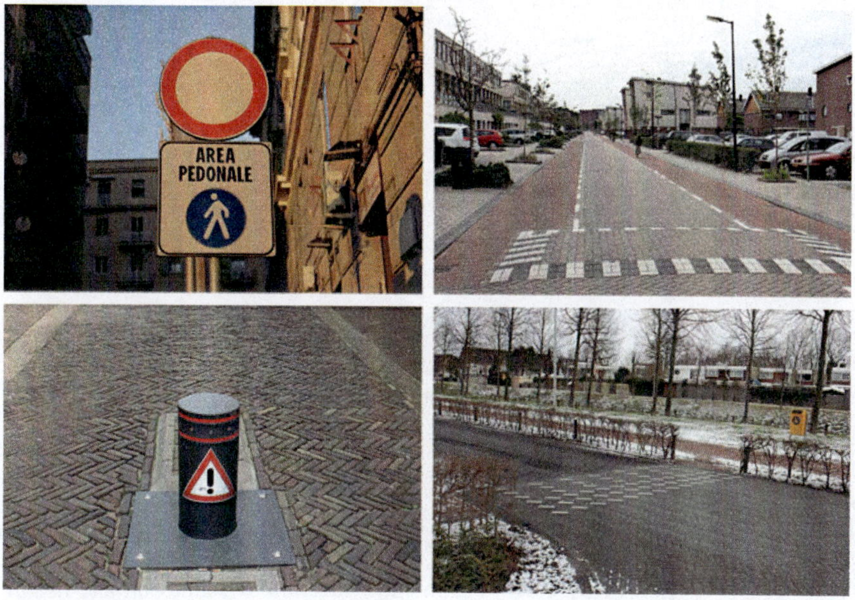

Fig. 2.4 Walkability increases when safety measures are introduced and include reduced speed limits, road bumps, and textured road surfaces

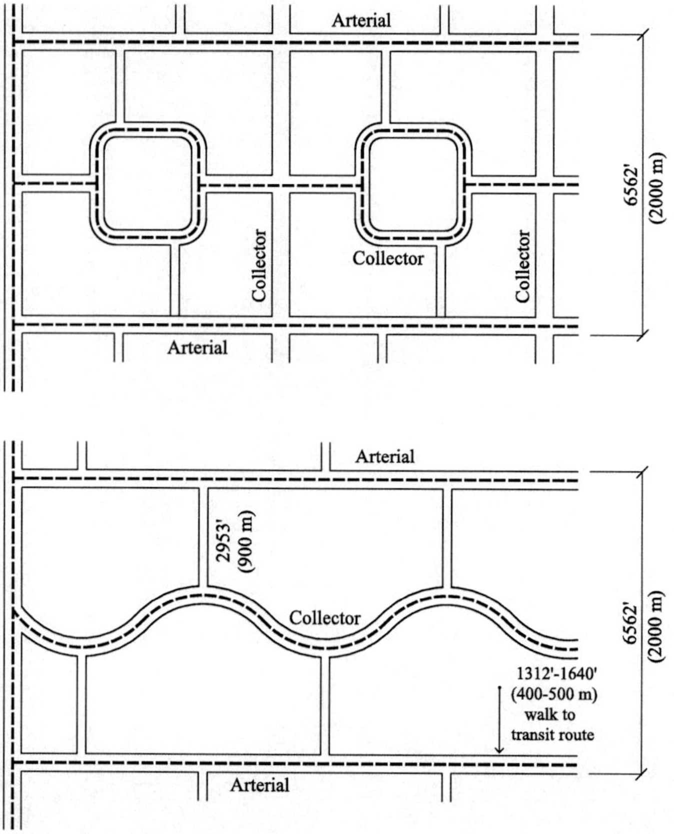

Fig. 2.5 The spacing and distance of bus stops from homes and each other are critical to consider when planning public transit

accessible to most inhabitants in less than a ten-minute walk from home. Dedicated vehicles to accommodate individuals with reduced mobility and seniors should also be considered in the implementation of public transit services.

2.1.3 Green Open Spaces

Open spaces in high-density environments have several environmental and health benefits. They can help mitigate the urban heat island effect which contributes to increase of the temperature in urban environments (Hammoud et al. 2018). Raised urban temperatures contribute to increasing energy usage, air pollution, and heat caused illness. Since cities produce 75 percent of total emitted CO_2, forests are one of the most efficient ways to mitigate greenhouse gas emissions, absorbing 35 percent of emitted CO_2 (Hammoud et al. 2018). Additionally, there are very clear links between access to nature and improved mental and physical wellbeing. Access to the natural environment is linked to improved cognitive function in children and adults. Adults, for example, are three times as likely to be physically active when living in residential areas with large green areas (Wells et al. 2010).

Open spaces require careful planning in high-density residential developments where a mix of private and public outdoor space and a variety of scales is recommended (Fig. 2.6). For homeowners, having a private outdoor space is associated with leisurely activities, and is commonly viewed as an extension of the home itself. In denser urban developments, private outdoor spaces can consist of front or back yards, patios, decks, balconies, or roof terraces. On the other hand, shared green spaces are important to the liveability, public engagement, and experiencing nature (Coolen and Meesters 2012). Public open spaces can be divided into a variety of sizes, promoting different uses—for gatherings from more intimate to larger scales. A series of several smaller green spaces distributed throughout a dense neighborhood is recommended to achieve the above noted goals (Coolen and Meesters 2012).

Communal outdoor spaces in high-density neighborhoods can facilitate social interactions and be bounded by buildings to maintain a sense of privacy and security. There should be choice and a clear distinction between places for dynamic and casual activities (Fig. 2.7). Parks can provide venues for sports and playgrounds which

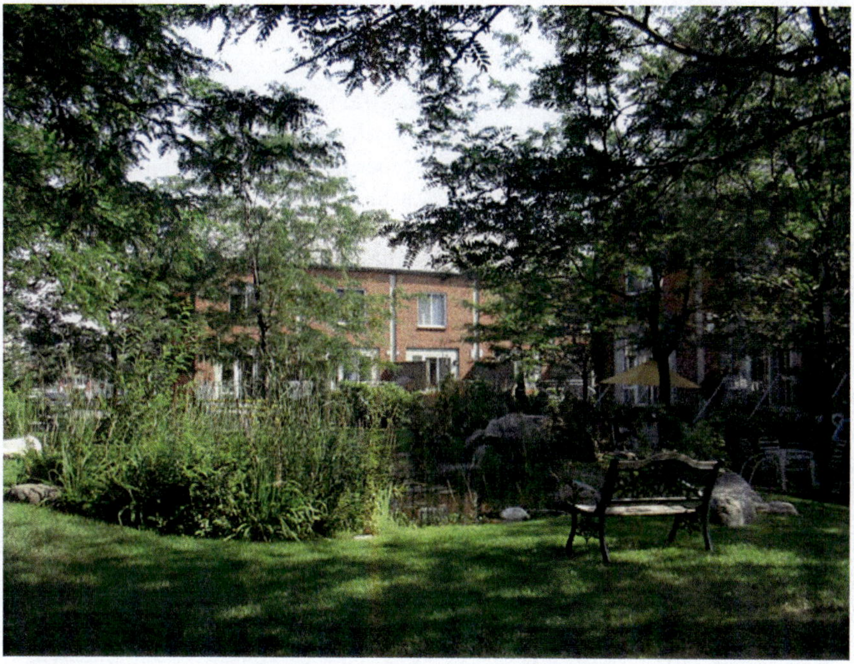

Fig. 2.6 Open spaces require careful planning in high-density residential developments. A mix of private and public outdoor spaces at a variety of scales is recommended

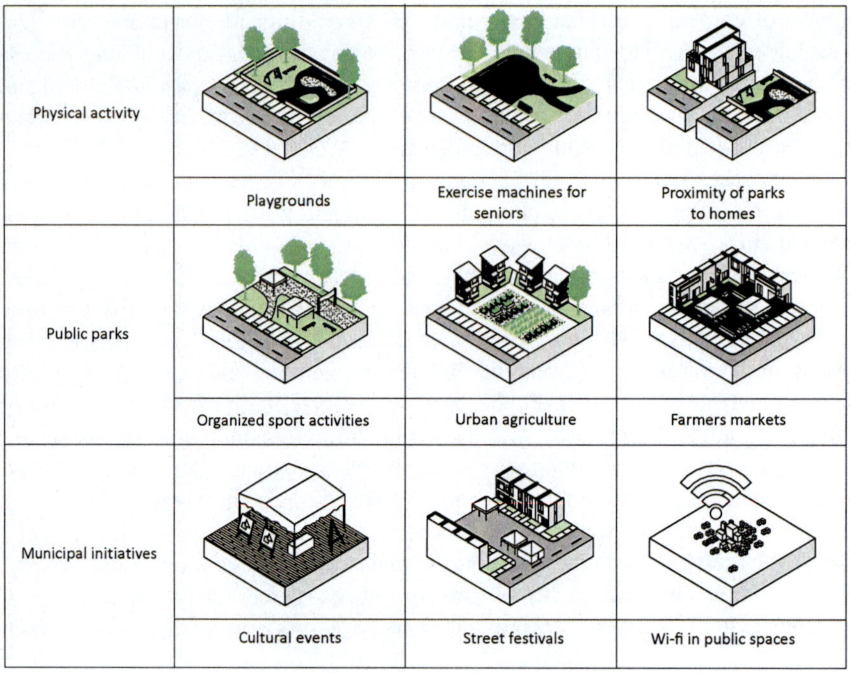

Fig. 2.7 Key aspects that make outdoor spaces welcoming and engaging

facilitate high levels of physical activity (Oka 2011). Passive areas for socialization, picnics, and reading should be considered as well, ideally located in contained areas further from roads. Continuous green landscapes help to maintain biodiversity and green corridors can be used to link parks to achieve this continuity. Planners should integrate sustainable ecological systems into green spaces. According to Faehnle et al. (2015), such systems may include urban agriculture for nutritional benefit, solar and wind energy, micro-climate regulation, air purification, and water collection.

2.2 Higher Density Dwellings' Forms

The chosen form of a dwelling is one of the greatest determinants of its sustainability. One of the main considerations in designing sustainable housing developments is energy consumption. This can generally be divided into two main categories: embodied and operational energy. *Embodied energy* refers to the energy used in the production of the building's materials, while *operational energy* refers to the energy consumed in the operation of the building, such as appliance use, heating, and cooling. Both embodied and operational energy consumption can be optimized with the right selection of a dwellings form. Fuller and Crawford (2011) found that the average per capita energy use associated with housing in Australia rose by approximately 337 percent between 1950 and 2000. Additionally, the size of the Australian average house tripled between 1950 and 2016, despite a general reduction in number of household members (Hoang and Vandal 2017). Therefore, in the design of higher density sustainable housing, having more energy efficient dwelling's types is critical. This section explores macro design aspects of high-density housing including typology, footprint, volumetric arrangement, attachment, and ground relation.

2.2.1 Typology and Footprint

There are several medium to high-density housing types which may be considered by designers to lower cost. Building multi-level homes increases cost and material savings by reducing the required foundation and roof areas while maximizing floor area. A *duplex* refers to a multi-level structure that is vertically divided into two units, saving up to 50 percent of roof and foundation area. Similarly, a *triplex* is a multi-level structure divided into three independent units, saving up to 66 percent of roof and foundation area compared to three detached, single-story residences. Additional saving will be obtained in four to six story tall apartment buildings which may or may not include an elevator and a high-rise apartment building.

In terms of carbon footprint, higher density dwellings are typically more energy efficient. Obrinsky and Walter (2016) found that single-family detached homes consume the most energy per household, attached homes in small multifamily developments consume less energy per household, and units in large multifamily buildings consume the least energy of the three (Obrinsky and Walter 2016). Using prefabricated construction methods can further reduce a building's embodied energy and material waste (Hoang and Vandal 2017).

The type and footprint of a dwelling affects land use and therefore its density and overall sustainability (Fig. 2.8). Commonly, the footprint of a dwelling can be either square or rectangular. While a square footprint makes subdivision of the floor plan easier, more rectangular, narrow-front designs typically permit higher density developments. The unit's footprint may also be determined by the number of bedrooms the designer wishes to include. Often, the width of rectangular dwellings will be the outcome of the number of bedrooms on either end of the home (Fig. 2.9).

The internal organization of a dwelling should aim to maximize space and flexibility for the occupants. Increased ceiling heights can help make units with smaller dimensions feel more spacious, as inhabitants are more likely to notice lower vertical heights than horizontal dimensions (Hoang and Vandal 2017). As noted by Fuller and Crawford (2011), the average household often moves several times over the course of many years into increasingly larger homes. Therefore, to avoid a move, providing opportunity for expansion and designing interiors to be flexible is essential. Strategic placement of entrances, stairs, circulation, and utilities can enable later modification of spaces between single and multifamily dwellings to facilitate multigenerational long-term occupancy.

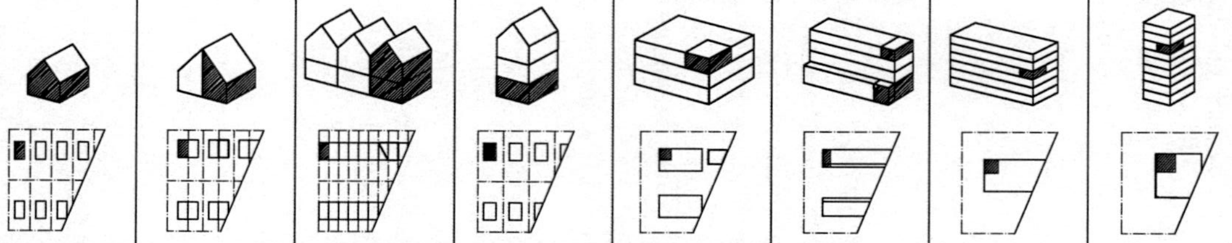

Dwelling type	1 Single detached	2 Semi detached	3 Row house	4 Triplex	5 3-story walk-up apartment	6 Combined apartments & row houses	7 Slab block apartment	8 High rise point block apartment
Isometric								
Plot plan								
Dwelling units/acre (dwelling units/hectare)	8 (20)	14 (35)	19 (47)	21 (52)	65 (160)	84 (207)	90 (222)	120 (296)
Floor area ratio % open space	0.24 76%	0.38 81%	0.56 72%	0.60 80%	1.36 55%	1.92 62%	1.78 62%	2.62 87%
Unit area in square feet (unit area in square meters)	1200 (111.5)	1200 (111.5)	1200 (111.5)	1200 (111.5)	800 (74.3)	800 & 1200 (74.3 & 111.5)	800 (74.3)	800 (74.3)

Fig. 2.8 The footprint of a dwelling effects its density levels which in turn effects the surrounding land use and therefor the project's overall sustainability

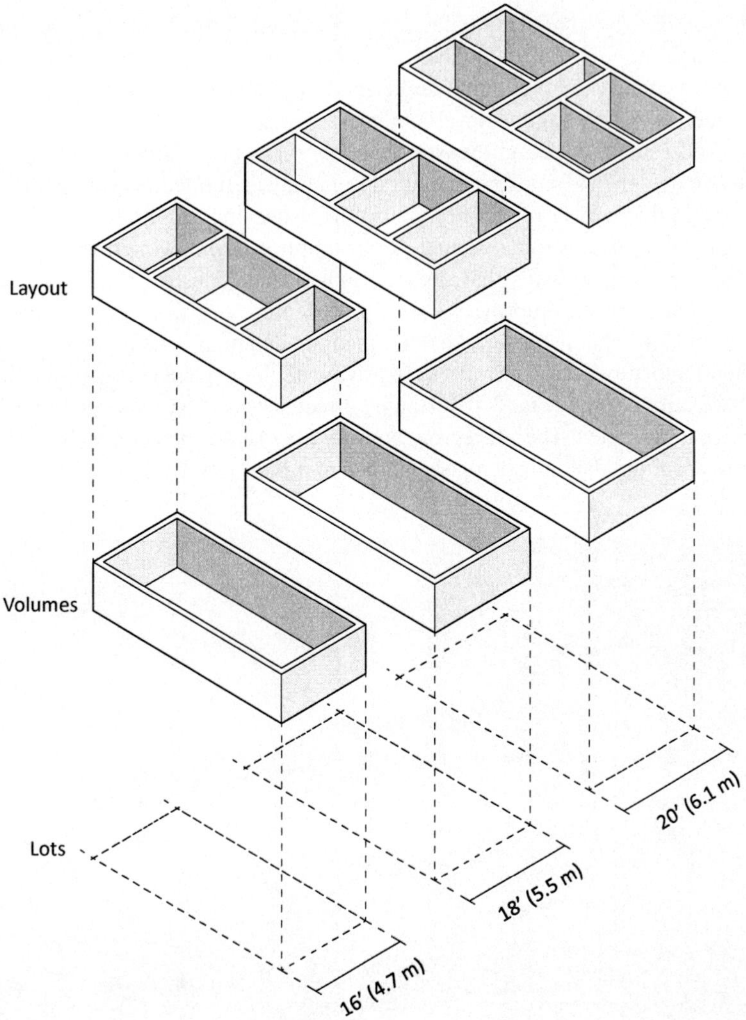

Fig. 2.9 The width of rectangular dwellings is based on the number and arrangement of bedrooms

2.2.2 *Attachment*

Attached dwellings are a more sustainable choice in high-density communities than detached houses since they improve material and energy efficiency. Heat loss is significantly reduced because developments with attached dwellings will have overall less envelope than equivalent sized detached dwellings, helping to reduce energy use. Operational energy consumption in low-rise attached dwellings can be up to 20 percent lower than detached dwellings of equivalent size (Wright 2010). Dwellings can be attached either as semi-detached units or in rows, typically composed of four to eight houses. Mixed housing types can be achieved in row housing developments by increasing the width of units on either end, to create larger multifamily homes. Row houses can also be made into duplexes or triplexes, to accommodate a variety of household sizes within one development.

Row houses and other high-density dwelling types bring a unique set of considerations for designers. Exposure to natural light may be limited in attached narrower dwellings, as the number of exterior walls diminishes opportunity for sun exposure. Inadequate sunlight may increase energy consumption during wintertime. Additional care must be given to walls that separate units. Their construction should reduce noise transmission and improve safety by preventing fire from rapidly spreading between units. Row housing, unlike other medium and high-density alternatives, has the potential to allow each unit access to private outdoor space in the form of a front yard, backyard, or both (Fig. 2.10). Having direct access to a private garden is often preferable to residents. Therefore, row housing may be an appropriate, high-density option to accommodate this demand (Coolen and Meesters 2012).

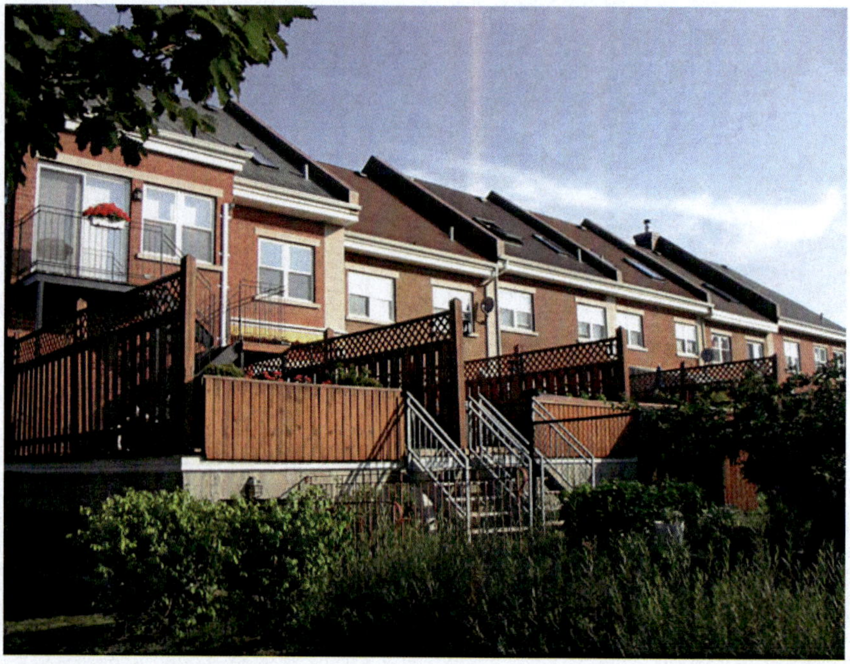

Fig. 2.10 Row housing allows each unit access to private outdoor space in the front, rear, or both

2.2.3 Foundations

There are several types of foundations which may be used in medium and high-density residential developments with varying levels of energy efficiency and costs (Fig. 2.11). The simplest option is the *slab-on-grade*. It requires the least concrete and can typically be constructed in a single pour. The thickness of a slab-on-grade is determined by climate and structural requirements. Another alternative is a crawl space foundation, which may be used when excavation is rendered difficult by the ground condition. *Crawl space* foundations are created when an elevated floor level is raised on a foundation of poured concrete, concrete blocks, piers, or piles. The slab-on-grade and crawl space constructions are the most efficient in terms of material use and greenhouse gas emissions, as they require the least amount of concrete.

The *basement* is the third type of foundation which requires the most concrete, excavation, and formwork. All of which contribute to increased embodied energy and greenhouse gas emissions through the required construction process. As a result, if reducing cost and carbon footprint is sought, it is environmentally favorable to forego the construction of a basement. However, there are certain cases which justify the addition of a basement level. A basement may facilitate increased residential density, in the case of a duplex where each unit is given one to two levels, or a triplex with three independent units. Additionally, a basement may be dug out but left unfinished, to be completed later, allowing for indoor expansion to accommodate an evolving household. Different types of ground relations are therefore favorable in different circumstances to maximize the sustainability of medium and high-density residential developments.

2.3 Denser Dwellings' Design

In designing compact dwellings, designers are challenged to create a comfortable, liveable interior space in a smaller area along with the challenge of maximizing energy efficiency and limiting material waste. This involves rethinking standard elements of larger homes to better suit the objective of a high-density environment. This section discusses micro design aspects which shape the character of high-density dwellings including dimensions, access and egress, and roof design.

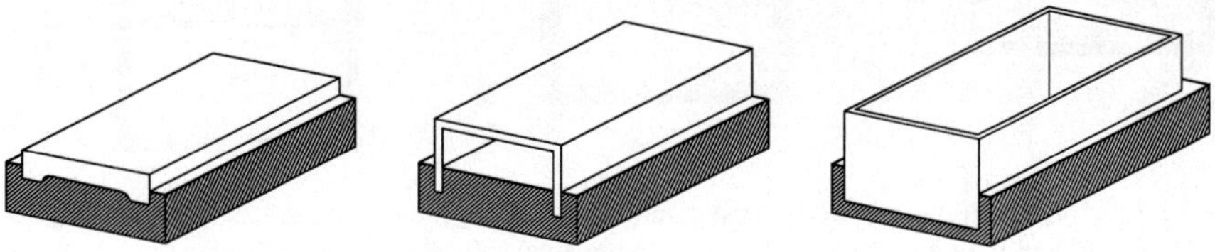

Fig. 2.11 The common ground relations of dwellings are slab-on-grade (left), crawl space (middle), and an in-ground basement foundation (right)

2.3.1 Dimensions and Orientation

The design of high-density dwellings requires strategic use of space to maximize the sustainability and livability of a residence. The primary consideration in determining the dimensions of a dwelling are the regulatory setbacks of the property. Minimizing the front setback can increase the residential density. The length of a unit is also determined by the chosen housing typology and the overall density of a project. Designers may comfortably fit all amenities in a unit as narrow as 3.7 m (12 feet) wide with strategic design and organization of space. The length of a unit can range from 9.8 to 12.2 m (32 to 40 feet) to comfortably fit a kitchen, dining area, living area, and service requirements (Fig. 2.12).

The dimensions of a unit are determined by the interior organization of rooms. Conventionally, as outlined by the Americans with Disabilities act, the minimum habitable size of a living room is 14 square meters (150 square feet), making the minimum room width dimension 3.7 m (12 feet) (Hoang and Vandal 2017). A 4.8 by 12 m (16 by 40 foot) floor plate is highly flexible. A home of this size does not typically require internal load-bearing partitions, which maximizes flexibility for providing optimal interior organization and future adaptability. These dimensions also typically permit modular prefabrication, which results in material and transportation savings, reducing the embodied energy of a project. The minimum habitable room height is 2.4 m (7.9 feet), however increasing the ceiling height of a denser unit creates larger volumes with smaller floor areas (Hoang and Vandal 2017). High ceilings also contribute to a sense of spaciousness. Additionally, using multiple vertical dimensions, such as a lower ceiling height in entryways, distinguishes spaces making smaller units more comfortable (Fisher-Gewirtzman 2017).

When taller and narrower homes are designed, special consideration should be given to sunlight access, particularly in row houses which have fewer facades. In the northern hemisphere, sunlight primarily enters through the southern façade, which

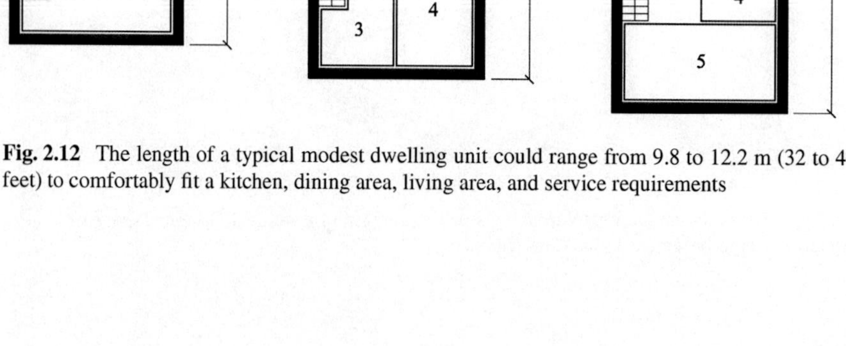

Fig. 2.12 The length of a typical modest dwelling unit could range from 9.8 to 12.2 m (32 to 40 feet) to comfortably fit a kitchen, dining area, living area, and service requirements

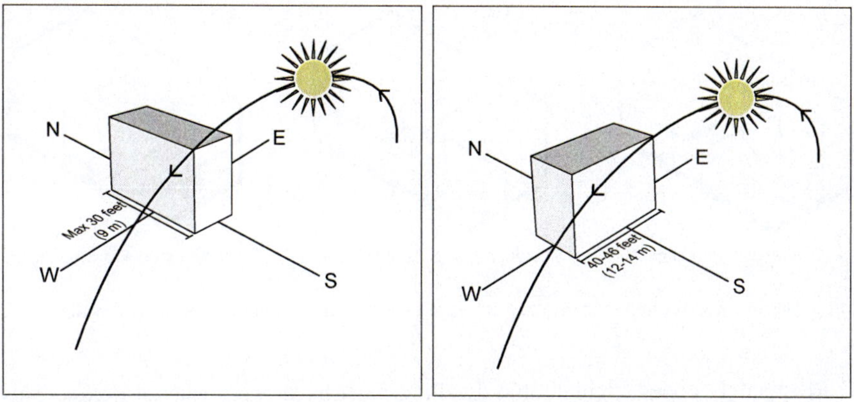

Fig. 2.13 The amount of daylight that enters the core of a townhouse will be determined by its orientation. A well-fenestrated east–west oriented end unit will have the most light

may affect the chosen dimensions of a residential development. In north–south-oriented rowhouses, designers should consider increasing the width and decreasing the depth of units to maximize the amount of sunlight that reaches the middle of the dwelling. Conversely, in east–west oriented units, an increased length can be more easily accommodated because sunlight enters in both directions (Fig. 2.13). Sunlight access is an important consideration for the energy performance of a building and should therefore be considered in determining the dimensions of a unit.

2.3.2 *Access and Egress*

Modes of access and egress should be carefully designed to ensure the safety and comfort of inhabitants in high-density dwellings. The main entrance establishes the identity of a dwelling and defines the boundary between the outside and inside, or public and private. In narrower units, the primary entrance is often best suited along one of the end walls (Fig. 2.14). This limits the disruption of interior spaces in units with reduced floor width. When it comes to duplexes and triplexes, entrances at the ground floor can be located at grade to simplify accessibility. Upper units can be. Accessed through common or private stairs located on the interior or exterior of the dwelling. Added volumes to the front or rear of attached houses can add variation in appearance and access configurations. Similarly, staggered units or recessed entrances can be used to create a more established boundary between units from each other and the street. In smaller units, entryways with reduced heights can create a defined subspace, giving residents a sense of transition into the main living space (Fisher-Gewirtzman 2017). These techniques however increase the surface area of the envelope, potentially increasing heat loss therefore should be sufficiently insulated if used.

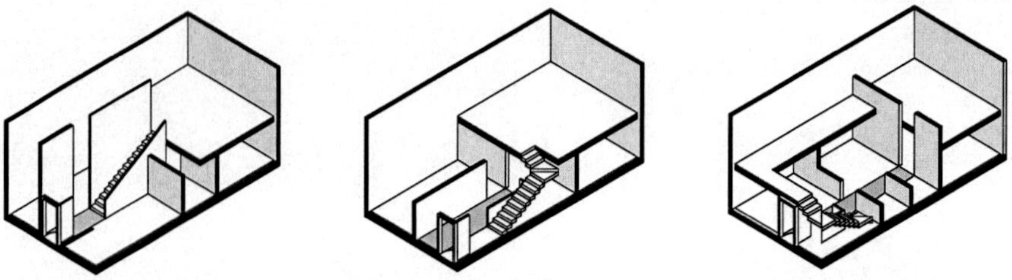

Fig. 2.14 In narrower units the primary entrance is best located along one of the end walls

Most safety codes stipulate that dwelling must have two means of egress. This can become a design challenge when residential density increases. The rear egress of a townhouse oftentimes provides access to a private or communal outdoor space. The transition between these functions should be designed carefully to maximize comfort and livability. For duplexes and triplexes, designers may introduce a private balcony on upper levels, leading to stairs, providing private outdoor space for residents while accommodating a secondary exit. When it comes to higher density apartment buildings, it becomes more of a challenge to facilitate safe emergency exit pathways for inhabitants. One proposed solution as high-density towers continue to grow vertically is the skybridge. Skybridges serve the purpose of creating more than one horizontal plane of connection, by linking high-rise towers at an above ground level. They provide alternate means of circulation through cities, making access and egress simpler for residents of high-rise buildings. Skybridges can improve safety by providing additional means of evacuation through a linked tower, making it easier for inhabitants to escape the building rapidly (Chow et al. 2013). Access and egress are important design considerations for both the safety and liveability of high-density residential developments.

2.3.3 *The Roof*

The design of the roof should also be uniquely considered in high-density developments. Roofing style, material, color, and form should be dictated by local temperatures, wind, sunlight, snow load, and rain to maximize efficiency (Fig. 2.15). A *flat roof* is the easiest to construct. Similar to a typical floor, joists are placed and covered with roofing materials. A flat roof can provide opportunities for future expansion if the designer ensures that the structure is sufficient to support another floor in the future. Flat roofs however permit less space for insulation, which can diminish a building's energy efficiency (Turner 2017). Proper roof insulation is key for energy efficiency because heat rises, if a roof is not sufficiently insulated it will result in significant heat loss. Additionally, flat roofs may collect more debris, and require sufficient drainage systems to prevent moisture-related damage.

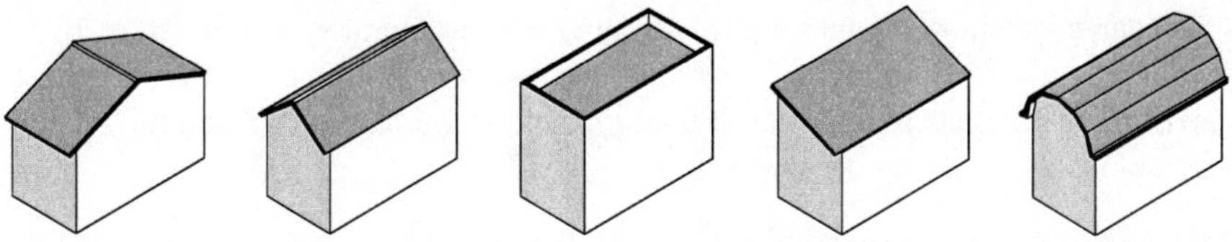

Fig. 2.15 The factors affecting the chosen roofing style, material, and color are commonly dictated by local building codes, and weather phenomena such as wind, sunlight, snow load, and rain

Pitched roofs can be constructed using prefabricated trusses. Off-site prefabricated trusses significantly reduce material and energy waste, improving construction efficiency. Pitched roofs allow for snow and rain to drain off, reducing the chance of weather-related damage and increasing the roofs service life. Wind, however, must be taken into consideration in the design of pitched roofs. Higher slopes experience a greater wind load, and therefore require greater structural reinforcement (Turner 2017). A higher slope may be worth the additional effort, however, if it is designed to be habitable. To provide natural light and ventilation to a livable attic space, windows or skylights can be used. Windows located on the roofs gable ends are recommended to maximize energy efficiency, as skylights and roof dormers are associated with more heat loss. Alternatively, passive ventilation can be achieved through the placement of vents on opposite ends of the roof to facilitate cross ventilation (Turner 2017). In general, designers should avoid more complex roof structures as more joints and surface area increases material use and potential for heat loss.

The application of these strategies will vary depending on the type and density of housing being designed. In row housing configurations, one continuous roof structure across dwellings is generally favorable because it limits the number of joints, which may lead to water damage or heat loss. In this case, the line of the roof should run perpendicular to the walls joining units. In larger apartment developments, designers may consider the opportunity for a rooftop terrace. A communal rooftop space can facilitate additional community interaction in high-density areas.

Implementation of a green roof is an additional design feature which makes a building more sustainable. A *green roof* is any type of roof that is designed to grow plants (Fig. 2.16). The structure starts with a waterproofing system, root barrier, and drainage system, onto which a layer of soil or aggregate is placed where plant species can be grown (Turner 2017). Green roofs help to naturally cool the building by reducing the amount of direct solar gains, the soil also provides a layer of insulation which regulates the internal temperature year-round (Clarke 2017). Green roofs can take the form of flat or gabled roofs depending on the budget and desired appearance. They may require some additional structural reinforcement. However, with the right choice of plants and organic material, some green roofs can weigh 120 kg/m^2, which can be supported by a relatively lightweight structural system (Clarke 2017). Overall, green roofs are highly beneficial because they promote biodiversity, limit water runoff, and help to improve air quality, making them worth considering in new developments.

2.3.4 Mixed Use Residences

The common tendency in planning low-density communities is to segregate residential and non-residential activities. Such practices encourage reliance on private cars and the building of an extensive road network. Proximity and accessibility to basic services such as groceries, schools, childcare centers, and medical clinics are essential. Locating them within reach of every home and having residential and non-residential functions in the same building will make them accessible and usable by pedestrian and cyclist. Moreover, such establishments need to be designed for changes and adaptability to economic or functional emerging realities when they take place. Mixing amenities with residences is therefore a key to successfully creating well-functioning sustainable and affordable communities.

As for implementation, the mix of land uses is governed by zoning regulations and bylaws that dictate how and where buildings and other civic functions such as parks be located, the types of permitted buildings, lot sizes, parking requirements, heights, and setbacks from the streets to name a few. Those regulations also specify the preferred types of amenities forbidding, for example, a polluting factory near dwellings. Zoning of what and how many amenities can mix with residences commonly differ from one community to another since the density, and the population would also be different.

Mixed communities also help with greater preservation of a site's natural features and a reduction of forest clearance. Creating green open spaces for passive or active recreation use is great for inhabitants of all ages. Furthermore, public transit can be integrated according to the lay of the land. The design of the streets can follow the site's curves and slopes, and bus stops can be located near trees to provide shade during summer (Raman 2010).

Mixed dwellings also support those who work from home, a population comprising a substantial sector of today's workforce given the power of digital communication. Mixed communities can in themselves offer these residents varied and stimulating lives, reducing their commutes to city centers. As well, growing demographic contingents including singles, childless couples, seniors, and young, career-oriented professionals are creating demand for non-traditional housing, and have lifestyles suitable to mixed developments.

Methods related to a neighborhood's *spatial organization* focus on horizontal and vertical separation of space and cluster organizations (Raman 2010). Horizontal separation of space divides land into areas dedicated by use and places all areas within walking distance from the home. Typically, commercial space would be located along street-level, with residences placed behind the stores. A method related to horizontal

Fig. 2.16 A green roof is any type of roof that is designed to grow plants. The three different types are extensive (top), semi-intensive (middle), and intensive (bottom)

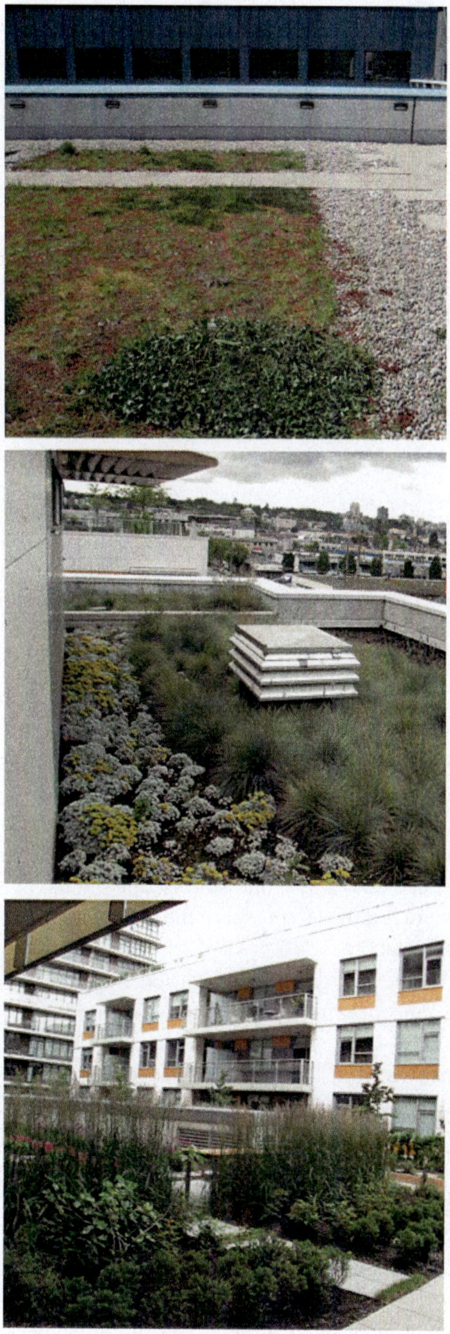

separation of space is to group buildings within a district, known as *cluster combination* (Raman 2010). Streets, parks, plazas, and squares separate individual buildings, forming urban districts within a single neighborhood. Both of these horizontal models require more land than a vertical separation of space, which entails making distinct separations within a high-rise building, segregating the various functions within.

If a building is intended to contain retail, office, and residential functions then it is advisable to arrange the work as follows: retail occupying the ground floor, offices above, and residential within the remainder. Also, it is important that each function have its own, separate entrance. As to parking, it is advisable to accommodate retail parking to an on-street location while residential and office parking is better placed behind or underneath.

2.4 Weaving-In a Denser Building in Melbourne, Australia

The 2 McIntyre Drive project by MGS Architects is composed of 69 apartments in Altona, a suburb of Melbourne in the State of Victoria, Australia, and is notable for its integration of high-density housing in a low-rise surrounding context (Horrocks 2013; Ray 2014). It reinterprets the suburban vernacular over multiple levels to create a connection that extends from the street to the main entrance (Fig. 2.17).

This project is the only apartment complex in an area characterized by single-family dwellings. Funding for the development came via the Social Housing Initiative through the Commonwealth's Nation Building economic-stimulus plan. In accordance with the objectives outlined by those funding the project, the building provides independent living accommodation for residents with disabilities and includes communal spaces, community gardens, and private open space. Parking spots, driveways, and streetscape plants, in addition to several existing mature trees, were retained to preserve the established character of the site which borders a council reserve.

The design process started by establishing a "U"-shaped footprint to create a connection from the street to the main entrance. The center of the "U" is a gathering space—a pleasant entranceway in which to wait for a friend or while away an hour or two. It also provides a natural point of visual interest for those gazing out their windows. The architects developed this space as a large courtyard reminiscent of the perimeter block plan used in European cities. It includes a communal garden tended by residents, as are the front yards of the ground floor tenants to create a pleasing, green facade.

It is worth reflecting on this building as a reinterpretation of affordable housing's relationship to both physical context and social status. Identity and a sense of place were intentionally encouraged in the design as an attempt to create an engaging communal environment. Yet, the designers were conscious of the fact that simply transplanting architectural values into a building that houses those in need, without sensitivity to context, risks a certain tackiness. It was critical to them that the design process considers planning, scale, material, color, and form in accordance with the

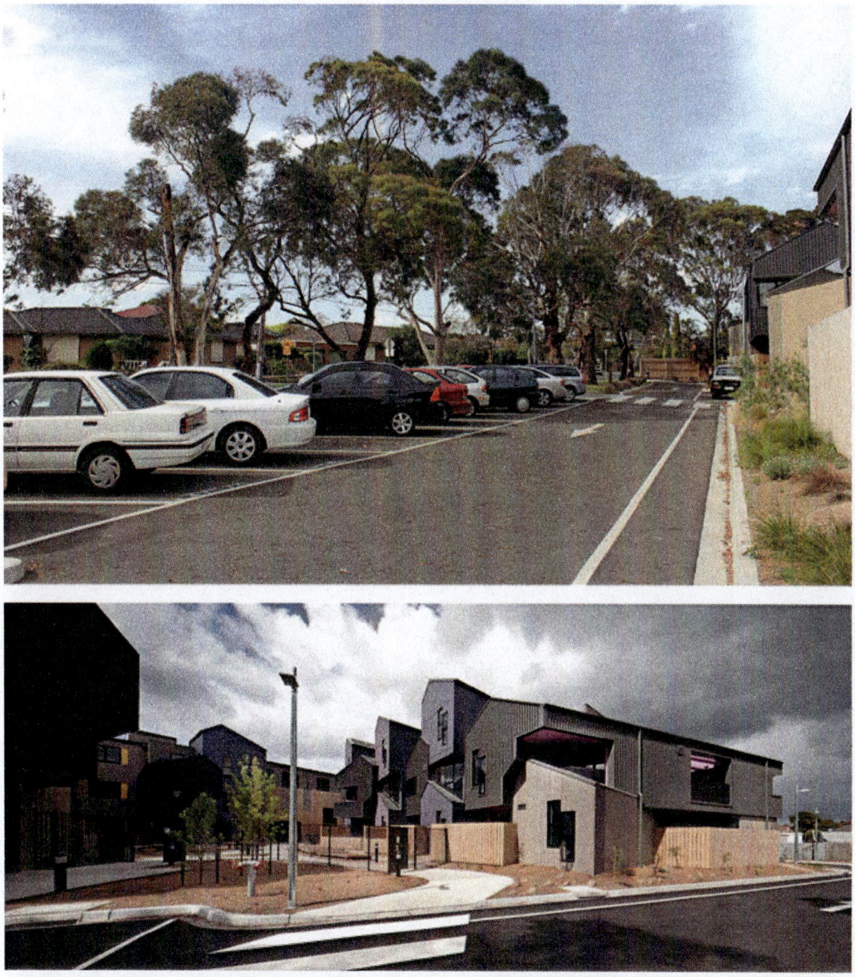

Fig. 2.17 The 2 McIntyre Drive project is composed of sixty-nine apartments in Altona, Melbourne, Australia. The project is notable in its attempt to integrate a higher density housing development into a surrounding suburban neighborhood

project's all-important social dimension (Fig. 2.18). An example of this approach is the nature of the design's generously landscaped courtyard, which serves as the focus of the visitor and resident experience. It offers retreat, recreation, and an opportunity for social interaction, vital to those in the community, all while consciously integrating the building into the suburbs.

The design considered the location and arrangement of a group of existing two-storey buildings that had previously served as nursing homes. In fact, to retain a sense of established order, the designers kept the alignment of the previous building footprint to the south. They also established planters, driveways, and parking along

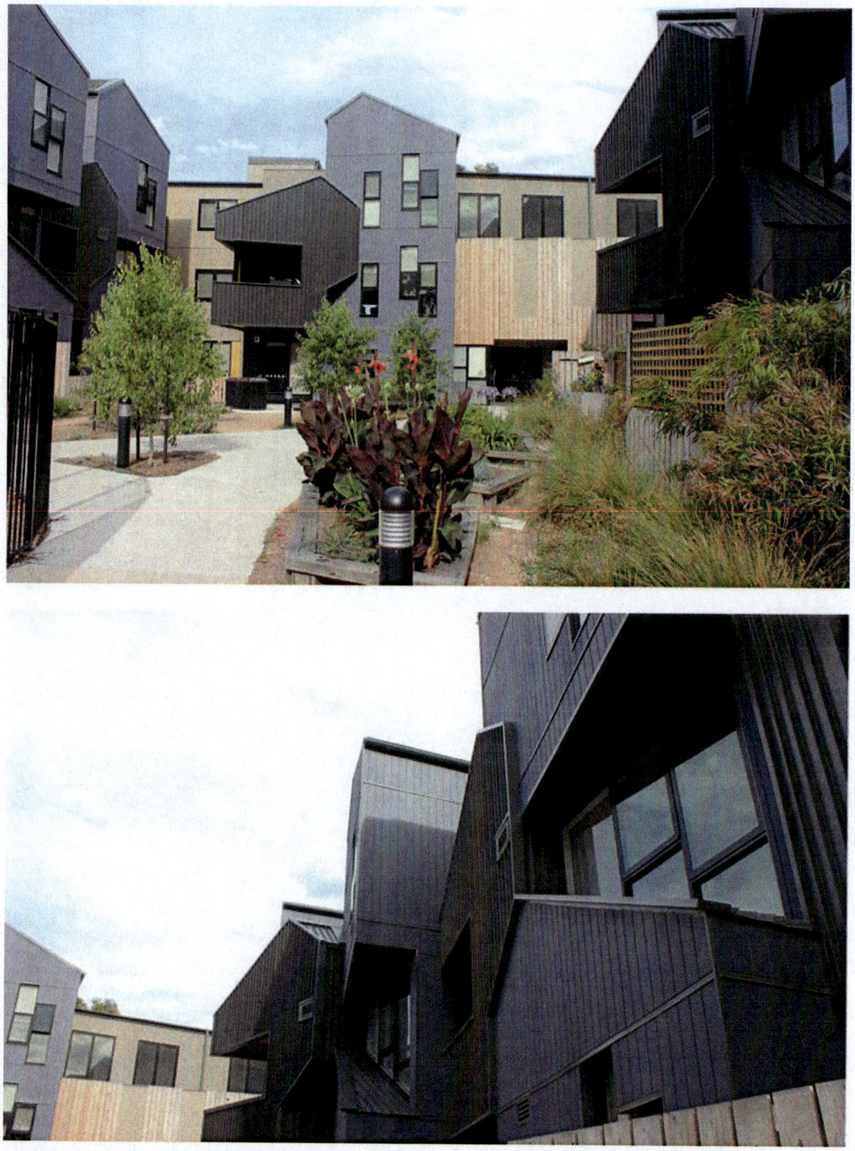

Fig. 2.18 The design process of the 2 McIntyre Drive project considered planning, scale, material, color, and form, all in accordance with the project's all-important social dimension

McIntyre Drive. The generous setbacks to the street and to the boundaries with neighboring residential properties were also retained from the previous structure. The project is respectful of the amenity of the suburb. It mediates streetscape-related concerns of scale and height by way of a mindful variation in building mass, and through experimentation in color and material texture as well as streetscape planting and a number of existing mature trees that were retained in an effort to preserve the established character of the site.

2.5 A Sustainable Apartment Building

APT1 is a sustainable apartment building based on a flexible modular configuration which was conceived for high and mid-density communities (Fig. 2.19). The building is designed to innovatively address diverse needs of modern households, accommodating a range of family compositions, ages, lifestyles, and budgets. The design aims to reduce the environmental impact of construction practices by using prefabricated, standard elements to produce rapidly assembled, flexible designs (Figs. 2.20 and 2.21). Sustainability measures consider the structure's relationship with its surrounding environment, helping to promote energy efficiency. The design employs a combination of architectural and urban planning strategies for high-density communities.

The apartments are composed of a single or a selection of modules, designed to be flexible according to the demands of the site and its occupants. Rectangular modules range in size, from 11 to 13 m (36 to 44 feet) long; half-sized modules, or

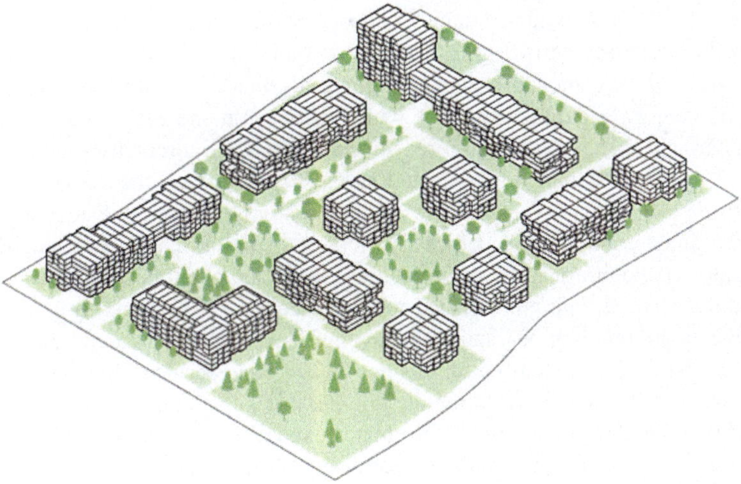

Fig. 2.19 The APT1, a sustainable apartment building, was conceived for high and mid-density communities

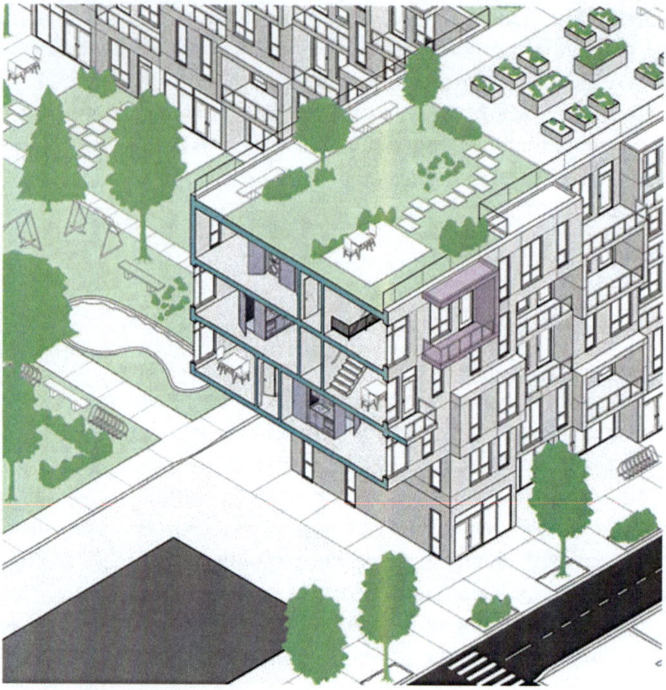

Fig. 2.20 A section of APT1

modules with integrated hallways and balconies are all variations on the standard module. Multiple modules can be merged to create larger units. Interior plan options are available to accommodate a range of household structures, from small studios to three-bedroom family units (Fig. 2.22). Various types of attachments can be used to coordinate the arrangement of units: the end-to-end connection with a central corridor between modules, a lateral connection with a hallway on one end, or the staggering of units around a shared central core. Specially designed circulation modules are provided to ensure adequate access and egress depending on the demands of the space to include elevators, stairs, fire exits, and storage areas. The attachment of modules can be configured to permit a variety of spatial arrangements, height, and site relations depending on the zoning requirements and client needs.

Modules' facades are embellished with a selection of elements to combat the repetitive, monotonous appearance of many high-density prefabricated designs. The ground level spaces can be used for businesses and other services, helping to create mixed-use neighborhoods. Additionally, the flexible design allows for a variety of configurations within neighborhoods to create mixed residential densities, open green spaces, and well-connected circulation pathways, all improving walkability.

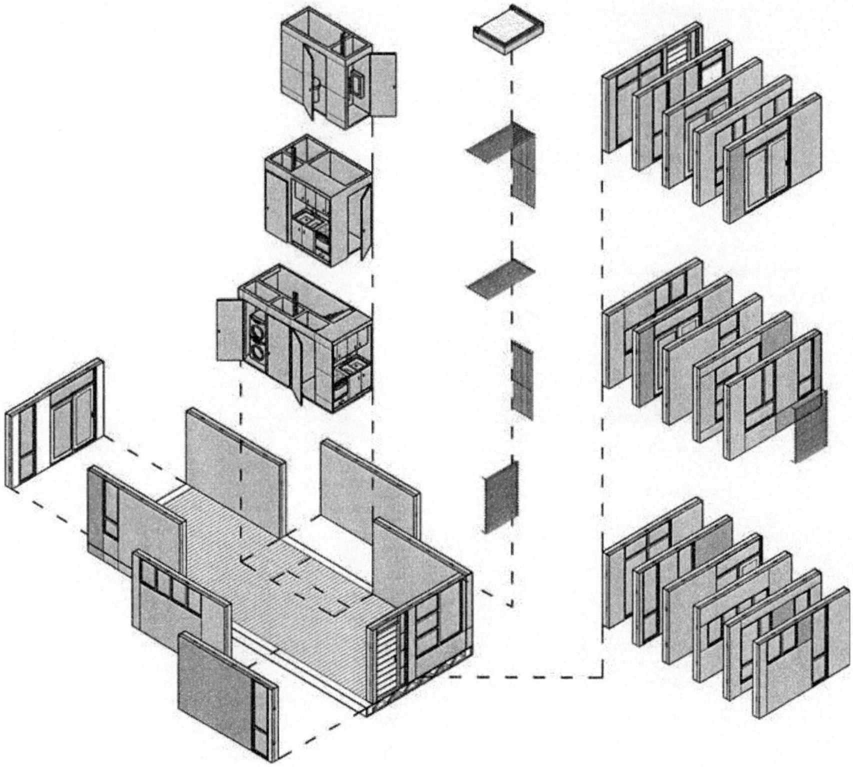

Fig. 2.21 Exterior panels and the utility pods used in assembly of APT1

Sustainable measures help to improve the energy efficiency of apartments; wind and sun protection are provided with strategically located coniferous and deciduous trees respectively (Fig. 2.23). Photovoltaic panels are used for active solar gains, while strategic design of the façade makes use of passive solar gains, helping to regulate the internal temperature. Moreover, a green roof helps to mitigate the heat island effect while providing a space for community gatherings and activities.

APT1 was conceived for a community constructed from prefabricated, modular units, and provides an adaptable system for sustainable, high-density developments (Fig. 2.24). Its design accommodates a high degree of flexibility, helping to create more walkable, diverse, and mixed-use neighborhoods.

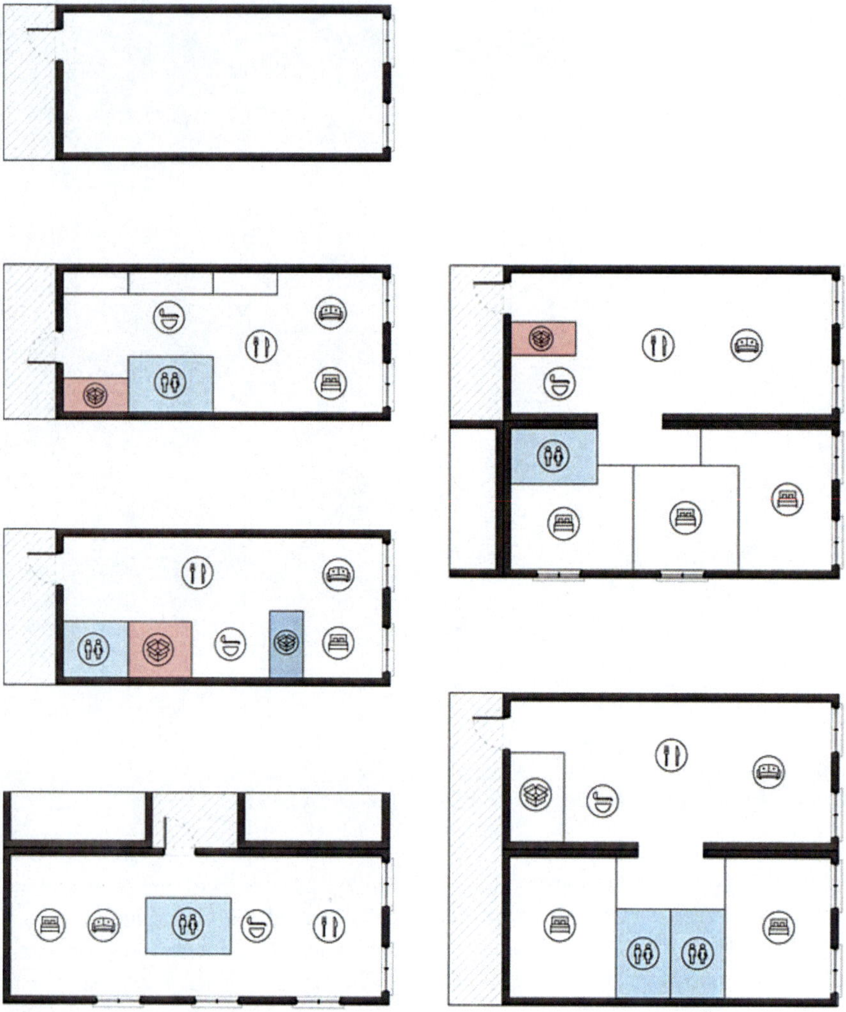

Fig. 2.22 Typical modules and dwellings' type in APT1

2.6 Final Thoughts

New, innovative solutions for sustainable urban environments are rapidly developing as more designers turn to high-density solutions to solve environmental issues and affordability. It is estimated that by 2050, the world's urban population will increase by 2.5 billion people (Wong et al. 2016). With new high-rise building solutions, urban density can be brought to unprecedented levels. Wong et al. (2016), propose a new way of imagining cities as three-dimensional matrices, where built planes above the ground level create rich urban environments enveloped by nature. Balling (2016)

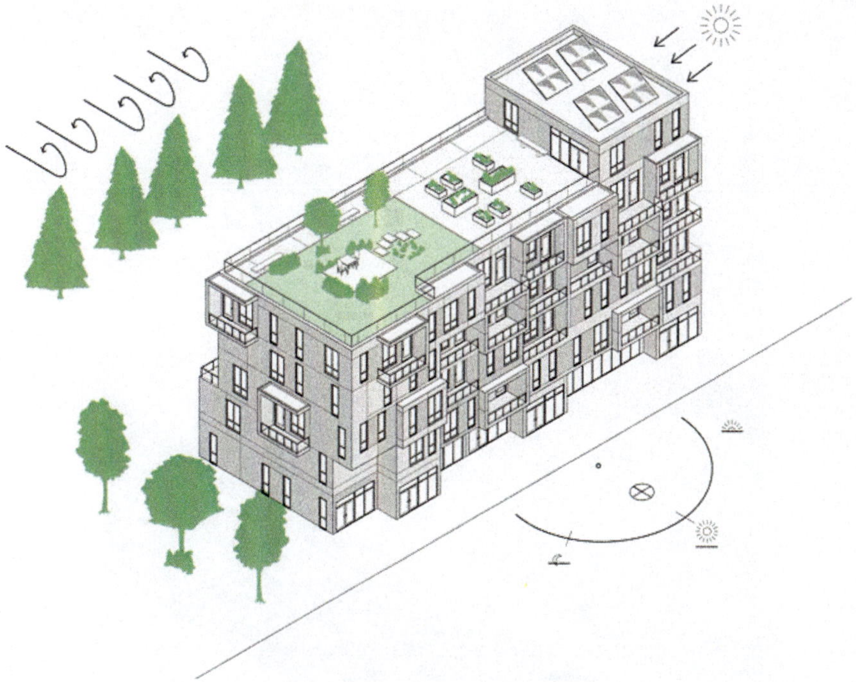

Fig. 2.23 APT1 was designed to consider and respond to the site's environmental aspects

discusses a model for a car-free interconnected city designed to house 100,000 people, entitled the *Greenplex* which strives to optimize the allocation of space in one highly dense megastructure. The Greenplex takes the notion of polycentric cities one step further by placing mixed-use buildings in the center, highly connected to residences and other services through multi-level skybridges (Balling 2016). These are just some of the many innovative models for future cities which prioritize sustainable urban development in very high-density contexts. Often, these projects blur the lines between engineering, planning, and architecture as they become more complex in nature.

Questions for a Follow-Up Discussion

1. What are the key features of planning communities for active mobility?
2. How will you prioritize the main aspects of higher density dwellings' forms?
3. What aspects of higher density dwellings' forms contribute to their sustainability the most?

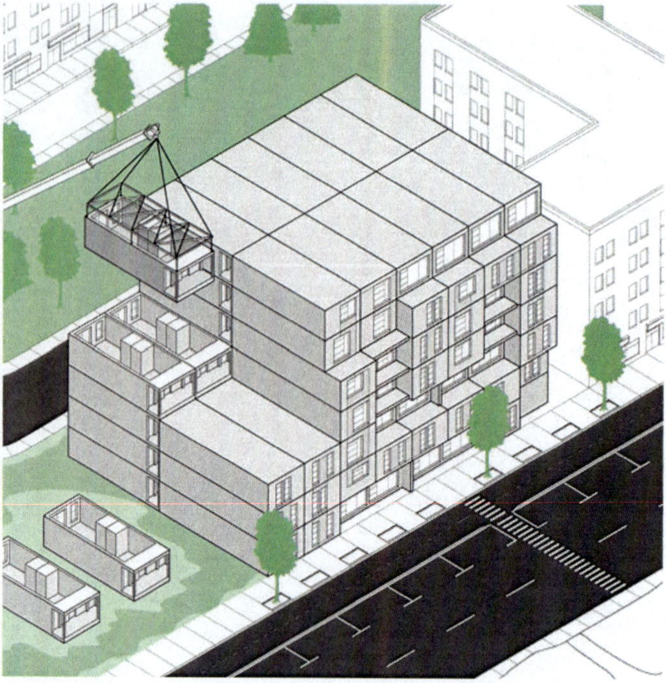

Fig. 2.24 Construction of the prefabricate APT1

References

Balling RJ (2016) Architecture/design: a car-free, polycentric city, with multi-level skybridges and inter-building Atria. CTBUH J 3:28–33. http://www.jstor.org/stable/44154343. Accessed 27 Sep 2022

Bharne V (2011) Humanizing high-rise urbanism: design strategies and planning tools. CTBUH J 4:18–23. http://www.jstor.org/stable/24192542. Accessed 27 Sep 2022

Bunbury A (2015) Designed to last: long live sustainable housing. ReNew: Technol Sustain Futur 132:64–67. https://www.jstor.org/stable/renetechsustfutu.132.64. Accessed 27 Sep 2022

Chow WK, Wood A, Li J (2013) Reduction in evacuation time for tall buildings through the use of skybridges. J Arch Plan Res 30(2):146–166. http://www.jstor.org/stable/43031086. Accessed 27 Sep 2022

Clarke D (2017) Green roofs. Sanctuary: Mod Green Homes 38:82–84. https://www.jstor.org/stable/90000611. Accessed 27 Sep 2022

Coolen H, Meesters J (2012). Private and public green spaces: meaningful but different settings. J Hous Built Environ 27(1):49–67. http://www.jstor.org/stable/41487473. Accessed 27 Sep 2022

Faehnle M, Söderman T, Schulman H, Lehvävirta S (2015) Scale-sensitive integration of ecosystem services in urban planning. GeoJournal 80(3):411–425. http://www.jstor.org/stable/24432627. Accessed 27 Sep 2022

Fisher-Gewirtzman D (2017) The impact of alternative interior configurations on the percieved density of micro apartments. J Arch Plan Res 34(4):336–358. https://www.researchgate.net/publication/328555089_The_impact_of_alternative_interior_configurations_on_the_perceived_density_of_micro_apartments. Accessed 28 Sep 2022

Fuller RJ, Crawford RH (2011) Impact of past and future residential housing development patterns on energy demand and related emissions. J Hous Built Environ 26(2):165–183. http://www.jstor.org/stable/41261677. Accessed 27 Sep 2022

Hammoud M, Baker W, Scheeren O, Parakh J, Hean CK, Lochhead H, Murray P, Azad MK, Pasquarelli G, Wood A, Hellmuth W, Wimer R, Lubin J, Xu Y, Thapar C, Chang A, Fiala R, Hardwick T, Lumpkin J, Murray W, Halliday P, Parry E (2018) CTBUH 2018 Conference special: polycentric cities: the future of vertical urbanism. CTBUH J 4:44–51. https://www.jstor.org/stable/26614165. Accessed 27 Sep 2022

Hoang M, Vandal A (2017) Micro-MACRO living in the global high-rise. CTBUH J 3:32–37. https://www.jstor.org/stable/90020855. Accessed 28 Sep 2022

Horrocks T (2013) "Come together", Monument 20th anniversary issue pp 58–63

Lee S, Koschinsky J, Talen E (2018) Planning tools for walkable neighborhoods: zoning, land use, and urban form. J Arch Plan Res 35(1):69–88. http://www.jstor.org/stable/45215820. Accessed 27 Sep 2022

Lindsay M, Williams K, Dair C (2010) Is there room for privacy in the compact city? Built Environ (1978–) 36(1):28–46. http://www.jstor.org/stable/23289982. Accessed 27 Sep 2022

Marckmann B, Gram-Hanssen K, Christensen T H (2012) Sustainable living and co-housing: evidence from a case study of eco-villages. Built Environ (1978–) 38(3):413–429. http://www.jstor.org/stable/23290271. Accessed 27 Sep 2022

Obrinsky M, Walter C (2016) Energy efficiency in multifamily rental homes: an analysis of residential energy consumption data. J Sustain R Est 8(1):2–19. https://www.jstor.org/stable/24876479. Accessed 27 Sep 2022

Oka M (2011) Toward designing an environment to promote physical activity. Landsc J 30(2):280–298. http://www.jstor.org/stable/43324379. Accessed 27 Sep 2022

Raman S (2010) Designing a liveable compact city: physical forms of city and social life in urban neighbourhoods. Built Environ (1978–) 36(1):63–80. http://www.jstor.org/stable/23289984. Accessed 27 Sep 2022

Ray E (2014) "Living on air," The age, Friday Sep 13 p 18

Steiner F, Simmons M, Gallagher M, Ranganathan J, Robertson C (2013) The ecological imperative for environmental design and planning. Front Ecol Environ 11(7):355–361. http://www.jstor.org/stable/43187631. Accessed 27 Sep 2022

Trubka R, Newman P, Bilsborough D (2010) The costs of urban sprawl—Predicting transport greenhouse gases from urban form parameters. Environ Des Guid: 1–16. http://www.jstor.org/stable/26150802. Accessed 27 Sep 2022

Turner L (2017) A roof over your head: choosing the right roofing materials. ReNew: Technol Sustain Futur 138:52–60. https://www.jstor.org/stable/renetechsustfutu.138.52. Accessed 27 Sep 2022

Wells NM, Evans GW, Yang Y (2010) Environments and health: planning decisions as public-health decisions. J Arch Plan Res 27(2):124–143. http://www.jstor.org/stable/43030900. Accessed 27 Sep 2022

Wong MS, Hassell R, Yeo A (2016) Garden city, megacity: rethinking cities for the age of global warming. CTBUH J 4:46–51. http://www.jstor.org/stable/90006409. Accessed 27 Sep 2022

Wood S (2015) The look and feel of a place: character, community, and the compact city. J Arch Plan Res 32(1):23–39. http://www.jstor.org/stable/44113096. Accessed 27 Sep 2022

Wright K (2010) The relationship between house density and built-form energy use. Environ Des Guid: 1–8. http://www.jstor.org/stable/26150782. Accessed 27 Sep 2022

Chapter 3
Quality Affordable Dwellings

Abstract The widening gap between household income and house price has become a significant barrier to homeownership. In many nations, high cost of land, infrastructure, labor, and materials have put homes beyond the reach of first-time home buyers and renters. These factors have heightened the need for and importance of affordable dwellings. Affordable housing is commonly defined as dwellings where owners or renters do not spend more than 30% of their income on shelter expenses. Affordable design may require downsizing, but not lowering the quality and standards of the dwelling and community. This chapter examines and outlines cost reduction strategies on macro and micro scales, with an emphasis on cost saving through urban planning, volumetric arrangement, lot sizing, housing shape, and material selection.

Keywords Density · Floor to area ratio · Lane parking · Lot configuration · Narrow lots · Open spaces · Vertical design · Volumetric design

3.1 The Housing Affordability Challenge

According to the US Department of Housing and Urban Developments (HUD), a household who spends more than 30% of their income on housing is *cost burdened* and anyone who spends more than 50% is *severely burdened* (Larrimore and Schuetz 2017). Furthermore, the term *affordability* also reflects on people's quality of life. When occupants spend a disproportionate amount of their income on housing, they do not spend enough on items such as nutritious food, healthcare, and other expenses that nurture their physical and mental wellbeing (Larrimore and Schuetz 2017). The underlying rationale is that homebuyers will be able to adhere to their financial commitment and still have means for other life's necessities.

One of the main barriers to affordability came with the turning of homes into investment vehicles which in many cities has led to an oversupply of luxury housing and a lack of affordable homes (World Finance 2021). The general rise in cost also coincides with increased housing demand, as a result of immigration, and population growth, primarily in and around urban areas. Furthermore, other rising costs such as

A. Friedman, *Fundamentals of Innovative Sustainable Homes Design and Construction*, The Urban Book Series, https://doi.org/10.1007/978-3-031-35368-0_3

labor, material, and energy have also contributed to the overall increase (Dahms and Ducharme 2022). Figure 3.1 shows the amount that homeowners or renters pay for a dwelling over the recommended 30% in Montreal, Canada.

Since homeownership can be a means of wealth accumulation, there are advantages to building equity through mortgage amortization in comparison to renting a unit. Over time, depending on location, the property value can appreciate and, as a result, the owner's equity increases (World Finance 2021). These circumstances have led to the daunting reality for today's entry-level earners putting homeownership beyond their reach. The impacts of the global housing crisis have negative effects not just on affording shelter but also on the livelihood of underprivileged occupants. According to Litman (2022), improving access to affordable housing can significantly increase people's economic opportunity by also improving access to education, employment, and health care.

Parallel to rise in cost, contemporary demographic changes and market trends are pressuring home builders to adapt their offerings. The transformation of the North American family—as described below—did not only affect society's demographic make-up, but it had many indirect links to the rising need for affordable housing. After World War II, home builders regarded their clients as a homogenous block. The common household was made up of a working father, a housewife, and their children. This family structure, with its space needs and lifestyle, influenced designs offered by architects and builders. The three-bedroom single-family detached home with a parking arrangement became a recognized feature across the North American continent. Housing solutions for families who did not fit this social model, it was assumed, would be found in apartments.

Decades later, from the 1960s onward to the turn of the century, this demographic make-up changed. Proliferation of birth control, along with new lifestyle and cultural tendencies gave rise to a new societal composition. The share of single-household and single-parent families grew. The number of seniors also increased as society got older. These new households, with a single breadwinner, could not master the means to purchase a home. They could also no longer rely on publicly funded housing since federal and local authorities had limited their participation in sheltering lower income households. Many had been left to fend for themselves, often relying exclusively on

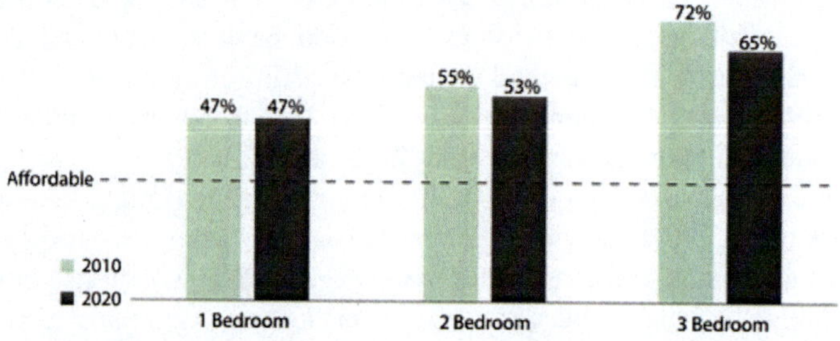

Fig. 3.1 The amount that homeowners or renters spend on lodging over the recommended 30% in Montreal, Canada

rental accommodations whose stock dwindled in many cities. Acquiring a home for those who had means became an even greater challenge, as the product itself swelled in size and as a result, became out of reach.

Most recently, the COVID-19 pandemic has also raised people's interest in low-rise dwelling types with private outdoor space. Despite their high cost, families are looking for single-family homes and avoiding dense and compact apartment complexes. When regarding affordability, one needs also to consider the relationship between life stages and housing needs. Need for a home and the use of its space cannot, therefore, be regarded in a static, linear fashion. Ongoing transitions and shifting priorities, a result of accumulating wealth, and aging, will all affect acquisition and use of homes.

3.2 Urban Planning Strategies for Affordability

In residential projects cost reduction begins with large-scale planning measures that will be highlighted in this section. As outlined in Chap. 2, one of the key strategies in achieving affordability is through densification. Denser developments reduce land cost per unit and contribute to sustainability by preventing urban sprawl (Litman 2022). *Density* can therefore be regarded as an index of a project's affordability level. It measures how many dwelling units per hectare (acre) are placed on the site to share land and infrastructure costs. These costs per capita will be the highest in low-density areas and the lowest in high-density. With increased density, there will also be declines in construction and maintenance costs of streets, parks, water, sewage, and solid waste disposal (Litman 2022).

Floor to area ratio (FAR) is the ratio of a building's total floor area to the size of the plot upon which it is built (Kurvinen and Saari 2020). As mentioned above, since land is becoming scarcer in urban areas, maximizing the ratio of floor area to perimeter becomes an increasing priority in achieving affordability. With the same floor area and height, simplifying a building's configuration, minimizing the number of corners, overall building perimeter, and exterior wall surface, will also lead to cost reduction and energy savings (Kurvinen and Saari 2020). In addition, when more units are attached, density increases, and more land is saved (Fig. 3.2). The leftover land can then be used to construct more dwellings thereby reducing the cost of land for each unit or be left public and/or green. Aspects of the relation between lot configuration and affordability will be outlined below.

3.2.1 Lot Configuration

A dwelling's typology, lot sizes, the placement of homes on them, and the urban configuration of a community will greatly influence the cost of each unit. In affordable housing project design, many conventional planning strategies must be altered to

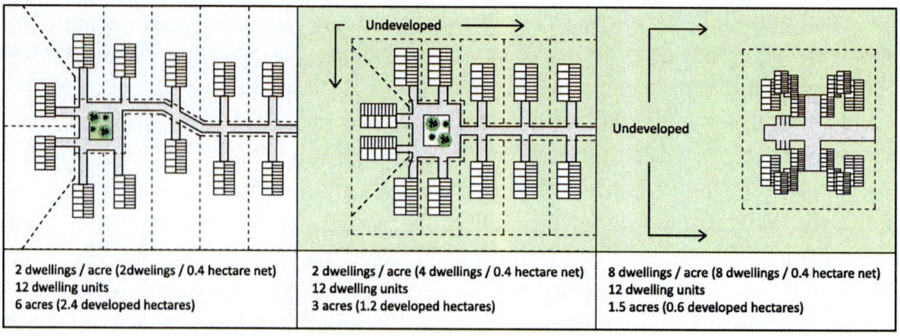

Fig. 3.2 When more units are attached, density increases, and more land is saved. The leftover land can then be used to construct more dwellings thereby reducing the cost of land for each unit

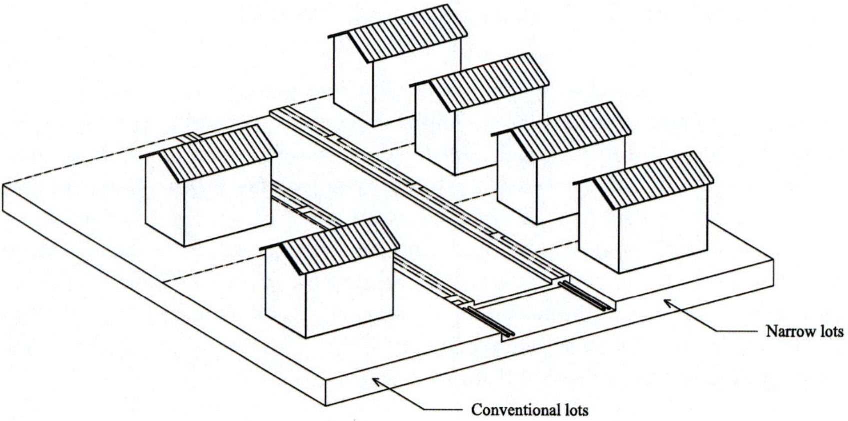

Fig. 3.3 Narrow lots are known to yield significant cost reductions of land and infrastructure

reduce costs. It is an interconnected process where the relationships between the above-mentioned aspects are all considered.

In an affordable community, lot size and configuration should be those that most increase density (Kurvinen and Saari 2020). Narrow lots will yield significant cost reduction of land and the infrastructure (Fig. 3.3). These lots can be designed to include *narrow center lots* with attached homes or *narrow angled lots*, to fit into a site's corner. To consider different types of living configurations, the end lot of each row can be wider to accommodate a larger unit. This will not only ensure the maximization of the space but provide a home for larger households (Southworth and Ben-Joseph 2017).

The *Zero-lot-line* is a land-saving strategy that was designed to mitigate high land costs and the demand for affordable detached single-family housing (Development Services Department 2019). In this arrangement, to reduce lot size, the house is built along one side of the property line to offer a private driveway or garage as well as

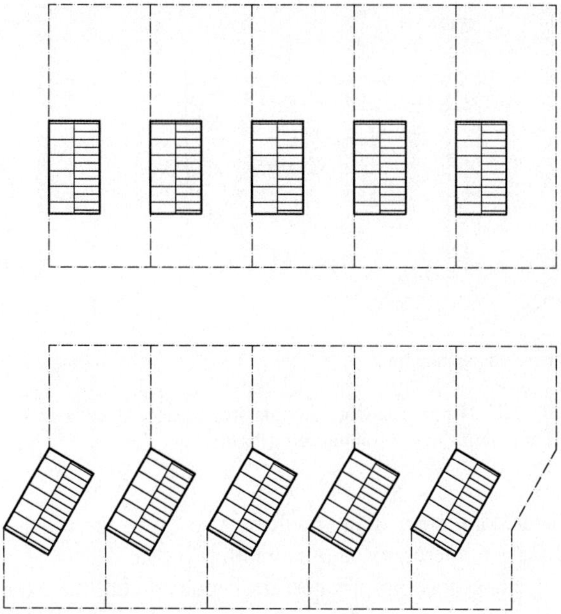

Fig. 3.4 The zero-lot-line is a land-saving strategy designed to mitigate high land prices and the demand for affordable detached single-family housing

green space at the rear (Chen 2021) (Fig. 3.4). This approach supports current desires to live closer to one's place of work by using willingness to settle for a smaller yard in exchange for access to public parks and amenities and affordability advantageously (Kramer and Sobel 2014).

3.2.2 Streets and Parking

Some 30% of residential site's area is commonly allocated to streets and parking to greatly determine the overall and unit costs (Litman 2022). The road network should, therefore, not create an isolated neighborhood but be integrated with nearby communities (Fig. 3.5). Road design and construction specifications are largely dependent on expected use. When a heavy volume of traffic is experienced, streets will sustain more wear and tear and therefore be costly to maintain. Therefore, prioritizing pedestrians, and cyclists' paths and public transit systems in affordable housing projects will reduce traffic load and the need for wide streets and as a result cost (CTOD 2014).

In addition to cost saving, the introduction of narrow local streets will also force motorists to slow down. Reduced speed limits support the location of pedestrian and cyclist paths as part of or closer to roads, unlike collector roads where a distance must be maintained. Furthermore, in narrow roads with low traffic volume and speed, it is not necessary to provide two moving lanes. If two cars are parked across from each other, there will still be room for one vehicle to pull over and allow the other to pass.

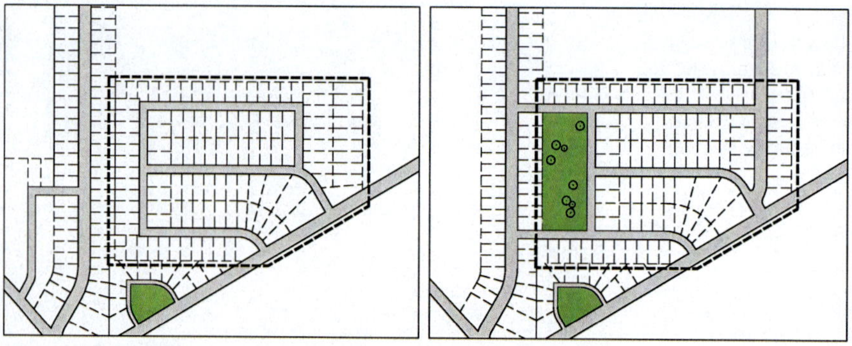

Fig. 3.5 The road network should not create an isolated neighborhood but be integrated with nearby communities to save on unnecessary driving

Therefore, cost savings will be a result of the precise interplay between the street function, hierarchy, and its width (Fig. 3.6).

When housing private cars is considered, the expensive nature of underground and above ground common covered parking structures and people's inability to pay for them, has resulted in having lot parking that consumes vast amounts of land and are often becomes an eye sore. Therefore, compact, and dense parking strategies need to be implemented to maximize buildable areas while simultaneously meeting the needs of drivers (Fan 2020; Kurvinen and Saari 2020) (Fig. 3.7).

Another parking option is at the home's rear with access from a lane. According to Huang et al. (2010), alley spaces have been seen as a waste of space since unlike the past, in contemporary suburban subdivisions deliveries are made from the front. Therefore, cities are eager to do without them to avoid the costly responsibility of maintaining both streets and lanes (Fialko and Hampton 2011). Yet, integrated lane parking stands to make better use of these spaces by being an affordable, parking alternative and place for informal social interaction (Fialko and Hampton 2011).

3.2.3 *Open Outdoor Space*

As housing density increases, so does the functional and psychological importance of open spaces between dwelling units (Brown et al. 2018). It is, therefore, crucial that open space will not be regarded as an afterthought but be designed to foster a sense of communal identity while providing privacy. In affordable housing this can be achieved by having public amenities such as parks and green areas near homes where people can congregate (Brown et al. 2018). Green space is of utmost importance, particularly for small homes and apartment buildings, where residents have little or no backyard. It can be arranged in variety of configurations as illustrated in Fig. 3.8.

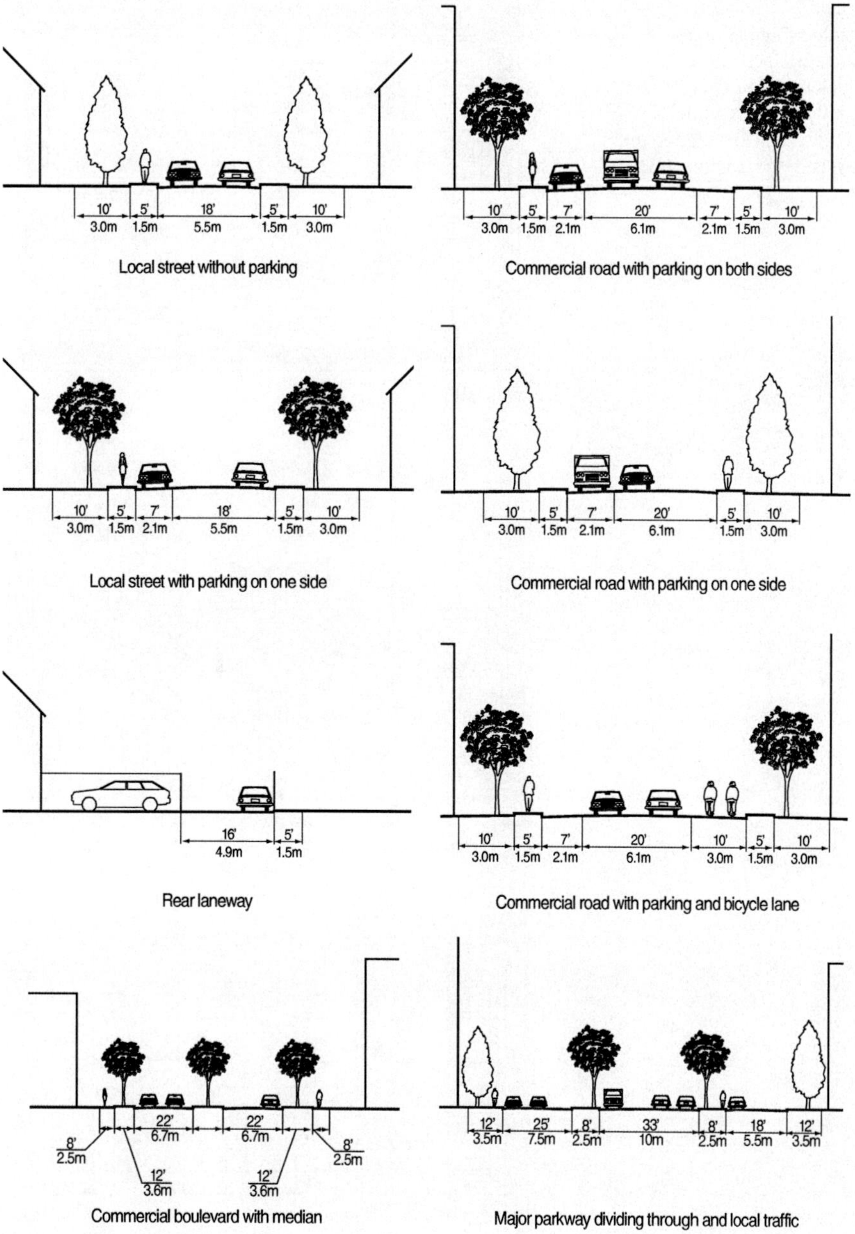

Fig. 3.6 When planning roads, cost savings will be a result of the interplay between the street function, hierarchy, and its width. Alternative width like the ones shown here can be considered

Fig. 3.7 Compact, land-saving parking strategies need to be implemented to maximize buildable areas while simultaneously meeting the needs of those with vehicles

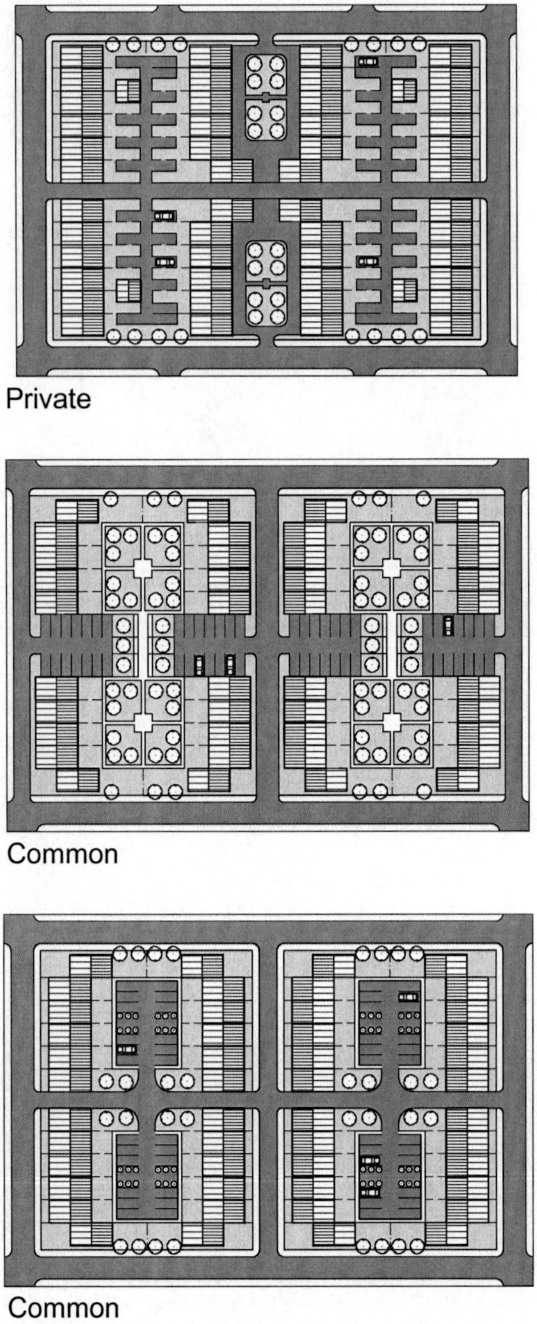

The *green belt* design, for example, regards the natural characteristics of the site as an open system (Euroleague 2014). The scheme is highly suitable for developments based on ideas of living in organic settings. Open spaces should have natural features and be accessible from all homes.

In places with population of mixed ages, accommodating the needs of various mobility levels is a necessity. Play areas for the youth such as soccer fields, courts, jogging, and cycling paths all need to be available for active recreation. Spaces for passive activities such as sunbathing, reading, and simple social get-togethers also need to be provided. Linking the various spaces to form an outdoor system while providing varying degrees of privacy in each is a fundamental principle of planning open spaces in communities.

In affordable neighborhoods, open spaces can be organized hierarchically, based on their order as public, semi-public, and private areas (Polat 2021) (Fig. 3.9). Rooted in ancient settlements, this system combines public plazas with courts and yards. The design is visibly more flexible and allows for a variety of different areas within a relatively short distance of each other by accommodating diverse activities and interactions between community members. Spaces are provided for large gatherings and intimate private encounters. Despite the notion that hierarchy leads to formality, the pattern can also take on a more freeform design in higher density residential configurations.

While neighborhood parks are to be shared and enjoyed by all community members, smaller communal areas can as a more private and intimate setting for small clusters of dwellings. These communal areas help to increase affordability by creating semi-private areas and thereby reducing the need for large private yards. Generally, the smaller the private area for each home is, the larger the communal areas should be. Therefore, integrating large communal gardens could offer a wide range of health benefits and greener interactive clusters.

The "exposure-effect" of community gardens on health and wellbeing has shown that the incidence of various chronic and non-communicable diseases is reduced by the availability and size of green spaces (Brown et al. 2018). Community gardens offer dwellers an opportunity to enjoy green space for food production as well as social interaction (Brown et al. 2018). In this context, the preservation of green space supports the concept of conservation design. This practice uses part of a property while leaving the remainder largely naturally undisturbed or rehabilitated (Arendt 1996). Since the vegetation in open spaces is conserved, landscaping costs for green areas are significantly reduced and, in some instances, eliminated.

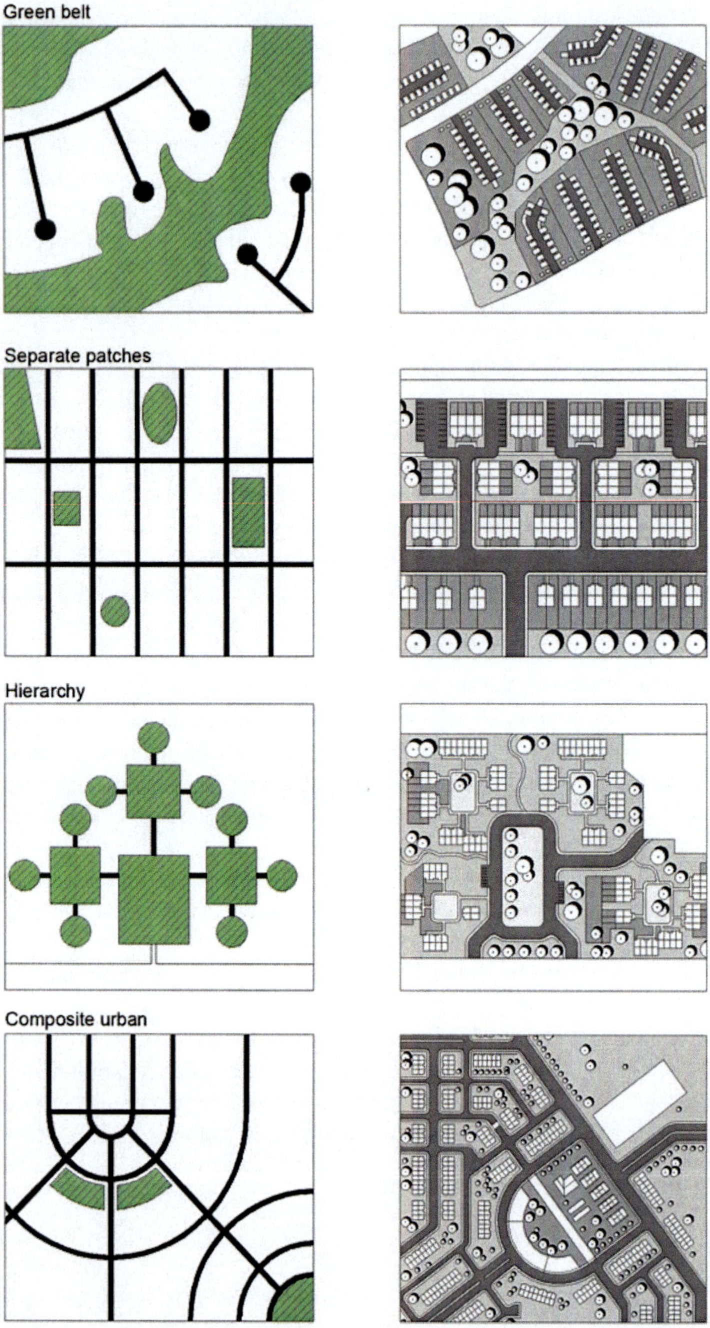

Fig. 3.8 Types of open space distribution in a community can consist of green belt, separate patches, hierarchy, or composite urban patterns

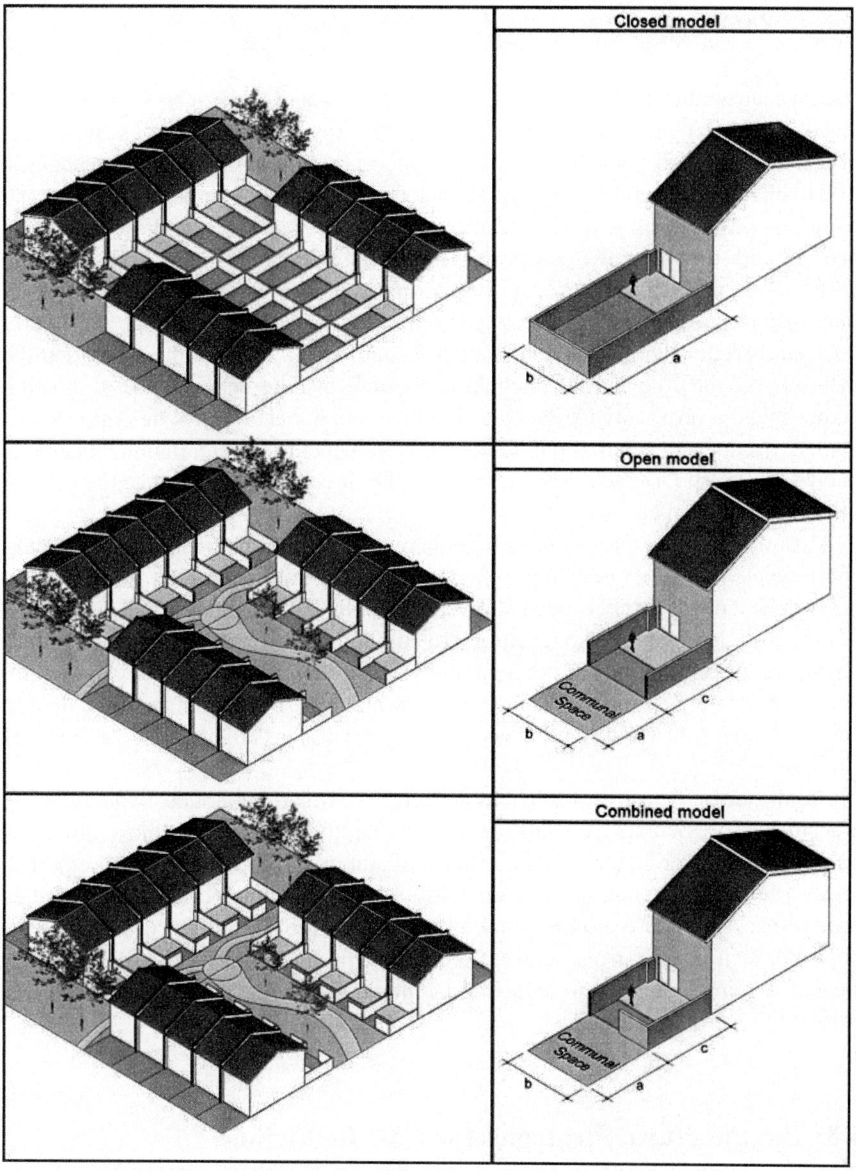

Fig. 3.9 In affordable communities, open spaces can be organized based on their order as public, semi-public, and private areas

3.2.4 Infill Housing

Vacant land within built areas offers opportunities and challenges to initiators of affordable housing projects. Easy linking with nearby roads and utilities and contribution to the rehabilitation of a rundown neighborhood are some valuable advantages. The likelihood of encountering contaminated soil, designing on odd-sized lots, and facing not in my back yard (NIMBY) sentiments are some of the drawbacks. The need to curb urban sprawl, adopt denser planning strategies, and lower dwelling costs makes infill housing worth considering especially in cities that have experienced rapid urban sprawl having led to traffic congestion, pollution, and poorly planned land development (Mohammadi-Hamidi et al. 2022). Others broaden the definition of infill housing to include major refurbishing or the reuse of existing homes in close proximity to amenities (Kramer and Sobel 2014). When infill development occurs near existing public transit, employment, and entertainment centers, known as Transit Oriented Development (TOD), it can also help reduce the need to drive which lowers emissions.

Most design guidelines suggest that the overall form of new infill homes should approximate that of its neighbors. Therefore, the overall height and roof line of a development needs do not exceed the average height of its neighbors, and the volume of the dwelling ought to also be similar to adjoining structures. Placing a large, tall townhouse in a community of predominantly small, single-family homes can visually dominate the neighborhood. Placing infill townhouses in communities that already have townhouses to avoid creating visual and social discord might be a recommended strategy.

Front, side, and backyard setbacks are often the first thing noticed when looking at a streetscape. Setbacks for infill homes should, therefore, reflect those of neighboring structures or be the average setback of adjoining properties to ensure proper scale within the neighborhood (Fig. 3.10). While guidelines often allow smaller front setbacks for courtyard row homes than for detached dwellings, setbacks should not differ significantly from adjacent properties. In this case, front yards would be smaller, and it is important to reduce apparent massing of the home as it approaches adjacent properties.

3.3 Architectural Strategies for Cost Reduction

Cost reduction strategies also include the home's configuration and performance. Energy consumption is, to a significant extent, proportional to house size and accounts for its operating cost. Therefore, downsizing a house will not only save on construction but will lead to reduction in energy consumption. Below are some specific strategies.

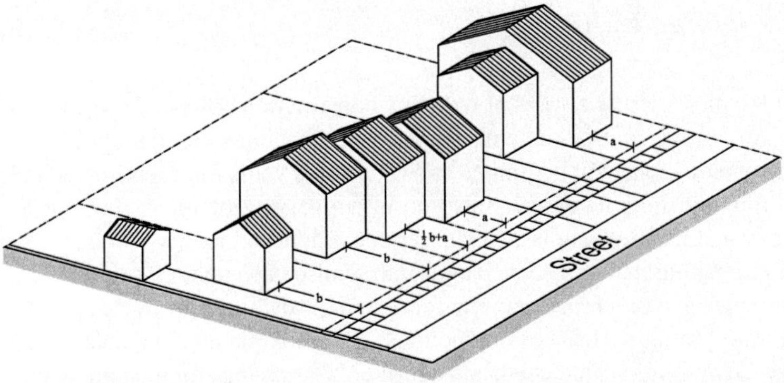

Fig. 3.10 Setbacks for infill homes should reflect those of neighboring structures or be the average setback of adjoining properties to maintain proper neighborhood's scale

3.3.1 Floor Staking

Having a taller rather than wider design is one of the most efficient ways to maximize the floor to area ratio (FAR) and affordability (Magay et al. 2018). While the vertical distribution of a unit's floor area will have the greatest impact on land use efficiency and density, it can also have substantial effects on the use of building materials, and, to a large measure, energy efficiency. The cost of a two-story square house, for instance, is lower than a one-story with equivalent area, since it has half the foundation and roof area. Floor to floor heights will also have an impact on the number of materials needed, especially the building envelope.

Vertical design is particularly beneficial when building in dense urban areas, and as mentioned above, would be a key strategy when building infill housing. The vertical design is also favorable to the stacking of wet rooms. This provides a more efficient and streamlined plumbing system throughout the dwelling and makes construction and long-term accessibility to the plumbing system easier.

3.3.2 Plan Simplification

One of the most practical ways of reducing materials needed and heat loss and as a result cost is by simplifying the unit's configuration. A building with a complex form often has more corners and perimeters which are costly to build. Simple configurations generally require less cutting and fitting of building materials and as a result less material wasted (Esteves et al. 2018). Therefor the shapes that designers should opt for are basic ones; rectangles and squares not only require less land but are easier to construct (Belniak et al. 2013).

3.3.3 Joining Units

One of the most effective ways of reducing construction costs and energy consumption is by joining units into semi-detached or rowhouse configurations, since the number of elevations will be limited to two or three walls for a semi-detached structure. Grouping units together is also an effective way of improving construction efficiency and usually results in a shorter construction time per unit. As for energy consumption joining four detached units into semi-detached, for instance, reduces the exposed wall area by 36%. Grouping four units as rowhouses provides an additional 50% energy savings. Heat-loss reductions of approximately 21% can be achieved when two dwellings are attached, and a further 26% savings for the middle unit when three or more dwellings are joined as rowhouses (Fig. 3.11).

Attached units will also reduce the amount of land required to build on. As noted above, when density increases, cost per unit declines as well, joining units therefore serves an important role in achieving affordability. The rowhouse configuration is more appropriate to adapt in its form, dimensions, and overall geometry and is therefore more economical (Donn and Thomas 2003).

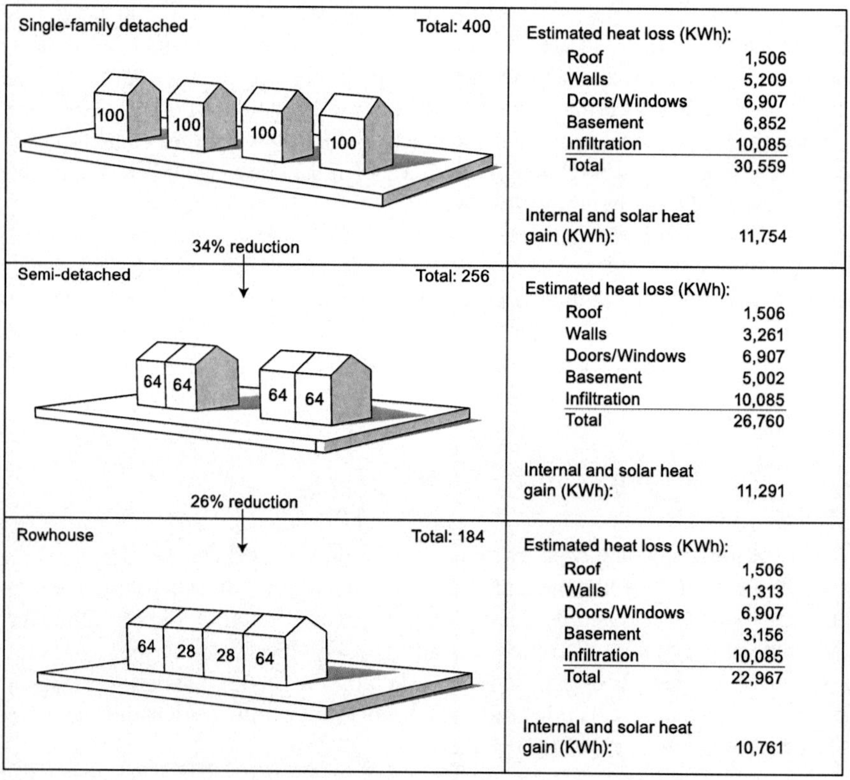

Fig. 3.11 Joining four detached units into semi-detached reduces the exposed wall area by 36% leading to considerable energy saving

3.3.4 Ground Relation

Having a basement is common in North American dwellings. When they are strategically designed, basements can lower construction costs and provide additional livable space. Recent improvements in construction technology contribute today to improving livability in basements. Better damp-proofing techniques have reduced the humidity that was formally associated with these excavated spaces and advanced heating and ventilation technologies can ameliorate air quality. Construction methods have also been developed to keep footings down, raise the lower floor, and insert large windows, thereby letting in more light.

There are several basement usage options available to designers. One of the most common uses of basements is as an indoor parking garage. In an affordable home it is a costly choice since valuable space will have to be devoted to a service function rather than living space. Another option is to leave the basement unfinished for future expansion. The children's bedroom can be located on the upper level and after occupancy as means become available, the residents can finish the space. Completion of the basement at a later stage may coincide with the family's life cycle. As children get older, they are likely to seek more privacy and bedrooms can be built for them at a lower level.

3.3.5 Chosen Materials

The chosen materials and how they are incorporated also greatly determine the overall cost of a project. The sourcing of materials, the materials themself, and their supply are all significant factors that need to be addressed before building. In Canada, for example, building with lumber is more affordable in comparison to using brick or concrete (McNeill Lalonde and Associates 2021). Wood is also lighter to transport and sustainably biodegradable when it is no longer needed (McNeill Lalonde and Associates 2021).

Building materials available offer a wide range of sustainable choices to the home designer. Careful research is required to identify and select materials and components that are environmentally sound, energy efficient, and perform efficiently over the life of the building. Using poor quality windows, or reducing insulation values, for example, is, in the long run, a poor strategy for saving, since the cost of future upgrades can be higher than the cost of installing high-quality materials from the outset.

Finishing materials account for a large percentage of the construction cost, and here one could take a different approach. It is possible to achieve significant savings through the use of less-expensive materials where these can be easily and practically replaced with better quality products in the future. Vinyl tile, for example, is the least-expensive flooring, and can be easily recycled and replaced with other products later. The same applies to hollow-core doors, light fixtures, and interior paint.

3.3.6 Modular Design, Dimensioning, and Efficient Framing Practices

A simple and effective way of reducing cost and waste is through careful dimensioning of the building to accommodate the modular configuration of materials. At the most basic level, designing within standard dimensions for structural wood framing members such as studs, joists, and plywood could result in substantial savings. It has been estimated that in a typical detached home, the use of general dimensions for stud spacing 405 mm (16 inch) module placing and dimensioning windows accordingly and locating partitions to line up with the structural studs may altogether save a ton of lumber in an average home. Designing for 2.2 m (4-foot) modules and 610 mm (24 inch) stud spacing alone can reduce lumber use by 8% (Friedman 2001). Providing for efficient details at corners and intersections of exterior walls and interior partitions doubles these savings. With more careful planning and material selection, the same principle could be implemented to accommodate interior finishes such as drywall and floor tiles. Cost savings are achieved not only through efficient use of materials, but also through reduced labor requirement, since less cutting and fitting is required.

3.4 An Affordable Community in Vijfhuisen, the Netherlands

Bloembollenhof is an affordable housing development in the town of Vijfhuisen, in the Netherlands. Designed by the firm S333 Architecture & Urbanism in proximity to Schiphol international airport, it is linked via an express train service to Amsterdam's city center. Fifty-two housing units have been placed into simple yet interesting and non-repetitive forms in the development. Houses range from large and detached forms to smaller social-housing blocks and are scattered freely, catering to several incomes and lifestyles (Fig. 3.12). VINEX is a spatial-planning method put in place by the Dutch government to control urban growth. This approach dictates that land be owned by the government and developed by the municipal planning authorities. Strict guidelines are to regulate the community's design and the site must consist of at least 30% affordable housing.

Bloembollenhof resembles a rural village, retains dense urban characteristics, and appears as a cluster of low-density housing typologies. The architects aimed to create a "regular irregularity" in their plan. This allows for the housing clusters to appear less formal, fostering the sense of a spontaneously built community as opposed to one that is strictly regimented. Houses have been placed in ways that frame communal open spaces and play areas for children. Residence views often overlook these courtyards, further connecting individual units to the broader community, and creating opportunities for parents to supervise their children from home. Further, the placement of houses also creates shortcuts between major streets and a pedestrian-oriented network of transportation routes.

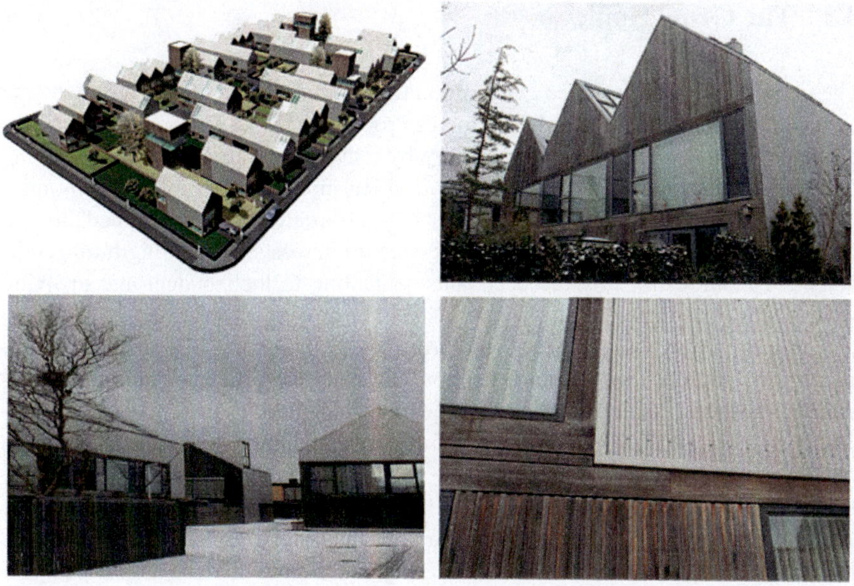

Fig. 3.12 The fifty-two housing units in the Bloembollenhof affordable development in the town of Vijfhuisen, in the Netherlands have been planned as a simple, dense yet non-repetitive form

Residences were designed to make their expansion easy and affordable. The thought process used in design and construction processes was that if a home is purchased prior to construction, future occupants can be consulted to achieve the living arrangements they desire and meet their budget. As a result, spaces vary, some with double-height ceilings, dormer windows, skylights, and various other additions. Generally, spaces in the house are left open, allowing for the later addition to suit changing needs. Residents can also purchase units attached to theirs, make wall alterations, and increase their total living area.

Houses are made in simple geometric forms yet incorporate several amenities. Roof terraces and patios have been cut out of the main mass, making up for the small private gardening area. Some rooms have been raised to the first-floor level, improving diagonal street views. The residences also house car-parking, minimizing the presence of automobiles on the street and eliminating the need for costly parking infrastructure. To lower costs, houses are clad in hardwood and inexpensive corrugated steel.

In sum, to achieve affordability the Bloembollenhof community development has applied low-density housing typologies to high-density living. Using small plot sizes and a scattered scheme, the development has created a dense environment that maintains many suburban characteristics.

3.5 The Grow Home

The Grow Home was designed and built on the campus of McGill University by the School of Architecture's Affordable Homes Program (Fig. 3.13). The main objective of the design was to improve affordability by reducing the footprint and achieving resource efficiency while addressing the needs of modern households. The home's name derives from a strategy which suggests that affordability can be attained through progressive building. It was designed to reduce initial costs by enabling future additions and modifications by the occupants—adapting to their budget and evolving space needs (Friedman 2001).

The narrow townhouse, with a footprint of 4.3 by 11 m (14 feet by 36 feet), allows for a high residential density, improving land use efficiency and preventing urban sprawl. The small area size of 93 square meters (1,000 square feet), for a standard, two-story unit, allows for the use of quality and durable materials and is more cost-effective to heat, cool, and maintain. The simple configuration of volumes also reduces construction costs and potential for heat loss. Additionally, the structure's dimension was chosen with modularity to limit materials waste and accommodates prefabrication using panelized or modular methods should there be an interest.

The Grow Home is offered with options for the foundation to include slab-on-grade, an unfinished basement, and a basement garage. The choices for the exterior

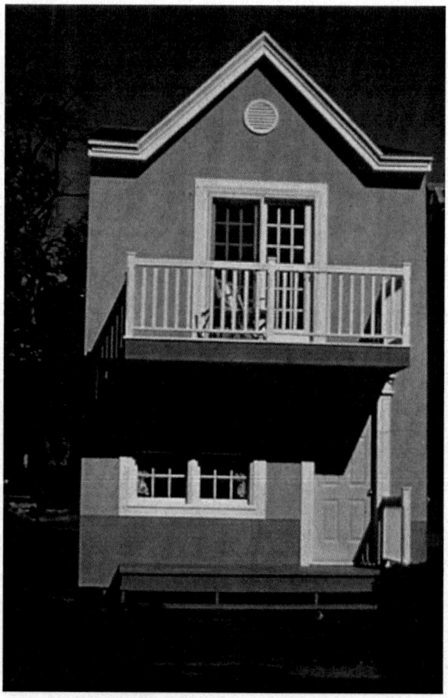

Fig. 3.13 The Grow Home on the campus of McGill University

include various balconies, porches, and roof shapes (Fig. 3.14). A range of interior and exterior finishings are available as well. The base model is finished with modest materials, with no basement level and a simple roof shape. Various alternative models include different configurations at higher costs.

The main floor includes a kitchen, a bathroom, a set of stairs in the middle of the unit, and a living room at the rear. While the first floor is fully finished, the upper is left unpartitioned creating a loft-style space, up to the owners to subdivide. Providing an unpartitioned space allows occupants more flexibility with respect to the organization of the floor. In the future, the second floor could be partitioned to create two separate bedrooms and a bathroom (Figs. 3.15 and 3.16). Since its introduction in 1990, the Grow Home concept was adopted and built by developers across Canada and the world. It demonstrated how designers can prioritize affordability by giving buyers the option to finish their home themselves while creating flexible and adaptable design (Figs. 3.17 and 3.18).

3.6 Final Thoughts

When designing quality affordable dwellings, one needs to recognize that the footprint of a home will be the outcome of the following functions. A dwelling with a square footprint will commonly require a wider lot than a house with a narrow rectangular footprint and will therefore be more expensive. Vertical designs make for the most efficient use of space since more stacking results in the need for less construction material. The cost of a two-story square house, for instance, is lower than a one-story with equivalent area since it has half the foundation and roof area.

When a building has several projections, and the upper floor extends beyond the lower, for example, costs are bound to rise. Projections therefore need to be introduced where they are most needed. When two dwellings are joined as semi-detached, the lot area can be reduced by as much as 18% and the exterior wall perimeter costs are reduced by a third. Joining units in a row will result in a 33% savings in lot area and street length, and 70% savings in the exterior wall perimeter for a middle unit.

The basement, when considered, offers an opportunity to create an independent accessory unit. The basement's unit may have its own entrance and can be used as a source of supplementary income for the household that might reside above. The design of a single-story dwelling is likely to require a larger lot as all the functions are congregated on a single level. As the number of units who use the same footprint mounts, each dwelling's share in the cost of land, infrastructure, and foundation declines.

Questions for a Follow-Up Discussion

1. What are the key urban planning strategies to achieve affordability?
2. What are the main architectural design strategies for cost reduction?
3. What were the key approaches to cost reduction in the design of the Grow Home?

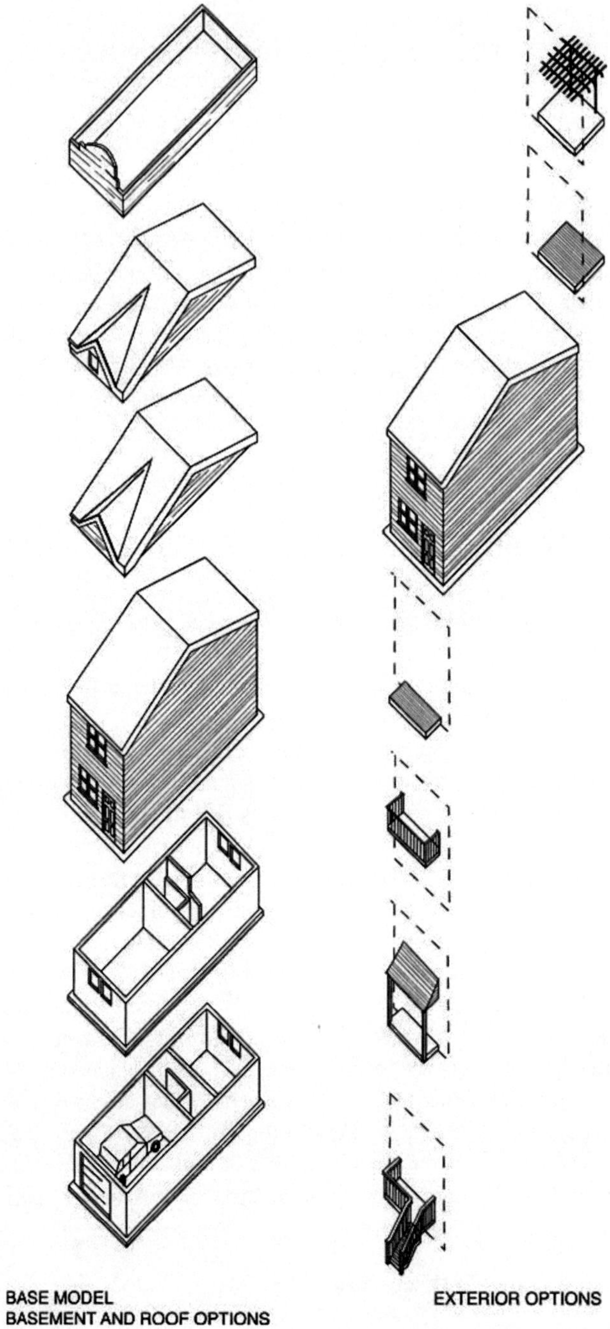

Fig. 3.14 The Grow Home is offered with options for the foundation including slab-on-grade, an unfinished basement, a basement garage. The choices for the exterior include various balconies, porches, and roof shapes

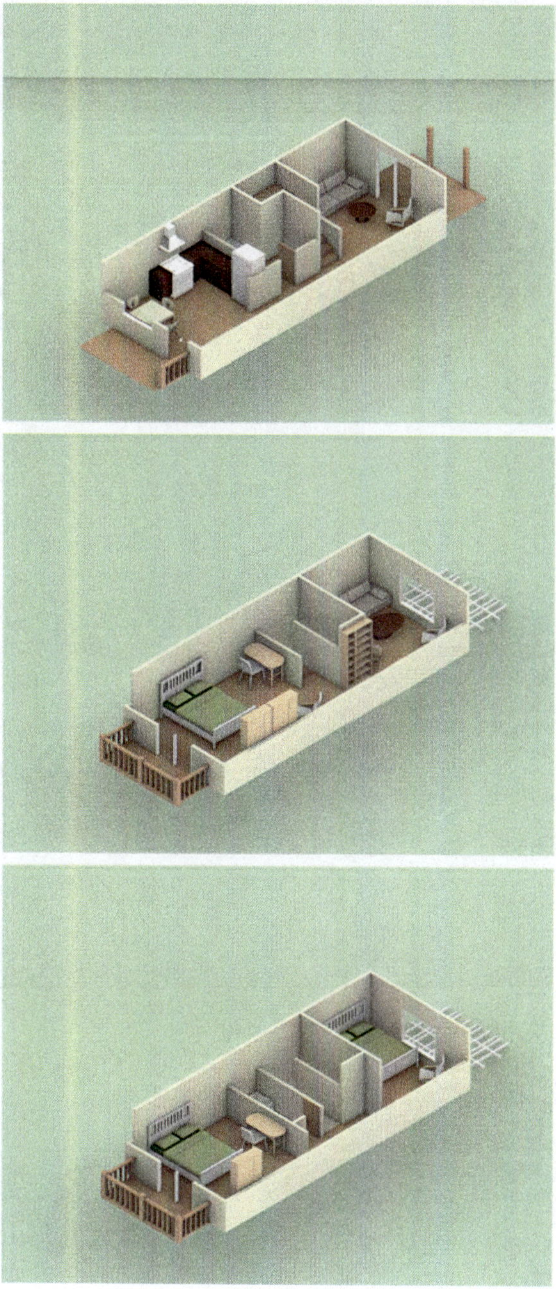

Fig. 3.15 The finished lower floor (top), the unpartitioned first floor (middle), and the finished first floor (bottom)

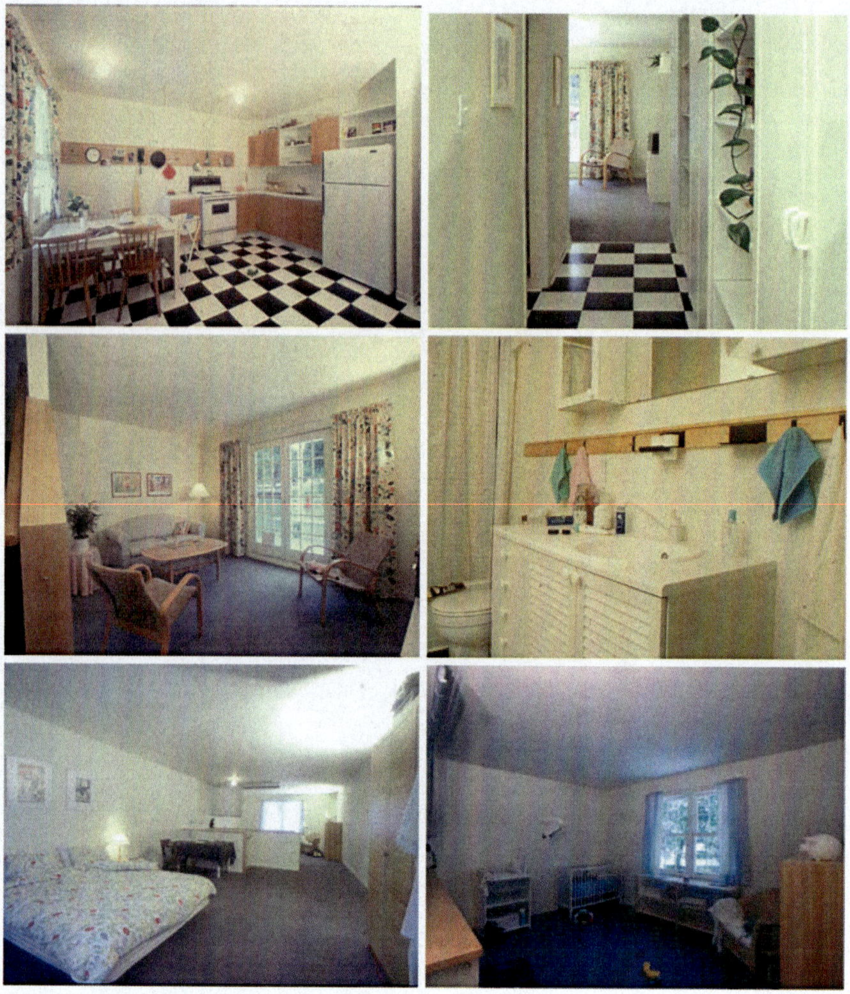

Fig. 3.16 Interior views of the Grow Home

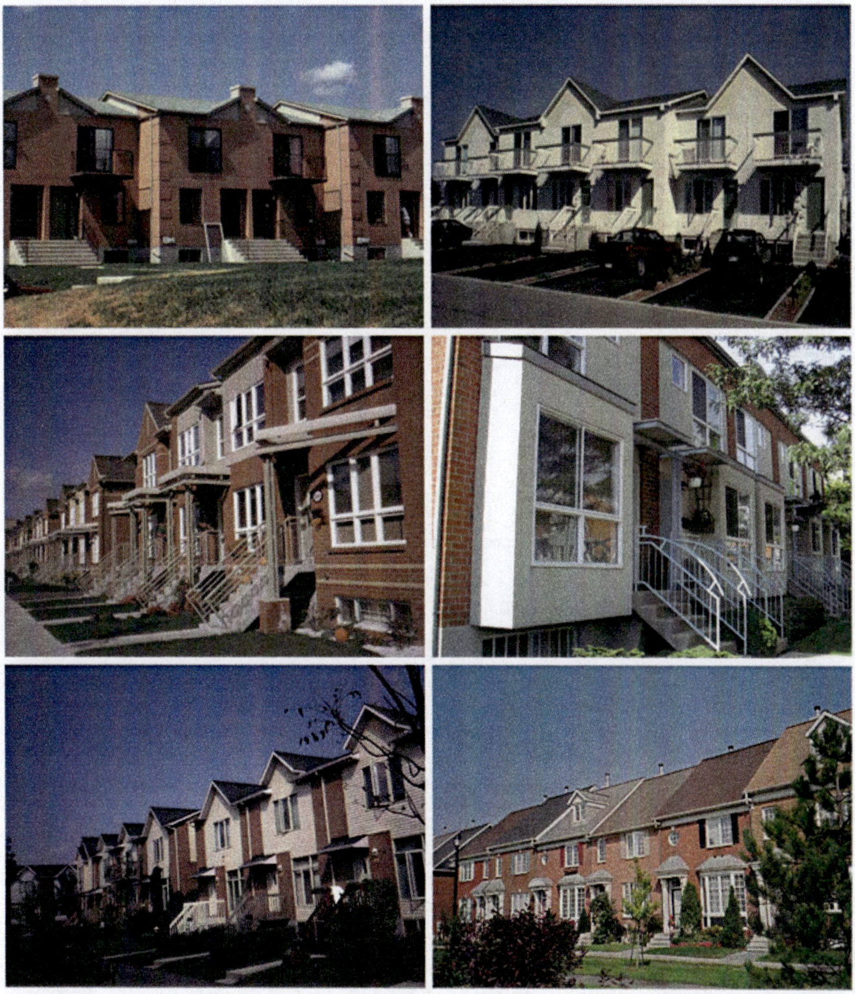

Fig. 3.17 Built Grow Home communities

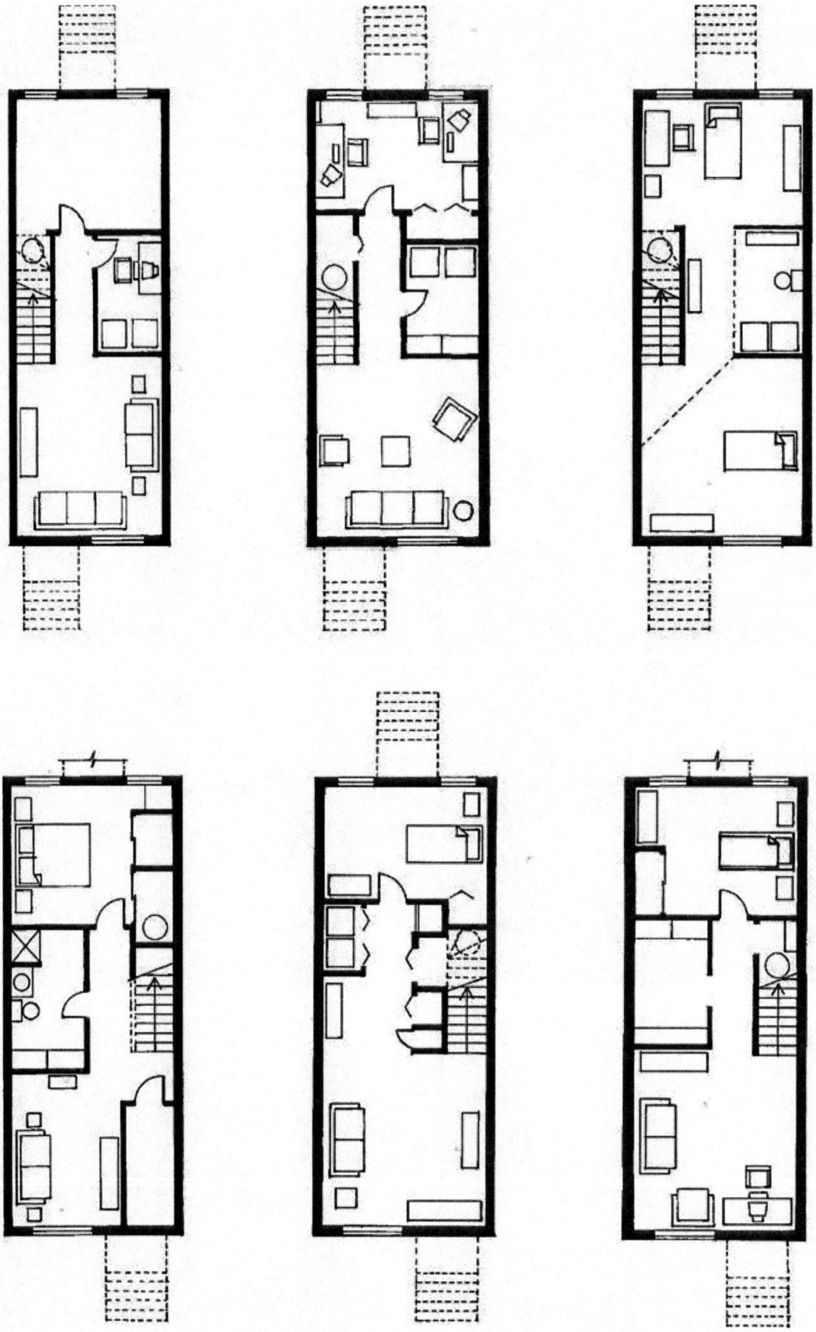

Fig. 3.18 A sample of basements that were originally unfinished and were later completed by the occupants

References

Arendt RG (1996) Conservation design. ACRD. https://www.acrd.bc.ca/cms/wpattachments/wpID188atID760.pdf. Accessed 28 Sept 2022

Belniak S, Leśniak A, Plebankiewicz E, Zima K (2013) The influence of the building shape on the costs of its construction. J Financ Manag Property Constr 18(1):90–102. https://www.emerald.com/insight/content/doi/10.1108/13664381311305096/full/html. Accessed 28 Sept 2022

Brown H, Proust K, Newell B, Spickett J, Capon T, Bartholomew L (2018) Cool communities-urban density, trees, and health. MDPI. https://www.mdpi.com/1660-4601/15/7/1547. Accessed 28 Sept 2022

Center for Transit-Oriented Development (CTOD) (2014) Creating connected communities—HUD USER. HUDuser. https://www.huduser.gov/publications/pdf/Creating_Cnnted_Comm.pdf. Accessed 28 Sept 2022

Chen J (2021) What is a zero-lot-line house? Investopedia. https://www.investopedia.com/terms/z/zero-lot-line-house.asp. Accessed 28 Sept 2022

Dahms K, Ducharme A (2022) Housing affordability monitor—2021q4—NBC. National Bank of Canada. https://www.nbc.ca/content/dam/bnc/en/rates-and-analysis/economic-analysis/housing-affordability.pdf. Accessed 28 Sept 2022

Development Services Department (2019) Design manual for small and narrow lots. Fayette Ville Arkansas. https://www.fayetteville-ar.gov/DocumentCenter/View/15360/Small-Lot-Infill-Standards-Manual?bidId=. Accessed 28 Sept 2022

Donn M, Thomas G (2003) Designing comfortable homes. CDN. https://cdn.ymaws.com/concretenz.org.nz/resource/resmgr/docs/ccanz/ccanz_tm37.pdf. Accessed 28 Sept 2022

Esteves A, Esteves MJ, Mercado MV, Barea G, Gelardi D (2018) Building shape that promotes sustainable architecture: evaluation of the indicative factors and its relation with the construction costs. Architect Res 8(4):111–122. http://article.sapub.org/10.5923.j.arch.20180804.01.html. Accessed 28 Sept 2022

Euroleague for Life Sciences Summer School (2014) Designing the nature of the Green Belt Ila + Irub - Boku. Boku. https://boku.ac.at/fileadmin/data/H03000/H85000/H85200/2014_Summerschool/Summerschool_2014_projects_neuweb.pdf. Accessed 28 Sept 2022

Fan F (2020) IOPscience. In: IOP conference series: earth and environmental science. https://iopscience.iop.org/article/10.1088/1755-1315/474/7/072091. Accessed 28 Sept 2022

Fialko M, Hampton J (2011) National association of city transportation officials. NACTO. https://nacto.org/docs/usdg/activating_alleys_for_a_lively_city_fialko.pdf. Accessed 28 Sept 2022

Friedman A (2001) The grow home. McGill-Queen's University Press, Montréal

Huang HB, Mapes J, Newell J, Wolch J, Seymour M, Reynolds K (2010) The forgotten and the future: reclaiming back alleys for a sustainable city. The forgotten and the future: reclaiming back alleys for a sustainable city. https://css.umich.edu/publication/forgotten-and-future-reclaiming-back-alleys-sustainable-city. Accessed 28 Sept 2022

Kramer M, Sobel L (2014) Smart growth and economic success: investing in infill development. United States Environmental Protection Agency. https://www.epa.gov/sites/default/files/2014-06/documents/developer-infill-paper-508b.pdf. Accessed 28 Sept 2022

Kurvinen A, Saari A (2020) Urban housing density and infrastructure costs. MDPI. https://doi.org/10.3390/su12020497. Accessed 28 Sept 2022

Larrimore J, Schuetz J (2017) Assessing the severity of rent burden on low-income families. In: The fed—assessing the severity of rent burden on low-income families. https://www.federalreserve.gov/econres/notes/feds-notes/assessing-the-severity-of-rent-burden-on-low-income-families-20171222.htm. Accessed 28 Sept 2022

Litman T (2022) Understanding smart growth public service cost savings. Victoria Transport Policy Institute. https://www.vtpi.org/sgcp.pdf. Accessed 28 Sept 2022

Magay AA, Bulgakova EA, Zabelina SA (2018) Organizing vertical layout environments: a forward-looking development strategy for high-rise building projects. In: E3S web of conferences. https://www.e3s-conferences.org/articles/e3sconf/abs/2018/08/e3sconf_hrc2018_01016/e3sconf_hrc2018_01016.html. Accessed 28 Sept 2022

McNeill Lalonde and Associates (2021) Wood vs. concrete: how structural materials are chosen. MLA Canada. https://mlacanada.com/newsfeed/wood-vs-concrete-how-structural-materials-are-chosen#:~:text=Upfront%20cost%20and%20long%2Dterm%20value&text=As%20a%20natural%20resource%20that,more%20expensive%20to%20build%20with. Accessed 28 Sept 2022

Mohammadi-Hamidi S, Beygi Heidarlou H, Fürst C, Nazmfar H (2022) Urban infill development: a strategy for saving peri-urban areas in developing countries (the case study of Ardabil, Iran). MDPI. https://www.mdpi.com/2073-445X/11/4/454/htm. Accessed Sept 2022

Polat Z (2021) The identity and hierarchy of urban parks: planning to design and management. https://www.researchgate.net/publication/348716388_THE_IDENTITY_AND_HIERARCHY_OF_URBAN_PARKS_PLANNING_TO_DESIGN_AND_MANAGEMENT. Accessed 28 Sept 2022

Southworth M, Ben-Joseph E (2017) Reconsidering the Cul-de-Sac. Real Estate Wharton Business School. https://realestate.wharton.upenn.edu/wp-content/uploads/2017/03/513.pdf. Accessed 28 Sept 2022

The Federal Highway Administration (2006) Neo-traditional neighborhood design. The Federal Highway Administration. https://safety.fhwa.dot.gov/PED_BIKE/univcourse/pdf/swless16.pdf. Accessed 28 Sept 2022

World Finance (2021) Solving the global housing crisis. World Finance. https://www.worldfinance.com/infrastructure-investment/solving-the-global-housing-crisis. Accessed 28 Sept 2022

Chapter 4
Comfortable Small Interiors

Abstract Future sustainable homes need to be functional, comfortable to live in, affordable, address new lifestyle trends, and as energy efficient as possible in their construction and operation. Choosing a dwelling type is one of the fundamental building blocks of any residential design process. What to build will depend on need assessment or the initiator's familiarity with the site and the clients. In addition to cost considerations, zoning bylaws, demographic make-up, and neighborhood norms will influence the chosen prototype. While this appears as a tall order for designers, builders, and homebuyers, it offers opportunity to be creative and address issues that in the past were considered marginal. This chapter offers a path to the conception of such homes' interiors by looking at a broader design process. The chapter will investigate design of small spaces including zoning, access, circulation, and spatial configuration. Finally, flexible design, space-making strategies, and finishings will be explored.

Keywords Access and circulation · Comfort · Flexibility · Spatial configuration · Zoning

4.1 The Need for Smaller Dwellings

The need for smaller dwellings is related to contemporary motivating factors such as affordability, densification, as well as demographic and lifestyle changes. In past decades, the area of homes grew progressively despite a decrease in the average household sizes (Nelson 2018a). This trend changed in the mid-2000s due to increasing land prices and increasing urbanization tendencies (Vachon 2018). Some of the aspects that led to size reduction are discussed below.

As previously stated in Chap. 3, affordability is a principal concern for new home-buyers as the cost of housing in many nations continues to rise. In 2022, for example, the average price of a home in Canada reached $816,720, a 30 percent increase in two years and nine times the average household income (National Post Staff 2022). Smaller units therefor are becoming more favorable and, in many cases, the only option for many buyers and renters. For example, units in apartment buildings have

A. Friedman, *Fundamentals of Innovative Sustainable Homes Design and Construction*, The Urban Book Series, https://doi.org/10.1007/978-3-031-35368-0_4

fewer rooms and accommodate smaller households, and as a result, are more affordable (Nelson 2018b). As the cost of housing continues to rise, the average size of apartments in urban areas has shrunk resulting in an increase in micro, studio, and one-bedroom units. Vachon (2018) attributes this rise in micro units, that are defined as any unit below 46 square meters (500 square feet), to the unmet need of affordable housing in densely populated cities. Consequently, to provide more affordable housing options without sacrificing liveability and comfort, designers are utilizing innovative strategies to get the most out of smaller spaces.

As discussed in Chap. 2, densification relates to the increasing population density of urban environments (Vachon 2018). Since the mid-1940s, many nations are experiencing and coping with significant unsustainable urban sprawl. Yet, due to increasing land and infrastructure costs of single-family dwellings, developers are opting to construct apartment buildings which in number of major cities now account for about 60 percent of the new dwellings stock (Vachon 2018).

Demographic transformations and lifestyle changes have resulted in a need to rethink modern housing. For example, individuals are increasingly opting to live alone and in Canada, the number of single households has grown more than twofold, from 1.7 million in 1981 to 4 million in 2016 (Tang et al. 2019). This can be attributed to a number of factors relating to the economy, culture, and lifestyles to name a few. One of these factors being lower mortality rates a result of improved public healthcare allowing more people to live longer, alone, and age in place (Nelson 2018a). Additionally, higher rates of divorce, lower fertility, and increasing numbers of young adults, who delay or forego marriage altogether have resulted in smaller households (Nelson 2018a). These changes necessitate a rethinking of housing and a call to design for smaller, more flexible dwellings.

4.2 Macro Design Aspects of Small Spaces

Key elements of designing small interior spaces are efficiency and functionality. To achieve that, one needs to foresee how the space will be used. The designer will be required to prioritize certain functions and eliminate unnecessary waste. For example, formal dining and living rooms are often less used in contemporary homes. Open, multifunctional spaces can maximize efficiency in compact dwellings. This is a change from the traditional, low-density homes, in which functions are formally separated room by room (Fig. 4.1). Various cultural and generational perspectives impact comfort and preferences when it comes to interior spaces. As older generations are able to age in place, they prefer a mix of private and open space (Halbe and Mark 2018a). Whereas younger generations are generally more informal and value self-expression, comfort, and flexibility in their space (Halbe and Mark 2018a). Small interiors must be comfortable and accessible for individuals of all ages and varying levels of mobility. This section discusses macro design aspects, including effective zoning of rooms, spatial configuration, access, and circulation of smaller interiors.

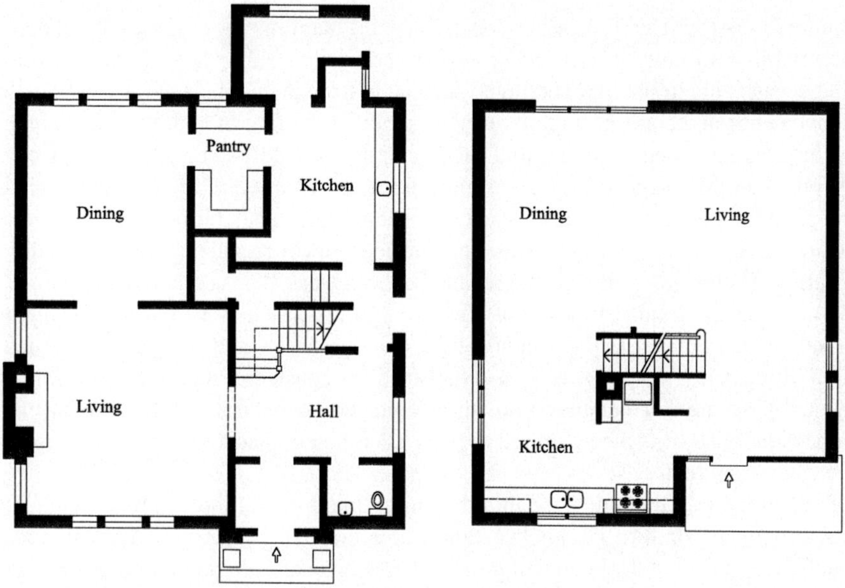

Fig. 4.1 Lower floors of a turn of the twentieth-century home and a contemporary residence. Over the years, enclosed individual functions were combined to form a single open space

4.2.1 Zoning Interior Space

The *zoning of a unit* refers to the designation of its functions into specific areas based on certain criteria and the household's activities. In larger homes, each room may have its own designated function, however, to create smaller interiors designers should carefully consider zoning to reduce unnecessary space. Modern, efficient housing developments require a variety of configurations to accommodate different households' needs and lifestyles, one standardized model can not sufficiently accommodate all modern households (Nägeli et al. 2016). Through the process of zoning, the designer can identify important criteria based on the occupants' needs to determine the most effective organization of space. The designer must first identify the necessary functions required of each space, these functions can then be grouped into various zones based on space availability and their categorizations (Du and Li 2013). A change in perspective from the traditional family home helps to envision smaller and more modern interiors by viewing the home as a series of areas for activity, rather than a collection of enclosed rooms, each with a prescribed function. This approach allows for more flexibility when zoning the various functions of the interior thereby maximizing the efficiency of a home based on its occupants' needs. Zoning can consider a variety of different variables including privacy, light, and utility, depending on the designer's priorities and requirements.

Zones can be categorized into public, semi-public, and private. *Public zones* are shared by household members and their guests (Fig. 4.2). Traditionally the living and dining rooms (both public) are best located near the main entrance, on the ground floor of a multi-story unit. *Semi-private zones* are primarily used by household members and occasional guests; they should be accessible from the public zones to maximize convenience and comfort. Finally, *private zones*, consisting of bedrooms and some bathrooms, are used by household members, and should be secluded if possible. Another useful categorization for zoning purposes is *day and night zones*, which consider the time of use for each space and natural light availability. *Day zones* may include living spaces, dining spaces, kitchen, and work areas, all require high-quality lighting. Therefore, daytime zones should be located based on access to natural light, often along the southern façade of the home. Bedrooms may be considered *night zones,* which depend less on natural light and can be located in a more secluded area. Bathrooms, storage spaces, and hallways are considered *service zones*, which typically are neither benefitted nor disadvantaged by natural light. Based on the zones identified, it is then required to place each function accordingly to maximize efficiency and better use.

When there is limited space in a dwelling, functions can be combined without compromising comfort. The kitchen, dining area, and living space may all fall into the public realm (Fig. 4.3). In contemporary homes, kitchens are often used as settings for day-to-day activities and spaces for socialization as well as food preparation. A separate dining area may not be the most efficient use of space. Therefor combining

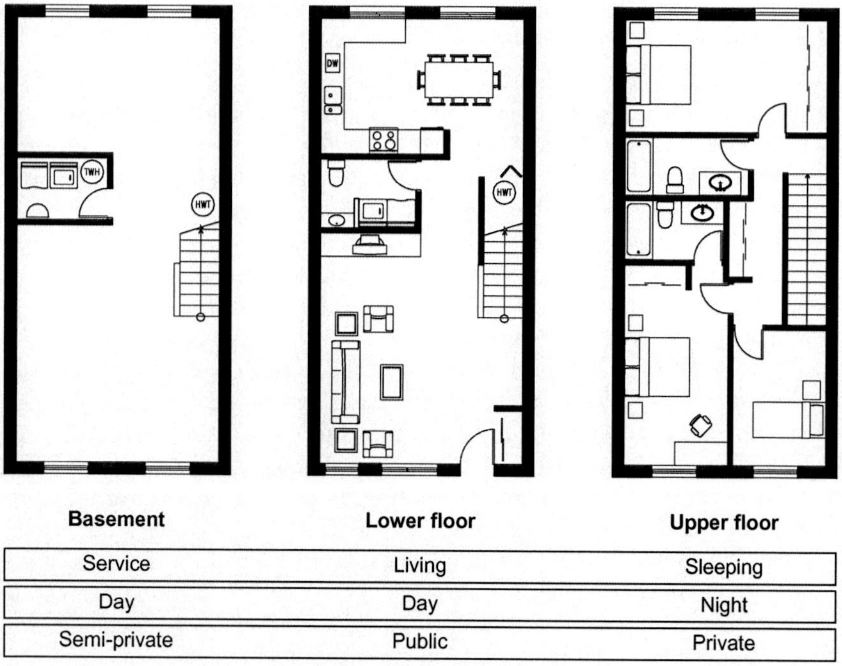

Fig. 4.2 A variety of identifications of spaces in a multi-level home as per their function, privacy, or time of use

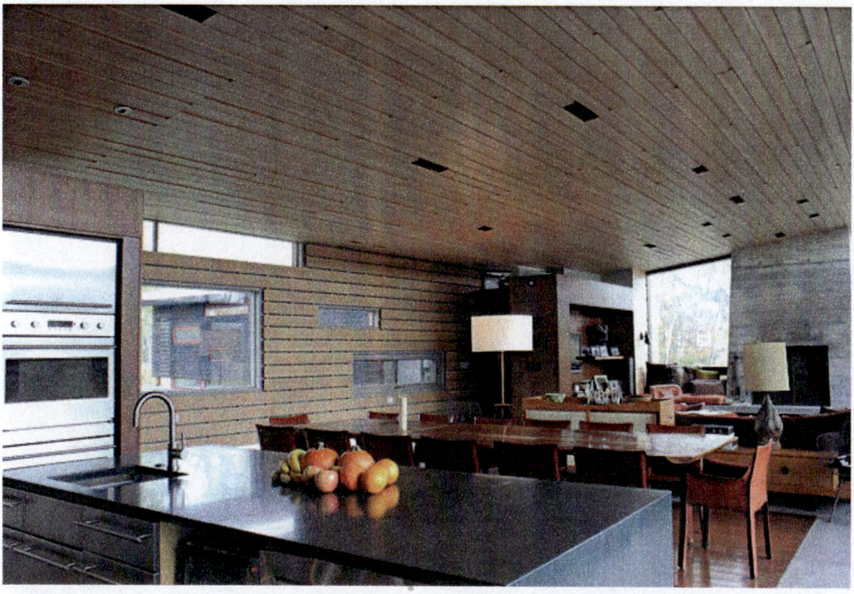

Fig. 4.3 The kitchen, dining area, and living space can be combined into a single open plan

the kitchen and dining areas into one space with the use of an extended kitchen table may satisfy a household's meal preparation and dining needs while also providing space for everyday family life. Workspaces can also be combined with living or bedroom spaces, and a laundry function can be combined with bathrooms or kitchens to eliminate the need for additional rooms. Naturally, these combinations will be implemented differently in each dwelling based on the household type and its needs.

4.2.2 Access and Circulation

Planning the access and circulation in a dwelling is important in defining its character and ensuring efficient use of space. The entryway is the transitional space between the exterior and interior and gives the first impression of the dwelling. The entrance should give a sense of the home, to make one feel comfortable and welcomed. For narrow houses, moving the entrance to one side of the front façade limits disruption to the interior spaces. Locating a window within view of the entrance immediately introduces the visitor to natural light and views to the outside, visually expanding the space (Fisher-Gewirtzman 2017) (Fig. 4.4). Additionally, using a lower ceiling height above the entrance frames the interior and creates a sense that the entryway opens up into the larger living space.

In small dwellings, the primary goal in the planning of movement should be to minimize the amount of space dedicated to circulation. When possible, circulation

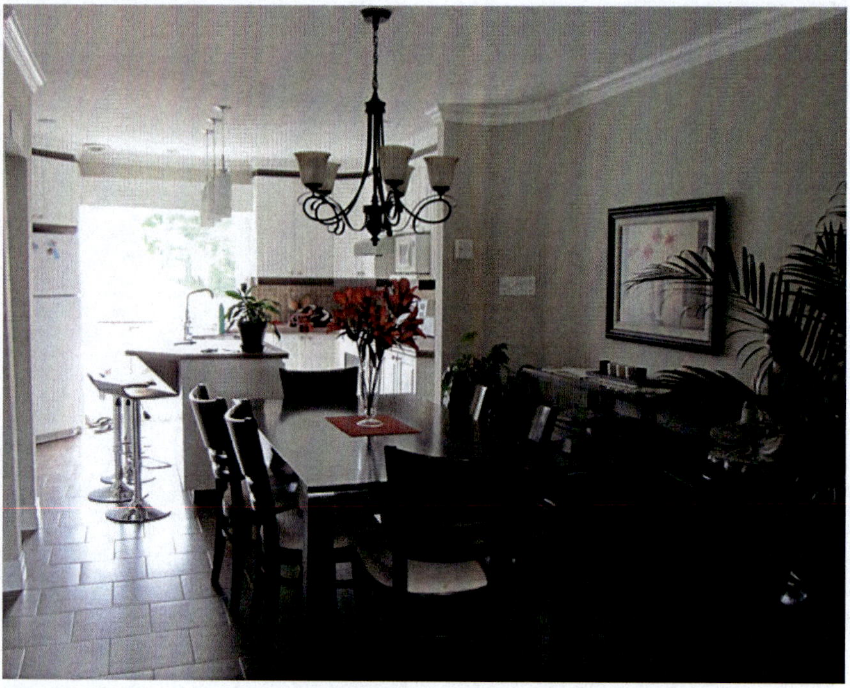

Fig. 4.4 A long view into the exterior upon entry will make a small space feel longer

spaces should be combined with other functions, such as storage (Fig. 4.5). With corridors of 1.2 m (48 inches) in width, 0.3 m (12 inches) of shallow storage space can be created, in the form of closets, cupboards, or shelving. Circulation should be planned to ensure that one need not cross through one zone to reach another. Also, private zones should not be located immediately off public zones. Additionally, in the configuration of an open plan, it is important to ensure that adequate circulation can be achieved without disrupting activities. Circulation can none the less be used to frame interior views and influence one's perception of a small space. Planning for smaller rooms to be entered from an angular view creates an immediate impression of more space, by seeing the room by its longest dimension: the diagonal.

In multi-level dwellings there are several types of staircases which can be used for efficient vertical movement (Fig. 4.6). In narrow townhouses, it is generally recommended to place one straight run staircase close to the entrance, longitudinally along the shared wall. This arrangement limits disruption to interior space. L-shaped staircases can be fit into corners and can often be placed near the entrance, creating a natural division of the interior plan. U-shaped staircases create a central void, this can be used to create an open core through the dwelling, increasing natural light access and giving a sense of spaciousness, but can be the least economical in terms of leaving useable floor space.

Fig. 4.5 When possible, circulation spaces should be combined with other functions, such as storage

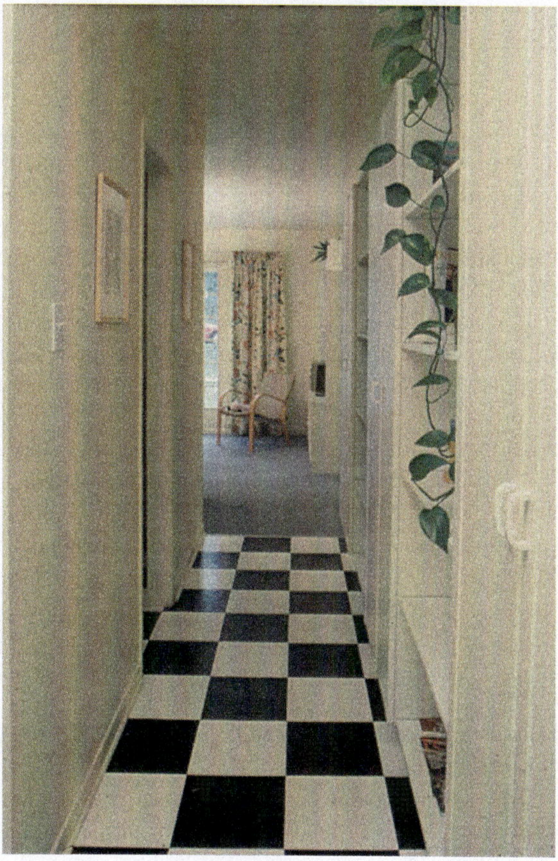

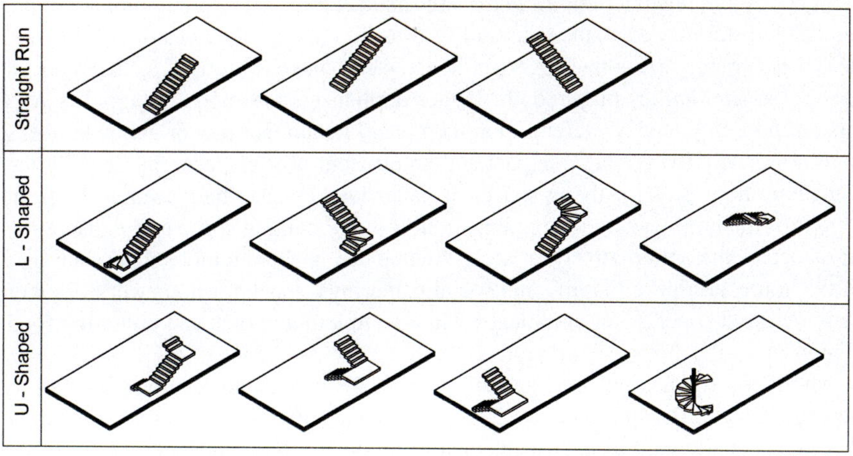

Fig. 4.6 Several types of staircases can be used for efficient vertical movement in multi-level dwellings

4.2.3 Spatial Configurations

The spatial configuration of smaller dwellings requires careful thought to maximize comfort and liveability. The main goal is to prevent a sense of enclosure which creates discomfort in small interiors. Perceived density relates to one's subjective response to the spatial configuration of a certain place. Without making changes to the area of a dwelling, improvements to its perceived density can result in more comfort for its inhabitants (Fisher-Gewirtzman 2017). Interior partition walls are the traditional approach to denoting areas of the home. However, in smaller dwellings this limits light access, uses valuable floor area, and makes spaces feel smaller. Open plan spaces which integrate multiple functions give the impression of more space than smaller, sectioned-off rooms. The use of an open plan for public zones and enclosed rooms for private zones reinforces the distinction of functional zones, which is particularly important in smaller units where zones are often near each other.

Instead of partitions, open plan configurations can use furniture, shelving, ceiling heights, and floor levels to denote space (Fig. 4.7). In general, taller ceiling heights can help to create a sense of spaciousness for residents of small dwellings (Hoang and Vandal 2017). The use of varying floor and ceiling heights can help to distinguish spaces for various functions in smaller interiors. Low ceilings create a sense of warmth and privacy while tall ones give a feeling of openness. These strategies can be used in different zones based on the designer's desired effect. Similarly, varying floor levels can help to distinguish separate functions within the home. In an open plan, certain areas such as the living space can be denoted with a small step down, creating a distinct space.

The spatial arrangement of functions will depend on the desired effect of the dwelling. In general, functions placed near the entrance have the highest traffic and are typically more associated with public zones. Whereas a degree of privacy is achieved by placing functions away from the main entrance. The predominance of the kitchen will vary based on the households' priorities and needs. Locating the kitchen near the entrance or in other high-traffic areas can make it more of a shared social space. Larger families may require bigger appliances and more storage, whereas smaller households may prefer a more compact kitchen. Narrow or galley kitchens permit less activity, but are compact and an efficient use of space, maximizing the remaining floor area for living spaces. Smaller kitchens can be placed at the front or rear of the unit, leaving the organization of the remaining space to the discretion of its occupants. Alternatively, a larger kitchen can become a more significant part of the home, permitting family and social gatherings, space for homework or other activities, and supervision of children while completing chores and preparing food (Fig. 4.8).

The living room provides a gathering space for household members as well as hosting guests, it is a space for social interaction and entertainment. Designers might then consider allowing access to a television without inhibiting other social activities.

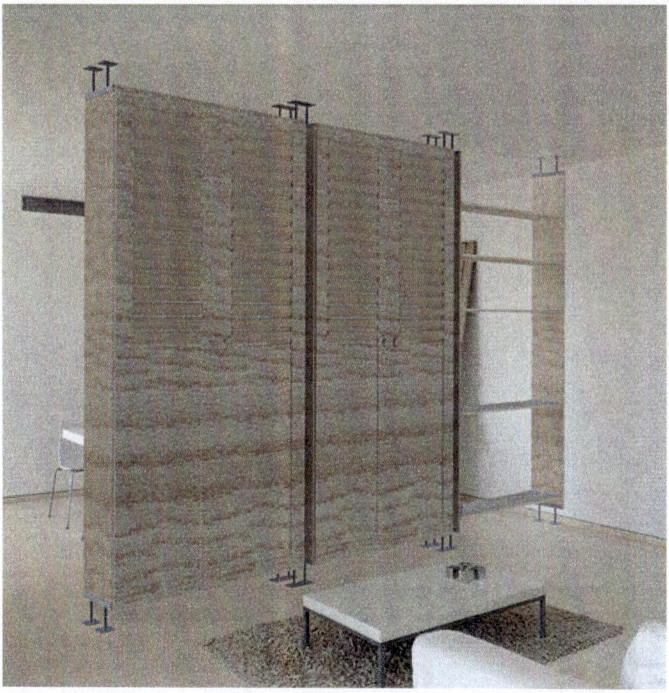

Fig. 4.7 Instead of permanent partitions, open plan configurations can use furniture or shelving to denote space

A large living space allows the occupants' flexibility to organize or subdivide the space according to their preferences and lifestyle. When placed toward the front of the dwelling, these areas can provide an inviting space for guests and socialization. Conversely, in a small unit, placing the living area near the rear with access to a backyard makes the outdoor space an extension of the living space. Similarly in an apartment, locating a balcony off the living space can have the same effect. Strategic use of openings can give the interior a sense of spaciousness. Large windows and glass doors in living spaces can open the interior up. Corner windows in particular help to break up the feeling of enclosed space. Furniture can be used to emphasize specific qualities in a small living space. Horizontality can be emphasized by using a low and long couch, giving the impression of a wider space (Halbe and Mark 2018b). Additionally, furniture can be used to create separate activity zones in small spaces without partitions, delineating the living space from the rest of an open area for example.

Bedrooms are always the most private zones, and their arrangement will vary greatly depending on the type of housing being constructed. In a townhouse or other multi-level dwelling, it is generally preferred to place bedrooms off the main floor, to distinguish them from more public zones of the dwelling. When located on the same floor as the main entrance, bedrooms placed away from the entrance can create

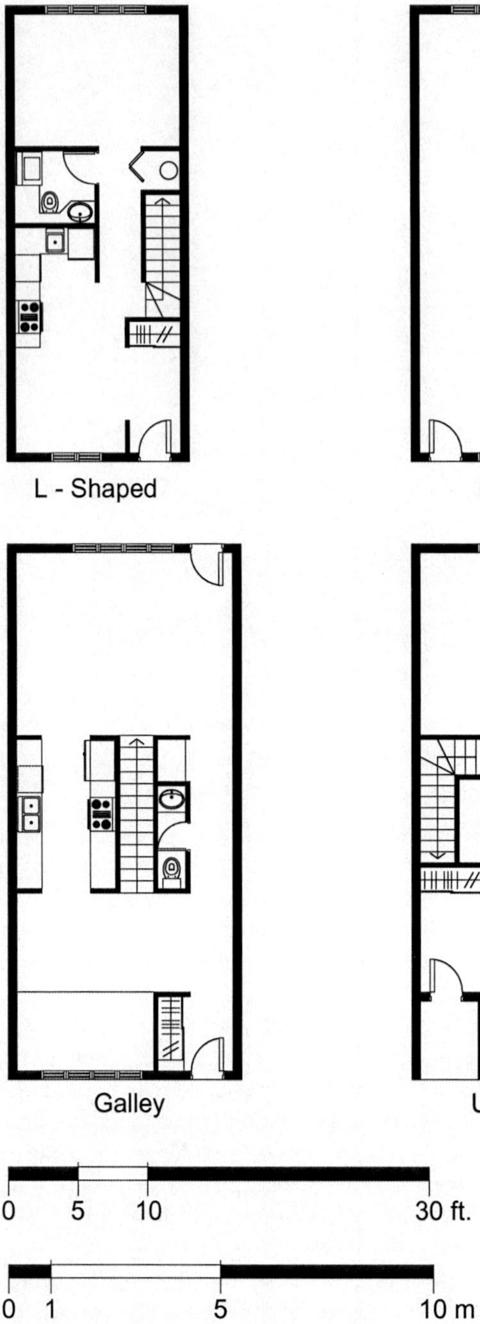

Fig. 4.8 Common kitchen layouts for small spaces

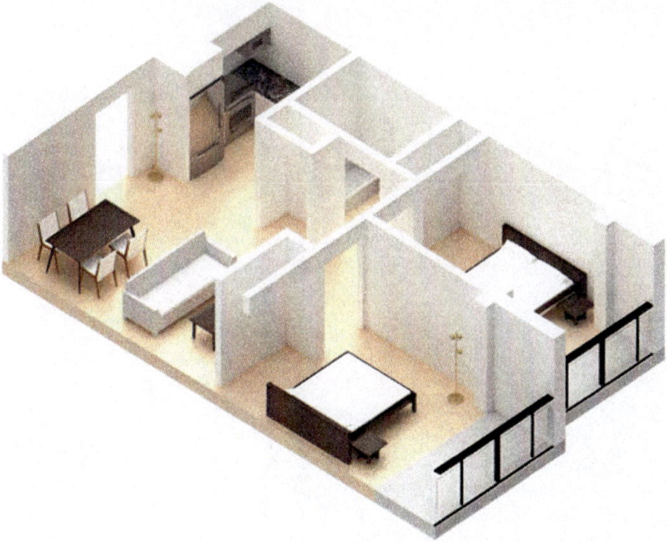

Fig. 4.9 When located on the same floor as the main entrance, bedrooms placed away from the entrance can create a sense of seclusion

a sense of seclusion (Fig. 4.9). In small units with tall ceilings, a *mezzanine* can be constructed over the kitchen or living areas, giving occupants a more private bedroom and distinguishing spaces through varying ceiling heights. Fisher-Gewirtzman (2017) found that the addition of a mezzanine space in a micro apartment unit was seen favorably by individuals for reducing the unit's perceived density, creating a more private bedroom space, and defining the living spaces below.

4.3 Micro Design Aspects of Small Spaces

Creating comfortable interiors requires designers to consider micro aspects that affect a small space's function. Seemingly, trivial aspects can have a significant effect on one's perception and comfort. One of the goals in designing sustainable housing is to facilitate long-term occupancy including planning for an uncertain future by maximizing liveability. This section discusses micro aspects of small space design including flexibility, space-making devices, and the use of light, colors, and finishings to create comfortable interiors.

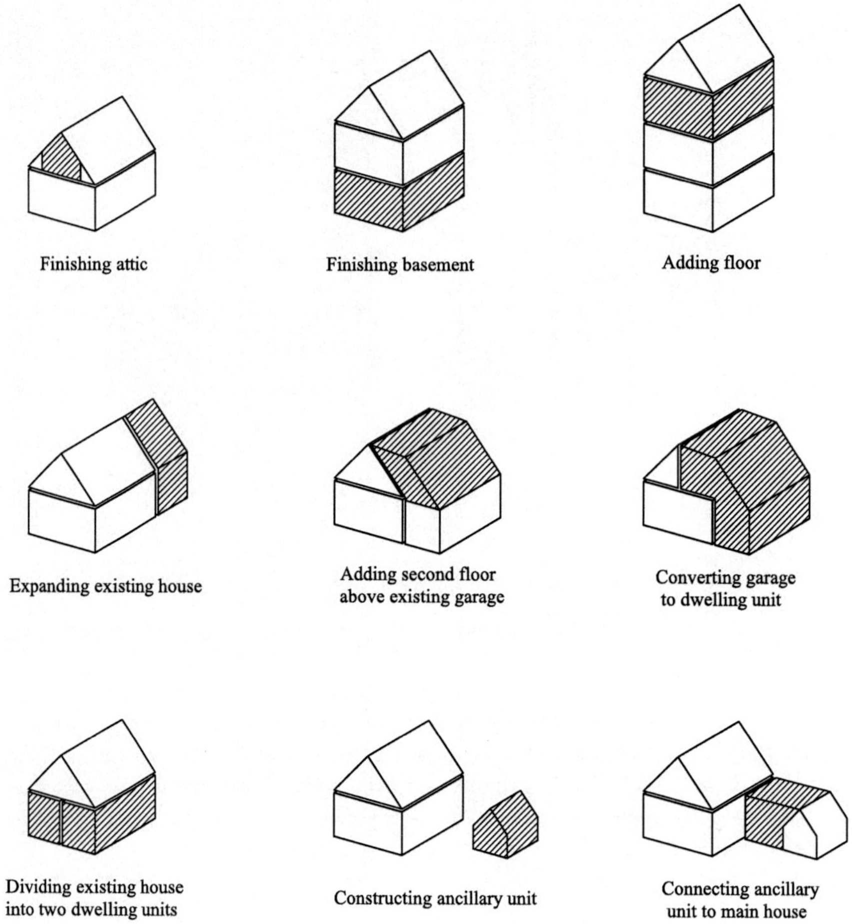

Fig. 4.10 Some common types of home adaptations involve subdivision of units into multifamily dwellings, additions, and modifying layouts to suit occupants' needs

4.3.1 Accommodating Flexibility

Planning for flexibility is essential to accommodating the evolving needs of modern households. This involves consulting with occupants on their space needs and facilitating post-occupancy modifications. The primary goal of flexible design is to plan for future uncertainty in order to prolong the lifespan and efficiency of buildings. There are a multitude of reasons that lead people to renovate their homes' interiors. The main factors often include change of household's size and newly acquired habits such as working from home. Some common types of adaptation involve subdivision of units into multifamily units, expanding dwellings, and modifying the layout to suit new needs (Fig. 4.10). In this regard there are certain strategies which can make dwellings more flexible to modifications.

As noted above, the easiest way to plan for flexibility is by creating an open plan. The use of an open plan allows for add-in growth, permitting residents to progressively alter their home according to evolving needs and budgets. In areas where more privacy is desired, freeing the interior space of load-bearing walls and columns allows for easy modification of the internal plan later (Fig. 4.11). Multipurpose rooms with sufficient proportions should be created by considering a variety of functions, including sleeping, entertainment, socializing, or working. Rooms between 3.7 by 3.7 m (12 by 12 feet) and 4.6 by 4.6 m (15 by 15 feet) are typically large enough to accommodate future adaptations. Service spaces such as bathrooms and kitchens are difficult to relocate once placed, therefore, careful consideration should be given to their locations to allow for adaptability of the adjacent living spaces (Dhar et al. 2013). Utilities including mechanical systems, electrical systems, ductwork, and plumbing should be easily accessible to enable maintenance and modification when necessary. Additionally, placing utilities within partition walls should be avoided to simplify future relocation of partitions (Dhar et al. 2013). By following these design recommendations, dwellings can become highly adaptable to future change.

Prefabricated *demountable partitions* can be introduced to divide spaces and enclose rooms while maintaining adaptability (Fig. 4.12). They can be assembled, disassembled, and relocated based on the needs of the household. There are generally three types of demountable wall systems which can be considered for more interior adaptability. The first is a *mobile* or *operable system*, its panels have a sliding mechanism attached to ceiling tracks which allow panels to slide in and out of place depending on their desired use. This type of system is recommended for occupants who want to modify a space frequently. For example. to gain more privacy, enclose a small workspace from a more public area to limit distraction without permanently partitioning the space.

Portable partition systems are prefabricated panels that are secured to channels in the ceiling and floor. These are a more semi-permanent solution to partitioning space. Finally, *demountable systems* are walls constructed of prefinished gypsum attached to metal studs at specific intervals. They are easier and faster to deconstruct and modify than traditional wall partitions. For example, they can be used to create two small children's rooms, when one child moves out, the space may be expanded into one larger room. Each of these systems have varying degrees of permanence to suit the inhabitants needs.

4.3.2 Space-Making Strategies

There are a variety of strategies to make efficient use of small interiors (Fig. 4.13). Open-floor areas can be subdivided using furniture to facilitate easy modification and maximize space efficiency. Moveable closets and shelving units, as well as couches, benches, and tables, can effectively subdivide an open plan while maintaining a sense of openness by blurring the visual boundaries between areas, making the unit feel more spacious than it would with partitioned areas. Additionally, in small spaces

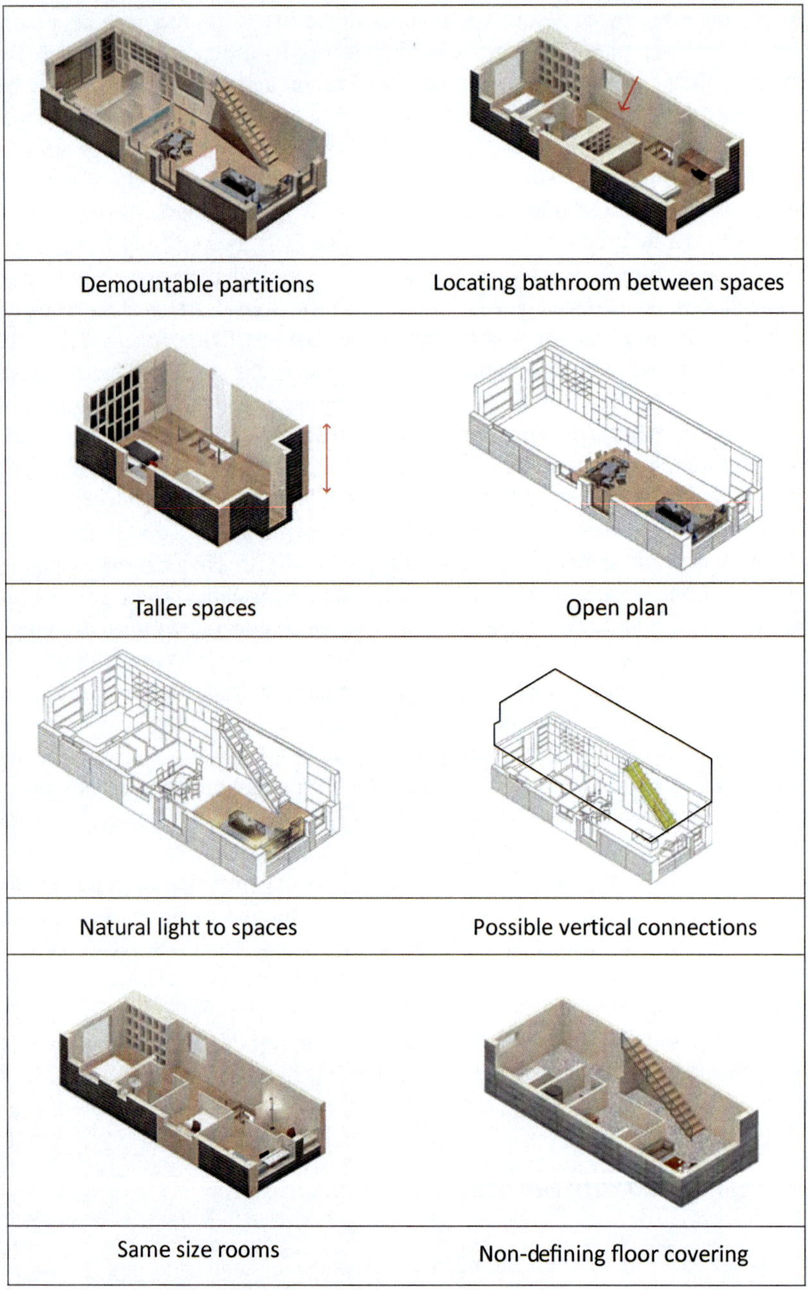

Fig. 4.11 Design strategies that facilitate future flexibility in small dwellings

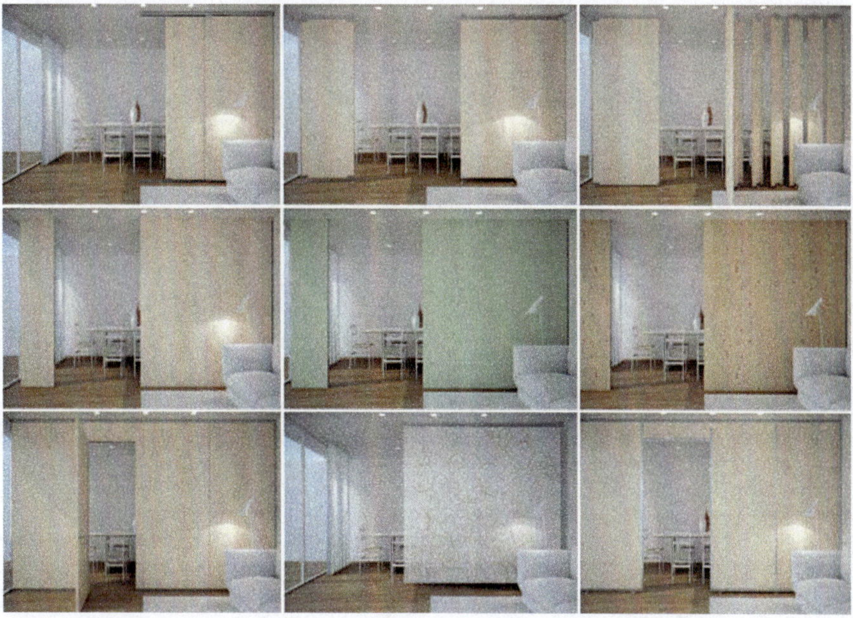

Fig. 4.12 Prefabricated demountable partitions can divide spaces and enclose rooms while maintaining adaptability

pocket doors should be considered as an alternative to hinged doors to maximize the available floorspace. They require less clearance than swinging doors and remain hidden within the wall cavity when opened.

Creating adequate storage space is one of the primary concerns in designing smaller dwellings. Freestanding storage such as closets and wardrobes can take up valuable floorspace and restrict movement. Built-in storage should be considered in small spaces to aid in limiting clutter. There are several ways to improve storage space without compromising floor area. *Residual spaces* are the small leftover areas after the main functions have been located in a home. In small dwellings, these locations can be areas under or on top of the stairs for example (Fig. 4.14). Open spaces underneath staircases can be used to create shelving units, small benches, or even compact workspaces to make the most of each space. Adding space upward should be the first method of adding storage without encroaching on floor space. With average North American room heights falling between 2.44 and 2.74 m (8 and 9 feet) tall, providing significant opportunities for storage space (Halbe and Mark 2018d).

Increasingly, compact, adaptable furniture units are being created to help maximize the efficiency of small spaces. Functional adaptable furniture can be easily modified to meet the dynamic needs of a household in particular spaces (Du and Li 2013). In bedrooms, this could include *trundle beds* which store additional mattress frames underneath to slide out, creating additional beds. Other examples include bed

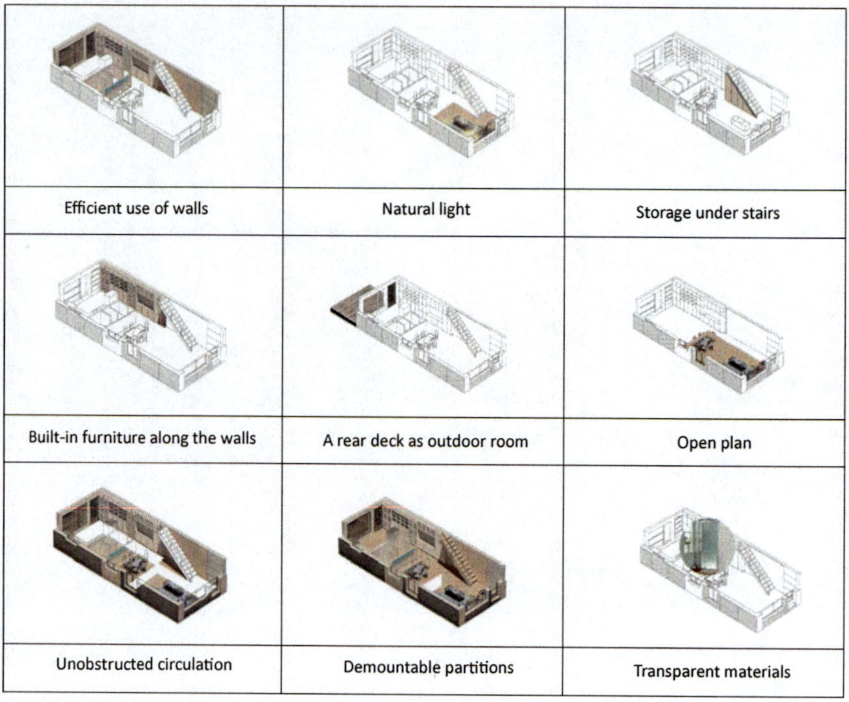

Fig. 4.13 Strategies to make use of small interiors and make them feel bigger

frames with storage drawers underneath for clothing and other items and *murphy beds,* built-in bed frames that use hinges to be stored vertically against a wall when not used, leaving space for a desk, seating area, among other functions. Similarly, in kitchen or dining areas, hinged tabletops can be folded out to minimizing their footprint when not in use (Fig. 4.15). Benches can feature low shelving units or fold open to be used as a hidden storage space. When there is not adequate room for a permanent workspace, built-in cabinetry and shelving units can feature a tabletop space which folds out and back into the wall when necessary. More on storage design will be outlined in Chap. 12.

4.3.3 Light, Colors, and Finishings

Careful attention to the use of lighting, color, and finishings in a dwelling can help to create more comfortable interiors for inhabitants. First, it is important to understand how psychological response to color impacts one's perception of a space. Colors can give the impression of advancing or receding, floating, or being grounded, and being lighter or heavier, among other sensations (Halbe and Mark 2018c). Wall colors also have a significant effect on our perception of the size of a space. Dark colored walls

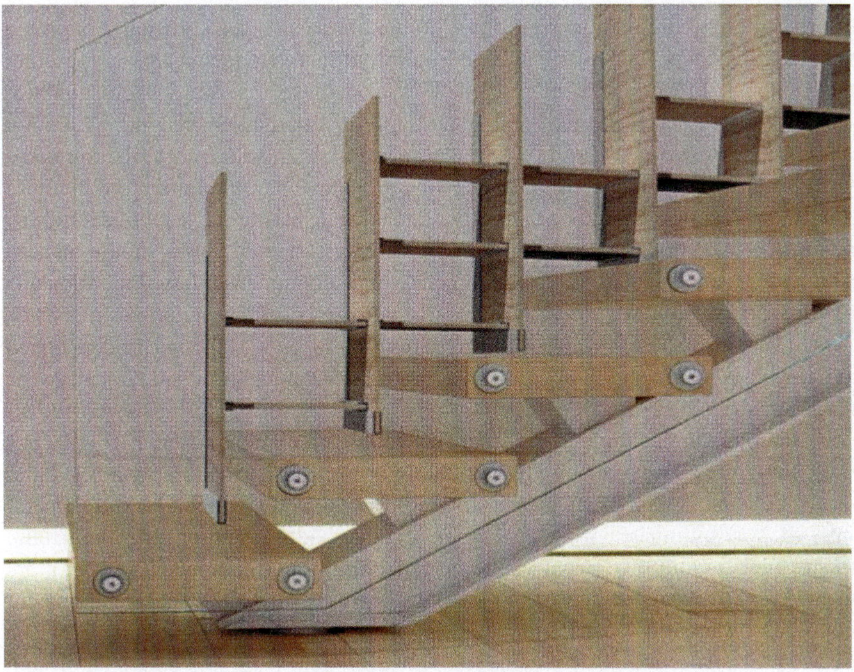

Fig. 4.14 In small homes, storage locations can be unused space on top of the stairs

Fig. 4.15 In a kitchen, hinged tabletops can be folded out to minimize their footprint when not in use

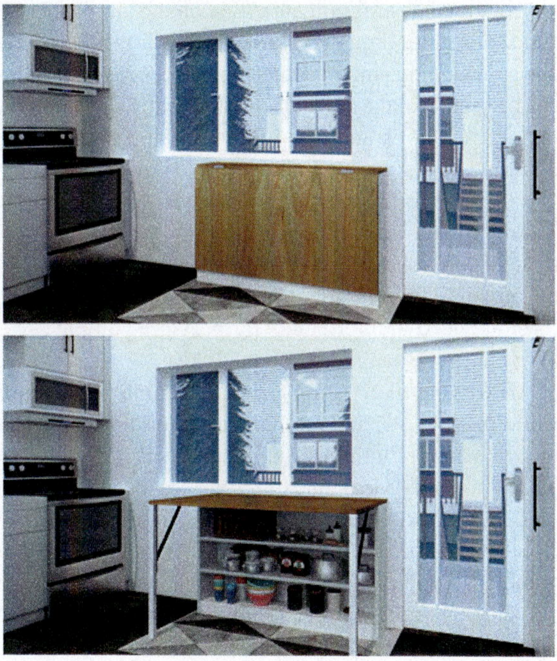

absorb light, making them appear closer to the observer. Therefore, darker colored walls can make spaces feel smaller and more intimate. Lighter walls on the other hand, reflect light, giving the impression of more openness. There are several qualities which influence our perception of color. *Hue* describes the wavelength that a color reflects, *brightness* measures how strong the color is, and *saturation* measures the purity of a color. Humans are most sensitive to changes in hue, therefore the hues of interior finishings should be carefully planned. The higher the level of saturation, the more visual attention a color draws, thus highly saturated colors should be used cautiously in small spaces. Each version of "white" contains very low saturation and high brightness, however small changes of hue can have a strong effect of a white painted room. Shades of white with low saturation of blue are typically considered cooler, while shades of white with low saturation of yellow or red are typically seen as warmer.

Light colors are key in making smaller dwellings feel more spacious (Fig. 4.16). Particularly in ceilings, because lighter colored ceilings give the impression of taller rooms. Finishing textures can have a similar impact to color. Rougher textured materials, such as unfinished wood and brick, create shadows and absorb more light. Meanwhile, smooth textures are better at reflecting light, making spaces feel more open. A small variety of colors and finishing textures in a small space can prevent monotony by creating visual character, highly saturated colors can effectively be used to create accent walls for example. However, very saturated colors and rough finishings should still be used sparingly in small spaces to maximize light and prevent a sense of enclosure.

Lighting has a significant effect on the atmosphere of a space. Providing an abundance of natural light makes homes feel larger and more open. The home's orientation and surroundings mainly determine access to natural light, however certain strategies can be used to make the most out of windows in small dwellings. Strategic planning of the size, quantity, and positioning of windows can help to bring in the most natural light and make spaces feel as open as possible. Windows overlooking interesting views are preferable in small spaces, to draw attention outward. Tall windows allow access to natural light while limiting views into the home from outside. Finally, frosted glass or window treatments can be used to give privacy without compromising on light in high-density environments. Sliding doors and Juliet balconies are recommended in micro units for providing ample access to natural light and a sense of openness (Hoang and Vandal 2017).

Mirrors are a very simple and affordable way to enhance natural light in the home while making a space feel larger. Carefully located mirrors can reflect daylight to improve natural light access in small spaces. Glass and other translucent furniture pieces are also ideal for small spaces because they allow light to pass through, making the space feel more open. The direction of natural light also impacts its effect on a space. In the northern hemisphere light from northern windows casts a cooler, bluer light on surfaces, whereas light from southern windows tends to be yellowish and warmer (Halbe and Mark 2018c). While the goal should be to maximize natural light in small spaces, bright sources of light can create glare causing discomfort, particularly when using screens. Giving people the capability to control

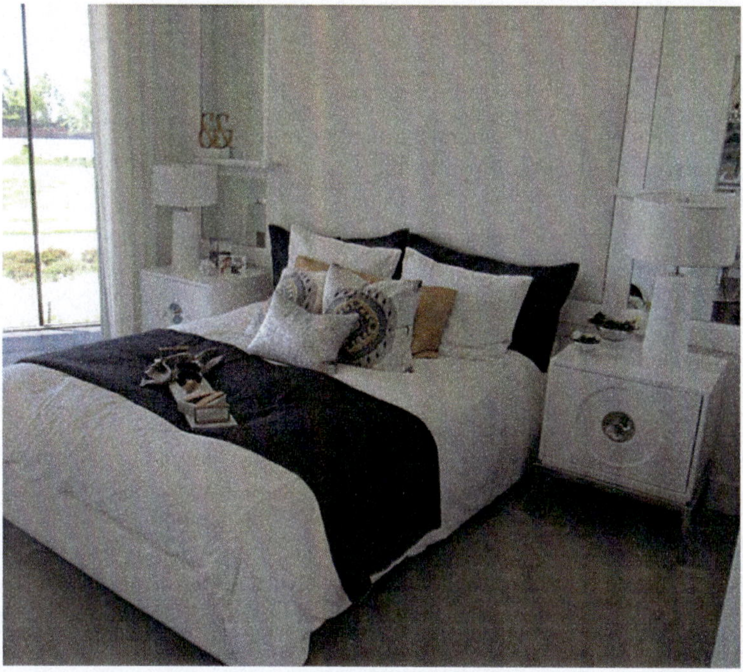

Fig. 4.16 Light colors, as used in this bedroom, are key in making smaller dwellings feel more spacious

their indoor environment improves their visual comfort and overall satisfaction with the space (Frontczak and Wargocki 2011). It is, therefore, appropriate to have window coverings allowing for occupants to control the amount of natural light they receive.

It is necessary to plan synthetic lighting that supports natural lighting and creates a comfortable interior. Interior artificial lighting comes in several categories of function: *ambient lighting* provides overall illumination, *task lighting* provides focused light for specific activities, and *accent lighting* which typically creates a warmer glow and adds depth to a space. Providing various types of light sources in a space helps to create a more comfortable environment for occupants. Implementing *circadian lighting*, which provides appropriate light based on the human body's circadian rhythm, has numerous benefits. The circadian rhythm helps to regulate many of the body's functions and is therefore a significant contributor to overall health and well-being (Halbe and Mark 2018c). Mindful planning of these details has a significant impact on the liveability, efficiency, and comfort of small dwellings.

4.4 Small Comfortable Interior in Zeeburgereiland, the Netherlands

The Netherlands has been facing a shortage of rental housing stock and a proliferation of derelict sites. Additionally, the number of single, high-potential people with their first jobs in their mid-20s to mid-30s is expected to grow to be larger than the population of Rotterdam by 2050. These people earn high enough incomes to disqualify them for social housing but not high enough to be in the free-rental sector. As a result, Heijman ONE in Zeeburgereiland, Amsterdam, the Netherlands was designed by MoodBuildersto to meet demands for quality and affordable rental housing that aims to solve multiple social issues including the one described above. At the same time, the 45 square meters (484 square feet) project focuses on the rental market to keep costs low yet display individuality (Hohenadel 2015).

Heijman ONE is a complete single-person home that can be placed temporarily in empty urban areas. Each unit is two storeys tall with all the necessary functions. The prefabricated home is a solution to derelict sites as installation can occur within a day until the land is ready to be developed. When a definitive building construction is about to start, the unit can be transported via truck to a new destination. Therefore, the temporary placement of the reusable home, with a lifespan of 25–30 years, becomes sustainable.

The home is equipped with a kitchen, bathroom, separate bedroom, large living room with open space, and a front door that leads out to a patio (Fig. 4.17). Once on site, the home is connected to existing utility facilities for water, sewage, and electricity. The units also have photovoltaic panels to generate energy and lower its carbon footprint. The home is even more energy efficient because of the solid wood skeleton, a recycled wood façade, and all-electric system. Since the home is built with preinstalled utilities, the building can be easily relocated to other sites.

The target residents are young adults who would otherwise have to settle for a small but expensive apartment, live with roommates, or move back in with family members. Consequently, each house is slightly varied so each property is clearly recognizable to give a sense of individuality and independence. The ground floor module links to the asymmetrical roof module which itself creates a diverse "roofscape". In addition, the two-storey home has a ceiling height of 5.9 m (19.3 feet) that allows a large amount of natural light into the house. The compact home is spacious enough for a large living space and a bedroom in the mezzanine that can accommodate a double bed. Heijman ONE ultimately created a dwelling that is more sustainable by being self-sufficient with the potential to be used by varying households.

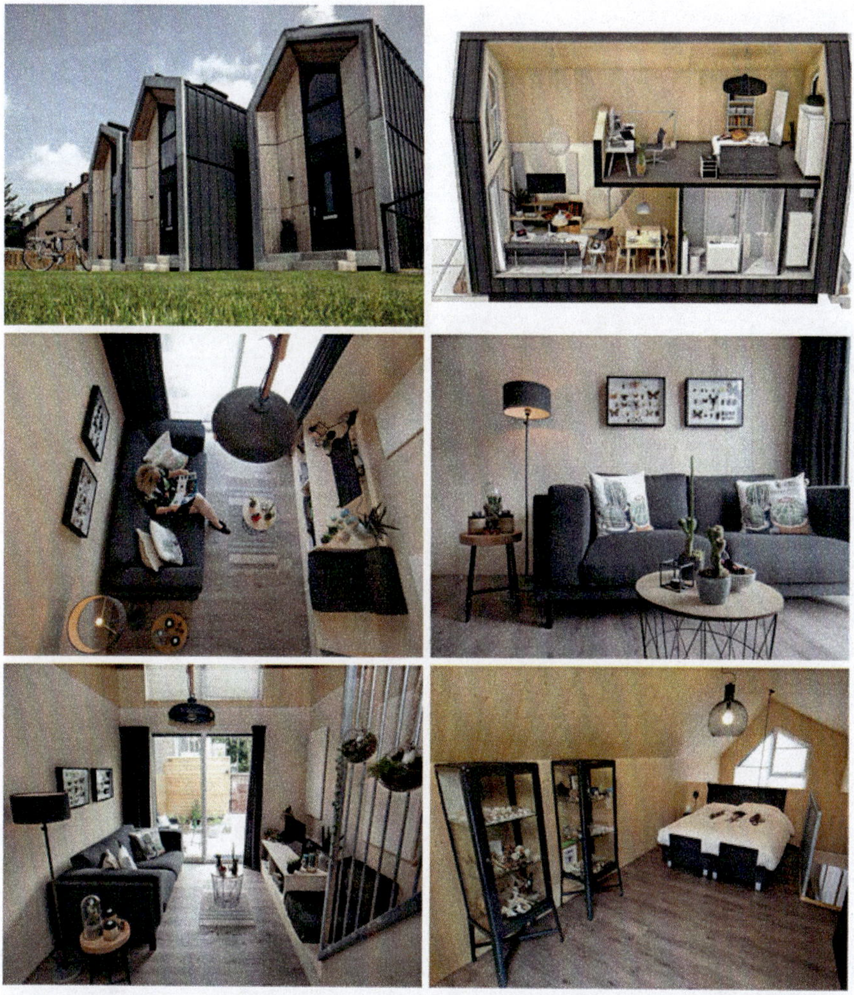

Fig. 4.17 The Heijman ONE home is equipped with a kitchen, bathroom, separate bedroom, large living room, and a front door that leads out to a patio

4.5 The Next Home

The Next Home was first built on the campus of McGill University as demonstration in 1996 (Fig. 4.18). The leading thought behind its design was that contemporary dwellings need to accommodate households with various spatial needs and budgets (Friedman 2002). Therefore, homes must be flexible and provide opportunities for choice and adaptability in their pre and post occupancy. Moreover, rising costs of housing and shrinking sizes of households call for the creation of smaller dwellings in denser communities.

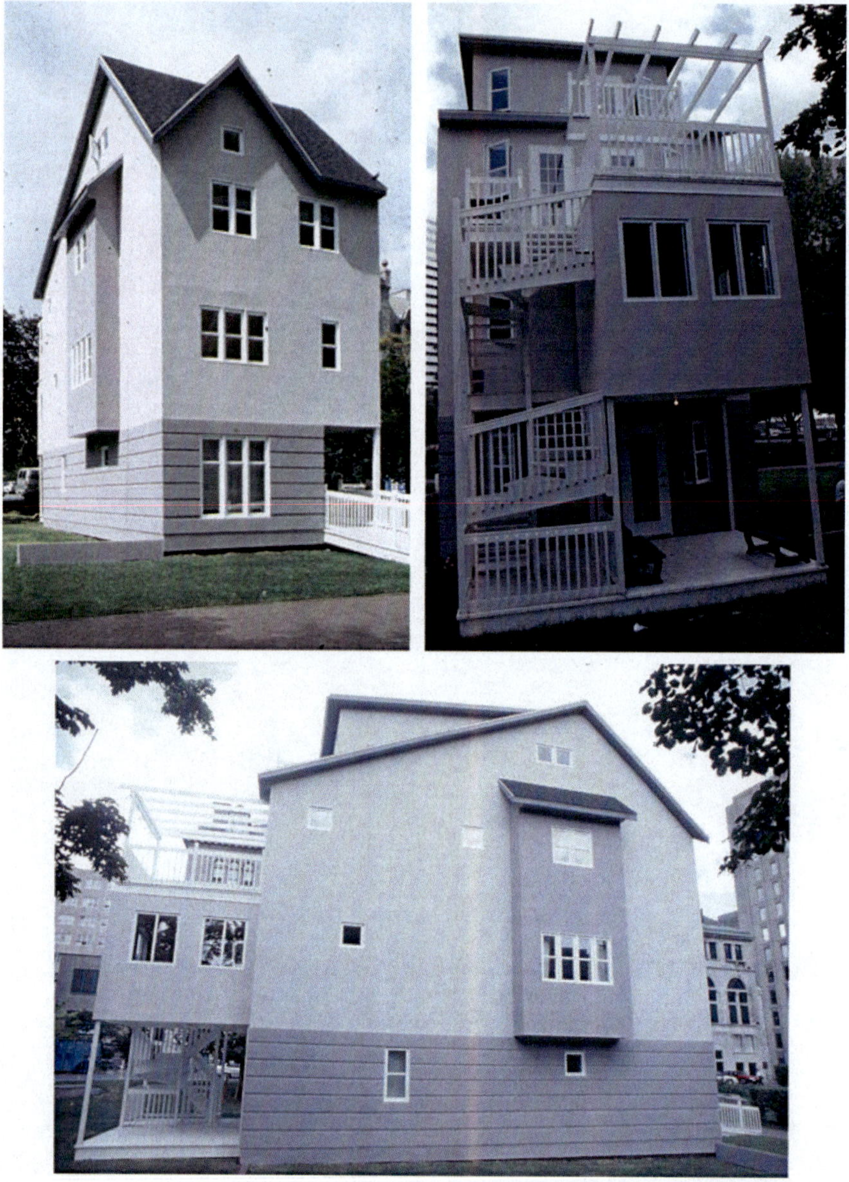

Fig. 4.18 Exterior views of the Next Home demonstration on the campus of McGill University

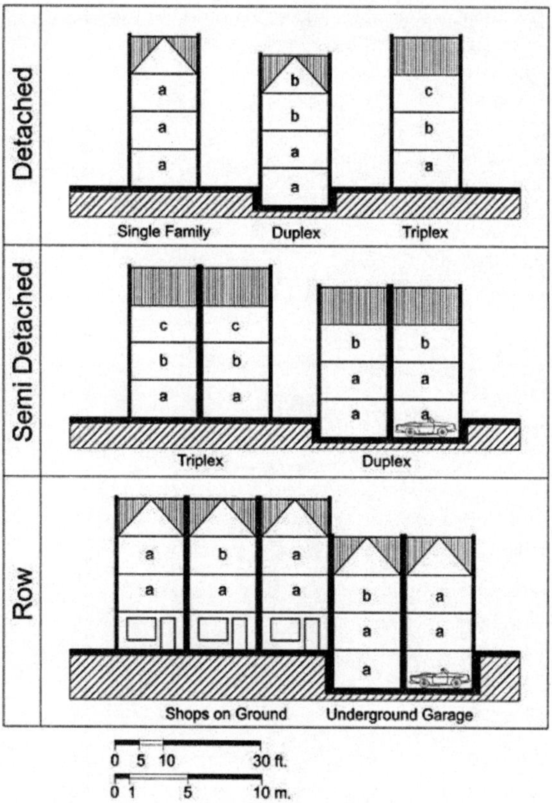

Fig. 4.19 The Next Home was designed to be built as a detached structure, semi-detached, or be part of a row

The Next Home was designed to be built as a detached structure, semi-detached, or be part of a row (Fig. 4.19). With footprints of 6.1 by 12.2 m (20 by 40 feet), the structure can become a single, two or three family unit where the interior of each dwelling can be made to accommodate different requirements.

Open web floor joists and a horizontal chaser for utilities' conduits allow bathrooms and kitchens to be located anywhere on each floor. The structures are designed to facilitate ongoing change by facilitating future additions of internal staircases for vertical connection, allowing for conversion between single and multi-level units. Using a menu, the buying process offers choice of internal components and their placement (Fig. 4.20). Buyers first select the number of floors they wished to purchase, and designers will then work with them to create a preferred interior layout and desired finishings according to their budget. Buyers will pay as per their chosen menu items which includes the length of interior partitions of their chosen layout.

To display the flexibility and range of choices available, the Next Home demonstration home presented scenarios for the arrangement of each floor based on three fictional households (Figs. 4.21, 4.22, 4.23, and 4.24). In the ground floor unit, a widower in his late sixties has a small home office, from which he operates a consulting firm. The kitchen and washroom are centrally located, separating the

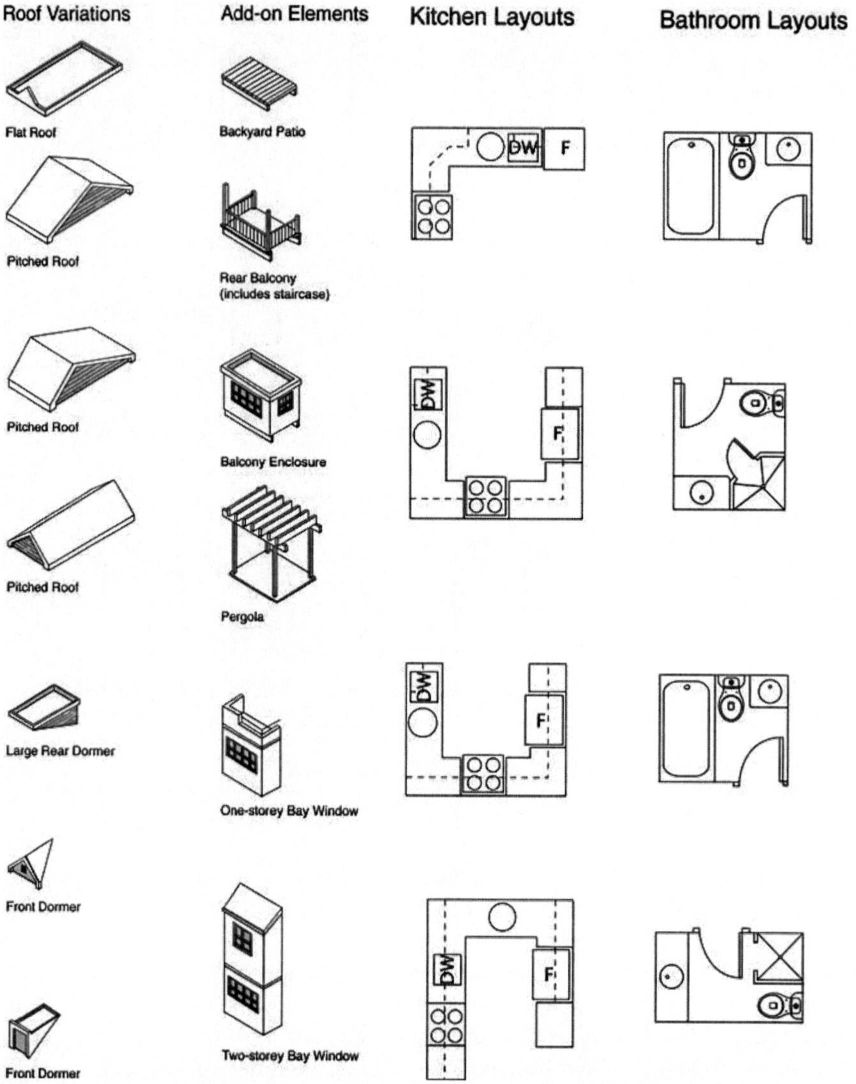

Fig. 4.20 Menu of external and internal components for builders' and buyers' choice

public zone of his office from his living spaces. Storage units partition the bedroom from the living area and walls are only used to separate the washroom and office entrance. Eventually, his office could be converted to a bedroom or living room and the unit is fully wheelchair accessible to facilitate aging in place.

A young couple who resides on the second floor were primarily concerned with affordability. They chose an open plan, using furniture to partition space, with the more public functions located toward the front of the unit. Having a small study space

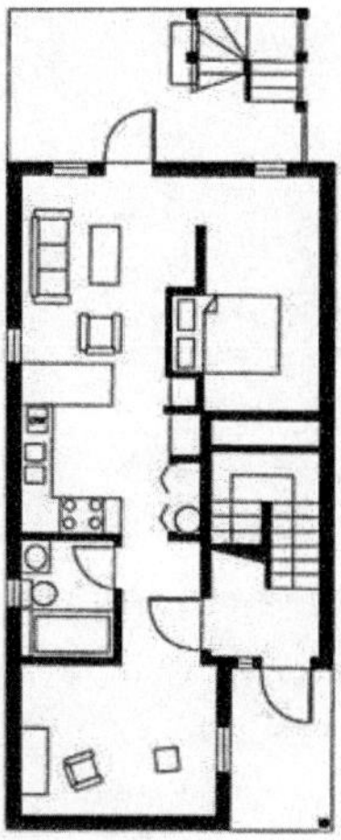

Ground Floor

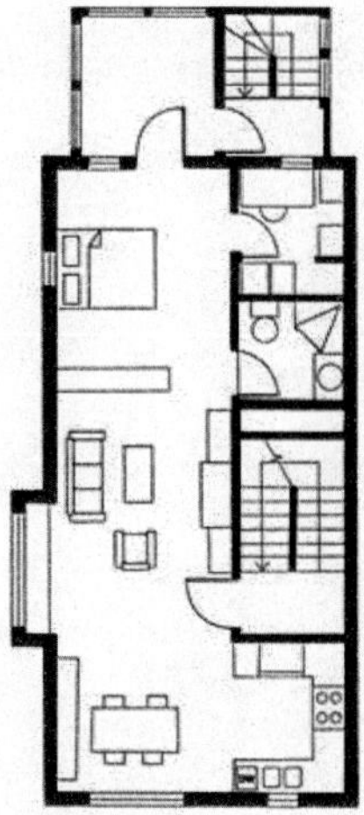

First Floor

The owner of the ground floor unit is a widower, age 57, who has worked as a civil servant for many years and has been offered an early retirement package due to budget cuts.

The owners of the second floor unit are a young couple without children who have been married for three years and who, until the purchase of their home, rented a one-bedroom apartment in a suburbna not far from the ciy center.

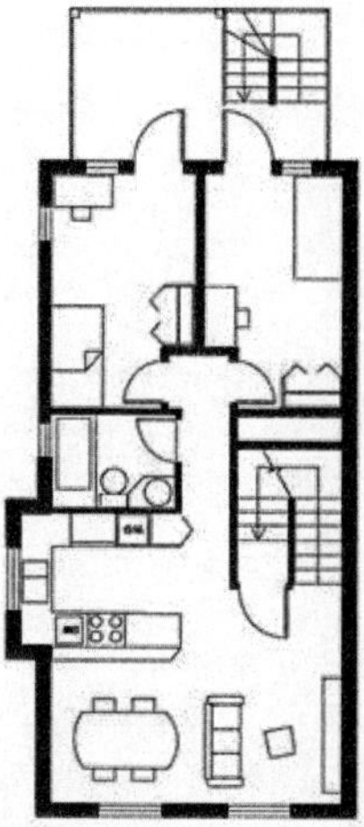

Third Floor

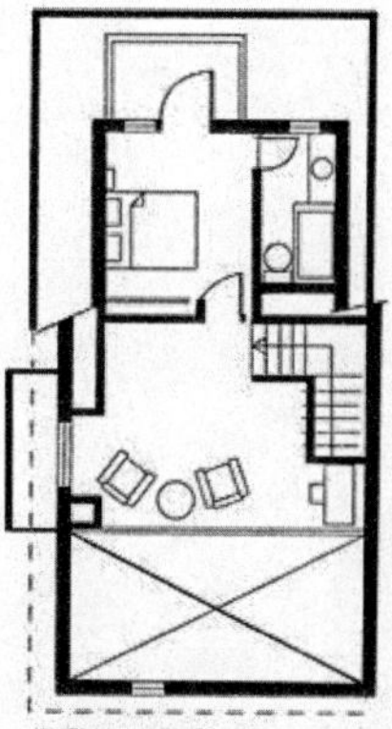

Mezzanine

The owner of the unit on the third floor and the mezzanine is a single mother, age 41, with two adolescent school children.

Fig. 4.21 Floor plans of the Next Home

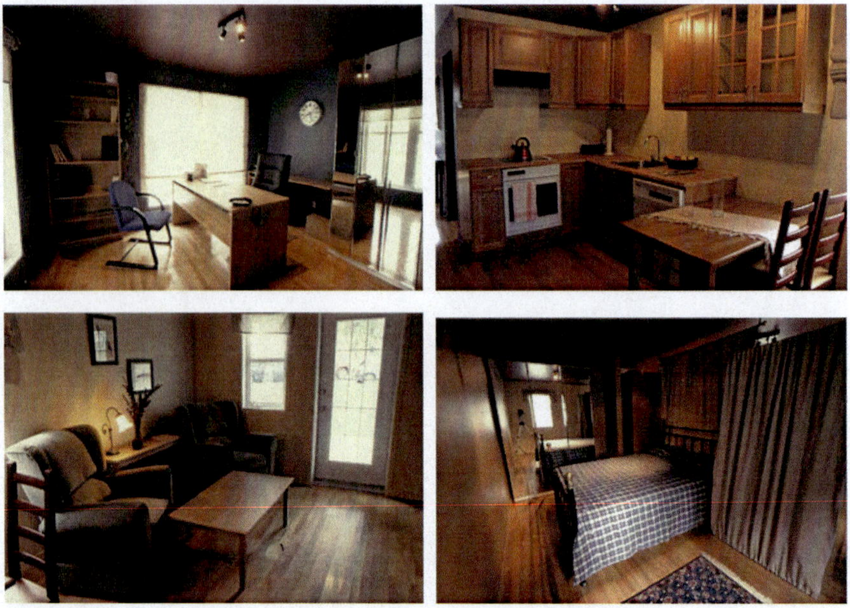

Fig. 4.22 The ground floor unit designed for a widower in his late sixties with a small home office

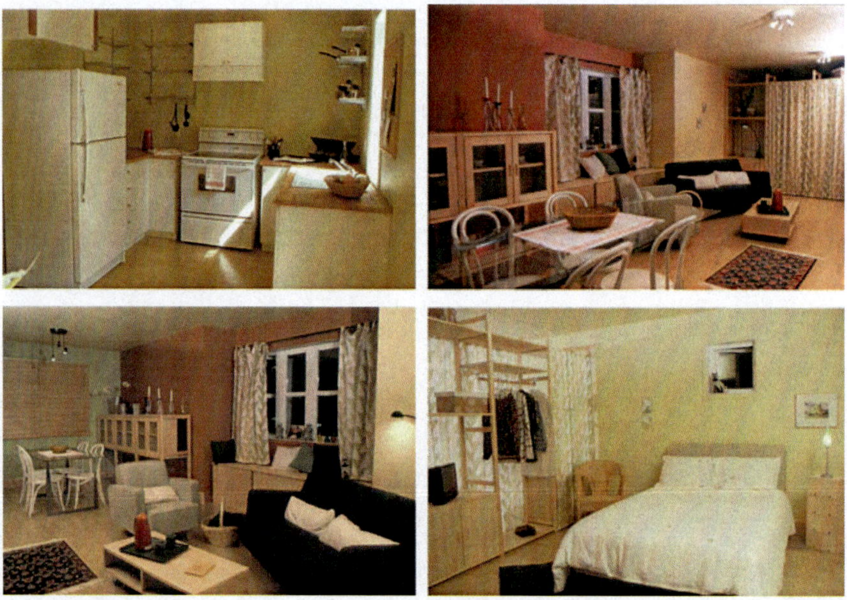

Fig. 4.23 A young couple resides on the second floor. They chose an open plan, using furniture to partition space with public functions located toward the front of the unit

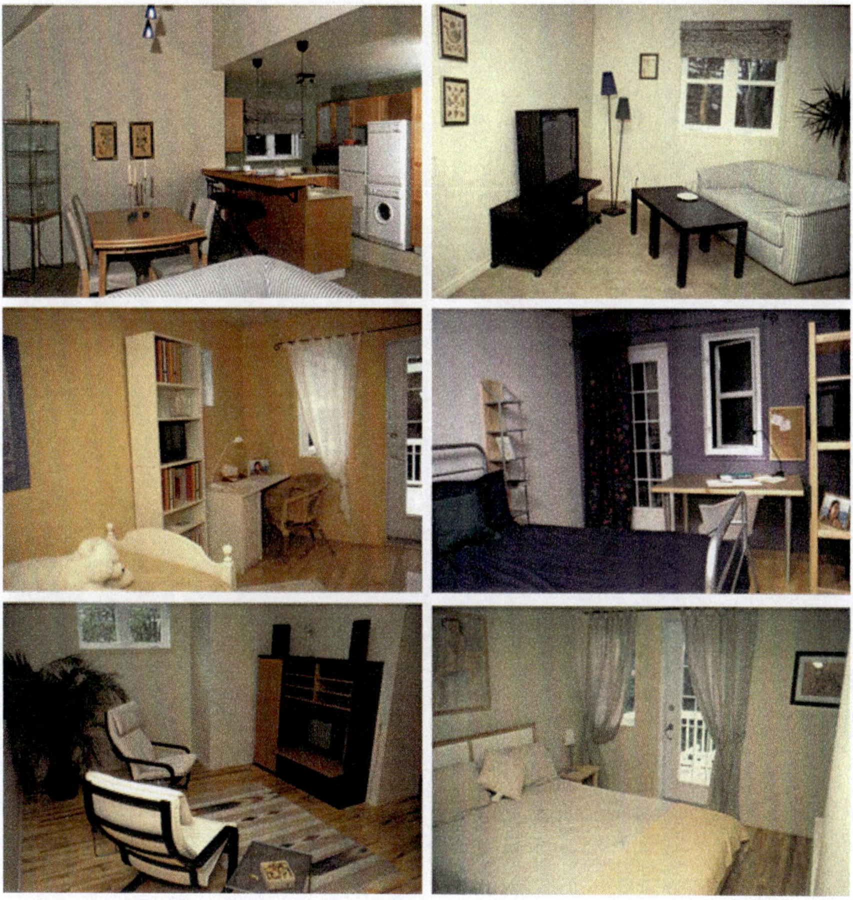

Fig. 4.24 The third floor designed for a single parent with two children. The roof at the rear of the structure is raised, allowing for a mezzanine above the third floor, used to create a separate living area, bedroom, and bathroom for the parent

was also a priority, placed along the rear of the unit alongside laundry and clothing storage, in an area that could be later converted to an enclosed room. Computer, telephone, and electrical cables were installed in special floor moldings around the perimeter of the unit, providing flexibility to modify or add receptacles in the future.

Finally, the third floor was designed for a single parent with two children. The roof at the rear of the structure is raised, allowing a mezzanine above the third floor, used to create a separate living area, bedroom, and bathroom for the parent. A larger kitchen with an area for bar seating was selected and the public zone with the living and dining area is located at the front of the unit. The two children's rooms are located at the rear of the unit. Demountable partitions were used, allowing the two bedrooms to be later converted into one living room when the children move out.

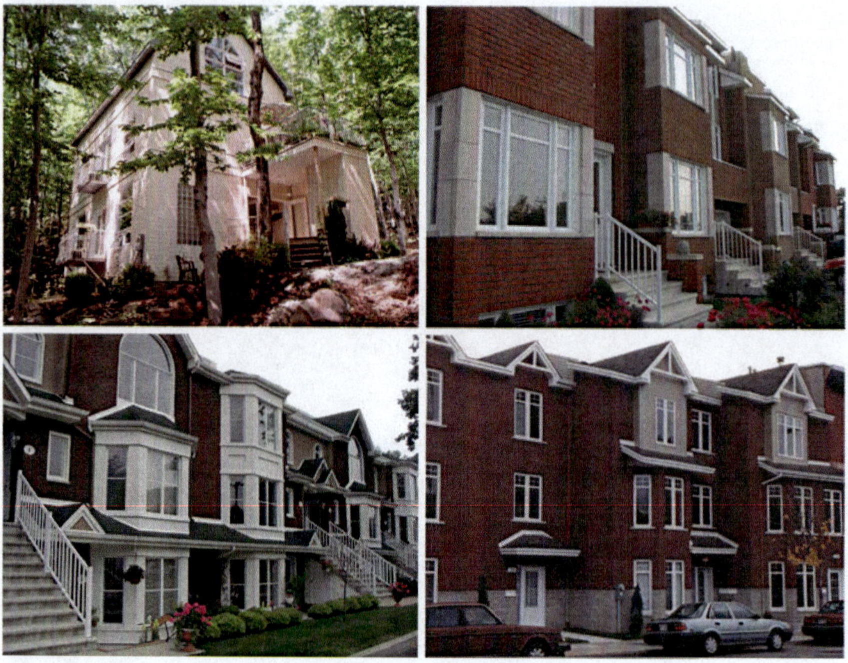

Fig. 4.25 The Next Home that was adopted and constructed by homebuilders

The Next Home that was adopted and constructed by homebuilders demonstrates how high-density developments with small interiors can be designed for a diverse range of households and facilitate longevity through adaptable design (Fig. 4.25).

4.6 Final Thoughts

As populations continue to grow in cities, the challenge of providing comfortable, dense, affordable housing will increase. Designers continue to work to create liveable and flexible interiors for modern households. As technology rapidly evolves, the introduction of smart technology into people's homes aims to streamline domestic life. Yuan et al. (2021) analyze the useability of smart kitchens in Chinese households for the future of residential interiors. Consumers today can find numerous pieces of smart furniture, companies claiming that they will simplify daily life and make the home more comfortable, including bed frames, bedside tables, desks, mirrors, and many others. While the world becomes more technology focused, fundamentally, designers of sustainable dwellings should design for the uncertainty of the future.

This involves creating flexible designs, which must be both long-lasting and adaptable to change, for a variety of non-traditional household types.

Questions for a Follow-Up Discussion

1. What social transformations led to a need for small dwellings?
2. What are the macro design strategies of small spaces?
3. What micro design strategies of small spaces?

References

Dhar TK, Hossain M, Rahaman KR (2013) How does flexible design promote resource efficiency for housing? a study of khulna, bangladesh. Smart Sustain Built Environ 2(2):140–157. https://doi.org/10.1108/SASBE-10-2012-0051. Accessed 5 Oct 2022

Du FL, Li T (2013) Selected, peer reviewed papers from the 2013 international conference on applied mechanics and materials (ICAMM 2013), November 23–24, 2013, Zhuhai, China. (2014). Study on architectural design for indoor space of small area house. Appl Mech Mater 477–478:1128–1131. https://doi.org/10.4028/www.scientific.net/AMM.477-478.1128 Accessed 5 October 2022

Fisher-Gewirtzman D (2017) The impact of alternative interior configurations on the perceived density of micro apartments. J Arch Plan Res 34(4):336–358. http://www.jstor.org/stable/44987241. Accessed 5 Oct 2022

Friedman A (2002) The adaptable house: designing for choice and change. McGraw-Hill, New York

Frontczak M, Wargocki P (2011) Literature survey on how different factors influence human comfort in indoor environments. Build Environ 46(4):922–937. https://doi.org/10.1016/j.buildenv.2010.10.021. Accessed 5 Oct 2022

Halbe SB, Mark RK (2018a) Chapter 1: why small living and for whom. In: Interior design for small dwellings. Routledge, pp 10–30. https://doi.org/10.4324/9780429506604. Accessed 5 Oct 2022

Halbe SB, Mark RK (2018b) Chapter 5: elements and principles of design for small dwellings. In: Interior design for small dwellings. Routledge, pp 82–106. https://doi.org/10.4324/9780429506604. Accessed 5 Oct 2022

Halbe SB, Mark RK (2018c) Chapter 6: color and light. In: Interior design for small dwellings. Routledge, pp 107–138. https://doi.org/10.4324/9780429506604. Accessed 5 Oct 2022

Halbe SB, Mark RK (2018d) Chapter 9: finding storage space. In: Interior design for small dwellings. Routledge, pp 189–201. https://doi.org/10.4324/9780429506604. Accessed 5 Oct 2022

Hoang M, Vandal A (2017) Micro-MACRO living in the global high-rise. CTBUH J 3:32–37. https://www.jstor.org/stable/90020855. Accessed 5 Oct 2022

Hohenadel K (2015) These sleek pop-up rentals designed for broke millennials redefine prefab housing. Slate Magazine. http://www.slate.com/blogs/the_eye/2015/01/23/heijmans_one_are_movable_rentals_for_young_single_person_households.html. Accessed 5 Oct 2022

Nägeli W, Tajeri N, O'Donovan JR (2016) Small interventions: new ways of living in post-war modernism. Birkhäuser, part of Walther de Gruyter GmbH. Retrieved May 28, 2022, from https://mcgill.on.worldcat.org/oclc/1061074982. Accessed 5 Oct 2022

National Post Staff (2022) A standard house costs almost twice as much in Canada as it does in the U.S. National Post. Retrieved from https://nationalpost.com/news/a-standard-house-costs-almost-twice-as-much-in-canada-as-it-does-in-the-u-s. Accessed 5 Oct 2022

Nelson A (2018a) Chapter 2: once we were small: traditional and contemporary homes. In: Small is necessary: shared living on a shared planet. Pluto Press, pp 21–43. https://doi.org/10.2307/j.ctt1zk0mpz.8. Accessed 5 Oct 2022

Nelson A (2018b) Chapter 3: apartment living in cities. In: Small is necessary: shared living on a shared planet. Pluto Press, pp 44–71. https://doi.org/10.2307/j.ctt1zk0mpz.9. Accessed 5 Oct 2022

Tang J, Galbraith N, Truong J (2019) Living alone in Canada. Statistics Canada. Retrieved from https://www150.statcan.gc.ca/n1/pub/75-006-x/2019001/article/00003-eng.htm. Accessed 5 Oct 2022

Vachon M (2018) The ever-shrinking condo. Can J Urban Res 27(2):37–50. https://www.jstor.org/stable/26542035. Accessed 5 Oct 2022

Yuan X, Deng F, Liu Z (2021) 14th international symposium on computational intelligence and design (ISCID) Hangzhou, China 11–12 Dec 2021. In: Design of Chinese smart kitchen based on users' behavior. essay, IEEE, pp 301–305. https://doi.org/10.1109/ISCID52796.2021.00076. Accessed 5 Oct 2022

Chapter 5
Attractive and Energy Efficient Facades

Abstract Exterior appearance is an important aspect that people will consider when buying a home. Allowing personalization of the home's exterior is an added incentive and likely to be the way of the future as more people will live in high-density urban settings. Creating a mass customized community where the dwellings have common underlying exterior design principles and where buyers are also permitted to have their own imprint on the façade is the trust of this chapter. The specific aspects to be investigated are homogeneity and diversity, principles of designing facades for choice, flexible exterior design according to layout, sustainable energy efficient design principles of facades, and offering a menu of choices for façade elements such as material choices, energy efficient windows, daylight, and ventilation.

Keywords Building envelope · Energy efficiency · Heat transfer

5.1 A Need for Diversity and Energy Efficiency in Building Envelopes

Facades play a significant role in the built identity of neighborhoods and the energy efficiency of dwellings. Having personalized and attractive exteriors is a key in establishing a sense of place (Fig. 5.1). In high-density residential developments, it is recommended to avoid overly repetitive façade designs that lead to homogeneity and monotony. Weinreb and Rofè (2013) found that perceived ugly, unpleasant, or ordinary architecture led people to experience more negative feelings while walking through a place. Whereas the presence of street focal points and windows on the ground floor level is associated with increased perceived walkability (Oreskovic et al. 2014). Additionally, high-density residential developments must serve a diversity of users, with various needs and preferences. Incorporating opportunities for choice and occupant personalization in dwellings will lead to more attractive facades and higher levels of satisfaction for the occupants.

The building envelope is also the barrier between the interior and exterior of the home, responsible for controlling the transfer of heat and providing shelter from outside elements. The energy performance and interior comfort of a building

A. Friedman, *Fundamentals of Innovative Sustainable Homes Design and Construction*, The Urban Book Series, https://doi.org/10.1007/978-3-031-35368-0_5

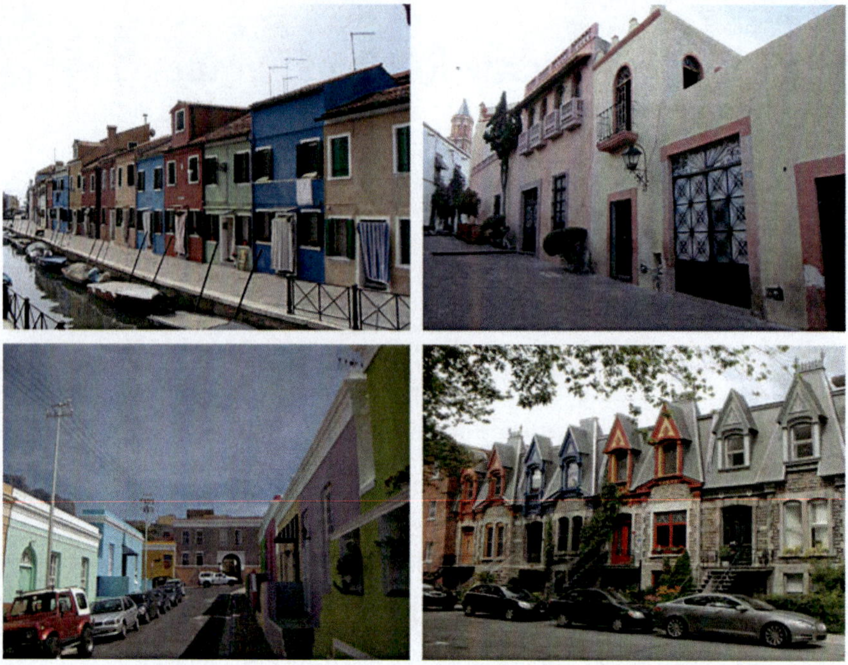

Fig. 5.1 Personalized exteriors establish a sense of place in Burano, Italy (top left), Santiago de Querétaro, Mexico (top right), Cape Town, South Africa (bottom left), and Montreal, Canada (bottom right)

are dependent on a well-designed façade. The thermal performance of materials, construction methods, access to daylight, ventilation, and energy-saving tools should all be considered in the design of the façade to create sustainable, comfortable dwellings.

5.2 Principles of Designing Attractive Façades

Since the façade offers the first impression of a dwelling, the designer needs to carefully consider its organization. There are a variety of strategies that designers can use to counteract monotony in high-density attached dwellings, the simplest way being through materials. Brick, for instance, has a timeless quality, comes in a variety of colors, and can be laid in patterns which create visual focal points. Wood siding, another vernacular option, can be painted to add color to a façade (Fig. 5.2).

Openings are the most important elements of the façade establishing the relationship between the home's interior and exterior. The opportunity for personalization is another aspect, as occupants experience higher levels of satisfaction with their homes when given control over their surroundings. Consulting with future occupants

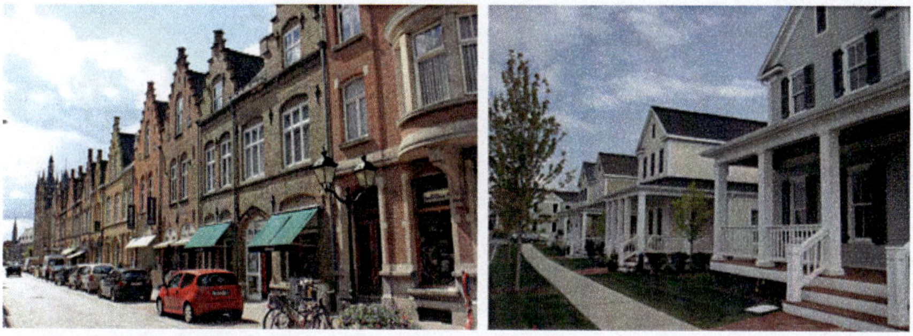

Fig. 5.2 Brick has a timeless quality, and it comes in a variety of colors (left). Wood siding, another vernacular façade option, can be painted to personalize buildings (right)

on the specifications of their unit when it comes to its exterior, helps to create more successful, less monotonous façades. This section discusses the logic of facades' design, principles of designing for choice, and a menu of design choices.

5.2.1 Logic of Façade Design

According to Herzog et al. (2017), the façade controls the home's degree of permeability to light and air. Parts of the façade can be classified into two categories: dynamic and static. *Static parts*, consisting of the opaque areas including the walls and roof, are responsible for sheltering and regulating the internal environment. Whereas the *dynamic parts* are the windows and openings that provide access to daylight, views, solar gain, and ventilation. The architect is responsible for finding the right balance between dynamic and static components of the envelope—both must be well-designed to provide comfortable and attractive dwellings. The designer needs to first recognize a site's challenges and opportunities in order to organize a façade that responds to conditions, orientation, neighboring buildings, and views. At times, it is also important to draw from the architectural character of the surrounding area to establish a sense of place if the home is proposed in an established neighborhood (Fig. 5.3).

Windows are one of the main determinants of the visual character of the façade. There are three main strategies in the organization of windows: systematic repetition, random order, and composition. *Systematic repetition* is the standard application of windows across the façade, the same types and sizes of windows are placed consistently without regard for the internal composition of the unit or clients' preferences. This approach gives the development of a standardized appearance, creating a sense of monotony, and limiting the personal identity from unit to unit. *Random placement*, the second option, relies primarily on the internal organization of the unit and occupants' preferences regarding the placement of openings. This option

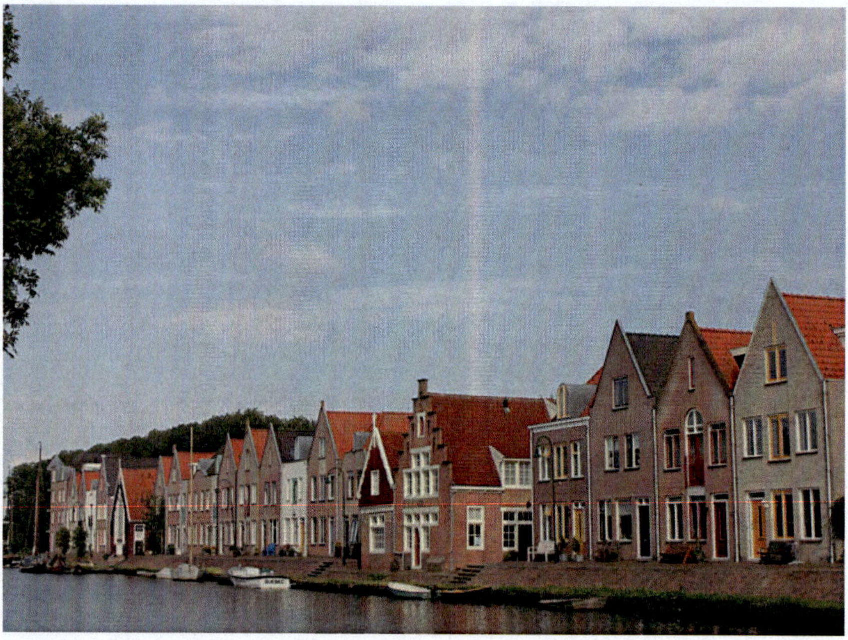

Fig. 5.3 In Edam, the Netherlands, the designers of contemporary buildings draw from the architectural character of the neighboring structures to maintain the original sense of place and appetence

allows for the most freedom, enhancing the individual identity of each unit and sacrificing the overall unity of the development. *Composition* combines the above two concepts to create a unified façade which still expresses the individuality of each unit (Fig. 5.4). Composition presents a selected variety of windows, openings, and other façade elements with a code of parameters set by the builder and designer. This ensures that the unit's facades share a consistent logic without creating a monotonous environment.

5.2.2 Principles of Designing for Choice

Designing for choice uses the composition approach to allow a variety of façade options while maintaining a consistent overall character of the building. Designers can offer clients a carefully selected choice of façade features to facilitate personalization and variety while maintaining a sense of continuity throughout the building envelope. For example, in row housing developments, buyers may choose between a front door with a porch or a bay window on the ground floor, a balcony or large window on an upper floor, and a dormer on the roof. This option for customization creates more diverse and interesting neighborhoods while enhancing occupants'

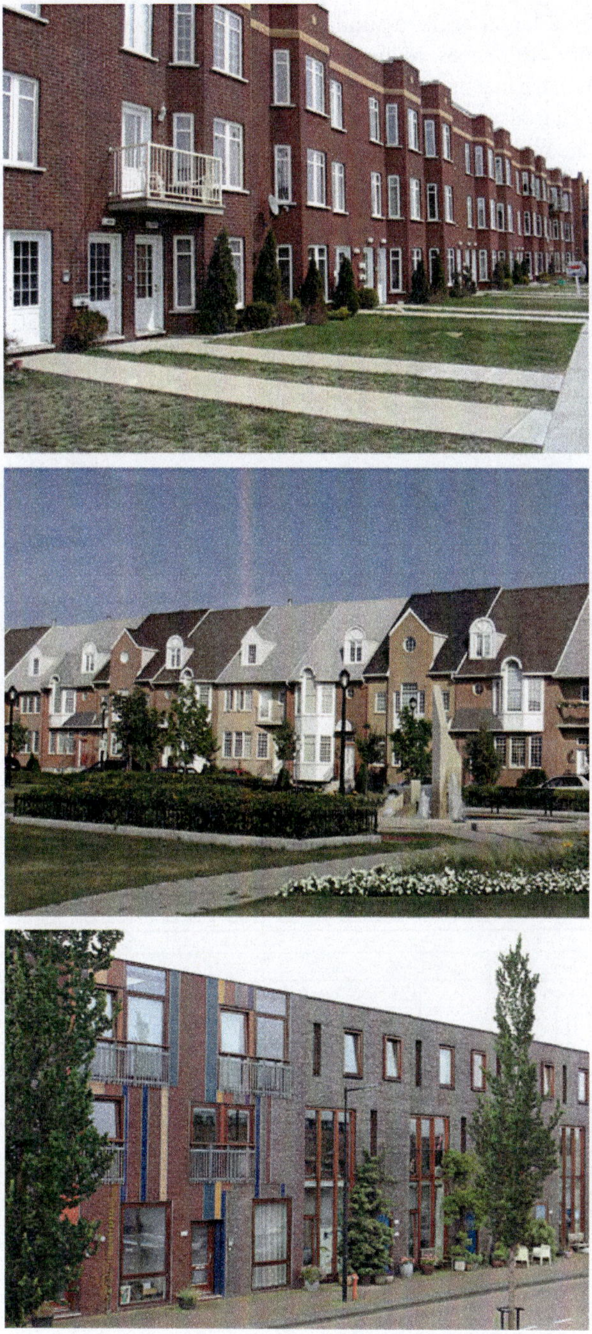

Fig. 5.4 Three main strategies can be used in the organization of windows: systematic repetition (top), random order (middle), and composition (bottom)

satisfaction. On a larger scale, a similar approach is taken in municipalities which enforce masonry ordinances, requiring commercial and residential buildings above a certain scale to use masonry facades (Deng 2010) (Fig. 5.5). The use of consistent, high quality materials throughout a neighborhood helps to establish its character and signifies an interest in architectural value and aesthetics, creating a positive impression of the place (Deng 2010).

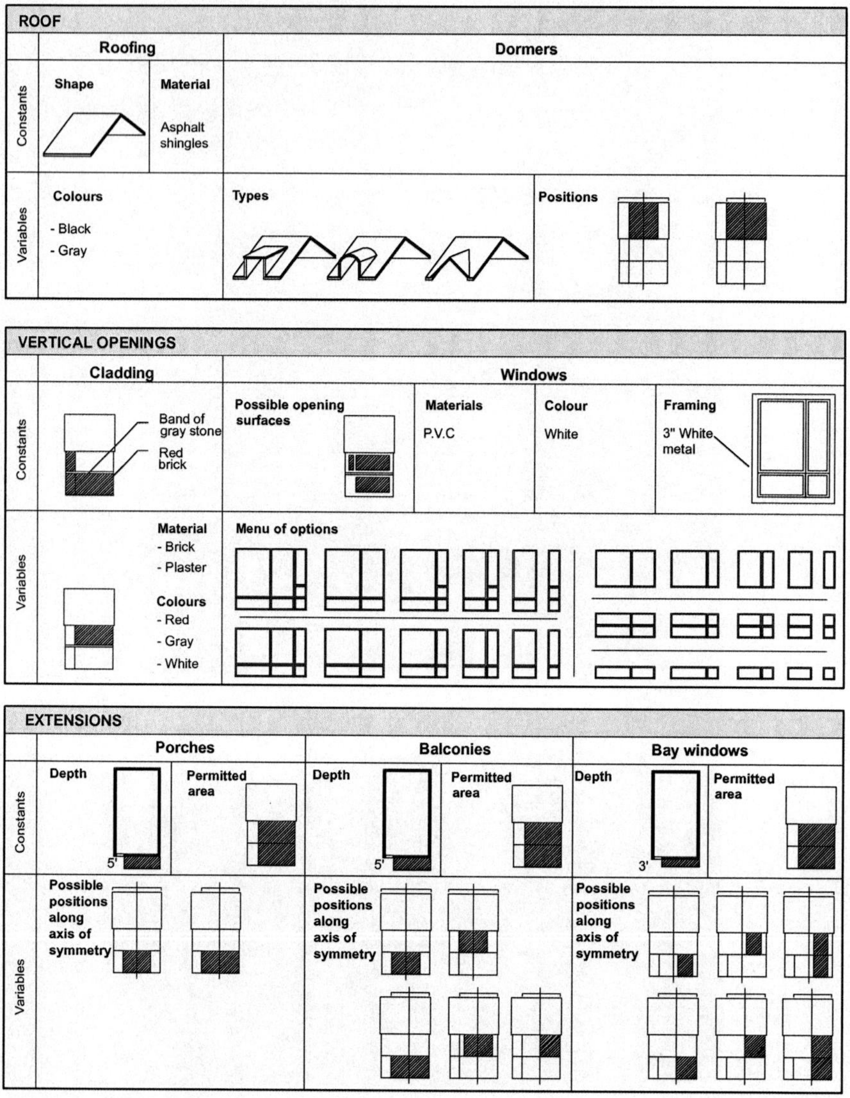

Fig. 5.5 By using written and illustrated guidelines for roof shapes and window sizes municipalities can regulate local ordinances

One methodology to provide choice and adaptability to facades distinguishes between *Opening zones* and *Infill components.* Opening zones are areas where infill components, such as various types of windows, projections, and doors, can be placed. The designer and builder determine the dimensions and location of opening zones on the façade. The variety of options for opening zones contain both infill (windows and doors) and opaque (exterior finishings) components (Fig. 5.6). One strategy would be to dictate a standard level for openings of various dimensions as per the specifications of each unit. The zone on the ground level is unique as it locates the entryway and potential commercial activity, its particular components will vary based on the home's relation to the street. Finally, bays and projections can add character and variety as façade elements. For example, the designer may stipulate that those projections can only be placed in certain zones with specified dimensions.

Designing for flexibility of the façade is important in allowing for future expansion or modification of the dwelling. When expanding an existent dwelling, any new addition must not inhibit the function of the existing envelope, particularly access to natural light and ventilation. Additionally, the opening zones can be structurally built to facilitate the modification of infill components. By enabling the modification of infill zones, units' facades can be modified to accommodate changing household structures. For example, if a multi-level single-family unit is converted into a multi-family building, a small balcony can replace windows on upper floors, to give each unit access to outdoor space.

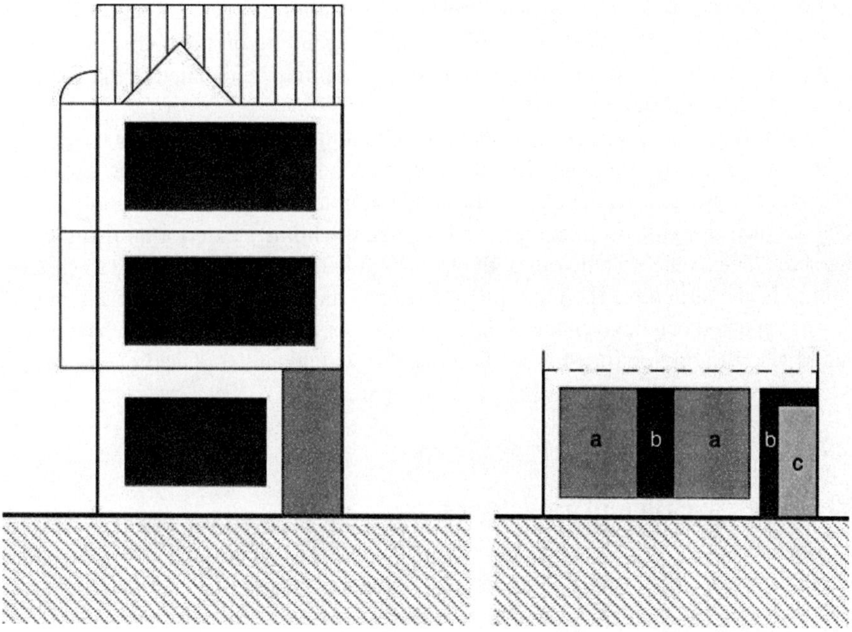

Fig. 5.6 A method to provide builders and homebuyers with choice and adaptability of facades is to distinguish between opening zones (left) and infill components that would fit in them (right)

5.2.3 Menu of Design Choices

Once the façade's opening zones have been determined, the specific types of components offered to clients and specifications are outlined by the designer. A designer can prepare a selected menu of façade options which are architecturally and aesthetically consistent, to establish the character of the community (Fig. 5.7). For example, square or rectangular windows with consistent framing and external features could be chosen based on the interior layout of the unit. The size and positioning of windows offered may be determined by the function of the interior. An example of this aligning at least one window or infill component with an edge of the main entrance helps to unify the ground floor entryway within the greater composition of the façade. As well, sill heights of 95 cm (3 feet) are ideal for placing countertops and other furniture underneath in a kitchen or dining room window. For commercial functions that might exist on the ground floor however, lowering the sill height is more suggested for function. Additionally, storefronts should be spaced in five-meter (16.4 foot) segments to create a stimulating environment for pedestrians (Adhitya 2017). A variety of opaque elements, which occupy the areas between infill components, can be offered. Options may include choices of specific panels, or standard dimensions and orientations of cladding elements. Finishings may be categorized into four main groups: masonry, recycled by-products, wood, and metal. The choices and specifications of finishings offered will depend on the designer's desired effect. A menu of options can be provided for the roof construction as well depending on each unit's specifications. Choices can be provided between an attic space, cathedral ceiling, mezzanine, or flat roof during the design stage, depending on the code established by the designer. For instance, creating a pitched roof may be dictated in the code, to give units a more residential scale.

Finally, types of finishings, trim, and ornamentation can be specified in the design code. The design of ornamentation and projections should avoid the two extremes of monotony and superficiality. Any ornament applied should be solid and durable because surface-applied ornamentation can give the home's exterior an inauthentic character. The element at the top of the front façade is key in establishing a sense of identity in the structure. Having a pitched roof as an anchoring element which can then be repeated can balance the façade with various elements (Fig. 5.8). The menu should also provide options for shading devices which can be placed on any level, to give users more control over their indoor environment.

5.3 Energy Efficient Envelopes

A building's envelope is one of the main determinants of energy efficiency, its primary purpose is to regulate the internal environment and provide protection from external elements. Poorly designed and constructed building envelopes can result in excessive heat transfer in and out of the structure, resulting in high energy costs and

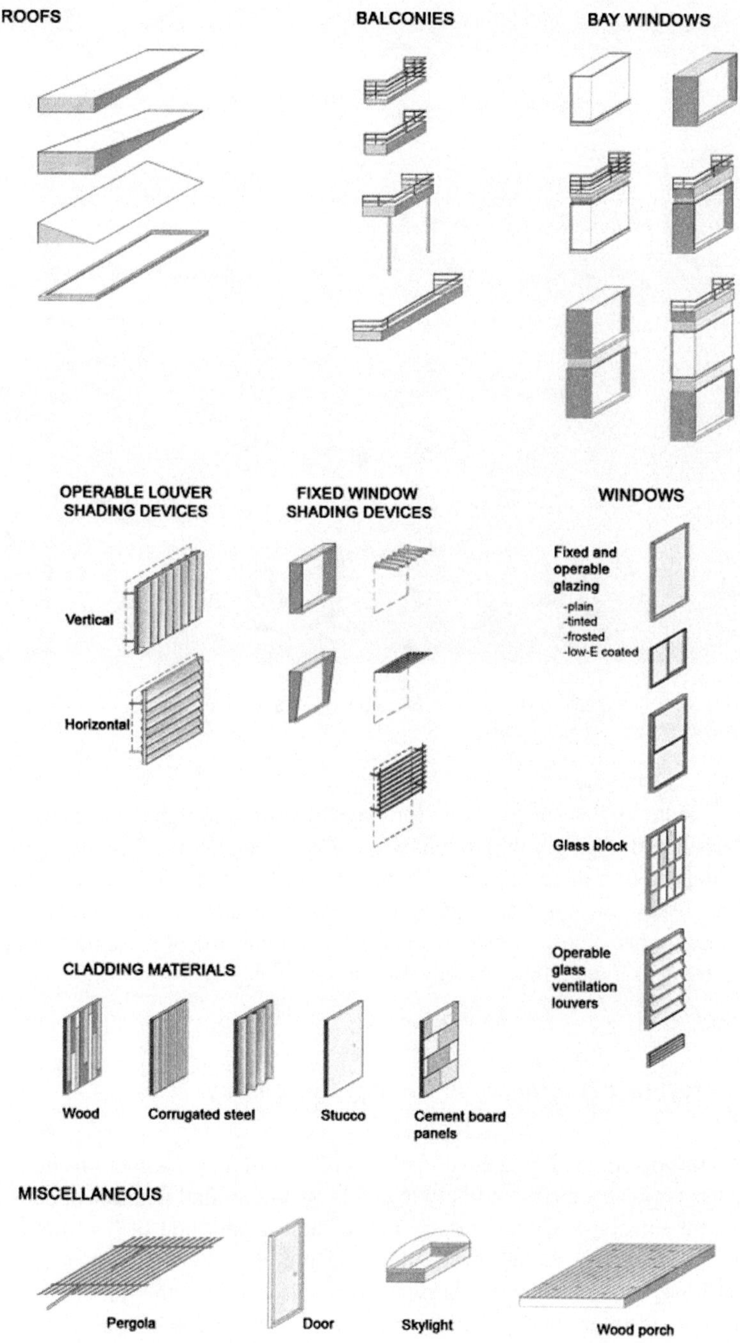

Fig. 5.7 A menu of façade components establishes the community's overall visual character

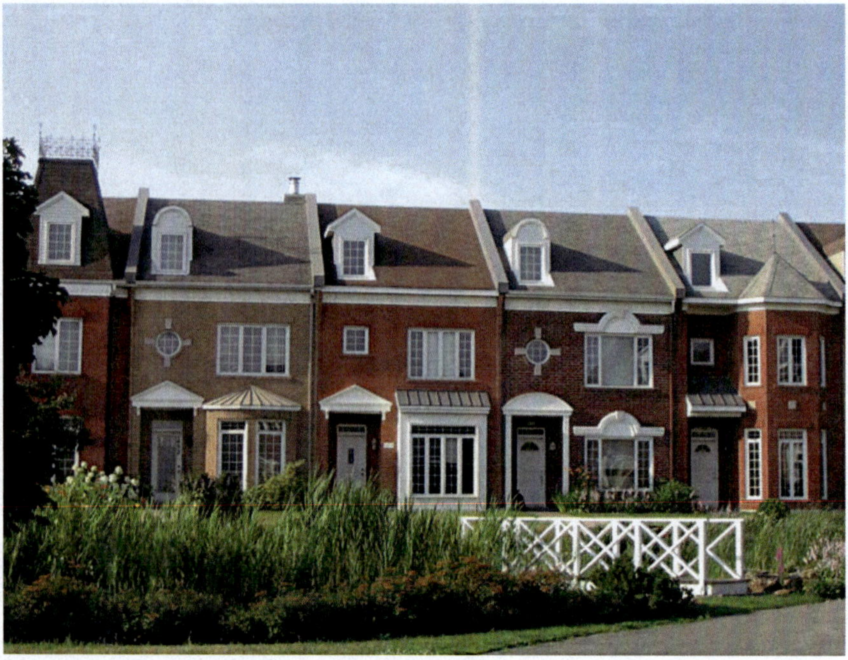

Fig. 5.8 Having a pitched roof as an identical feature on all the homes can act as a unifying element of a façade with various components

decreased indoor environmental comfort. Heat transfer through the envelope occurs when there are differing internal and external air temperatures. The degree of heat transfer depends on the characteristics of the building envelope such as windows, materials, insulation, and form (Zemella and Faraguna 2014). This section discusses the key design principles, choice of materials, and construction methods related to creating energy efficient building envelopes.

5.3.1 Design Principles for Energy Efficiency

Heat loss through the building envelope is mainly caused by conduction and convection. These processes are most likely to occur on and around window openings and edges of the envelope. Conduction refers to the transmission of heat through the materials themselves—if they have a high degree of conductivity, heat can easily transfer through the envelope. Thermal bridging describes this process in which heat transfers through more conductive materials, which can account for up to 10 percent of total heat loss in the envelope. This often occurs when the areas around structural members are not properly insulated. Convection refers to the transmission of heat resulting from the movement of air, this type of heat transmission is often caused by air leakages in the façade. In a typical dwelling, about 25 percent of heat loss can be caused by air leaks in the envelope. This uncontrolled movement of air reduces the energy efficiency of the home and can lead to moisture damage and mold growth. The most leakage generally occurs where building materials are joined, often around window and door frames, and at the intersections of walls, floors, and the roof (Fig. 5.9).

There are a variety of construction strategies which can be used to minimize heat transfer and moisture infiltration through the envelope. Areas where the envelope is interrupted by doors, windows, joints, and other openings have the most potential for air leakage and moisture infiltration. Placing a continuous polyethylene moisture barrier is necessary to prevent mold growth and deterioration of building elements.

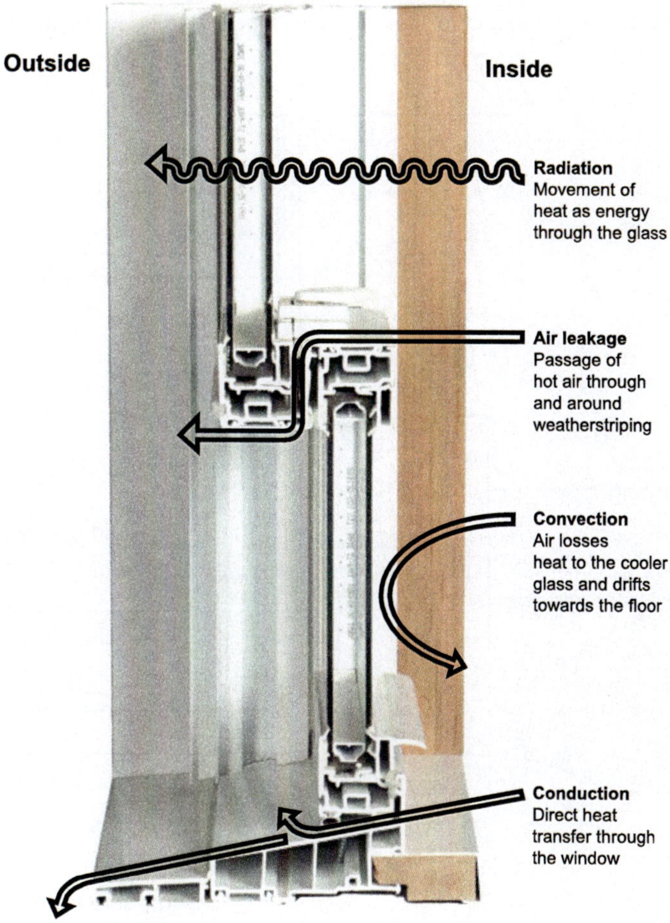

Fig. 5.9 The most air leakage from buildings generally occurs where materials are joined, often around window and door frames

Additionally, while all areas of the building envelope require insulating, additional insulation should be placed in areas where the envelope is interrupted by openings or edges. While basement spaces can provide valuable floor space for occupants, they can be a significant source of energy loss for a dwelling and are susceptible to moisture infiltration and cracking. Alternatives such as *slab-on-grade* or *crawlspace* foundations are less prone to heat loss and are thus more energy efficient. Additional insulation should be placed around the edges of the foundation where more heat is lost due to thermal bridging (Fig. 5.10). The roof also generally requires significant insulating to limit heat loss because of its large surface area, particularly in cold climates, as heat rises, a poorly insulated roof can result in significant energy losses. The maximum amount of insulation according to budget and unit size should be used to improve energy performance.

Designers should explore alternative methods of wall construction which can improve the energy efficiency of building envelopes. One option is a double stud

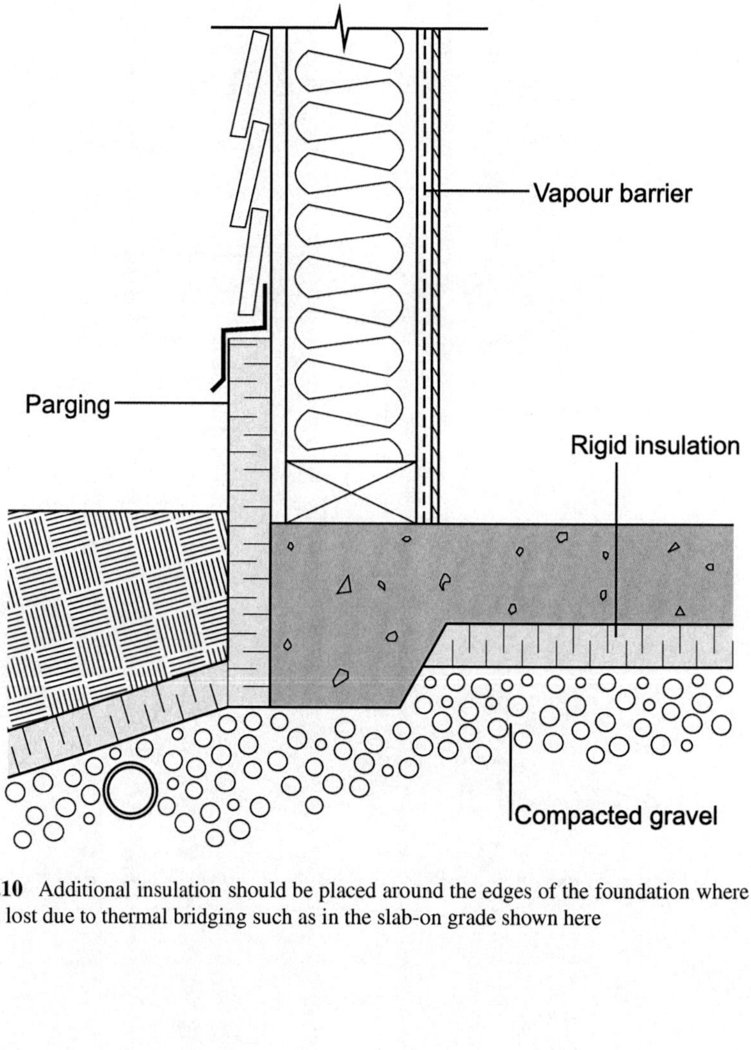

Fig. 5.10 Additional insulation should be placed around the edges of the foundation where more heat is lost due to thermal bridging such as in the slab-on grade shown here

wall, featuring one load-bearing wall and an additional non-load-bearing wall. Both walls are filled with insulation and separated by an additional insulating layer. This can greatly improve the energy performance of a dwelling. However, it requires more material and occupies valuable floor space in smaller units. A standoff wall is a less costly alternative which allows for additional layers of insulation to be installed between the load-bearing wall and the finished interior. There are also more efficient envelope framing methods which can help to conserve materials and energy (Fig. 5.11). Using bigger components with wider spacing allows for larger sections of continuous insulation to be placed, helping to improve its efficacy.

Richman et al. (2010) propose a double buffer zone wall method to improve the thermal performance of envelopes in cold climates. The typical wood-frame residential façade in cold climate regions is composed of a masonry wall, an air cavity, rigid insulation, a wood stud cavity, batt insulation, a vapor barrier, and gypsum board (Fig. 5.12). The proposed alternative introduces a ventilated airspace in between the masonry wall and rigid insulation. It was found that this method of construction helped to preheat ventilation air in residential buildings, reducing the energy required to heat the building (Richman et al. 2010). In high-rise buildings, double skin facades can be used to improve energy efficiency. Saroglou et al. (2020) found that on average 50 percent less energy was consumed for cooling purposes in high-rise buildings that had a double skin façade. *Double skin facades* consist of three components: the internal façade, an air cavity, and the external façade—this allows for natural ventilation systems to be introduced, which reduces energy spent for cooling purposes and creates a more comfortable interior (Saroglou et al. 2020).

5.3.2 Choice of Materials

When choosing façade materials to limit heat transmission, thermal mass and thermal resistance are important properties to consider. Materials such as concrete and masonry generally have high thermal mass—meaning these materials absorb heat from solar radiation throughout the day and slowly release it into the home when it cools down at night (Hampton 2010). Thermal mass is used to regulate the internal temperature of a building, it is most effective when there is a dramatic fluctuation between day and night temperatures. In more temperate climates, thermal mass can still be used as a strategy to regulate temperature but requires more strategic planning.

Materials with high thermal mass must be placed in areas where they can absorb heat from the winter sun throughout the day, if they do not receive adequate sunlight, they can have the opposite effect by absorbing and releasing warm indoor air (Hampton 2010) (Fig. 5.13). Conversely, if they receive too much sun in the summer, they can increase the internal temperature (Hampton 2010). Thermal resistance describes the ability of heat to flow through a given material, represented by the *R*-value, it is commonly used to evaluate the insulating properties of materials (Moe 2012). There are a variety of types of insulation commonly used to raise the thermal resistance of the building envelope, including batt, blanket, loose fill, foam, and

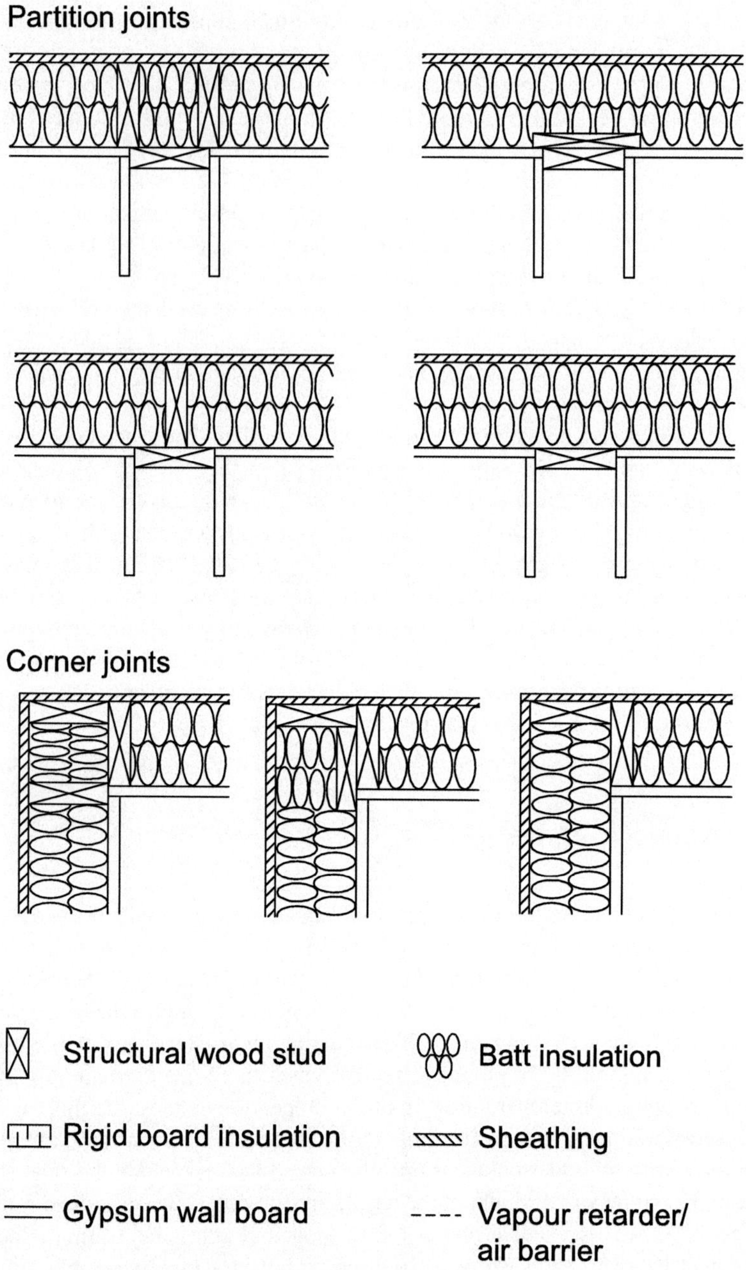

Fig. 5.11 Efficient envelope framing methods can conserve materials and energy

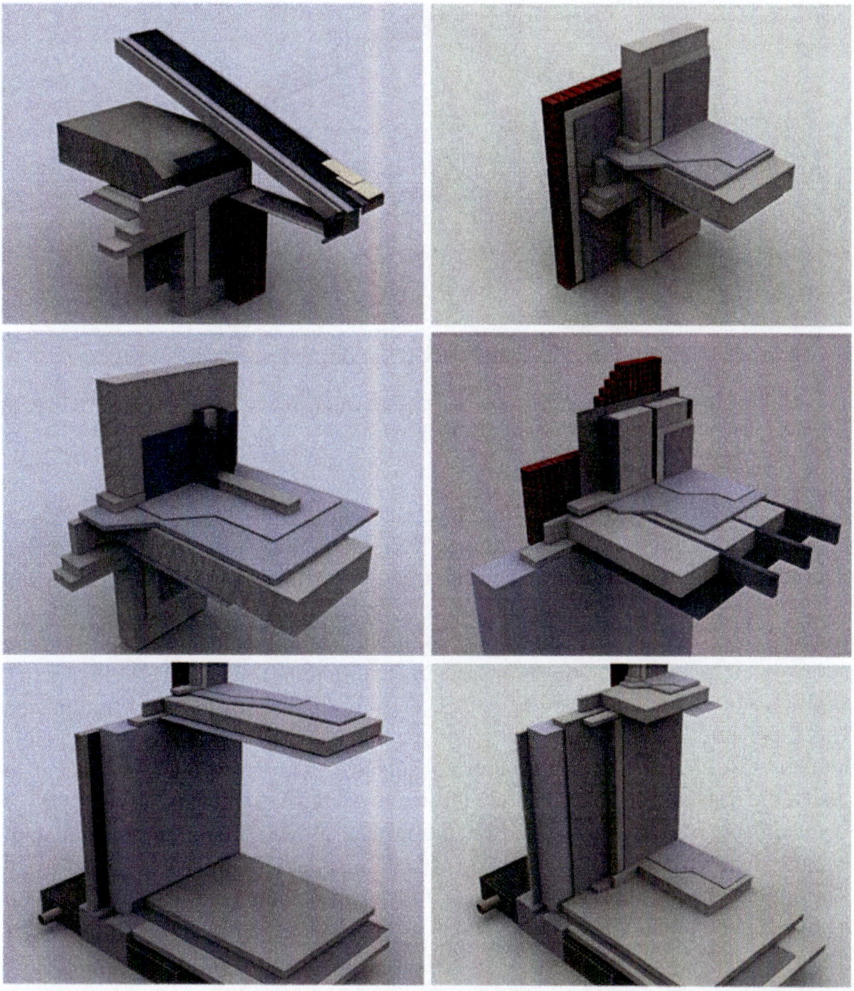

Fig. 5.12 The typical wood-frame residential façade in cold climate regions is composed of a masonry wall, an air cavity, rigid insulation, a wood stud cavity, batt insulation, a vapor barrier, and gypsum board as shown here in roof, wall, floor, and foundation

reflective. Within different types of insulation, air pockets help to limit heat transfer through the material. Therefore, overfilling cavities with insulation often limits its efficacy. Batt insulation is a common, affordable choice. Cellulose insulation is made of recycled paper; however, it is prone to having uneven coverage, reducing its *R*-value (Hampton 2010). Insulating paint is a newer product, consisting of a ceramic-based paint which can reduce heat transfer by up to 40 percent (Hampton 2010). Rigid insulation has the highest *R*-value for its volume, it is therefore most effective for insulating floors and the exterior layer of the envelope.

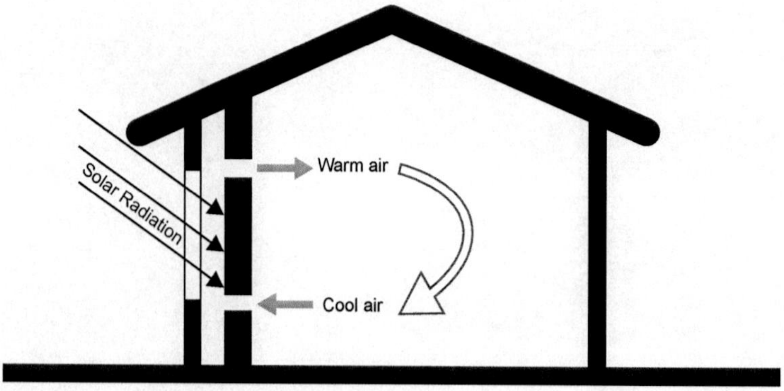

Fig. 5.13 Materials with high thermal mass can be placed in areas where they can absorb heat from the winter sun throughout the day and release it overnight

There are several sustainable building material options that can reduce waste and significantly improve energy efficiency. Some popular residential construction materials are engineered lumber and light-gauge steel, used for joists, beams, and flooring. *Engineered lumber* makes use of smaller trees, helping to preserve old growth forests, it requires less wood volume than traditional timber and is more energy efficient to produce than concrete or steel (Green Building n.d.) (Fig. 5.14). Lumber can be glue-laminated and stacked using finger-jointed wood. Laminated veneer lumber uses thin layers of wood glued together. *Light-gauge steel* is a viable alternative to lumber and has several advantages. It is lightweight, consistent in quality, fire resistant, and manufactured from recycled materials. *Building-integrated photovoltaics* (BIPV) can be used on the façade, allowing building envelopes to generate their own energy by making use of solar radiation. There are a variety of different types of photovoltaic panels which can be employed based on budget, location on/of the building, and levels of sun exposure. The most common use of BIPV systems in residential buildings is with grid-connected rooftop systems, they can occupy between 7 and 15 square meters (75–161 square feet) of roof area (Snow and Prasad 2011). A house with high energy efficiency can save up to 80 percent of energy consumption using BIPV systems thereby reducing dependence on grid electricity and greenhouse gas emissions (Snow and Prasad 2011).

5.4 Energy Efficient Windows

Windows are often considered the weakest links thermally of a building envelope due to their lower *R*-values, which refers to their insulating capabilities, than other façade elements and because they are also particularly prone to air leakage and moisture infiltration. Well-designed windows, however, are essential in providing natural light and ventilation to homes. Therefore, the careful design of windows is

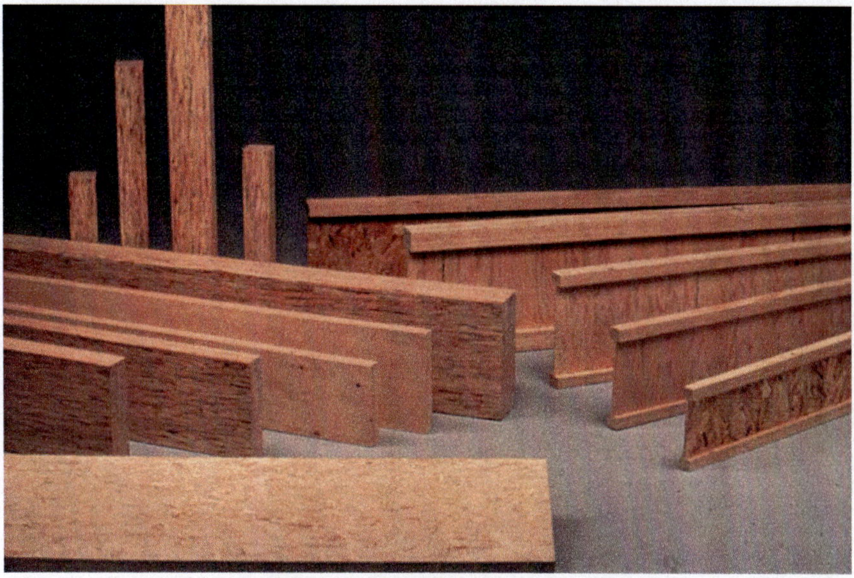

Fig. 5.14 Engineered lumber makes use of smaller trees to help preserve old growth forests. It requires less wood volume than traditional timber and has a lower carbon footprint than concrete or steel

required to create energy efficient envelopes without sacrificing comfort. This section discusses key elements of efficient window design including window composition, heat transfer, daylight, natural ventilation, and energy-saving features.

5.4.1 Window Composition and Heat Transfer

There are three main components which should be considered for energy efficiency when planning windows: the frame type and material, glazing unit, and spacer bar (Fig. 5.15). The glazing unit is composed of a frame and glass panes. There are two separate framing components: the *casing*, which is the outer frame anchored to the wall, and the *sash*, which is the inner frame enclosing the glazing unit. The window frame can account for up to 20 percent of the heat loss in a window unit, it is susceptible to heat loss by both convection and conduction. Therefore, the selection of the proper window frame is essential in designing energy efficient windows.

Wood is an effective insulator, making it an energy efficient material choice for a window frame. It can be easily damaged by moisture however and must be well-maintained to prevent deterioration. Aluminum and vinyl window frames on the other hand are much more durable but are good conductors and therefore result in greater heat loss. Wood frames can be used in combination with a protective aluminum or vinyl coating to make use of its thermal qualities while being protected

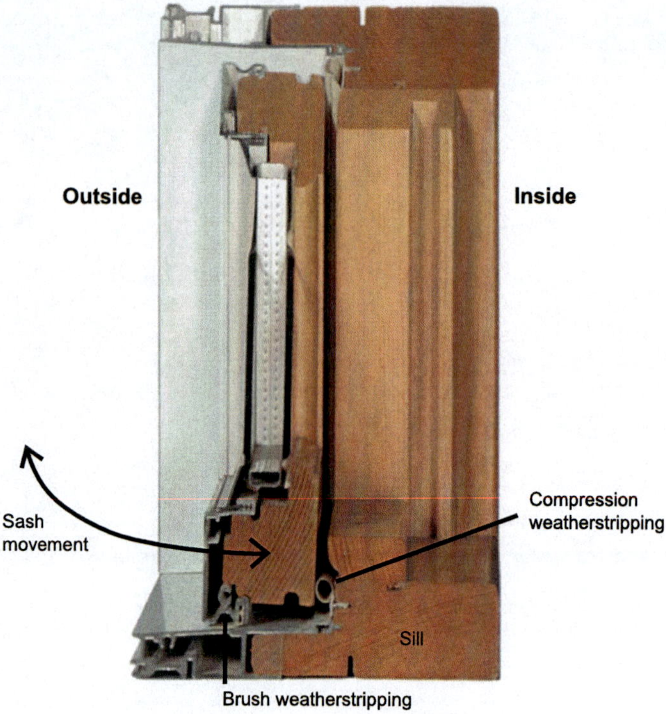

Fig. 5.15 The three main components to consider for energy efficiency when selecting windows are the frame on its material, glazing unit, and the spacer bar

from damage. Alternatively, aluminum frames can be built with a thermal break, containing insulating materials to reduce its conductivity. Window frames are also very susceptible to air leakage, this can be reduced with careful installation and the selection of appropriate frames. While operable windows are often preferable for natural ventilation purposes, they are more prone to air leakage than fixed windows.

There are various types of operable windows including waning, hopper, casement, sliding, single hung, and pivoted (Fig. 5.16). In general, a more linear length of window joints allows for more air leakage. Operable windows with pivoting components can be the most efficient because they make use of compression seals minimizing air leakage. To limit air convection, joints must be made as airtight as possible by using sealants and gaskets on all fixed components of the window. The spacer bar is used to prevent air and moisture from infiltrating the glazing unit itself. It is composed of metal and sealed by an organic compound. Insulated spacer bars can also be used to limit conductive heat loss around the edges of the glazing unit. The glazing unit itself is susceptible to heat transfer due to solar radiation, and conductive heat transfer due to the low *R*-value of glass.

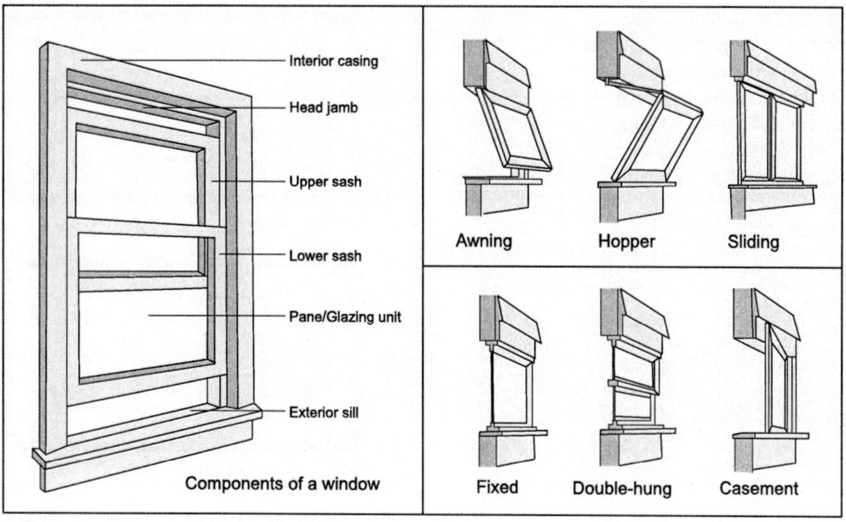

Fig. 5.16 There are various types of operable windows including awning, hopper, casement, sliding, single hung, and pivoted

5.4.2 Daylight and Natural Ventilation

Daylighting refers to the practice of positioning windows to provide natural light to interior environments. Access to natural light helps to create more comfortable, interesting, and productive environments for occupants (Ander 2016). Access to sunlight is very dependent on the building's orientation and surroundings, however there are strategies to maximize daylight in each indoor environment. In general, a window that accounts for 20 percent of a room's floor area can provide sufficient daylight even without direct sunlight access. The shape, size, and positioning of the window will depend on the depth of the room in which it is located. Higher and taller windows allow for light to penetrate further into a space (Ander 2016). Therefore, taller windows can be advantageous in deeper rooms (Fig. 5.17).

When planning the organization of windows on the façade, designers should consider opportunities for natural ventilation (Fig. 5.18). Well-designed natural ventilation can provide significant energy savings and improve occupants' comfort levels. Candido (2011) argues that it is not feasible to rely solely on mechanical systems to regulate the internal environment of buildings, the ventilation strategy must be adapted to the local microclimate. Therefore, it is important to incorporate opportunities for natural ventilation. Operable windows can help to expel excessive heat, humidity, and unwanted odors, while allowing occupants more control over their environment. Requirements for the necessary operable windows for natural ventilation are stipulated by codes and regulations for the quality of the indoor environment; it is advantageous to place windows in accordance with wind patterns throughout the site, and to provide windows on opposite facades of the unit to facilitate a cross-breeze.

Fig. 5.17 The shape, size, and positioning of the window will depend on the depth of the room in which it is located. Higher and taller windows allow for light to penetrate further into a space

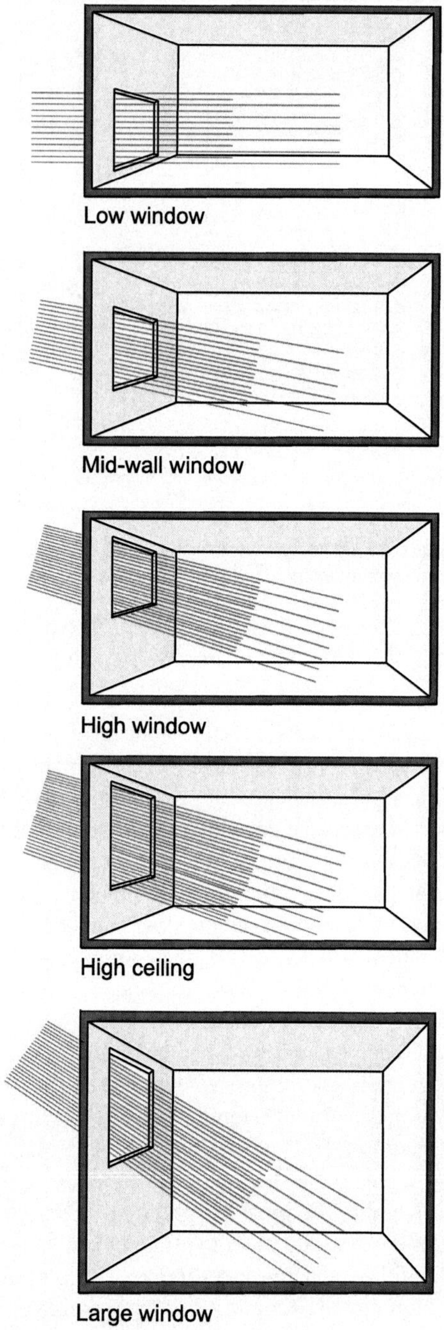

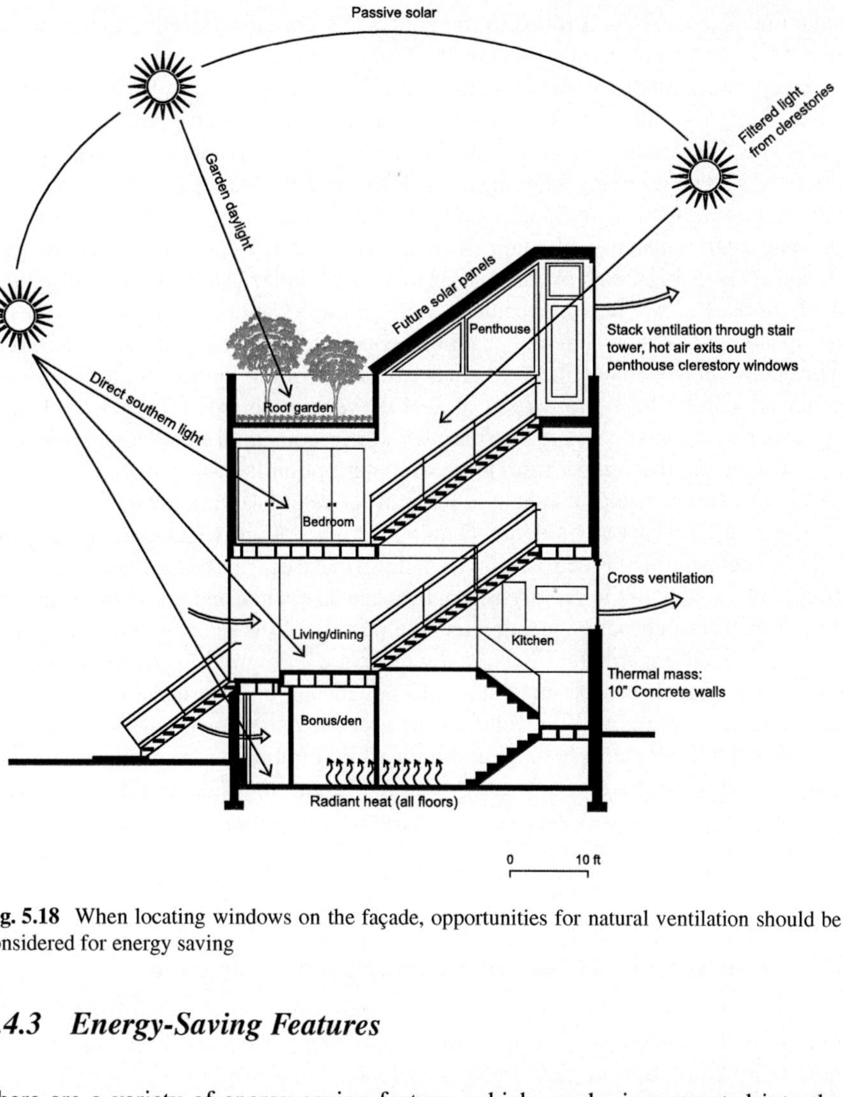

Fig. 5.18 When locating windows on the façade, opportunities for natural ventilation should be considered for energy saving

5.4.3 Energy-Saving Features

There are a variety of energy-saving features which can be incorporated into the design of windows to improve their thermal performance. Conductive heat loss across the glazing unit can be controlled by the number of glass panes and the airspace between them. High-performance glazing systems aim to admit natural light while limiting the amount of heat transmission, this can be achieved through a double paned or insulated glazing unit, often including a low-emissivity coating to help limit radiative heat gain (Ander 2016). Double or triple glazing is one of the best ways to reduce conductive heat loss in windows. By using multiple panes with sufficient space between them, the thermal resistance of the glazing unit is increased. Using a low-conductivity gas in between windowpanes is an additional measure to further improve the thermal resistance of a glazing unit. Radiative heat transfer caused by solar radiation can be controlled by applying a low-emissivity coating to the glazing unit, which reflects certain wavelengths of radiation outwards.

Overheating caused by summertime heat gain through windows can lead to significant energy use. Shading devices can be used to limit excessive heat gain through windows. The proper design of shading tools depends on the solar path across the building, neighboring buildings' shadows, and the time of year. There are a variety of exterior and interior shading devices which can be used depending on the particular demands of the unit, such as structural add-ons, louvers, blinds, and shutters (Fig. 5.19). Fixed overhangs can help to minimize overheating but also limit daylight access, whereas moveable devices allow for more control based on the occupants' preferences. Internal shading devices are typically more affordable and more conveniently maintained, whereas external shading devices tend to perform better at limiting heat gain (Ye et al. 2014). Venetian blinds reflect natural light upward, helping to illuminate the space with diffused natural light while mitigating overheating and discomfort from glare. Another option is insulated shutters on the window's exterior, which can help to limit all forms of heat transfer.

There are also several modern technological innovations which can improve the efficiency of windows. Electrochromic windows make use of using an electric current within the glazing unit to rapidly switch between an opaque and transparent surface. Similarly, thermochromic glazing alters the opacity of the glazing unit in response to the external temperature. Finally, on windows with high levels of sun exposure, photovoltaic panels can be integrated into the glazing units to make use of excess solar energy. Monocrystalline cells can be spaced apart allowing 4–30 percent of light through, giving access to natural light while limiting excessive heat gain and generating its own energy (Snow and Prasad 2011). Through the use of these features, the energy performance of the façade can be greatly improved without sacrificing on windows.

5.5 Diversity of Facades in Västra Hamnen, Sweden

After a 1970s recession in shipbuilding, the Western docks of Sweden's second-largest city, Malmö, were left largely abandoned. An urban renewal project of the district has aimed to revitalize the area and establish it as a model for sustainable urban living where walkability and shared streets will be made a central focus. With a master plan by Klas Tham, the community boasts 1,100 residential units. The opening of the Øresund bridge, linking Malmö with Copenhagen, and the construction of Santiago Calatrava's iconic Turning Torso, have established Västra Hamnen as an exciting new community and has become a symbol of Malmö's economic revival (Foletta and Field 2014).

In accordance with the intention to create sustainable developments, the project takes many measures to follow suit including diversity of housing types and facades' appearances (Fig. 5.20). It exercises an awareness in the choice of exterior construction materials, avoiding ones deemed hazardous to the environment and instead

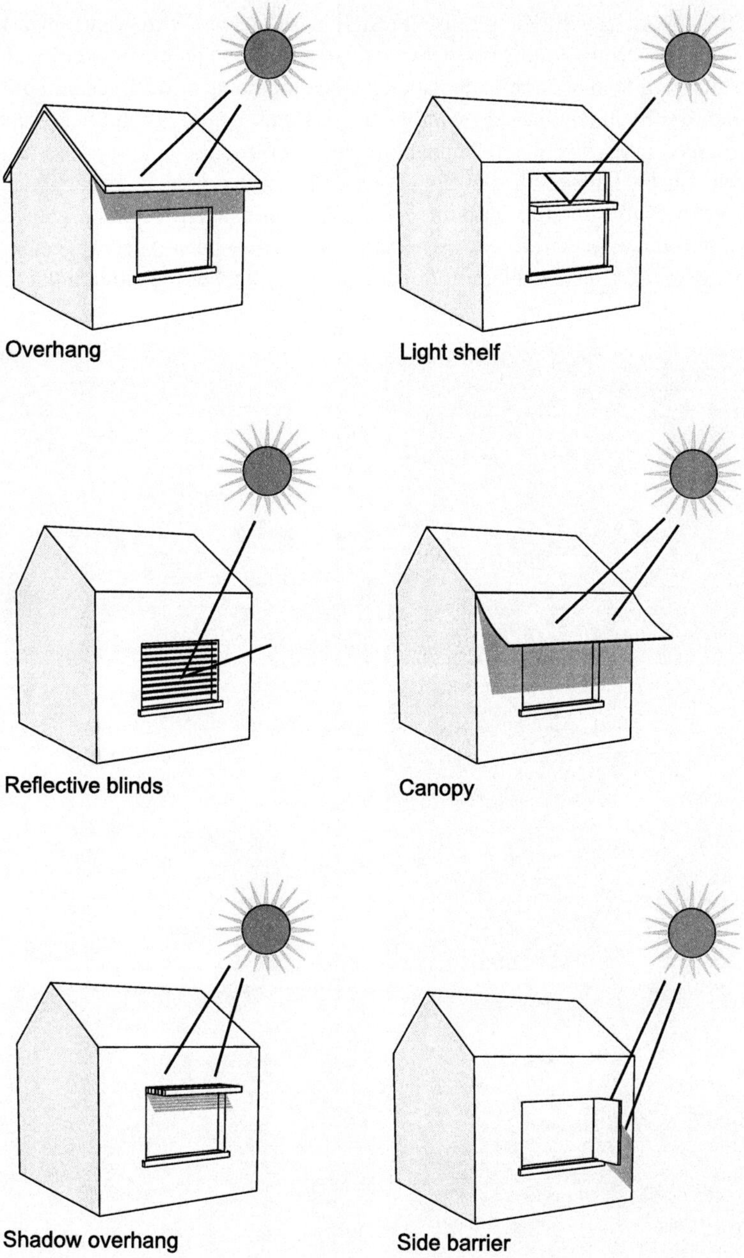

Fig. 5.19 Exterior and interior shading devices can be used depending on the demands and uses of the home, such as structural add-ons, louvers, blinds, and shutters

opting to support recycling and reuse through these means. With grants awarded by the Swedish government, this project invests in renewable energy sources and other devices that will lead economic and societal developments down a sustainable path. These sources include solar and wind energy, as well as biogas and heat pumps in which excess energy is collected and stored for later usage.

Within Västra Hamnen, the developers have placed an emphasis on car-free living. In the first neighborhood to be built, most streets are shared between pedestrians, cyclists, and drivers. Private car ownership has further been discouraged in other neighborhoods. To make walkability pleasant, green patches are either embedded in

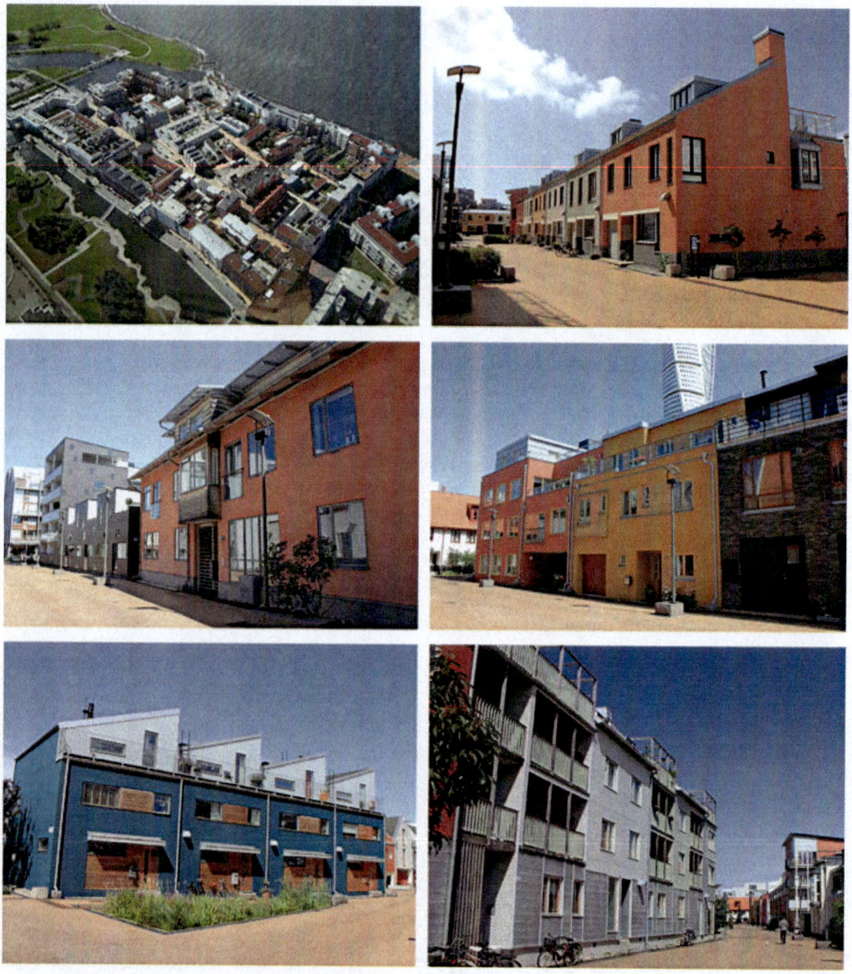

Fig. 5.20 To create a sustainable development, the Västra Hamnen project takes many measures to include a diversity of housing types, colors, and shapes

the road or in a shallow canal that runs alongside the street, and public art is integrated as well.

Cycling is another important form of transportation in Västra Hamnen. Malmö is considered one of the world's leading bicycle cities, with 420 km (261 miles) of cycle paths. Within Västra Hamnen alone there are 8.2 km (5 miles) of newly built laneways. The comfort of cyclists has been recognized, with resting rails at all traffic lights. Along the streets and near homes, amenities such as bike storage and sitting areas for seniors have been installed.

Special charges for private parking have been used to discourage car ownership. Parking is limited to 0.7 spaces per unit, as opposed to the Malmö average of 1.1. This space limit was too ambitious, however, and a greater than expected level of car ownership has forced a new multi-level parking garage to be built. Newer additions to Västra Hamnen have now been allotted an increased 0.75 spaces per unit. A green car-sharing service has also been established.

5.6 Affordable Prefab Home

The Affordable Prefab Home exemplifies a flexible system for sustainable prefabricated residential developments. It is meant to engage future occupants in the design stage, making use of technology to realize designs in a highly efficient manufacturing process—similar to the Grow Home and the Next Home as mentioned in Chaps. 3 and 4, respectively (Friedman 2001). Buyers have unique needs and design ideas for their homes, making their knowledge and participation valuable in the design process. Prefabricated homes can be assembled quickly and under optimal conditions, where construction can be broken down into a series of controlled and efficient tasks in a manufacturing plant. Whereas conventional construction methods using on-site assembly are more dependent on environmental factors, prone to error, and material waste. The manufacturing process of the Affordable Prefab Home therefore improves the energy efficiency of the building envelope (Friedman and Cammalleri 1992).

The technique of mass customization facilitates flexible designs while establishing economies of scale by using several small numbers of standard components to create a selection of many layouts and façade options for buyers to choose from. Components can then be manufactured at a large scale without creating repetitive and monotonous units.

The design process began by identifying the standard types of interior and exterior partitions which would be offered across all units (Figs. 5.21 and 5.22). A variety of different panels would then be created. Once the types of panels were chosen by the designer and builder, the offerings could be visualized for prospective buyers.

Mass customization allowed the buyers to work with the designers to customize the layouts and appearances of their homes to suit their preferences and budgets (Figs. 5.23, 5.24, and 5.25). A variety of plans were prepared in advance as well

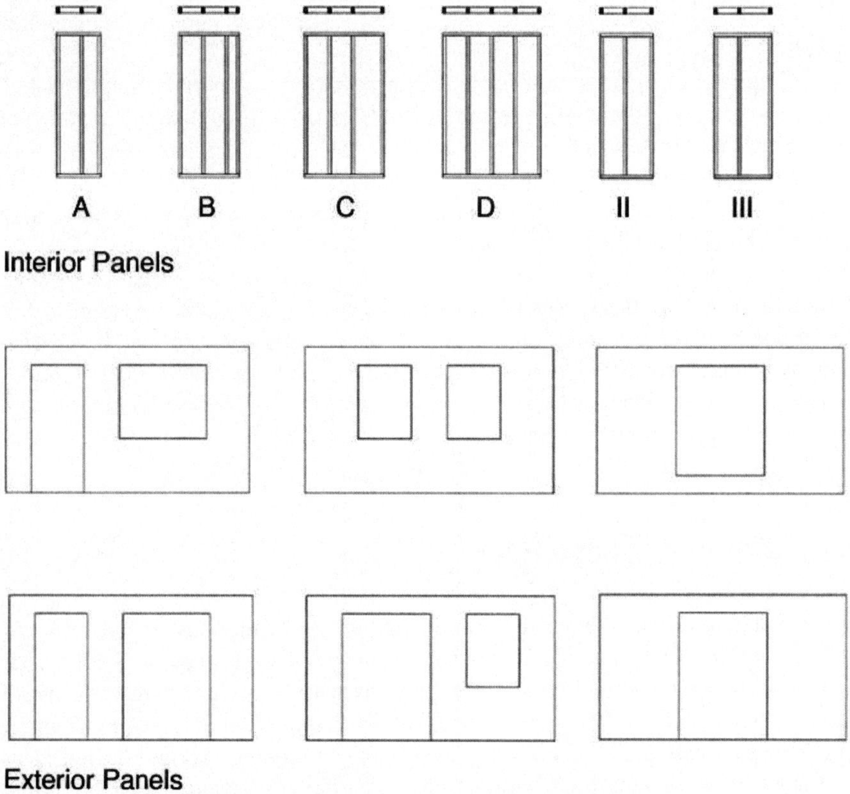

Fig. 5.21 The design process of the Affordable Prefab Home began by identifying standard types of interior and exterior partitions which would be offered and used across all units

which could be selected without modification. Once the interior arrangement was decided, the panels for the exterior envelope could be chosen.

It was recognized that giving occupants the ability to personalize their façade would be important, as it would impact their perception of their own home, as well as its presence on the street. Exterior panels had to accommodate a variety of internal configurations while offering buyers choices. Nine panel configurations were created for the facades: six for the front and back walls, and three for the side walls (Figs. 5.26 and 5.27). Depending on the size and number of attached walls, units could be built using anywhere between two to six exterior panels.

Once the selection process was complete, designs could be quickly realized in the manufacturing plant. The sales office would send information regarding the units directly to the technical office of the manufacturer, where a technician would determine the necessary components in each design. Tables to indicate the required amount of each panel for production would be created. A cutting station would then prepare

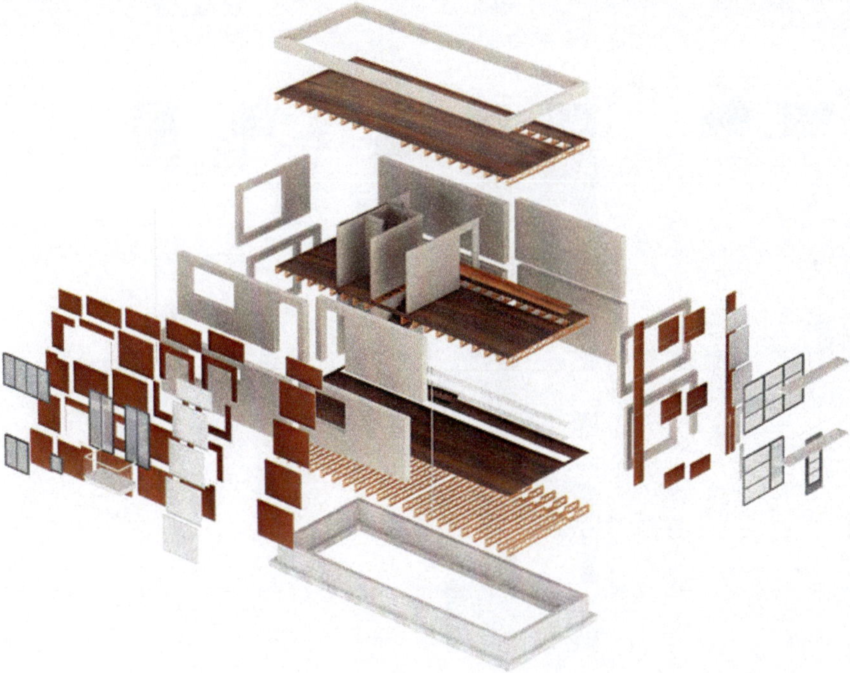

Fig. 5.22 Axonometric showing exterior and interior components of an Affordable Prefab Home

and cut the necessary lumber for each component, which would be transported to a semi-automated assembly station where a worker would construct the panels. The same types of panels could be mass-produced together despite being prepared for different units, making their assembly more efficient. Workers would add openings, insulation, and any other specified components in the manufacturing process. Finished panels would then be transported to storage, in preparation for shipment and the final on-site construction.

The design and construction of the Affordable Prefab Home demonstrate how flexibility and customization can be facilitated with a more efficient means of production. Prefabricated panels are affordable, energy efficient, and can be used in different combinations to accommodate the various needs of different households. This method of construction creates more sustainable building envelopes with opportunities for personalization (Fig. 5.28).

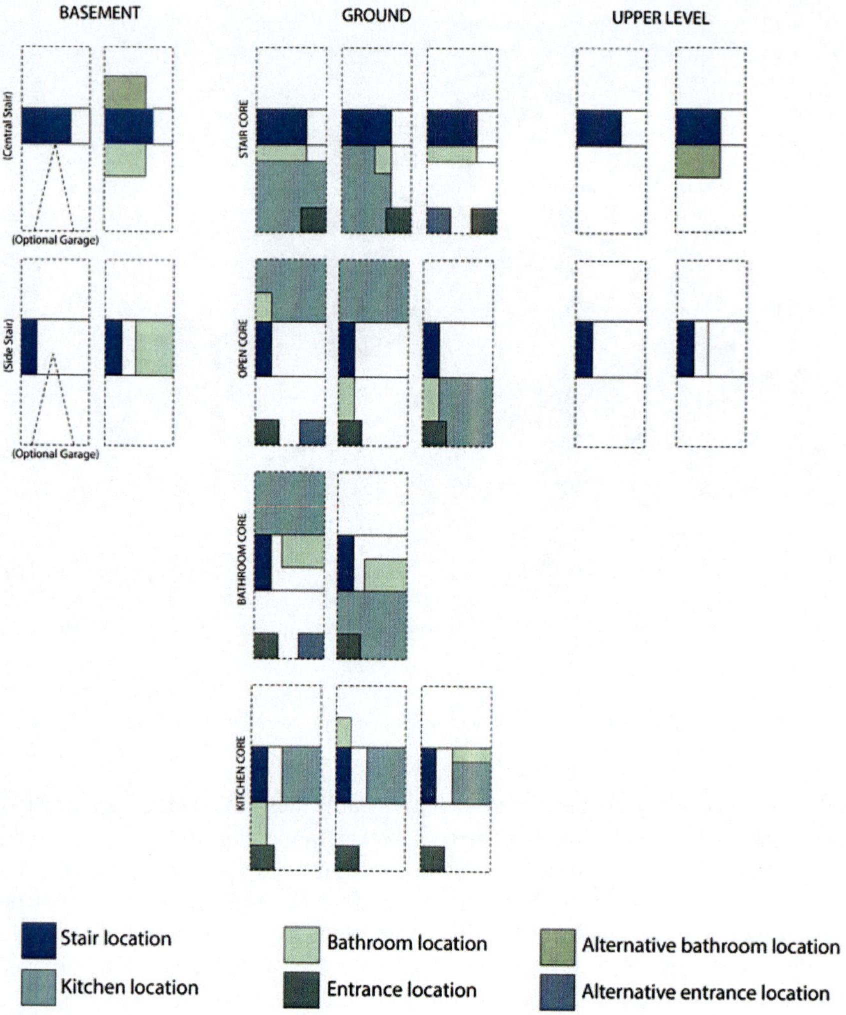

Fig. 5.23 Schematic representation of possible design layouts

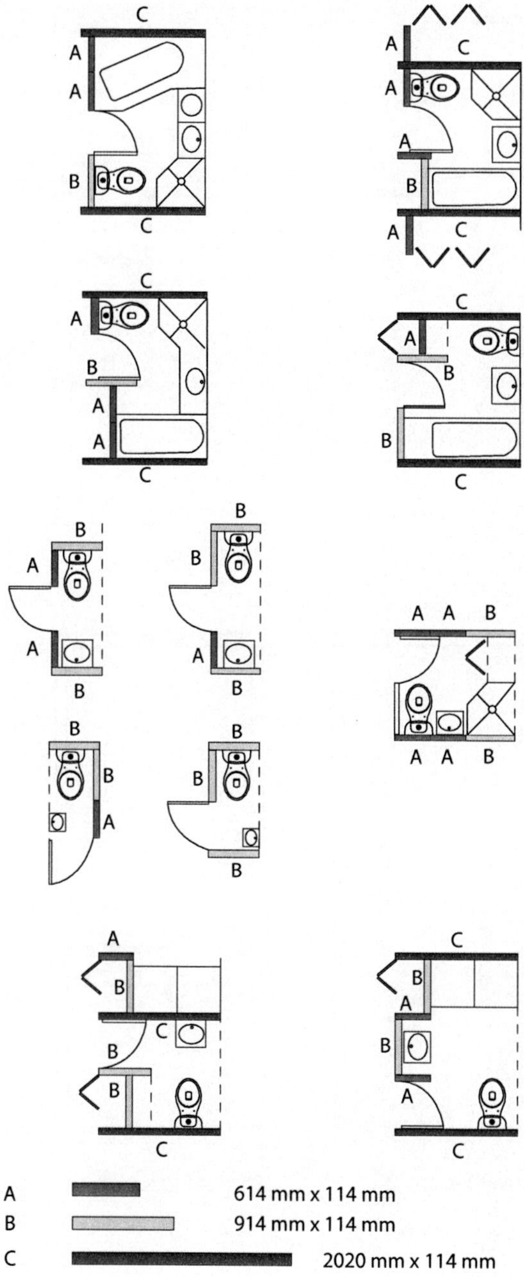

Fig. 5.24 Bathroom configurations using standard panels

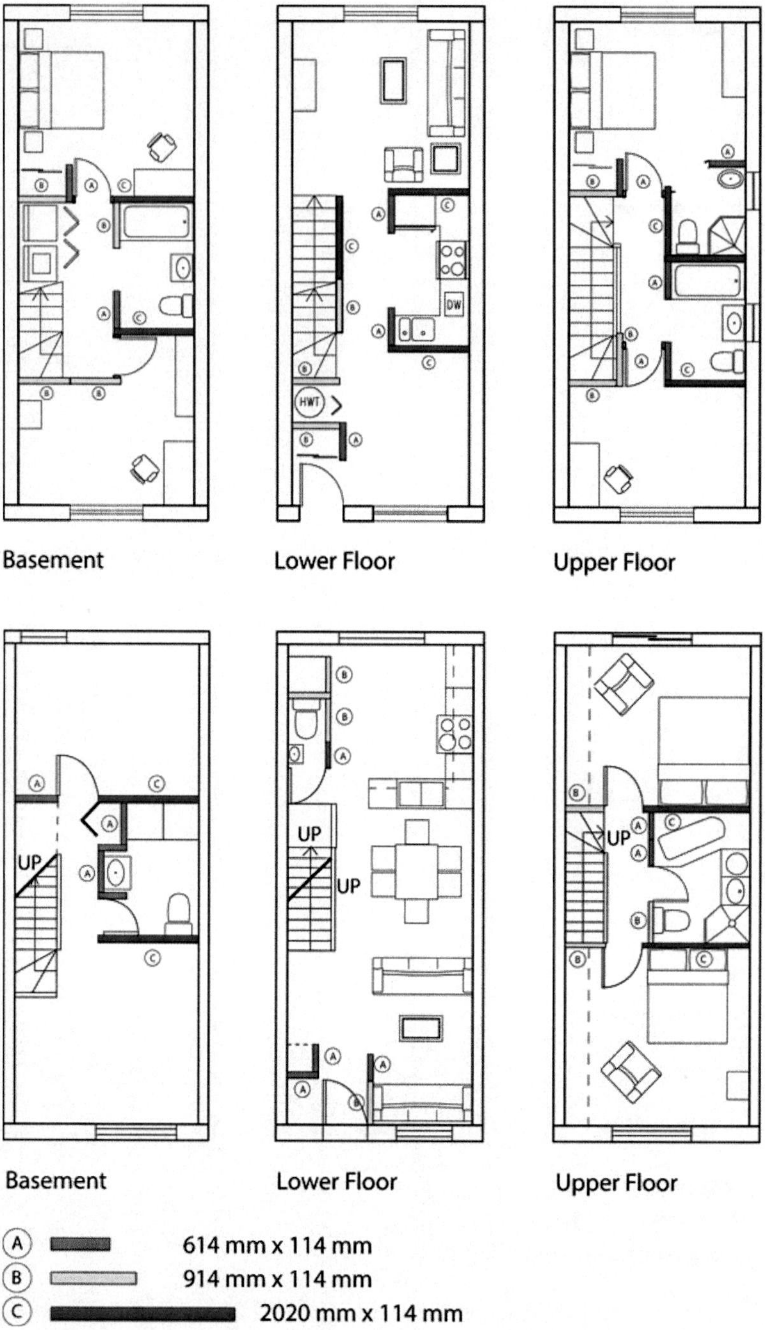

Fig. 5.25 Interior arrangement of two dwelling units using standard panels

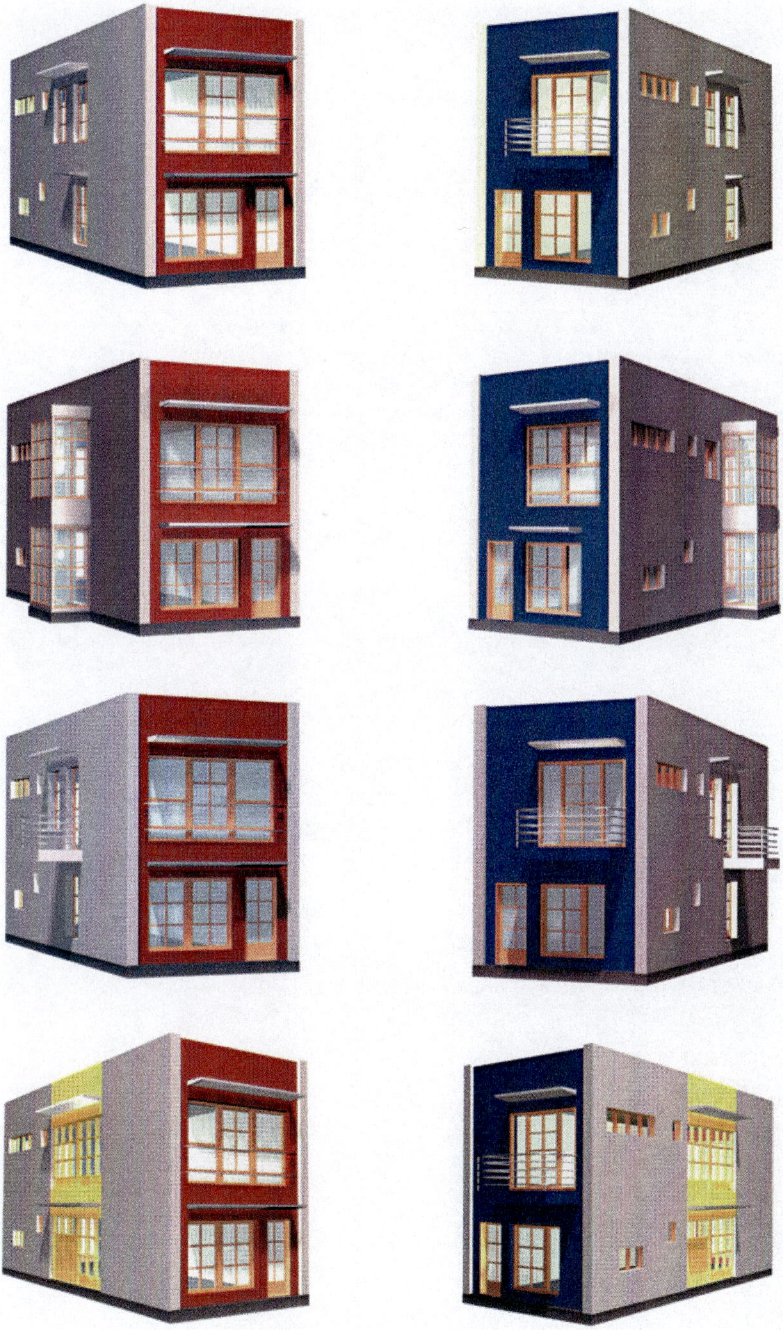

Fig. 5.26 Optional façades using standard panels

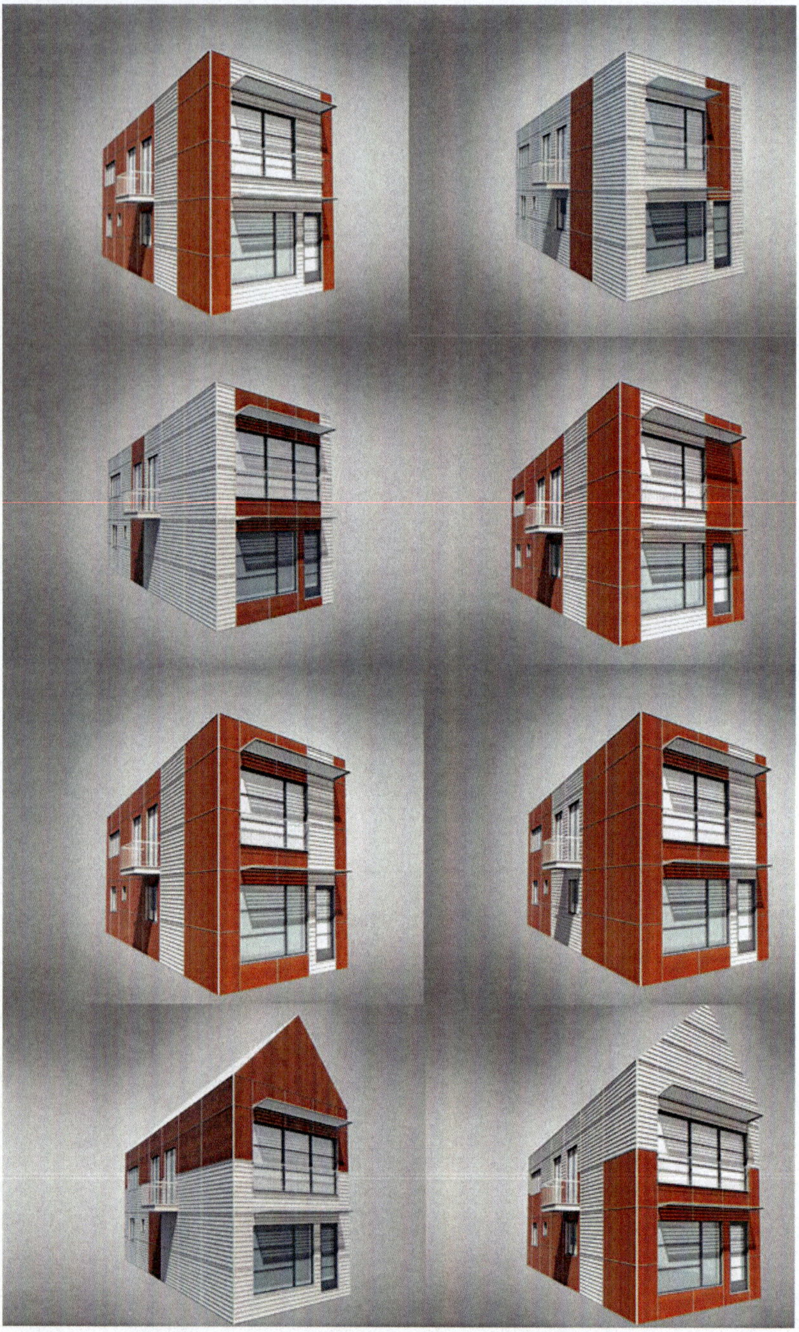

Fig. 5.27 Optional façades using standard panels

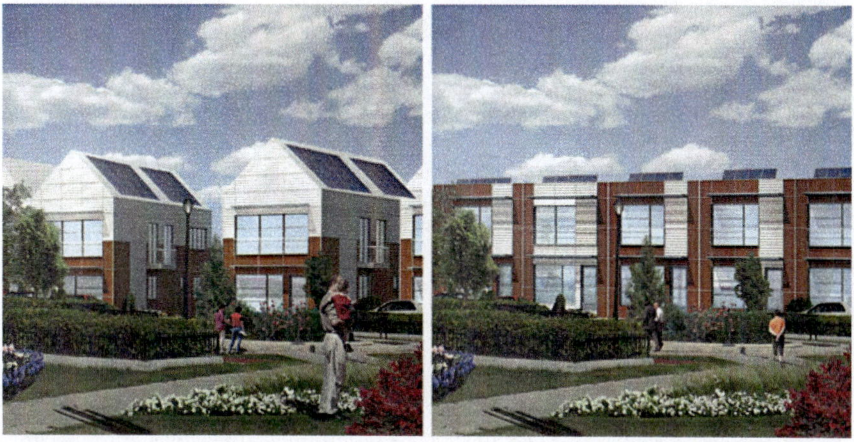

Fig. 5.28 Affordable Prefab Home Communities

5.7 Final Thoughts

As higher density residential developments become the norm in many areas, the demand for attractive and sustainable building envelopes will be intensified. A building's sustainability is primarily determined by energy efficiency, which depends on the thermal qualities of its envelope. Designers continue to experiment with innovative façade technologies which aim to improve the sustainability and liveability of high-density dwellings. Green facades and vertical farming are gaining popularity in urban environments to reintroduce nature to high-density areas. Yuan et al. (2019) found that with sufficient green coverage and the careful selection of plants, vertical farming can be used to improve the air quality in naturally ventilated buildings. Another new concept for sustainable façade design is the use of biomimetic facades, which derive inspiration from biological patterns to create innovative, energy efficient designs (Webb 2022). These are a few of the innovative models for energy efficient building envelopes which may become the future of urban architecture.

Questions for a Follow-Up Discussion

1. What are the key principles of designing attractive façades?
2. What are the main aspects to consider in the design of energy efficient envelopes?
3. What are the main criteria effecting selection of energy efficient windows?

References

Adhitya S (2017) A rhythmic approach to urban design. In: Musical cities: listening to urban design and planning (DGO-digital original). UCL Press, pp 26–33. https://doi.org/10.2307/j.ctv550cz9.10. Accessed 5 Oct 2022

Ander GD (2016) Daylighting. In: National institute of building sciences. Retrieved from https://www.wbdg.org/resources/daylighting. Accessed 5 Oct 2022

Candido C (2011) Adaptive comfort: passive design for active occupants. Environ Design Guide 1–5. http://www.jstor.org/stable/26151893. Accessed 5 Oct 2022

Deng L (2010) The value of regulating building facades: a case study of masonry ordinances in Chicago suburbs. J Arch Plan Res 27(4): 271–286. http://www.jstor.org/stable/43030912. Accessed 5 Oct 2022

Foletta N, Field S (2014) Europe's new low car(bon) communities. http://www.gwl-terrein.nl/files/artikelen/low%20carbon%20communities.pdf. Accessed 5 Oct 2022

Friedman A (2001) The grow home. McGill-Queen's University Press, Montréal

Friedman A, Cammalleri V (1992) Industrialization of narrow-front rowhousing using panelized wall systems. McGill University, Montreal

Green Building (n.d.) In APA—The Engineered Wood Association. Retrieved from https://www.apawood.org/green-building. Accessed 5 Oct 2022

Hampton A (2010) Thermal mass and insulation for temperate climates. Environ Design Guide 1–11. http://www.jstor.org/stable/26150778. Accessed 5 Oct 2022

Herzog T, Krippner R, Lang W (2017) Facade construction manual (Second, Ser. Construction manuals (english). Edition DETAIL. https://doi.org/10.11129/9783955533700. Accessed 5 Oct 2022

Moe K (2012) Insulating North America. Constr Hist 27:87–106. http://www.jstor.org/stable/44215886. Accessed 5 Oct 2022

Oreskovic NM, Roth P, Charles SL, Tsigaridi D, Shepherd K, Nelson KP, Bar M (2014) Attributes of form in the built environment that influence perceived walkability. J Arch Plan Res 31(3): 218–232. http://www.jstor.org/stable/44114605. Accessed 5 Oct 2022

Richman RC, Cianfrone C, Pressnail KD (2010) More sustainable masonry façades: preheating ventilation air using a dynamic buffer zone. J Build Phys 34(1):27–41. https://mcgill.on.worldcat.org/oclc/646281834. Accessed 5 Oct 2022

Saroglou T, Meir IA, Theodosiou T (2020) Façades: improving the energy efficiency of a mediterranean high-rise envelope. CTBUH J 2:20–27. https://www.jstor.org/stable/27034277. Accessed 5 Oct 2022

Snow M, Prasad D (2011) Building-integrated photovoltaics (BIPV). Environ Design Guide 1–15. http://www.jstor.org/stable/26151891. Accessed 5 Oct 2022

Webb M (2022) Biomimetic building facades demonstrate potential to reduce energy consumption for different building typologies in different climate zones. Clean Technol Environ Policy 24(2): 493–518. https://doi.org/10.1007/s10098-021-02183-z. Accessed 5 Oct 2022

Weinreb AR, Rofè Y (2013) Mapping feeling: an approach to the study of emotional response to the built environment and landscape. J Arch Plan Res 30(2):127–145. http://www.jstor.org/stable/43031085. Accessed 5 Oct 2022

Ye YY, Xu P, Mao JC (2014) Selected, peer reviewed papers from the 2014 2nd international conference on renewable energy and environmental technology (REET 2014), 19–20 August 2014, Dalian, China. (2014). Energy performance of external versus internal shading devices in residential buildings. Appl Mech Mater 672–674:546–549. https://doi.org/10.4028/www.scientific.net/AMM.672-674.546. Accessed 5 Oct 2022

Yuan C, Shan R, Adelia AS, Tablada A, Lau SK, Lau SSY (2019) Effects of vertical farming on natural ventilation of residential buildings. Energy Build 185:316–325. https://doi.org/10.1016/j.enbuild.2018.12.028. Accessed 5 Oct 2022

Zemella G, Faraguna A (2014) Evolutionary optimisation of facade design: a new approach for the design of building envelopes. Springer. https://doi.org/10.1007/978-1-4471-5652-9. Accessed 5 Oct 2022

Chapter 6
Innovative Construction Practices

Abstract A renewed interest in prefabricated housing has emerged along with recent development in innovative manufacturing. The negative effects of conventional building practices on the environment coupled with an urgent need for affordable housing has encouraged the home building sector to prioritize resource efficiency and conservation. The rapid rise of digital production, along with mass customization that adapts an end product to a consumer's needs, makes a long-term practice of off-site fabrication viable. Prefabrication's importance rests in its ability to provide flexible living models for various demographics, while optimizing resource consumption. This chapter introduces the fundamentals of prefabricated construction on its main types including mass customization, time-saving interior construction methods, and design for growth and adaptability.

Keywords Dry construction · Prefabricated homes · Mass customization · Modular construction · Resource efficiency

6.1 A Need for Resource Conservation and Affordability

Overconsumption of non-renewable resources has led to a call for resource preservation. Driven by rising ecological awareness, declining availability, and high costs of natural resources communities began to reexamine waste disposal methods and homebuilders to reconsider their current practices (Hoyt 2020). The challenges in conventional construction practices are among others the poor handling of demolition waste (CDW) (Behroozy and Keegel 2014). The need for resource efficiency can be understood through the concept of natural limits. Some of the resources that the construction industry relies on will eventually run out since they can only be replenished at a certain rate (EEA 2020) (Fig. 6.1). There is an urgent need to find ways to generate greater returns from the same number of resources by shifting to a circular production method where products are used as long as possible (EEA 2020). In addition, the demand for raw materials has increased globally in recent decades and the use of natural resources is expected to double between 2010 and 2030 (Ncube et al. 2021). Therefore, using resources efficiently and economically is

A. Friedman, *Fundamentals of Innovative Sustainable Homes Design and Construction*, The Urban Book Series, https://doi.org/10.1007/978-3-031-35368-0_6

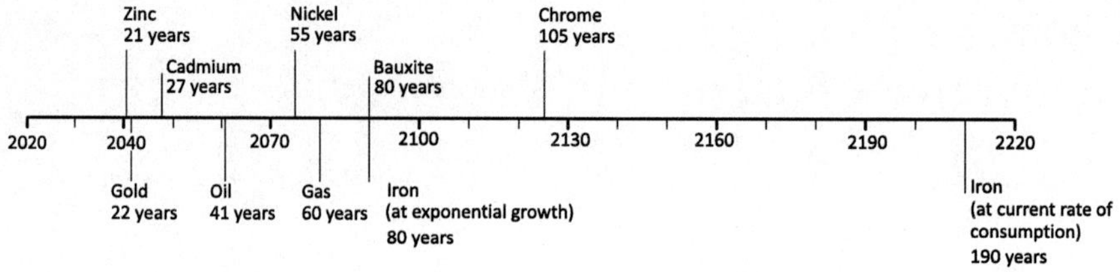

Fig. 6.1 Some of the resources that the construction industry relies on are projected to run out since they cannot replenish fast enough naturally to meet the projected demand

set to be a way to deliver more with less. Promoting resource efficiency also stands to increase the competitiveness of the industry, create jobs, stimulate innovation, boost sectors such as recycling and resource recovery, and help ensure more secured supplies (EEA 2020). It is with this mindset that this chapter suggests that an efficient construction system can benefit from prefabrication, which will facilitate assembling, disassembling, and recycling practices (Macieira and Mendonca 2016).

Housing affordability is another challenge that the housing market of many nations is facing. In Canada, the year 2021 saw a sharp increase in costs of housing ownership and rentals with median multiples of 10 or more in some markets (Cox 2022). This lack of affordable homes has led to a trend whereby young people stay at their parental homes longer (Hankin 2022). Another housing trend is the agglomeration of millennials in urban areas. Driven socially or due to the lure of more work opportunities, members of the millennial generation are moving to regions with a higher proportion of renters compared to homeowners, which further pushes rental prices up (Hankin 2022). The overarching effects of these phenomena contribute to the rising demand for innovative affordable and adaptable dwellings.

6.2 Fundamentals of Prefabricated Homebuilding

Prefabrication is the practice of assembling building components in a factory and transporting complete assemblies or sub-assemblies to the building site for construction (Editors of Encyclopaedia 2022). The rising interest in prefabrication can be attributed to their financial, environmental, quality, and time-saving advantages. Economies of scale can be achieved through the mass production of parts due to the specialization and standardization of pieces which reduce costs (Bertram et al. 2019; Larsen et al. 2019). Furthermore, industrialization enables manufacturing in controlled indoor settings (Fig. 6.2). This promotes precision during production in comparison to outdoor building sites especially in a challenging climate. Prefabrication enables production of high-quality elements while also considering workers' health and safety. Since fabrication is not weather dependent, it allows sensitive tasks, such as waterproofing or spray processes to continue regardless of weather

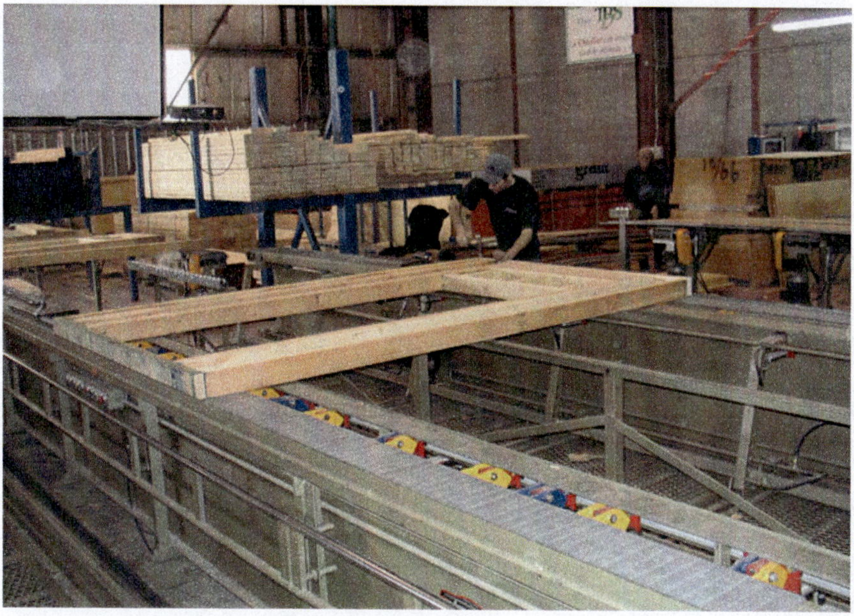

Fig. 6.2 A prefabricated production plant

conditions (Derananian and Donlon 2019). As a result, the improved work environment increases productivity and the product's quality. Once the prefabricated parts are completed, they can be inspected and amended in the plant if needed (Derananian and Donlon 2019). Furthermore, a reduction of on-site construction time lowers labor costs, and limits damage due to warping, rot, vandalism, and theft. The assembly of prefabricated parts usually requires a smaller team of workers and construction coordination is simplified as a result.

6.2.1 Main Types of Prefabrication Methods

The three main types of prefabrication are panelized, modular, and kit of parts. *Panelized* prefabrication is the most widely used and the end products are also referred to as 2D elements. Panels of different sizes, some only with framing and others with insulation and windows, are assembled according to plans to form the bearing structure of the building (Derananian and Donlon 2019). Panel systems applicable to wood-frame residential construction can be categorized as: (1) open-sheathed panels, (2) structural sandwich panels, and (3) unsheathed structural panels (Derananian and Donlon 2019). Unsheathed structural panels' greatest advantage is overcoming the inadequate workmanship which may be found in conventional construction sites without resorting to unfamiliar building techniques (Fig. 6.3). Furthermore, alternative timber products such as *cross laminated timber* (CLT), *dowel laminated timber* (DLT), and *glulam* are panel systems not widely used for multifamily housing. However, the 2021 version of the International Building Code is projected that the use of mass timber will be permitted in taller buildings (Hoyt 2020).

Modular construction refers to the factory fabrication of sections that consist of an entire house or a part of one also referred to as 3D elements (Fig. 6.4). Once manufactured and produced, the sections are sent to the site where they are hoisted into place by crane (Bertram et al. 2019). The maximum width of modules for road transport that does not require a special vehicle escort is typically around 4.8 m (16 feet). This either increases the cost of transporting larger units if escort is needed or limits the size of modules (Bertram et al. 2019). Using modules can cut construction time by 20–50% and costs by 20% (Bertram et al. 2019). A 3D approach to building has the potential for maximum efficiencies and time savings and is most suitable for affordable housing projects (Bertram et al. 2019).

Kit of parts consists of well-marked pieces, such as studs or windows that are shipped to the site for assembly (Fig. 6.5). This method can be understood as precut and labeled components that are sized for convenient handling that match the dimensions of shipping constraints. They are the tightest way to pack elements and therefore are recommended for shipping over long distances. The assemblies are conceived in a systematic way based on a certain increment, size, or shape (Howe et al. 1999). After rules of connection and increments are established, the number of possible shapes and appearances is limitless.

Regardless of the method, the use of prefabricated construction turns the building process into an "assembly line" style of work in a safe and controlled environment (Howe et al. 1999).

6.2.2 Other Prefabrication Typologies

Plug and Play homes are prefabricated, modular units which can be installed rapidly and made ready for immediate use by simply plugging-in their utilities (People's Architecture Office 2018). The term Plug and Play originated from the realm of electronics that refers to products installed by the user that can function immediately without professional help and can subsequently be unplugged and removed or repositioned with equal ease. With modern shipping techniques, Plug and Plays are dwellings that can be transported or relocated worldwide. Unlike other types of prefabrication methods, which call for some assembly, Plug and Play units require no site work and simply need to be connected to utilities once placed (Stinson 2018) (Fig. 6.6).

Plug and Play units need to be transported without causing external or internal damage. They also should be light enough to ship but well build to sustain maximum wind drag (People's Architecture Office 2013). Windows and doors present a challenge since they weaken the overall structure and are easily breakable. Units which are designed to be expandable fold into the structure and are therefore protected

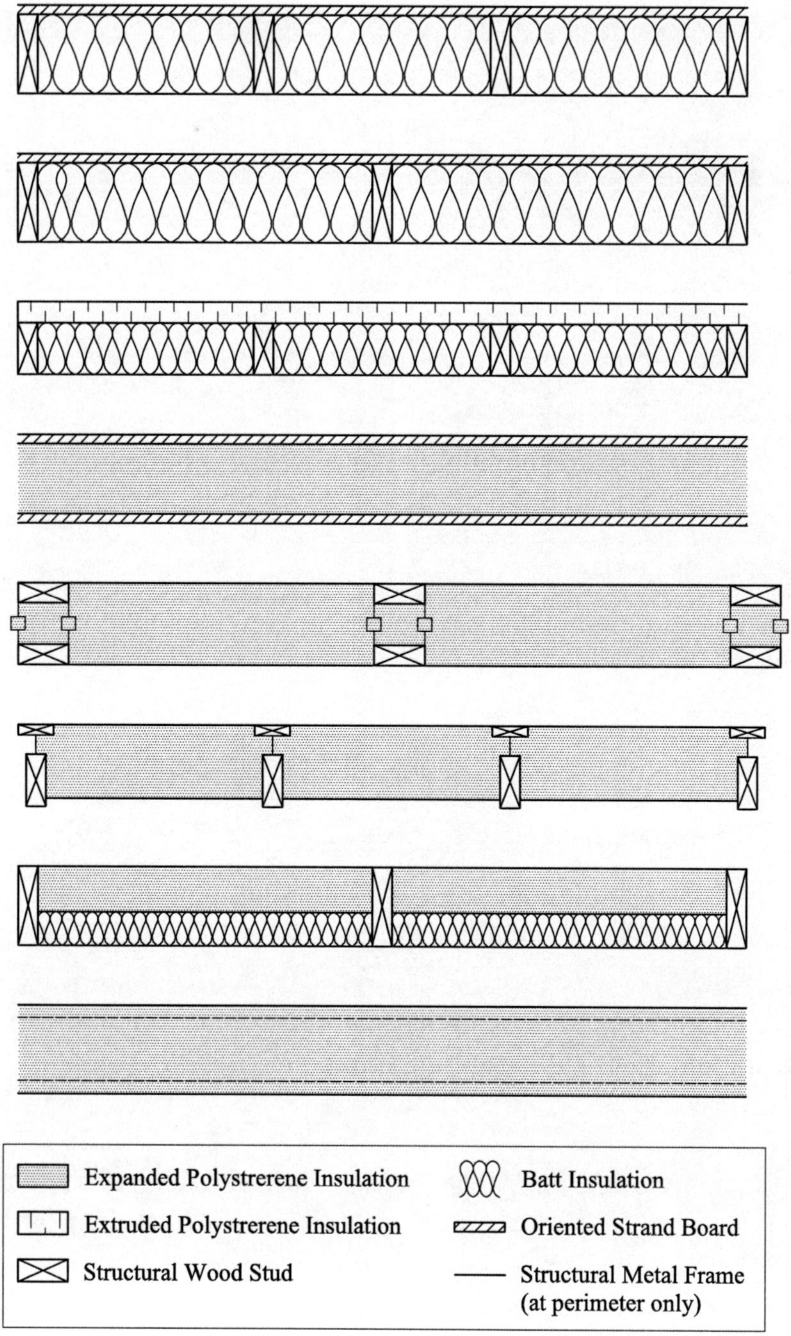

Fig. 6.3 Sections of prefabricated panelized wall systems

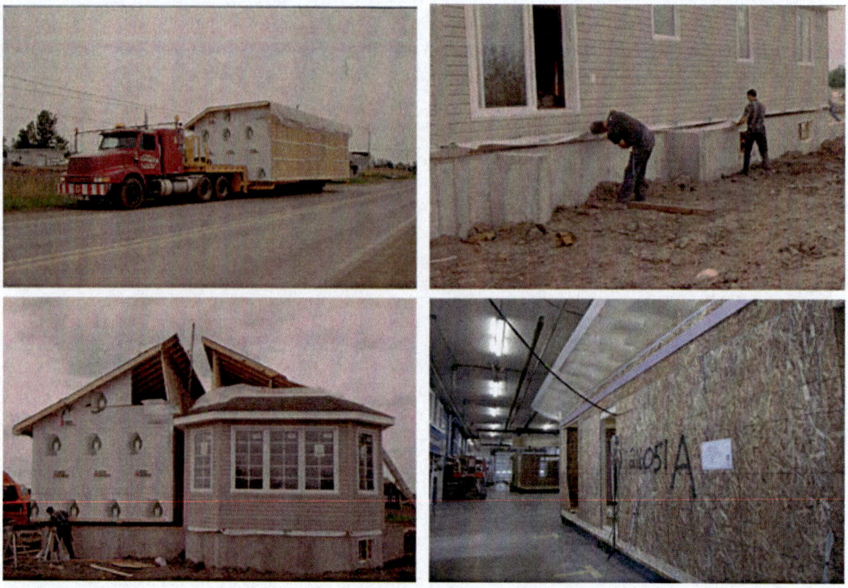

Fig. 6.4 Fabrication of modular home, delivery, installation, and on-site finishing

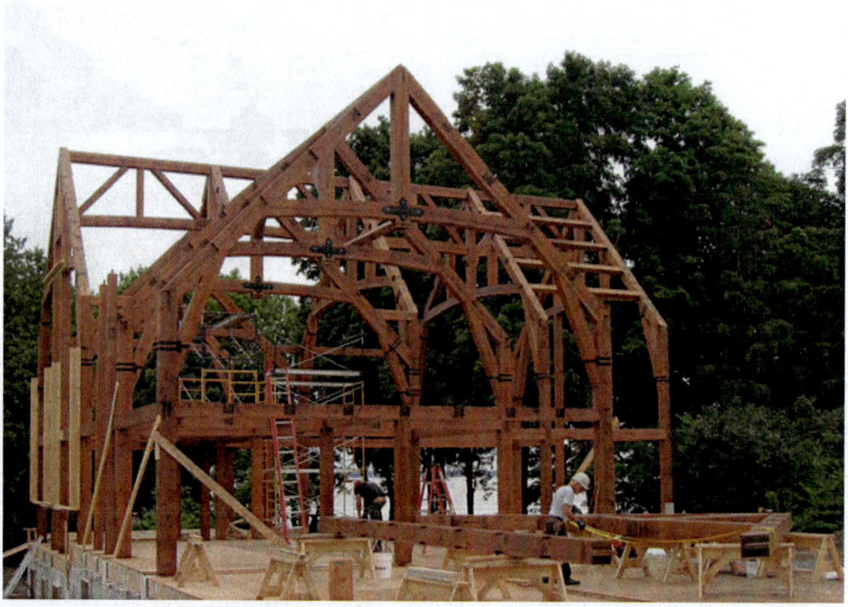

Fig. 6.5 Kit of parts consists of well-marked pieces, such as studs or windows that are shipped to the site for assembly

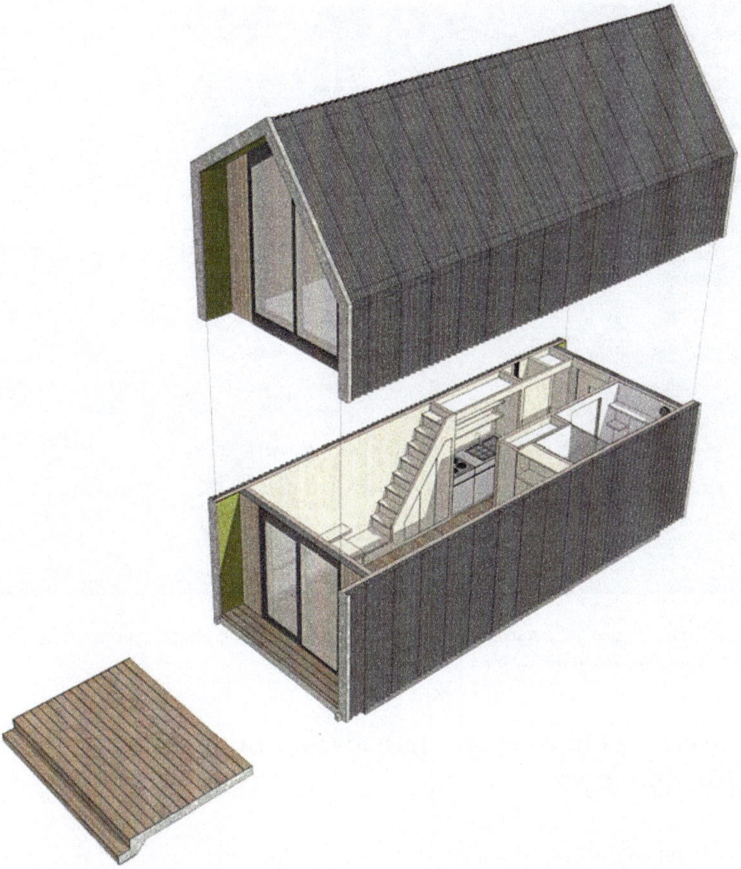

Fig. 6.6 Sections of a Plug and Play home

during transport. This feature, alongside mezzanines increases living space. Plug and Plays are generally small units such as accessory dwelling units (ADUs), therefore are less costly than conventionally constructed homes which creates an opportunity for designers to source higher quality materials, sophisticated technologies, and customizable features (People's Architecture Office 2022).

Mobile homes are a variation of the Plug and Play unit that are prefabricated in plants yet can easily be customize to the dweller's needs (Office of Mobile Design Corp 2021). Like Plug and Play units, mobile homes can be unplugged and repositioned with as much ease as there was during installation (Office of Mobile Design Corp 2021). Mobile homes are offered in different versions that vary in length, number of levels, and expandability with foldout or pop-up mechanisms (Fig. 6.7).

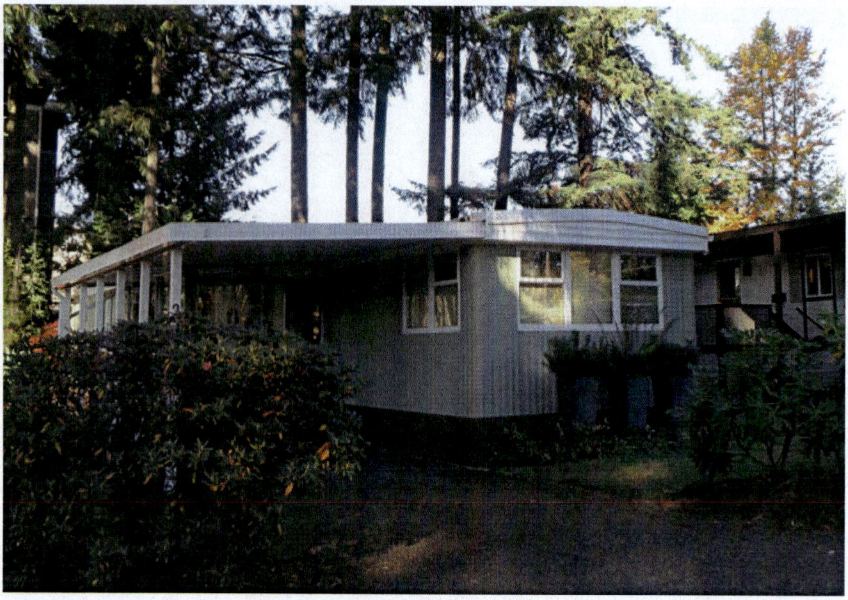

Fig. 6.7 Mobile homes are a variation of Plug and Play units that are prefabricated in plants, yet can easily be customize to the dweller's needs

6.3 Selecting Fabrication Methods and Building Communities

It is important to distinguish and understand the three main types of prefabrication methods that were outlined above in relation to a best-suited context (Howe et al. 1999). A panelized system is a recommended method for single-family projects on a challenging narrow infill urban site. They can also be used for an accessory dwelling unit (ADU) be it a garden pavilion, a retrofitted garage, or an annex to a building (Lessard 2018). Since panels are much smaller and lighter, they can be installed with simple equipment which is also easier to operate and transport (TechPrefab 2021; Bertram et al. 2019). Transporting panels is also simpler compared to modules (Fig. 6.8). Their flat, rectangular configuration, and lightweight make them easy to stack and fit onto smaller trucks, compared to the considerably larger vehicles needed to transport modules.

Building types such as dorms or hotels are better suited for modular construction (Hoyt 2020). Modular construction has an advantage over panelized systems for projects with a high level of repeatability and a high ratio of wet to dry functions such as kitchens (Bertram et al. 2019). Additionally, less manual on-site labor is required for modular construction which means even lower costs.

Construction of kitchens and bathrooms can be time consuming and costly in conventional construction practices. Building these parts as pods off-site may lead

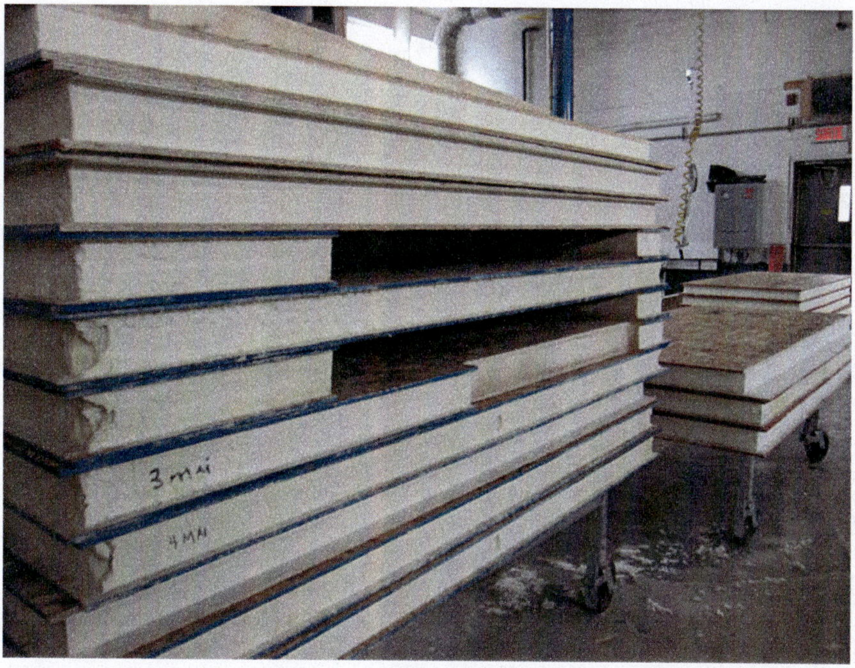

Fig. 6.8 Transporting stackable prefabricated panels is simpler and less costly compared to modules

to time and cost savings (Larsen et al. 2019). Bathroom pods for example, such as those built by Italian company Bathsystem (2019), represent a preferred solution for a variety of projects ranging from student housing, mid-level/luxury dwellings, low-rise hotels, hospitals, care homes, and military barracks. Through innovative technology, bathroom pods can be robust, with carefully selected materials to increase durability, strength, and design appreciation (Fig. 6.9). Their production can be checked for quality control by the product development and technical teams before shipment to the site. Similarly, modular kitchen pods are adaptable to all types of residential development projects. Clients can select various forms and shapes of pods to best suit their dimensions and taste (Bathsystem 2019).

Population growth and rapid urbanization are prompting a need for higher density environments ultimately leading toward interest in apartment living. In addition, in many nations, household size has been declining, which increased demand for small units, especially one and two-bedroom apartments (Lee 2021). On the supply side, as land costs increase, developers are fitting more dwellings on a single lot and building in existing neighborhoods. These challenges can be facilitated with the use of prefabricated technologies. Designers can introduce variability by offering a range of unit types that cater to clients with varying demographic compositions and lifestyles (Lee 2021).

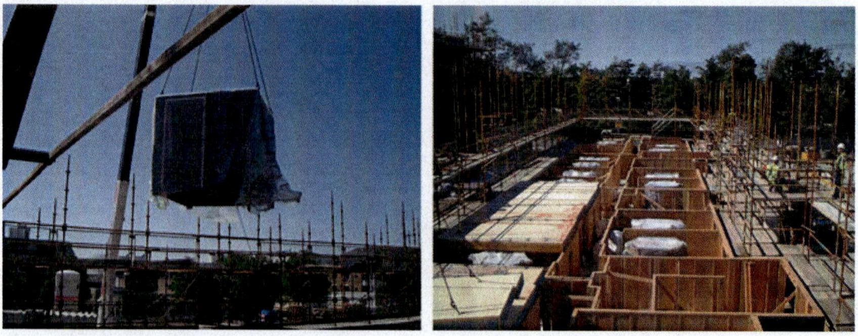

Fig. 6.9 A bathroom pod is being lifted and installed in a floor

In addition to apartments, in recent years prefabricated mass customized (MC) communities have emerged as an alternative to conventionally built subdivisions (Kashef 2009). Being built better, they can offer sustainability while satisfying a wide range of inhabitants' needs. MC is a manufacturing paradigm that enables customized and personalized design at a mass production cost. Its lower unit cost increases quality, shortens project time, and is considered highly relevant for the future of the homebuilding industry (Larsen et al. 2019).

Despite a common belief that mass production techniques at a community scale create uniform and monotonous units, MC uses these methods to produce units at a low cost while maintaining flexibility and adaptability to each occupant's need (Larsen et al. 2019). This is done by offering a range of unit types for different demographic cohorts. Furthermore, the integration of several unit types such as townhomes, and single-family residences into one community makes it more affordable due to economies of scale.

6.3.1 *Mass Customization and Digital Fabrication of Housing*

Mass customization is the production of individually customized goods and services. It relates to the ability to provide customized products or services through flexible processes, in high volumes and at reasonably low costs including housing. The concept emerged in the late 1980s and was viewed as a natural outcome of processes that have become increasingly flexible and optimized regarding quality and costs. In addition, mass customization, as an alternative, gave companies an edge in a highly competitive and segmented market.

The process of customization of housing can be considered as a multi-faceted process that involves consideration of various aspects, such as managerial and technical. It is a production strategy that aims to provide customers with individualized design at near-mass production efficiency. The starting point toward mass customization of dwellings would be mass production of one-of-a-kind products. Mass production companies work based on a fully developed model by making standard components to stock. These companies may decide to shift toward mass customization in response to market pressures and customer demand for a broader product portfolio (Kashef 2009).

Housing, unlike other products, requires a response to social and cultural issues when gathering clients' preferences. There have been valuable contributions to the process of mass customization of housing in diverse domains. On the one hand, there was an attempt to develop design systems that could be used by architects or homebuyers to customize dwellings on various levels, ranging from direct participation in unit layout to finishing material selection. On the other hand, efforts have been made to explore the potential of digital fabrication and how it could be deployed to produce mass customized housing. However, innovative computational techniques require an exploration of new opportunities in the process of mass customization of housing. Recent advancements in design and manufacturing technologies made the idea of mass customization possible.

The common application of customization within the prefabricated housing industry is to provide homebuyers with different alternatives regarding layout, finishing, and systems. For example, the homebuyer will be offered to choose between housing layout A, B, or C, then kitchen layout A, B, or C, an optional extra garage, or even an extra story. This approach has been defined in research as *multiple-choice housing*, which takes the form of printed catalogues and, recently, developed into interactive electronic catalogues. Electronic catalogues typically take the form of internet websites which offer users the ability to navigate and then modify the design on a computer. Such a trend is considered to be exhaustive, as the architect is required to formerly design all possible alternatives. As a result, in some cases the number of alternatives has to be kept limited to three to four options in order to avoid additional overhead costs. Moreover, offering many choices might be confusing to some homebuyers. Nevertheless, there are still chances that customers' desired design variation will not be offered, since alternatives are developed according to the architect's view, not the customer's demand (Larsen et al. 2019).

6.4 A Mass Customized Prefabricated Community in Ijburg, the Netherlands

The Floating Houses project in Ijburg, near Amsterdam, the Netherlands is a mass customized prefabricated community that was designed by Marlies Rohmer Architects & Urbanists (Fig. 6.10). The Dutch have a history of living close to the water and coping with the challenges that come with it. Consequently, houseboats and floating hotels are no strangers to the canals of the Netherlands. Lately, however, there has been a growing enthusiasm for living directly on the water—as opposed on

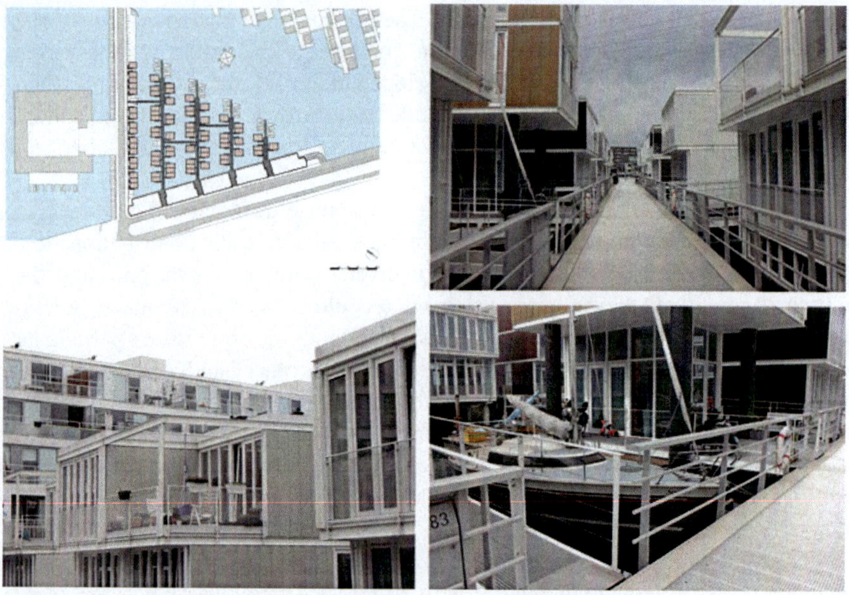

Fig. 6.10 The Floating Houses in Ijburg's project in the Netherlands is a mass customized prefabricated community

shore—because of rising sea levels (two-thirds of the population already live below the sea level), increases in precipitation, and shortage of land. The 10,652 square meters (114,657 square foot), 75-unit Floating Houses project capitalized upon this reality and aimed to lessen the housing pressures in the dense metropolitan areas of Amsterdam (ArchDaily 2011).

These 142 square meters (1,529 square foot) floating residences more closely resemble land-based houses than boats and are classified as immovable properties. Hence, they represent a neighborhood, laid out in a triangular shape derived from the diagonal slicing of the basin by the powerlines. Although living on the water is not a new concept, it is very different than building on land, and so must respect the unique properties of water. The houses are built in a shipyard about 65 km (40 miles) north of IJ Lake and tugged through canal locks. Due to the method of transportation, the canal locks restrict the houses from exceeding a width of 6.5 m (21 feet). Each structure is supported by a buoyant concrete tub that is submerged in water to the depth of a half story. On top, a lightweight supporting steel construction is built, which is fitted with wood paneling to create rooms and floors. In addition, the steel frame can be covered with brightly colored paneling which can later be changed by the owner for their preferred aesthetics or more privacy. Finally, to prevent the houses from floating away or into one another, they are anchored to the lake's bed by steel mooring plates.

In general, for all the units, the lowest story, which is partly submerged, contains the bathrooms and bedrooms. The kitchen and dining space, with a covered veranda, are located on the raised ground floor. The upper floor is comprised of the main living area which is connected to a cantilevered open-terrace deck. Despite the similarity in appearance, there are multiple possible layouts for each unit. Additionally, pre-designed extension packages give options for further customization, such as sunrooms, verandas, awnings, and more, which can be easily attached to the skeleton frame. Due to the height difference between the jetty, water, and front door—on the ground floor—they are bridged by a boardwalk that circles around the home and slopes down to the water.

6.5 Dry Interior Construction

Dry construction refers to a building process that avoids wet tasks such as plastering and painting (Fig. 6.11). The installed panels are built to completion and need only to be clipped onto vertical or horizontal metal elements, which reduces labor costs (Webster 2022). The use of water commonly accompanies the construction lifecycle, ranging from the extraction of raw material to the demolition phase at the end of building's life cycle (Macieira and Mendonca 2016). Therefore, dry construction systems replace the need for moisture-retaining materials such as plaster as it uses partitions that can detach easily and therefore made for disassembly. Not only is this construction method cost-effective, but due to newly introduced materials and construction technologies, it offers an architect creative freedom since the surfaces involved are almost infinitely malleable (Macieira and Mendonca 2016).

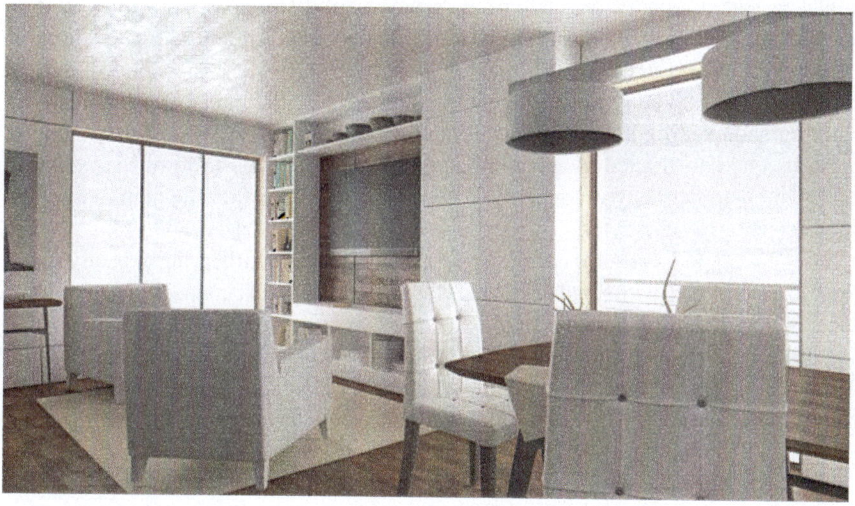

Fig. 6.11 Dry construction as seen here in the wall covering refers to a building process that avoids wet tasks such as plastering and painting

Dry construction strategies simplify flexibility and waste reduction (Macieira and Mendonca 2016). Fostering interior flexibility can increase the building's lifetime by extending its materials' life cycle and opportunity for the occupants to modify the home according to their evolving space needs and budget. In addition, components of dry construction are light weight, and easily portable to lower transportation and the manpower cost.

6.6 Design for Growth and Adaptability

Mobility and evolving space needs are a few of the challenges that occupants commonly must adapt to. As a result, design and construction methods need to be flexible to permit future growth and change. Prefabrication simplifies this by allowing adaptive customization, a result of a foreseeing change to meet users' needs (Gilmore and Pine 2014). This could be achieved for example through 3D computer modeling that allows users to custom design their homes and simulate visual walk-throughs before the construction of their dwelling begins (Larsen et al. 2019). Change should be exemplified in the form of disassembling or recycling building parts, rather than disposing them (Zairul 2021). These outstanding features support the circular economic framework through improved product quality and waste reduction (Zairul 2021).

Another benefit of prefabrication rests in their use as additions to existing homes. They offer advantages to cities, builders, and homeowners, including cost savings, potential for higher density, ease of construction, and less material waste. They can either be added adjacent to, or on top of an existing structure, but either method faces its own set of challenges (Fig. 6.12). A logistical consideration with additions is how to connect the new and the old, which differs between adjacent and vertical cases. For adjacent additions, a concern is to make sure the intersecting roofs of the addition and existing home shed water and snow properly. A sufficient foundation must also be constructed to accommodate the addition. This form of rear addition is less likely to be met with opposition because it does not alter the neighborhood's streetscape or the character.

For vertical additions, logistical concerns include the load-bearing capacity of the existing structure to withstand the added weight. Additionally, connections of utilities such as electricity, water, and heat may be more complicated for taller additions than adjacent one. When considering transportation and placement on site, it is essential that building materials are lightweight. Such materials include panels, steel frames, composite wood, and hybrid systems that can result in faster on-site construction leading to financial benefits (Hoyt 2020).

Another approach to maximize the advantages of standardization and customization is the mass production of customizable end products (Dollarhide 2021). This method is adaptable to the occupant's needs for interior changes over time. Under this system, homebuyers can easily alter their place of living after occupancy,

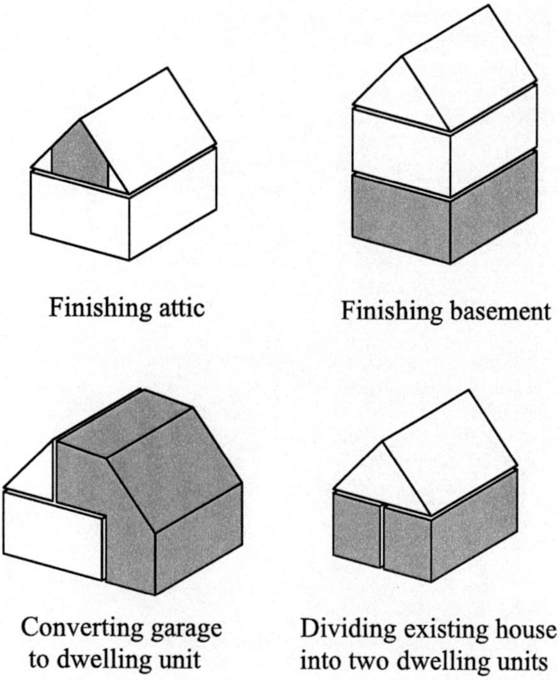

Fig. 6.12 Add-on and add-in methods in dwellings

through the selection of alternative end products which are mass-produced to work with a universal plug-in system. In addition to the economic advantages that mass customized communities offer, they can be designed for disassembly and recycling thereby being more sustainable.

6.7 The Pod Home

The Pod Home is a design concept for the construction of affordable, adaptable, and sustainable housing using prefabrication strategies. The design goal is to plan for the evolution of neighborhoods, through the addition of new homes and easy modification of existing ones with minimal cost and environmental impact (Fig. 6.13). The Pod Home design enables a variety of configurations to create diverse, high-density neighborhoods. Structures can be connected, and the number of floors increased to improve density. This strategy allows for the creation of single-family structures, duplexes, and triplexes. A selection of roof types, add-ons, and projections can be used to add diversity and variation to the built environment.

The design is named for its use of prefabricated pods, containing the unit's main utility functions to simplify construction. Pods are manufactured and shipped to the site, ready for installation. They are composed of a rigid outer structure with a flexible internal arrangement, depending on the unit's specifications (Fig. 6.14). Pods

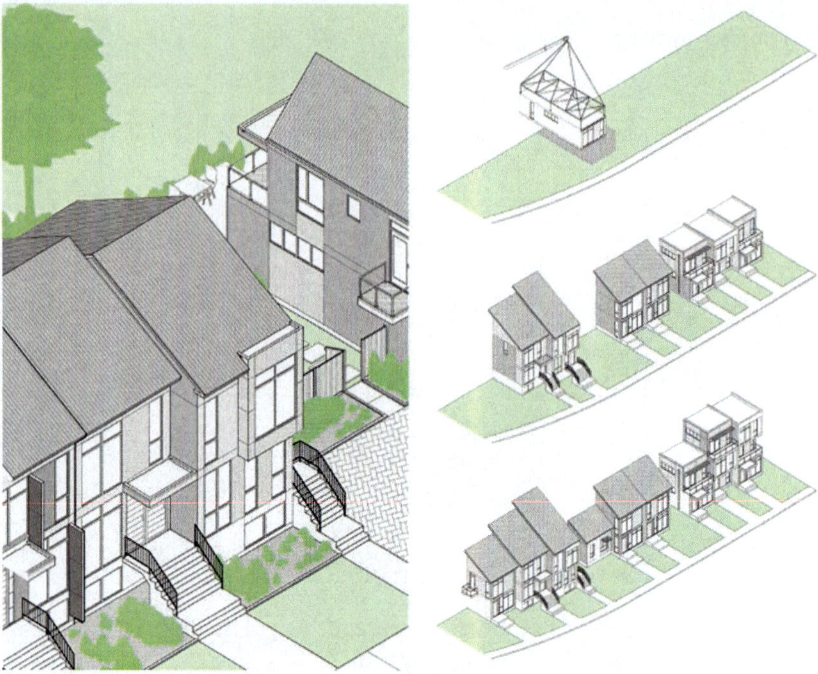

Fig. 6.13 One of the goals of the Pod Home is to plan for neighborhood evolution, through additions to and easy modifications of existing structures with minimal cost and environmental impact

can contain any combination of six main functions: mechanical space, storage space, bathroom, kitchen, laundry, and stairs. They are available in three different sizes: 0.9 by 2.4 m (3 by 8 feet) containing two to three functions, 1.5 by 2.4 m (5 by 8 feet) containing four to five functions, and 1.5 by 3.6 m (5 by 12 feet) containing four to six functions (Fig. 6.15). For example, in a smaller one-story unit, a single medium pod could accommodate the kitchen, bathroom, laundry, and mechanical space for the household (Fig. 6.16). Otherwise, a larger, multi-story unit may have one large pod containing the kitchen, bathroom, laundry, mechanical space, and stairs on the main level. With an additional pod containing a second bathroom, storage, or mechanical space on the second floor. This system allows for a high degree of flexibility based on the needs of each household while enabling homes to be efficiently constructed.

The internal layout of each unit is highly flexible. A variety of configurations can be created depending on the placement of pods within units, which can be located strategically to organize space within the home. The standard model has a footprint of 4.3 by 11 m (14 by 32 feet) and the smaller economy model measures 3.6 by 9.7 m (12 by 32 feet) (Fig. 6.17). Pods are designed to be easily placed anywhere within the structure. The grouping of wet functions and appliances reduces the amount of plumbing installation and mechanical space required; reducing construction time, costs, and materials needed. Once installed in the unit, the pods are connected to water

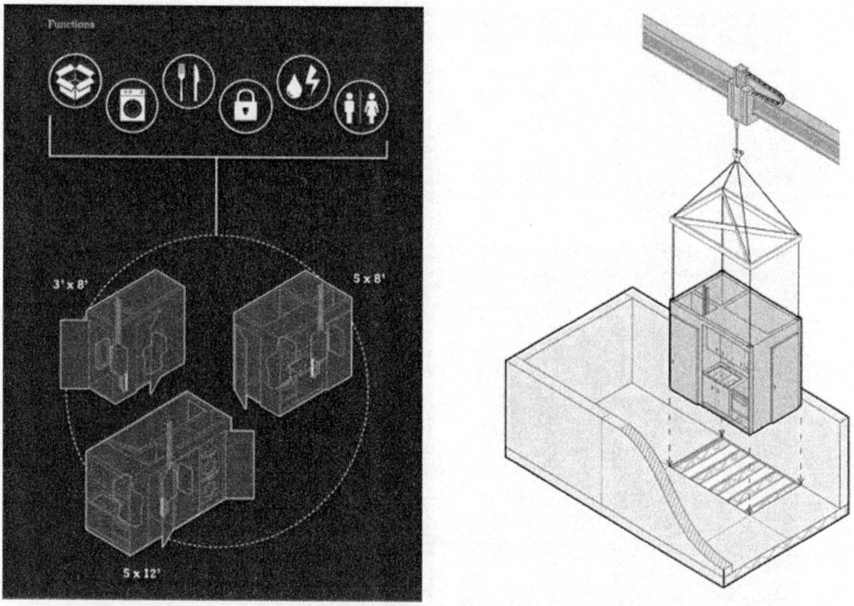

Fig. 6.14 Pods are manufactured and shipped to the site ready for installation. They are composed of a rigid outer structure with a flexible internal arrangement, depending on the unit's specifications

and power systems through chasers in the structure below. Chasers run lengthwise along a structural grid aligned with the floor joists and girders, allowing pods to be located anywhere on the grid.

Prefabricated partitions are designed for both the interior and exterior of the home to further simplify the construction of the Pod Home. Various types of walls are offered to accommodate different internal configurations and façade appearances. Exterior walls are offered with a selection of window and door placements. A menu of façade options provides an opportunity for personalization with a variety of doors, windows, balconies, cladding, and shading devices to be placed on the unit (Fig. 6.18).

The Pod Home design demonstrates how prefabrication can be used in residential developments to streamline the construction process without sacrificing any of the customization and flexibility offered by on-site construction (Fig. 6.19). The solution comes from the prefabricated pods, grouping together the units' main functions according to the demands of each household.

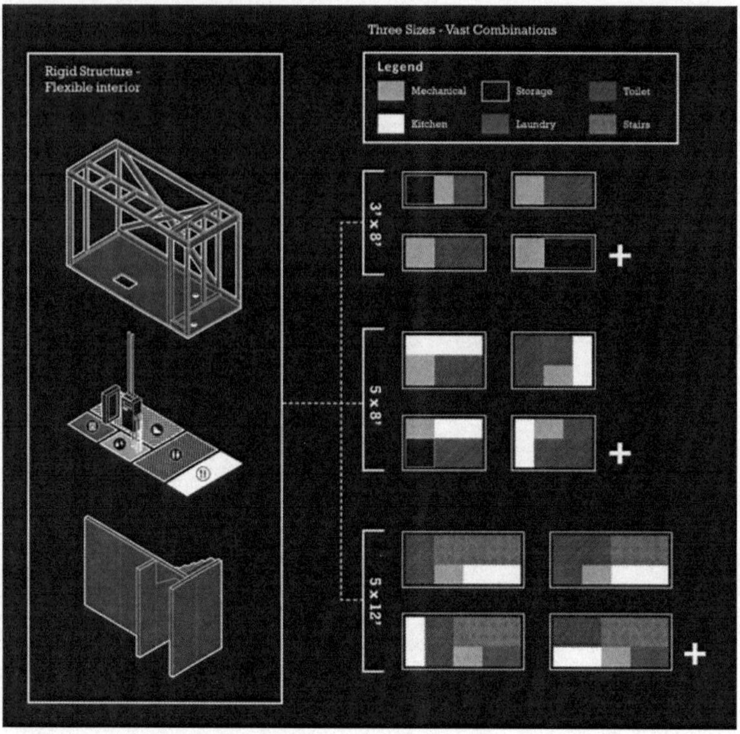

Fig. 6.15 Pods come in different sizes and can contain any combination of six main functions: mechanical space, storage space, bathroom, kitchen, laundry, and stairs

6.8 Final Thoughts

It is likely that demand for natural resources will increase and force innovative designs and construction methods to reduce consumption. The future of prefabrication and efficient construction will unfold on par with building information modeling (BIM) software (Zairul 2021). The software's ability to visualize and track various components in three dimensions makes it uniquely capable of merging with factory production. A single model could be used to extract material volumes, produce shop drawings for component fabrication, and test assembly scenarios. Findings also indicate that BIM is an efficient and effective method for measuring carbon emissions from the construction of new buildings (Hao et al. 2020). A second potential offered by BIM has emerged in the generation of part identification systems, enabling the direct transfer of coded information at the time of fabrication through barcodes or radio-frequency identification. This identification system could be critical in the efficiency of building's assembly and disassembly.

The major hurdles to overcome are increasing construction costs and as a result lack of affordable housing. In the future, the challenges of rising costs could be offset

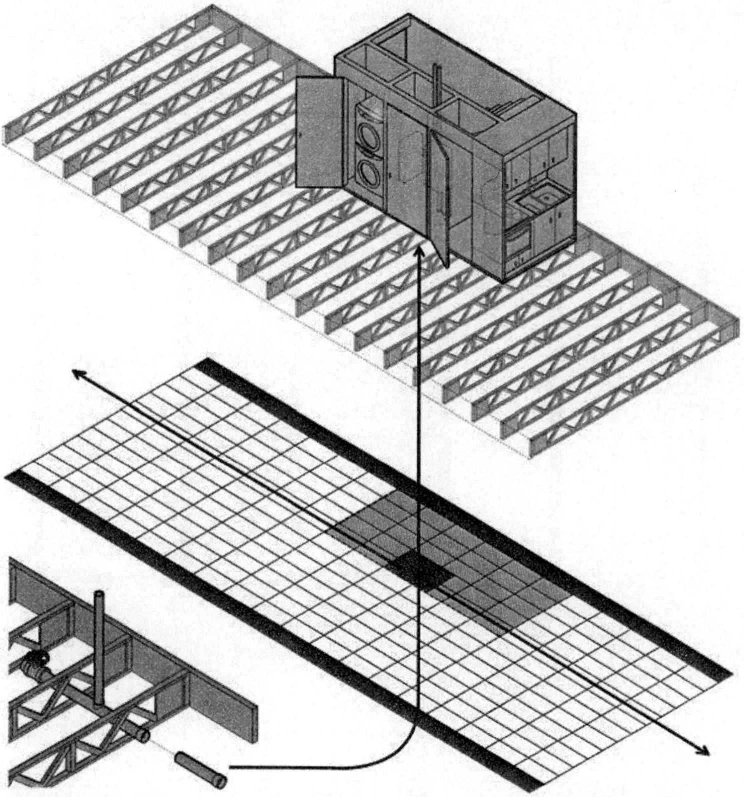

Fig. 6.16 Pods are designed to be easily placed anywhere within the structure. They are connected to the home's main utility systems that run in a chase along the dwelling's long dimension

by new engineering and construction prototypes which will unlock productivity gains and save money (Bertram et al. 2019). Furthermore, prefabrication methods such as modular construction are still very much outliers from the mainstream construction industry. Even so, there are plausible signs of what could be an attested broad-scale disruption in the making (Bertram et al. 2019). Prefabrication is already attracting competitors and offers an opportunity to revolutionize the real estate and construction sectors.

Questions for a Follow-Up Discussion

1. What are the key aspects that prompted a need for resource conservation?
2. What are the main types of prefabricated dwellings and when will you select each?
3. What are the advantages and disadvantages of dry interior construction?

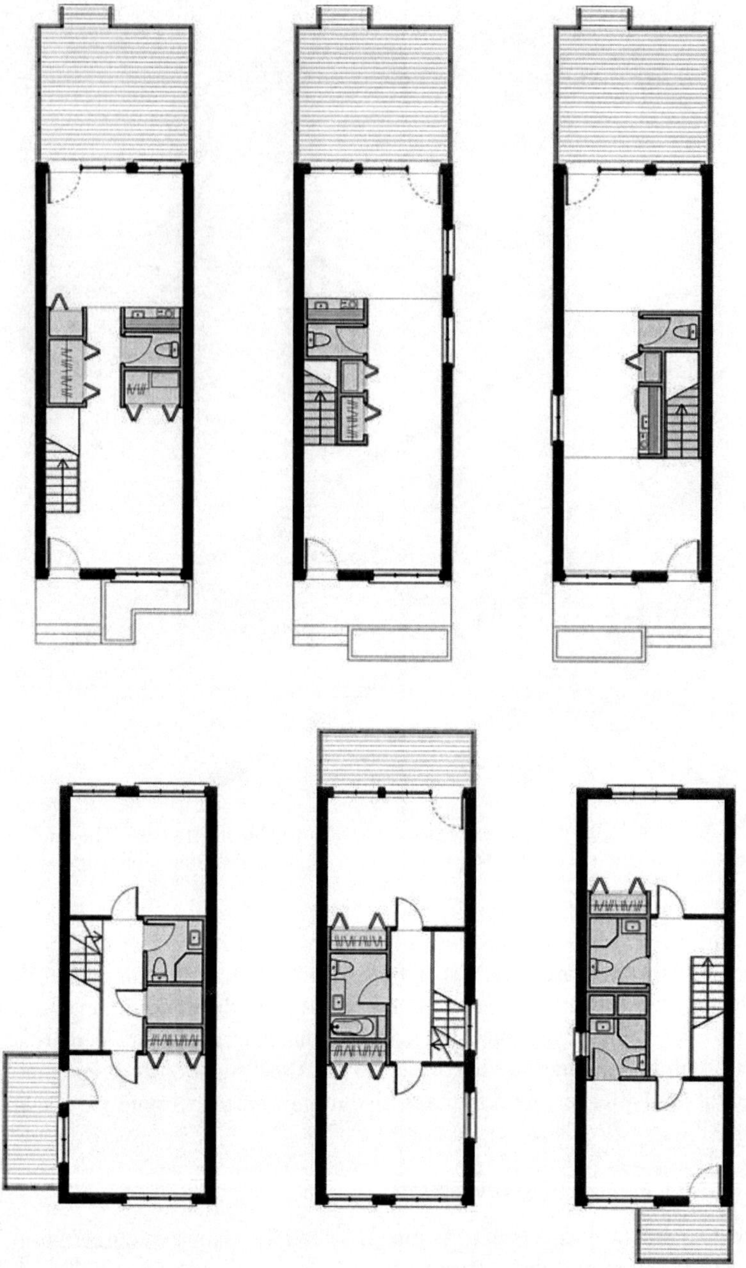

Fig. 6.17 Floor plans of a lower floor (top) and upper floor (bottom) showing variety of pods

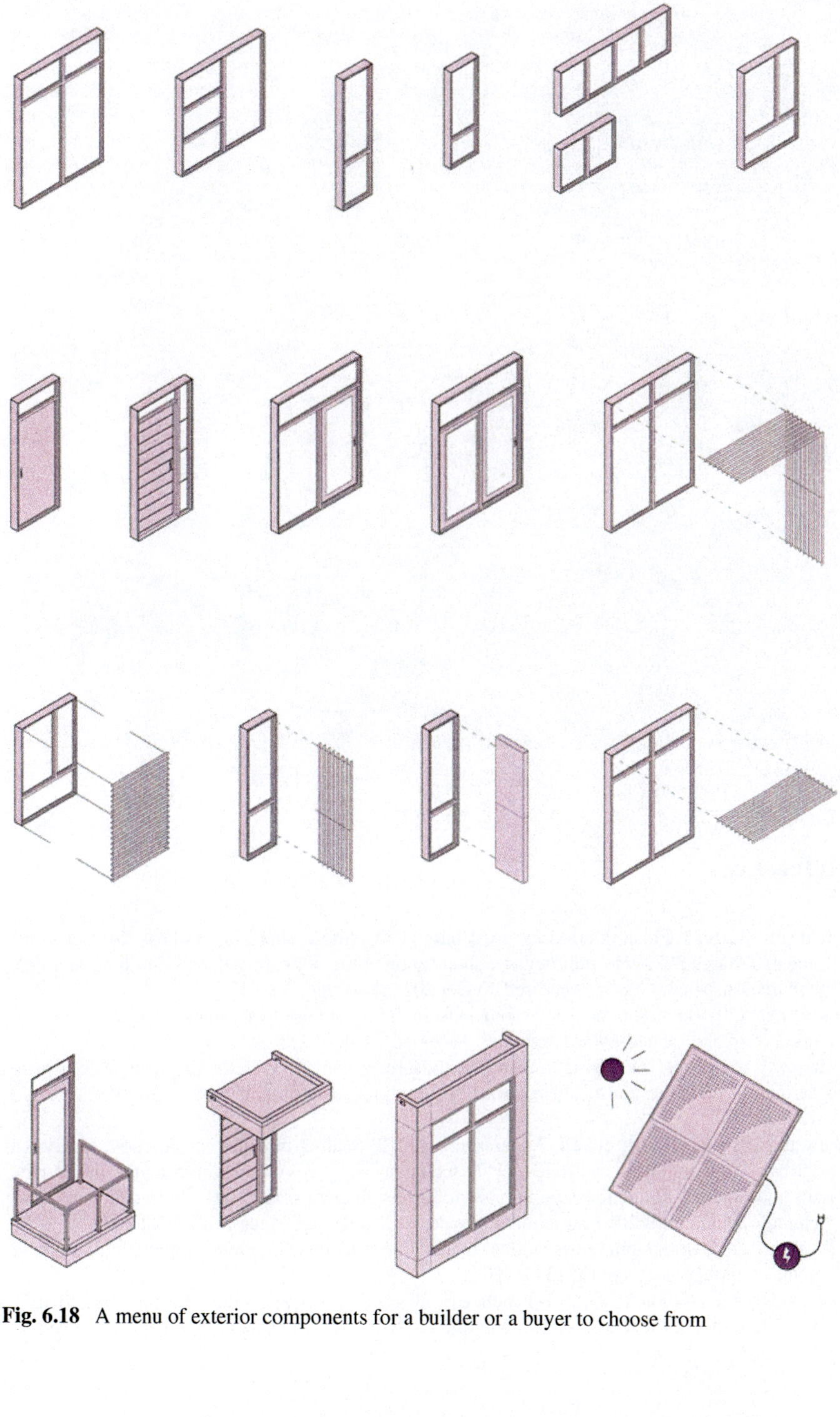

Fig. 6.18 A menu of exterior components for a builder or a buyer to choose from

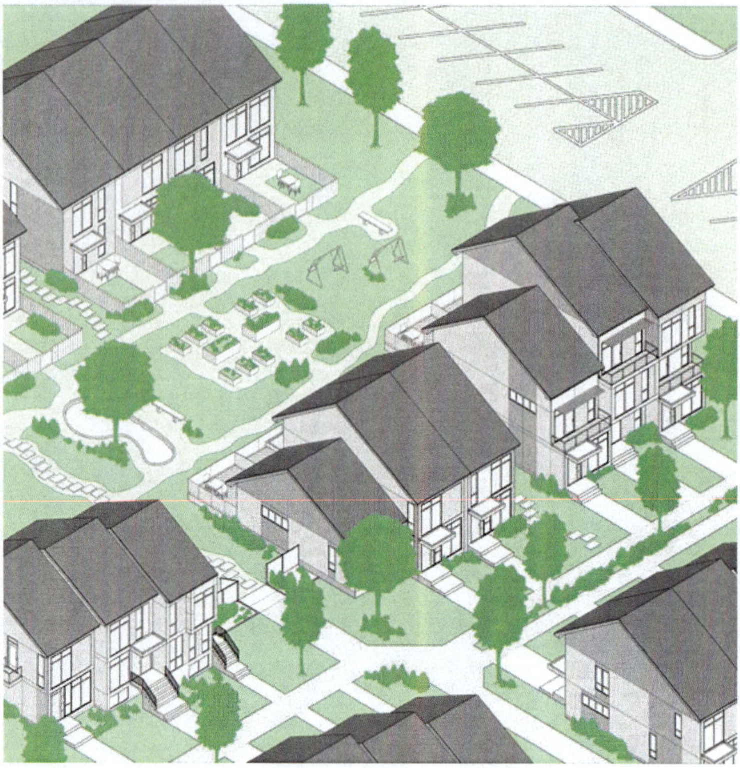

Fig. 6.19 A community made up of Pod Homes

References

ArchDaily (2011) Floating Houses in IJburg / Architectenbureau Marlies Rohmer. Last updated March 20, 2011. http://www.archdaily.com/120238/floating-houses-in-ijburg-architectenbureau-marlies-rohmer/. Accessed 13 Oct 2022

Bathsystem (2019) Prefabricated bathroom pods. In: Bathsystems https://www.bathsystem.com/en/products/prefabricated-bathroom-pods/. Accessed 13 Oct 2022

Behroozy A, Keegel TG (2014) Wet-work exposure: a main risk factor for occupational hand dermatitis. In: Safety and health at work. https://pubmed.ncbi.nlm.nih.gov/25516808/. Accessed 13 Oct 2022

Bertram N, Blanco JL, Mischke J, Andersson N (2019) Scaling modular construction. In: Global infrastructure initiative by McKinsey & Company. https://www.mckinsey.com/~/media/mckinsey/business%20functions/operations/our%20insights/voices%20on%20infrastructure%20scaling%20modular%20construction/gii-voices-sept-2019.pdf. Accessed 13 Oct 2022

Cox W (2022) Demographia international housing affordability. In: Demographia. http://demographia.com/dhi.pdf. Accessed 13 Oct 2022

Derananian AR, Donlon MK (2019) Panelized wall systems – joint detailing for Success – IIBEC. In: IIBEC. https://iibec.org/wp-content/uploads/2019-cts-derananian-donlon.pdf. Accessed 13 Oct 2022

Dollarhide M (2021) Mass customization. In: Investopedia. https://www.investopedia.com/terms/m/masscustomization.asp. Accessed 13 Oct 2022

Editors of Encyclopaedia BT (2022) Prefabrication. In: Encyclopædia Britannica. https://www.britannica.com/technology/prefabrication. Accessed 13 Oct 2022

European Environment Agency (EEA) (2020) Why is resource efficiency important? https://www.eea.europa.eu/themes/waste/resource-efficiency/why-is-resource-efficiency-important. Accessed 13 Oct 2022

Gilmore JH, Joseph Pine B (2014) The four faces of mass customization. In: Harvard Business Review. https://hbr.org/1997/01/the-four-faces-of-mass-customization. Accessed 13 Oct 2022

Hankin A (2022) 5 reasons millennials aren't buying homes. In: Investopedia. https://www.investopedia.com/news/real-reasons-millennials-arent-buying-homes/. Accessed 13 Oct 2022

Hao JL, Cheng B, Lu W, Xu J, Wang J, Bu W, Guo Z (2020) Carbon emission reduction in prefabrication construction during materialization stage: a BIM-based life-cycle assessment approach. In: Science of the Total Environment. https://www.sciencedirect.com/science/article/pii/S0048969720313838?via%3Dihub. Accessed 13 Oct 2022

Howe AS, Ishii I, Yoshida T (1999) Kit-of-parts: a review of object-oriented construction techniques. In: ISARC Proceedings. https://www.iaarc.org/publications/proceedings_of_the_16th_isarc/kitofpartsa_review_of_objectoriented_construction_techniques.html. Accessed 13 Oct 2022

Hoyt H (2020) Harvard Joint Center for Housing Studies|Joint Center ... In: Joint center of housing studies. https://www.jchs.harvard.edu/sites/default/files/media/imp/harvard_jchs_gramlich_design_and_construction_strategies_multifamily_hoyt_2020_3.pdf. Accessed 13 Oct 2022

Kashef M (2009) Sense of community and residential space ... – researchgate. In: Research Gate. https://www.researchgate.net/publication/41668003_Sense_of_Community_and_Residential_Space_Contextualizing_New_Urbanism_within_a_Broader_Theoretical_Framework. Accessed 13 Oct 2022

Larsen MSS, Lindhard SM, Brunoe TD, Nielsen K, Larsen JK (2019) Mass customization in the House Building Industry: literature review and research directions. In: Frontiers. https://doi.org/10.3389/fbuil.2019.00115. Accessed 13 Oct 2022

Lee H (2021) Are millennials leaving cities? Yes, but young adults are not. In: Are Millennials leaving cities? Yes, but young adults are not | Joint Center for Housing Studies. https://www.jchs.harvard.edu/blog/are-millennials-leaving-cities-yes-young-adults-are-not. Accessed 13 Oct 2022

Lessard G (2018) Principles and best practices – arpent. In: Larpent. https://www.larpent.ca/wp-content/uploads/2019/04/AccessoryDwellingUnitsPrinciplesAndBestPractices.pdf. Accessed 13 Oct 2022

Macieira M, Mendonca P (2016) (PDF) building rehabilitation with dry and wet systems ... In: Research Gate. https://www.researchgate.net/publication/305787506_Building_Rehabilitation_with_Dry_and_Wet_Systems_-_Embodied_Water_Comparison. Accessed 13 Oct 2022

Ncube A, Matsika R, Mangori L, Ulgiati S (2021) Moving towards resource efficiency and circular economy in the brick manufacturing sector in Zimbabwe. In: Science Direct. https://doi.org/10.1016/j.jclepro.2020.125238. Accessed 13 Oct 2022

Office of Mobile Design Corp Eof (2021) Nomad live & work-electric-truck. In: OMD. http://www.designmobile.com/#/nomad-live-work-electric-truck/. Accessed 13 Oct 2022

People's Architecture Office Eof (2013) In: Courtyard house plugin. People's Architecture Office. http://www.peoples-architecture.com/pao/en/project-detail/3. Accessed 13 Oct 2022

People's Architecture Office Eof (2018) SHANGWEI URBAN VILLAGE PLUGIN HOUSES. In: Shangwei Urban Village Plugin Houses/People's Architecture Office. http://www.peoples-architecture.com/pao/en/project-detail/27. Accessed 13 Oct 2022

People's Architecture Office Eof (2022) Our vision. In: Plugin House. https://www.paopluginhouse.com/our-vision. Accessed 13 Oct 2022

Stinson L (2018) Prefab 'plug-in houses' help revitalize dilapidated buildings. In: Curbed. https://archive.curbed.com/2018/12/18/18145769/prefab-plugin-houses-help-revitalize-dilapidated-buildings. Accessed 13 Oct 2022

TechPrefab (2021) Innovative construction: panelized vs modular. In: Tech Prefab. https://www.techprefab.com/post/panelized-vs-modular. Accessed 13 Oct 2022

Webster M (2022) Wet work definition & meaning. In: Merriam-Webster. https://www.merriam-webster.com/dictionary/wet%20work. Accessed 13 Oct 2022

Zairul M (2021) The recent trends on prefabricated buildings with circular economy (CE) approach. In: Cleaner engineering and technology. https://www.sciencedirect.com/science/article/pii/S2666790821001993. Accessed 13 Oct 2022

Chapter 7
Utilities Systems for Sustainability

Abstract In conventional residential construction, it is often difficult to locate and access utility conduits once installed on walls and floors. As a result, upgrades and repairs are complicated, time consuming, costly, and require unsustainable demolition. Alternatively, a single passage for all utilities can be created using an open-web floor joist system allowing for greater ease in upgrading and repairing systems. In addition, as part of the utilities an enhanced water management system will contribute to sustainability through water harvesting and recycling. Furthermore, landscaping techniques such as xeriscaping can reduce freshwater consumption. This chapter discusses the installation and sustainability of varying utility systems and plumbing, electricity heating, ventilation, and air conditioning (HVAC), and Wi-Fi networks in homes.

Keywords Black water · CIPP · Cisterns · Ductwork · Gypsum boards · Grey water · HVAC systems · Open joist systems · Water efficient technology · Water harvesting · Wi-fi · Vertical stacking · Xeriscaping

7.1 Evolution and Challenge of Utilities' Insulations

The genesis of plumbing systems dates back to ancient societies. In ancient Rome even before the turn of the century, fresh water was brought to the city using aqueducts and was subsequently collected in tanks and distributed through tunnels to baths, fountains, and public toilets (Delile et al. 2017). At this time, it was common for wealthy Roman families to have private sewage systems connected to their households. A later innovation, lead pipes, resulted in poisoning, causing numerous deaths (Delile et al. 2017) (Fig. 7.1).

Following the Industrial Revolution, plumbing systems made of cast iron and copper pipes were realized and installed exposed in the dwelling's interior (Hunter 1941) (Fig. 7.2). When electricity was invented in the 1700s and later brought into the private home in the late nineteenth century, electricians began concealing wiring in studs and floorboards which required notching. This work included tedious chiseling of wood and plasterwork and the realization that wires and pipes simply could not

A. Friedman, *Fundamentals of Innovative Sustainable Homes Design and Construction*, The Urban Book Series, https://doi.org/10.1007/978-3-031-35368-0_7

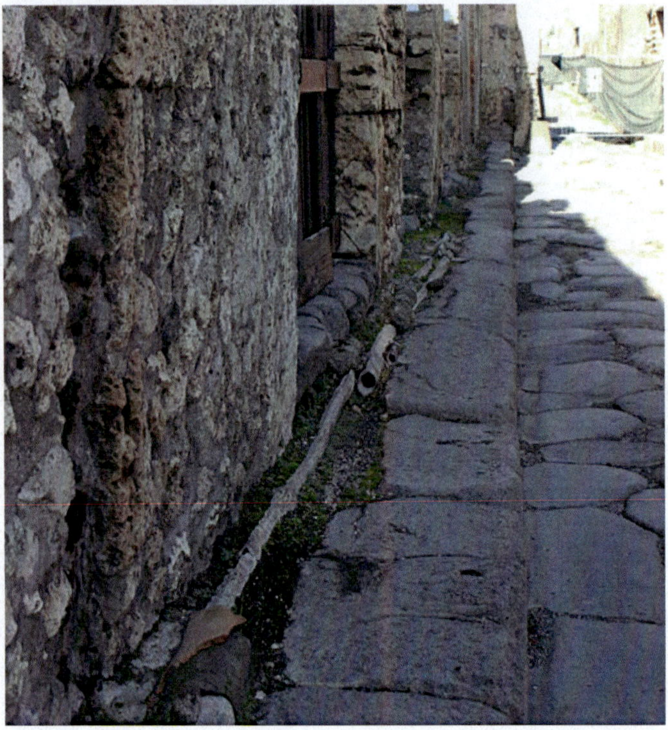

Fig. 7.1 Lead pipes along a road in Pompeii

run between certain points. At times it resulted in leaving wires exposed to create a safety hazard. With evolving regulations and better practices, plumbing and wiring were enclosed in the structure to create safe and tidy spaces.

In the late twentieth century, the introduction of Polyvinyl Chloride (PVC) replaced cast iron and copper pipes to simplify plumbing work. It was light and easily workable where each segment connected to one another using modular prefabricated fittings with adhesives to reduce installation complexity and time (Kantamneni et al. 2021). Finally, in the 1990s, the flexible Poly Coiled Tubing also known as BOWPEX pipes, made out of cross-linked polyethylene were introduced to further simplify work (Rubeiz 2004) (Fig. 7.3).

Other relevant innovations in the latter half of the twentieth century in this context include those of construction tool development. The use of power tools such as circular saws, jigsaws, power drills, sanders, grinders, and routers simplified many tasks including cutting, opening, and closing of walls and ceilings (States and Mcquaid 2022). Another significant innovation was the introduction of drywall that replaced the time-consuming plaster work with simplified construction work (Triforce 2015).

Despite these advancements, contemporary installation of utilities is still complicated and laborious (AHAM 2021). The running of wires, ducts, and pipes through

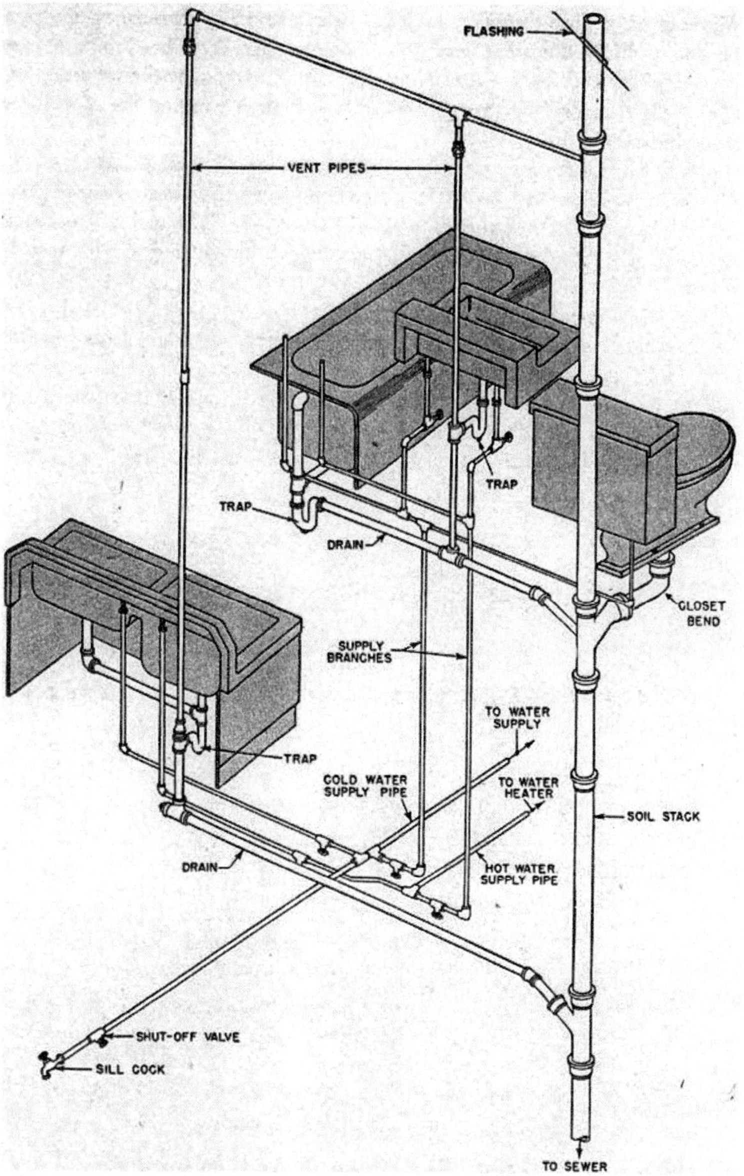

Fig. 7.2 A plumbing system made of cast iron and copper pipes

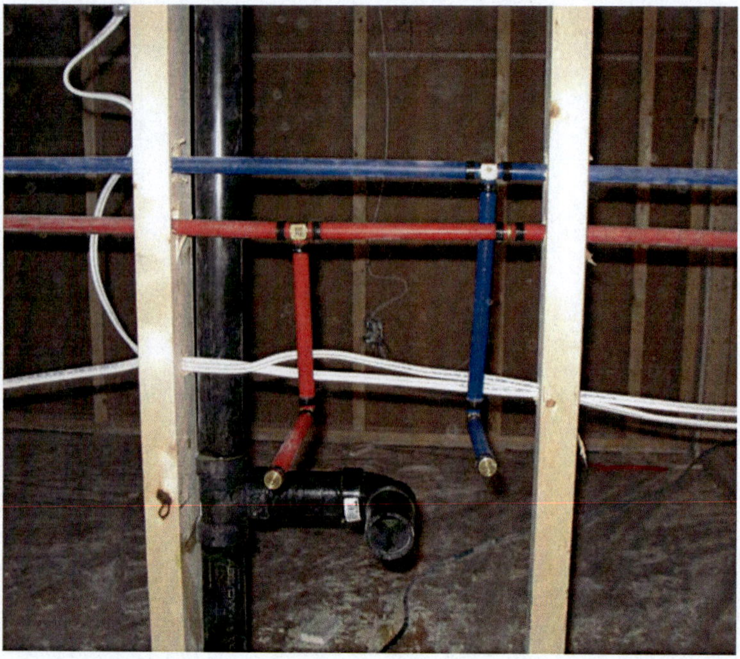

Fig. 7.3 Polyvinyl chloride (PVC) pipe for drainage and the flexible poly coiled tubing also known as BOWPEX for hot and cold-water pipes

partitions, floors, or ceilings requires knowledge of the house's structure and keeping track of where wiring is located (Fig. 7.4). Once a building is complete, adding utilities requires more costly labor, and materials, making the entire process cumbersome. Furthermore, not only does drywall often need to be demolished to access utilities, but other sections of the wall and ceiling must be torn down as well. Lack of documentation of original installation also makes maintenance extremely challenging especially when professionals other than the original are engaged (CAA 2022).

7.2 Plumbing and Electrical Systems

Prior to exploring new cost saving methods for plumbing and electrical systems let us begin by looking at conventional insulation methods of utility systems. Once a dwelling's rough structure has been constructed, installation of the home's plumbing begins. Three main components make up this system: water supply, water and waste removal, and the fixtures through which the water is moved. There is often a "main stack" which includes a vent pipe that runs in the center of the home and exits through the roof (Fig. 7.5). All the stacks eventually connect at the home's lowest point to be excreted via the house sewer to the municipal collecting system (Law 2022).

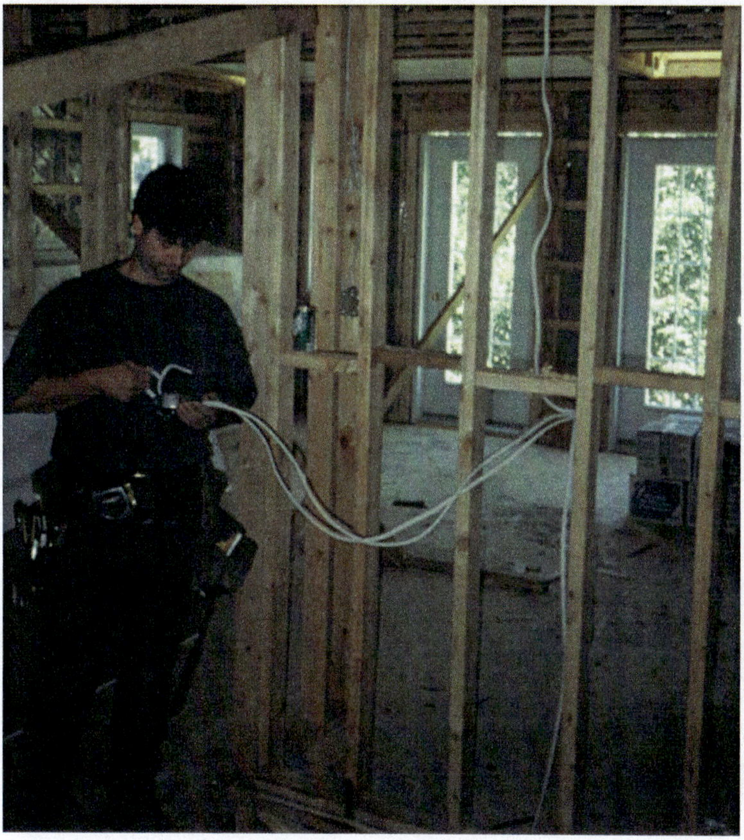

Fig. 7.4 An electrician installs electrical wiring in a home

Current plumbing designs and installations have a scattered nature and are often installed with little coordination with other utility systems. Due to the contemporary tendency toward circular building methods, home builders are changing the way plumbing systems are installed to have them more accessible, and more easily replaced and recycled. This can be achieved through better organization and strategic placement of the systems. In addition, accurate sizing of the system and allocating the appropriate pipes for their optimal function will also lead to cost savings.

A significant contributor to the simplification of plumbing system installation can be attributed to the introduction of *open web joist floor system* (Triforce 2020) (Fig. 7.6). Rather than suspending mechanical systems below the floor joists, an open web joist facilitates their installation within the floor (Triforce 2020). A single passage could be used to centralize the plumbing system, where there will be a horizontal canal in the floor where pipes, electric wire, and air conditioning ducts will be placed. The passage which is referred to here as a "chase" will allow utilities

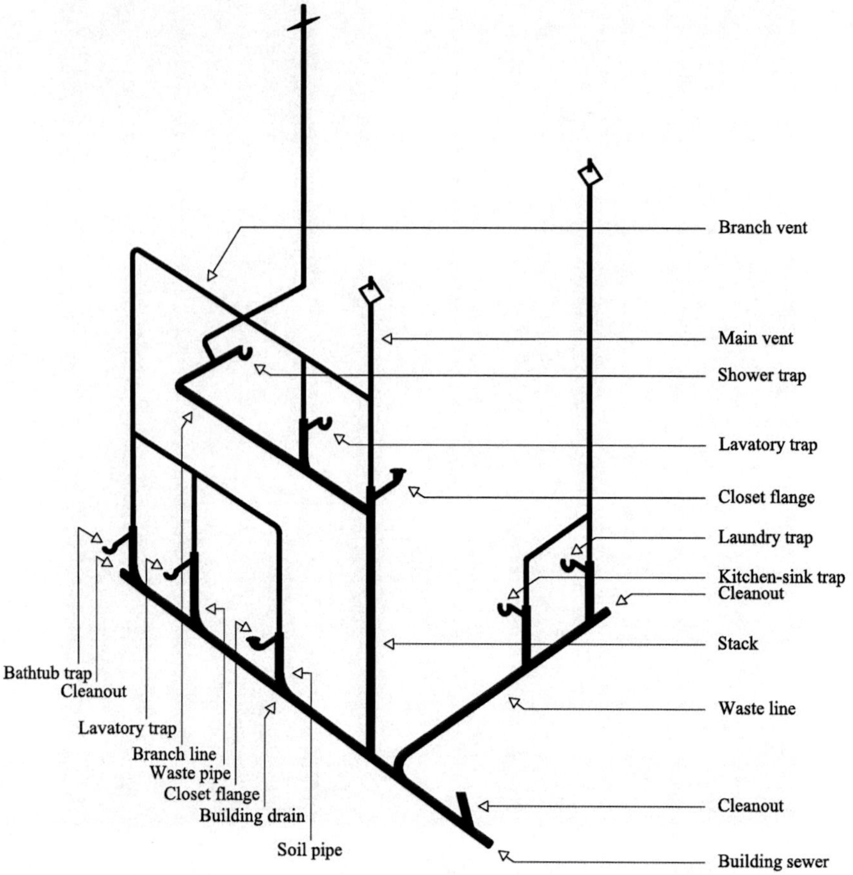

Fig. 7.5 A contemporary home's plumbing system

to easily pass through lower and upper floors for greater ease in locating, changing, and repairing them (Fig. 7.7).

Another time and cost saving design strategy is the stacking of a dwelling's wet functions (Fig. 7.8). Locating bathrooms and kitchens next to each other and on top of one another would make the installation of plumbing more efficient through centralization. Clustering the wet rooms will shorten the length of the conduits required to reach them (Clifton 2020; Line 2019). When two wet functions are located back-to-back, such as the kitchen and bathroom, the wall between them can be designed and build to contain the required pipes which will enable easier future access to said pipes (Clifton 2020). Another consideration should be to ensure that the same common wall should be wide enough to permit the passing of several drain and water pipes (Clifton 2020).

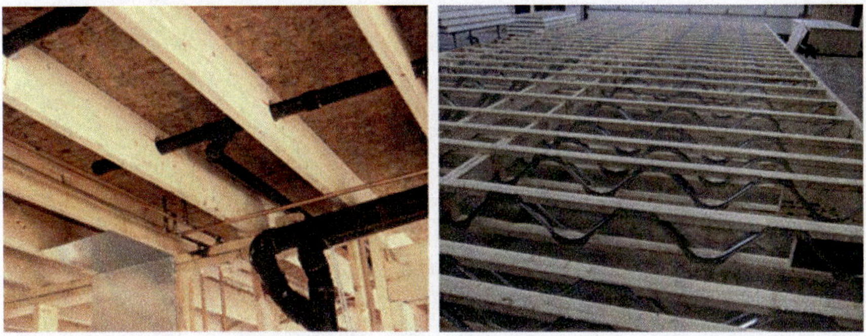

Fig. 7.6 The introduction of open web joist floor system (right) simplified the installation of utility lines compared with solid sawn joists (left)

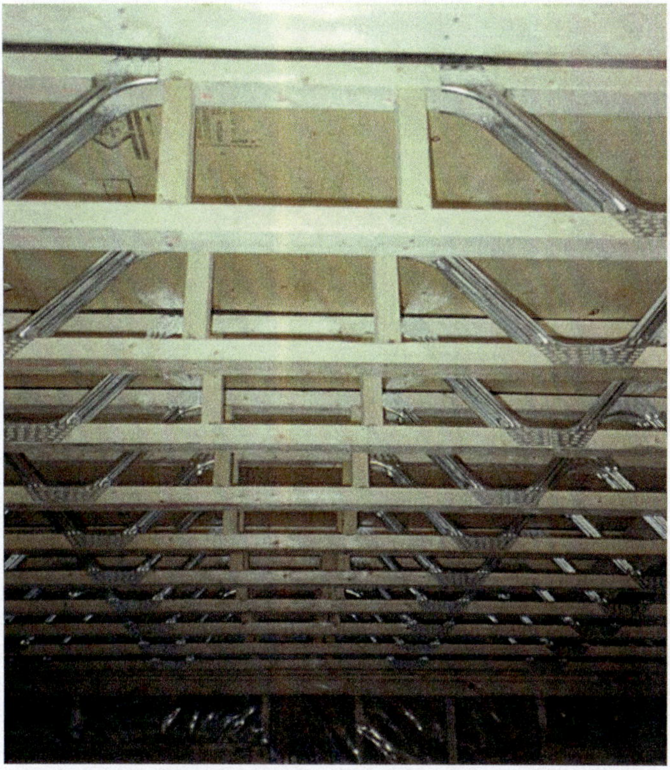

Fig. 7.7 The passage in the open web joist floor system (referred to as chase) allows utilities to easily pass through lower and upper floors for greater ease in locating, changing, and repairing them

Fig. 7.8 A significant time and cost saving design strategy is the stacking of the dwelling's wet functions

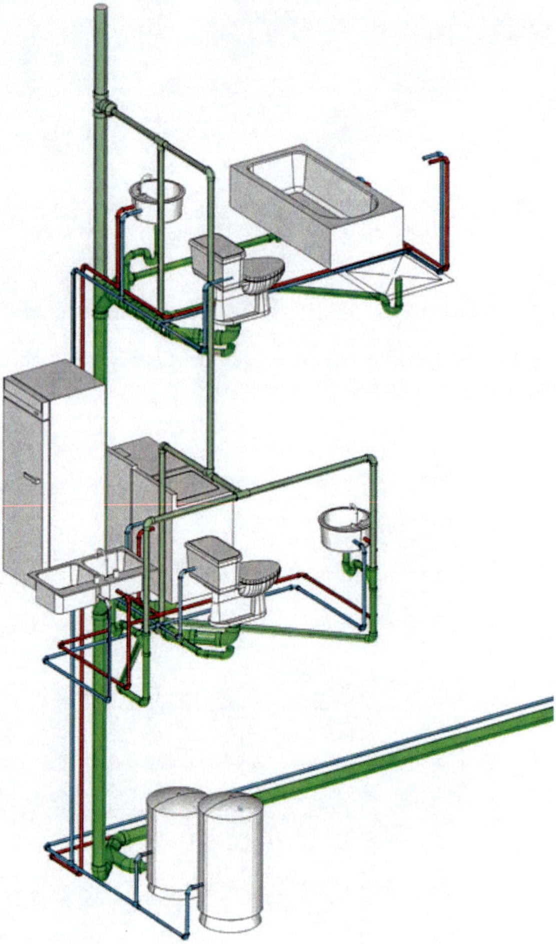

High-rise residential and multifamily dwellings pose many unique challenges given their size and orientation. Solutions include various manners of efficient installation and maintenance of plumbing. Cured-in-place pipes (CIPP) is a form of vertical pipe repair that eliminates the need to open walls to access damaged pipelines. CIPP is a restoration process that involves the insertion of a liner tube and the use of resin (Ji et al. 2020). Once in place, a curing process is initiated, which provides a tight fit against the existing pipe. Besides a quick installation process, CIPP is more durable and lasts longer compared to traditional pipes. Repairing and relining pipes, instead of completely replacing them, can prevent a need for them to be maintained for over 50 years offering major cost savings over the course of the product's life (Britnell 2015).

Another critical domestic utility system whose installation requires modification is electrical. Made up of wires, switches, and plugs the system carries electricity

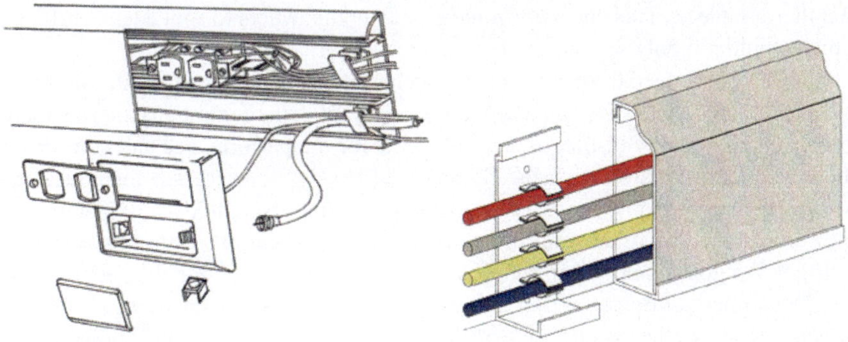

Fig. 7.9 By enclosing wires within a floor base molding they can be routed unobtrusively throughout residences

from a source, usually a nearby transformer, to an outlet in the home. The system's main purpose is to satisfy the user's electricity needs while complying with safety standards. Most electrical systems exist in walls however, to avoid removing a wall's covering to access these systems, a viable design would be to install wiring along floor baseboards. This would better organize the wiring, making it more accessible without it being a safety hazard. By enclosing wires within base moldings, they can be routed unobtrusively throughout residences (Fig. 7.9). This way, access points can be chosen and adjusted by users, simply by sliding the outlet along the molding. If cables require replacing or repair, they can be accessed without the cutting and refinishing of entire walls. Most baseboards are made of solid wood. But composite materials such as medium-density fiberboard (MDF) are sometimes used, since they're less expensive and resistant to mold, especially in the case where water intrusion is an issue (Canada NR 2014).

7.3 Water-Saving, Harvesting, and Recycling Systems

Due to global warming, water scarcity has become acute in many of the world's regions leading to water conservation initiatives. One of the best ways to ensure a reduction in water consumption in the home is to incorporate water efficient technologies in rooms such as bathrooms and kitchens. Within the bathroom, efficient flushing and showerhead technology are becoming common. These technologies include dual-flush, high-efficiency, pressure-assisted, and composting toilet waste (US Department of Energy 2010). The low flush water closet for example is a two-lever system where smaller quantities of water are used for liquids and greater quantities for solids. Redirecting sink gray water into the toilet tank is also an effective reuse method (Fig. 7.10). Similarly, regular checks for toilet leaks reduce the probability of water waste (US Department of Energy 2010). This can be done by placing food coloring in the tank. If color appears in the bowl without flushing, that indicates that there is a leak. This checkup would enable homeowners to spot a leak early on, which would not only save water, but reduce monthly utility bills.

Showering is one of the main contributors to water consumption, and shower facilities are directly related to water and energy saving. With the continuous improvement of consumers' quality of life, the requirements for the performance and comfort of showering facilities are increasing (Zhang et al. 2021). Showerheads manufactured before 1995 require 50% more water than new and improved models. Replacing old showerheads with the new and improved models, particularly with a flow rate of less than 2.5 gpm (gallons per minute), can lessen water usage for showers by 50% (US Department of Energy 2010). The most common low-flow showerheads consist of aerating showerheads and streamline-flow showerheads. Aerating showerheads combine air with water to form an even full shower spray, whereas streamline-flow showerheads form individual streams of water and create less steam.

The next major "wet room" is the kitchen. Selecting water efficient home appliances is key, as certain models consume more water than others during their operation. Dishwashers and washing machines need to be researched before purchase, not only for their energy efficiency but also for their water consumption. For laundry appliances, front loading machines are known to be more water and energy efficient than top loading machines. Sourcing water efficient appliances can be identified through ENERGY STAR certifications labels. Washing machines that meet the ENERGY STAR criteria require 30% less water and consume half as much energy as conventional washers (US Department of Energy 2010). A washing machine that does not qualify for an ENERGY STAR rating can use 64.3 L (17 gallons) of water with every load of laundry (US Department of Energy 2010). This equates to the same

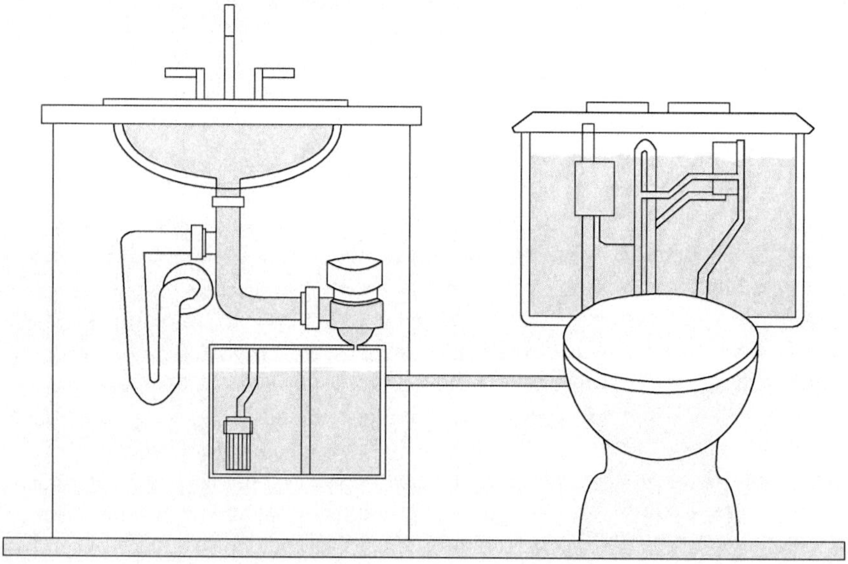

Fig. 7.10 Redirecting gray water from sinks into the toilet tank is an effective reuse method

amount of water used in a typical shower. Furthermore, dishwashers that qualify for the ENERGY STAR label use 18% less water and 10% less energy than conventional machines (US Department of Energy 2010).

Water harvesting is an ancient practice that has allowed communities to exist in semiarid and arid areas where natural sources of fresh water were limited. The process refers to the collection of rainstorm-generated runoff from a catchment to provide water for human use (Fig. 7.11). The water collected can either be utilized immediately or be stored in aboveground ponds or in subsurface reservoirs, such as cisterns (Hillel 2005). The collected water can be used for outdoor watering purposes or indoor water usage depending on one's location. Fresh water may also be supplemented by rainwater collected using a cistern for the rainwater harvesting system (Hillel 2005). Other typical uses of harvested rainwater include landscape irrigation, washing, and toilet flushing (US Department of Energy 2022). The harvesting system can be set up beside the home with the relevant system components and sizing, tailored to the available space and monthly rainwater available in the region where the home is located. In theory, this water could also be used for drinking if stored and treated since it is usually cleaner than grey water. However, pollutants such as dust, nitrogen, carbon dioxide, and sulfur dioxide, can form acid rain making the collected rainwater unusable. In any case, on-site collection strategies minimize stress exerted on storm sewers during periods of heavy rainfall. In municipalities that combine storm and sewer lines, rainwater can be used to propel and dilute black water which is an added benefit.

Besides rainwater, the home produces different types of water, such as black and grey by recycling this water, which will further reduce water waste. The recycling of grey water refers to the treatment and subsequent reuse of wastewater from sinks, dishwashers, showers, hand basins, baths, and washing machines (EPA 2022). The water is referred to as "grey" as it contains fats, oils, harmful chemicals, bleaches, and germs that can affect human health (EPA 2022). Untreated greywater can have impacts on water quality and public health due to the high bacterial concentration, nutrient discharge, biological oxygen demand, and high saline load (EPA 2022).

Greywater discharge can also be harmful to the natural environment it gets released into, because plumes may remain on the surface of aquatic environments and accumulate pollutants. The excessive nutrients from grey water can contribute to the overgrowth of algae, which can be problematic for vessels when grey water is released into marinas via sewers (EPA 2022). Therefore, the treatment and reuse of grey water has a multitude of benefits such as reducing the pollution that enters waterways, reducing pressure on wastewater treatment facilities, and reducing the need for the expansion of drinking and wastewater treatment infrastructure, which is energy and labor intensive. By installing a greywater system, homeowners would be able to reuse up to 60% of household water for irrigation and flushing systems (US Department of Energy 2010). Integrated greywater treatment particularly in water scarce-arid environments provides plants not only with water but even other required nutrients and organics (Pradhan et al. 2019).

Blackwater discharge is any waste from toilets or urinals. It is defined either as treated or untreated and contains disease-causing bacteria and viruses that could

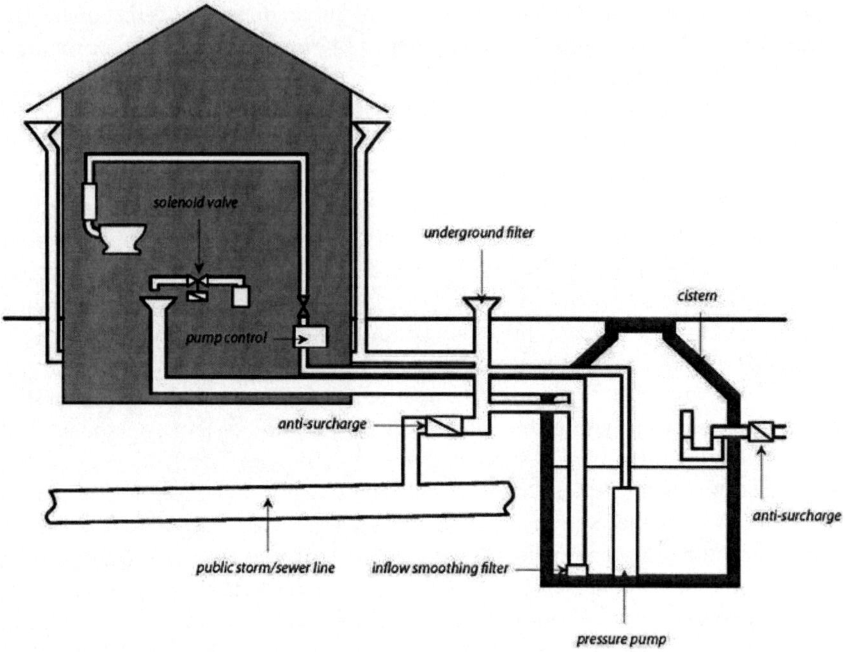

Fig. 7.11 Collection of rainstorm-generated runoff from a catchment to provide water for household use

harm humans if contacted (EPA 2022). The main difference between black and grey water is that black water is wastewater from bathrooms and toilets that contains fecal matter and urine (Global 2022). Depending on the water treatment code and regulations, water from kitchens and dishwashers can also be considered blackwater due to the contamination of pathogens and grease. Black water requires biological or chemical treatment and disinfection for reuse, whereas grey water requires little to no treatment, depending on how it is planned to be reused (Global 2022).

Blackwater has traditionally been difficult to treat due to the high concentration of pathogens, however technological improvements have led to the creation of an effective blackwater management system. *Aquacell* is an in-building water recycling system that collects blackwater from toilets and has undergone a systematic treatment process. It includes aerobic screening, biological treatment, ultrafiltration, ultraviolet disinfection, chlorine resistant protection, and nutrient removal (Aquacell 2019). The technological capability of the Aquacell exemplifies the core nature of what a circular economy should encapsulate by preventing the release of toxic water and instead producing usable water.

For homes with a backyard, *xeriscaping* is a form of landscaping that conserves water and reduces irrigation loads (Society 2011). A significant proportion of residential water use is attributed to irrigating lawns, which can be reduced by xeriscaping (Society 2011). In xeriscaping designers seek out ways to reduce the amount of water

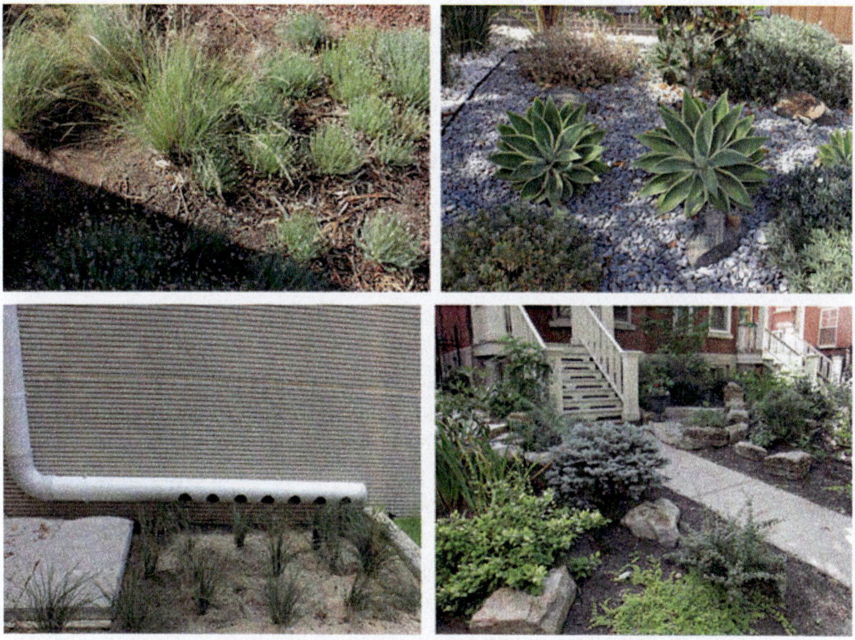

Fig. 7.12 In Xeriscaping designers' landscape to reduce the amount of water needed for irrigation

needed overall by maximizing the effectiveness of natural precipitation (Fig. 7.12). A good xeriscape design can even raise property values by drought proofing it. Xeriscaping studies the natural contours and drainage patterns of the land and identifies zones with different watering needs. This enables homeowners to better select appropriate vegetation and better allocate their water supply to avoid overwatering (Society 2011).

The optimal strategy in home water conservation is a combination of various water-saving techniques. Rather than merely reducing the water used, current technologies allow homeowners to use alternate water sources such as recycling gray and black water effectively and implementing xeriscaping. While some of these technologies have yet to be perfected, they can be integrated into the home's utility system from the start.

7.4 Installing Heating, Ventilation, and Air Conditioning (HVAC) Systems

HVAC encompasses all the different systems used for moving air between indoor and outdoor areas, along with the heating and cooling of homes. More efficient HVAC systems can lead to a significant reduction in power consumption. For context residential and commercial buildings consume over 40% of the total power usage in many developed countries (Seyam 2018). HVAC systems can be classified into central and local systems. Central or district HVAC systems are located outside of buildings in a central equipment room and deliver the conditioned air by a ductwork system. Two systems that are often set up as central systems are: heating and cooling panels and water-source heat pumps. Local HVAC systems on the other hand do not require ductwork and are mostly placed inside or adjacent to the living or working spaces and serve one single zone (Seyam 2018).

One of the requirements for centralized HVAC systems to operate is ductwork that transports air from the equipment throughout the building. The following principles help reduce the complexity of the ductwork system. Since hot air rises and cold air falls, for air conditioning to work properly, the large ducts that transport air back to the central unit, need to be installed high up against the wall of each upper floor, to capture warmer air and return it for cooling (Johnson and Menzie 1992). Efficient heating means installing a return at a low point on the first floor, to capture cooled air and returning it for heating. Ducts should be installed in a way, so they run as directly as possible from the basement to each register. This is due to the fact that more turns in the duct system slow down the flow of air and therefore slow down the time in which air will be delivered to the register (Johnson and Menzie 1992).

One of the main challenges of current construction practices is a lack of cohesive mapping. If the HVAC system needs to be modified or cleaned, swift accessing of the system can only be achieved with the use of the open web joist system to fit the ducts in them. However, the mapping of the duct path to the upper floors must occur before the cutting of any holes. Furthermore, ducts can run between the wall's studs and then turn into the space between two joists. Homebuilders should ensure that the studs and bays line up well enough to leave a clear path.

New and improved HVAC systems are being designed to require smaller ductwork and utilize less floor space for mechanical rooms (Murphy and Harshaw 2011). An example is active chilled beams (ACB) that consist of a fin-and-tube heat exchanger (Murphy and Harshaw 2011). They contain an integral air supply, where primary air passes through nozzles, which induces air from the space upward and through a cooling coil (Murphy and Harshaw 2011). The smaller ductwork that this system requires enables more air to be delivered to each room. Further, by increasing insulation, energy demand overall is reduced, which subsequently lowers the number of thermoelectric module units that need to be used. This then reduces the overall volume of wiring required in installing the system (Martín-Gómez et al. 2021). The future of improving HVAC systems requires good operational control by having them centralized, and easily accessible. This allows for faster identification of the system that requires modification or maintenance (Martín-Gómez et al. 2021). Big Data analysis will also provide smart energy audits for the efficiency of HVAC systems to support high prediction, which in turn will reduce the need for regular modifications to the system to prevent the need for the destruction of drywall and detailed mapping of the entire electrical and plumbing system needed to power the HVAC system (Martín-Gómez et al. 2021).

7.5 A Home with Well-Integrated Utilities

The Catskills House, in White Lake, New York, US designed by husband-and-wife firm J_spy Architecture as their family retreat in the Catskills Mountains (Fig. 7.13). Although the site is expensive, the couple decided to build only on a small portion of it, thereby cutting down carbon footprint and cost. The home is industrial looking and unassuming on the outside, yet welcoming and comfortable indoors (Lasky 2018).

The massing of the home is formed of different blocks. For the ground level, three concrete boxes combined. Two of which contain private and public functions, with the entrance tucked in between. A smaller third, block, serves as a mechanical room. Although the interiors of the living room are made of concrete as a continuation of the facade, the juxtaposition with large, floor-to-ceiling windows brings natural light and warmth into the home. A fourth metal box that rests on top to heighten the main living area provides spaciously high ceilings. Extending from this room is a patio with unobstructed views of the surrounding landscape.

Due to the remote location, the only utility to reach the site was electricity. For heating, the architects opted to use a more efficient system, deciding on a geothermal heat pump, connected to 122 m (400 feet) deep well that provides heating. This is joined with a layer of hydronic radiant tubes beneath the flooring to ensure even

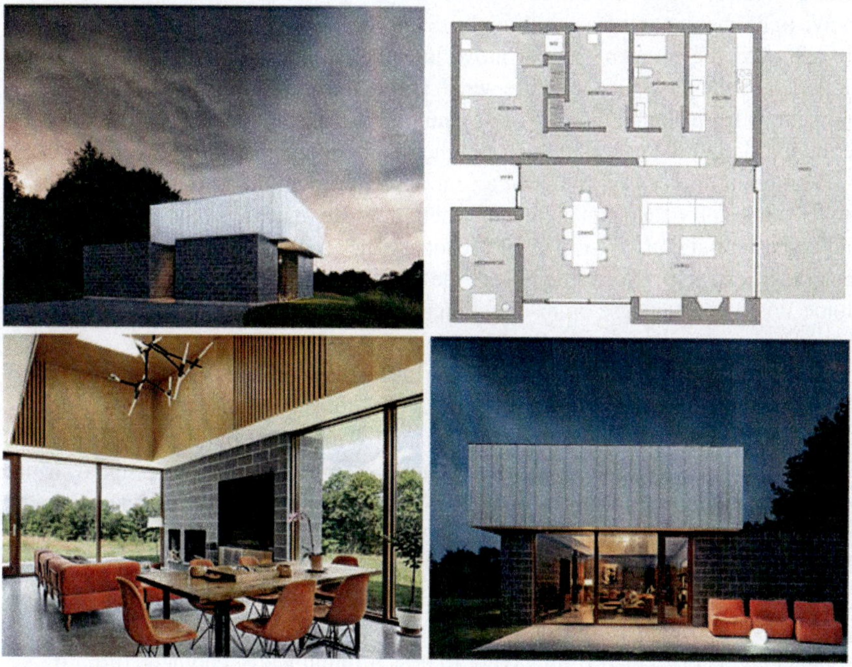

Fig. 7.13 In the Catskills House, in White Lake, New York, US, the only utility available on site was electricity. The architects decided to install a geothermal heat pump, connected to a 122 m (400 feet) deep well

warmth throughout the house. During the summertime, ducted system air conditioning is employed. Together these two systems drastically reduce electricity bill. The substantial overhang of the metal box also plays a part, reducing solar gain in the summer. The home is also equipped with high-performing insulation and a tankless hot water system to further cut down energy usage.

7.6 The Internet of Things (IoT) and Wi-Fi

The Internet of Things (IoT) and Wi-Fi have been crucial to contemporary domestic communication. Wi-Fi was first introduced in 1971 when the University of Hawaii developed the ALOHAnet (Boston University 2018). The ALOHAnet became the first wireless packet data network. It wasn't until 1999 when the trademark name of Wi-Fi was created by the newly formed Wi-Fi Alliance (Boston University 2018). The technology's capability has transcended profoundly, with a top speed of 1,300 Mbps (megabits per second), in comparison to its previous connection speeds of at most 11 Mbps (Boston University 2018). Wi-Fi uses radio waves to transmit data from a wireless router to enabled devices such as video game consoles, computers, smartphones, and cars to name a few (Verizon 2022). Most urban and suburban areas offer accessible 5G Home Internet services.

Wi-Fi has become a household staple, to the point where its setup and connectivity are achieved the moment people move into their homes. This demand is further propelled by the post-pandemic life which still requires a significant amount of work and school from home. The increased number of users and devices in the home setting has led to the necessitating of higher bandwidth capabilities (Verizon 2022).

Positioning the central router in a central location within a house, apartment, or office enables all occupants no matter where they are to always receive high-quality and consistent signal. Wi-Fi signal broadcasts in all directions, so positioning the router in a central location will improve the signal spread (HQ Editorial Group 2020). Home Wi-Fi products use omni-directional antennas, which radiate horizontally all around but are weaker vertically. Therefore, placing a wireless device on a table or shelf and keeping it elevated will better utilize the transmission from the omni-directional antennas (HQ Editorial Group 2020).

Placing the router in close proximity to a microwave is one of the most detrimental ways to interfere with Wi-Fi signals because it emits a very strong signal in the 2.4 GHz band (HQ Editorial Group 2020). Within the home, metal-based appliances such as refrigerators, and steel or aluminum furniture can create Wi-Fi dead zones as they can reflect and/or absorb the signals, thereby creating a barrier between the device and the router (HQ Editorial Group 2020). Designers need to plan the position of the wireless router away from said obstacles, so that the signal is not absorbed by metal and is prevented from having to pass through walls, cabinets, furniture, or other physical obstacles. Furthermore, refrain from placing a wireless device near water sources, and avoid heat and humid areas to prevent damage to the router (HQ Editorial Group 2020). The best rooms to keep the Wi-Fi box are in the living room and study room, whereas the kitchen, bathroom, and basement should be avoided.

7.7 Domus Ex Machina

Domus ex Machina was designed as a conceptual prototype with advanced manufacturing practices in mind and integration of new technologies to streamline its performance. Its design addresses four challenges of modern housing: the diverse sociodemographic makeup of households, the environmental impact of the construction industry, simplification of construction and utilities' installation, and finally affordability (Fig. 7.14). Analysis of lifestyle trends, high-quality design, and technological advances enable an innovative approach to the home's design.

The design aims to provide a comfortable interior that can accommodate a range of different household needs. Each has a footprint of 4.8 by 1.2 m (16 by 40 feet). Developments are comprised of three-story structures, built with either finished or unfinished basements, that can be configured into duplexes, triplexes, and single-family units (Fig. 7.15). Homes can be arranged as rows, semi-detached, or detached structures. A specially designed software allows buyers to visualize their future homes

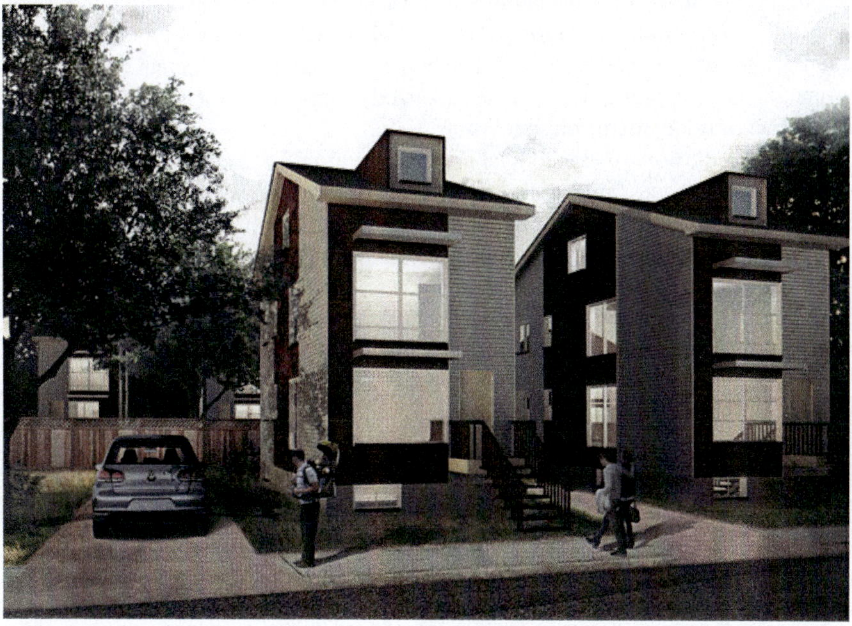

Fig. 7.14 Domus ex Machina was designed with advanced manufacturing practices in mind and integration of new technologies to streamline its performance

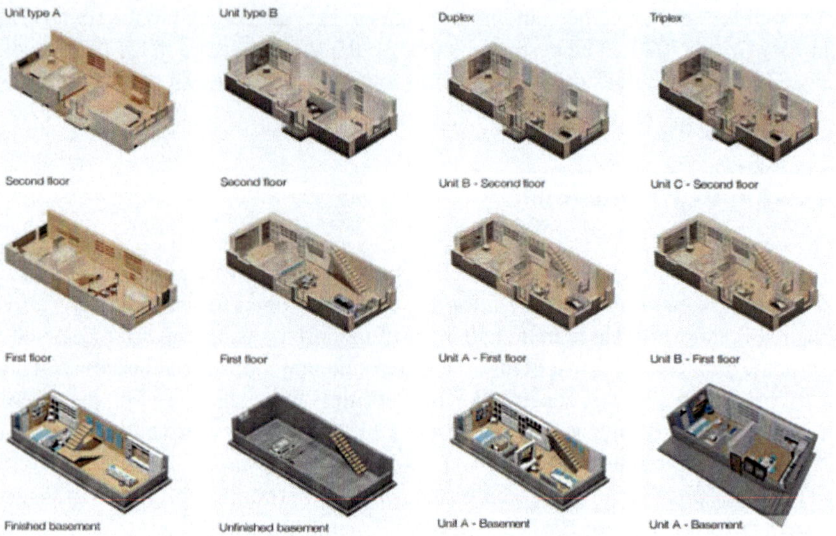

Fig. 7.15 Domus ex Machina is comprised of three-story structures, built with either unfinished or finished basements, that can be configured as single-family, duplex, and triplex units

and select different interior configurations and finishes. A selection of sustainable systems is available within the bathroom and kitchen offerings, including low-water flow toilets, grey water filtration systems, composters, recycling units, and high-efficiency washers and refrigerators. The home is designed for efficient construction by making use of prefabricated and panelized components for simplified on-site installation of floors, partitions, and roofs.

Systems in the home use a modular approach to simplify assembly and installation while improving accessibility for later modifications. Open web floor joists allow a single chase to run the length of the unit, passing all the home's utilities (Fig. 7.16). This system limits material waste while making it simple to locate wet functions and utilities for repair and adaptation later. The chase culminates at a wet wall, where the kitchen, bathroom, and laundry of the unit are located (Fig. 7.17). The wet wall is placed in a consistent location on each floor allowing for utilities to be stacked vertically, reducing material use (Fig. 7.18). Functions may be spread out across floors or shared on one wet wall depending on the unit's size (Figs. 7.19 and 7.20). For instance, a three-story unit may have the laundry function in the basement, a kitchen and bathroom on the first floor, and a larger bathroom on the second floor, all located in the same vertical position along the wet wall. In a one-story unit, all these functions would be located along the shared wet wall. This reduces the distance that plumbing travels to appliances, simplifying the installation process, and making utilities easy to locate in the future. Electrical wiring and internet connections are enclosed within accessible base moldings rather than within the walls themselves.

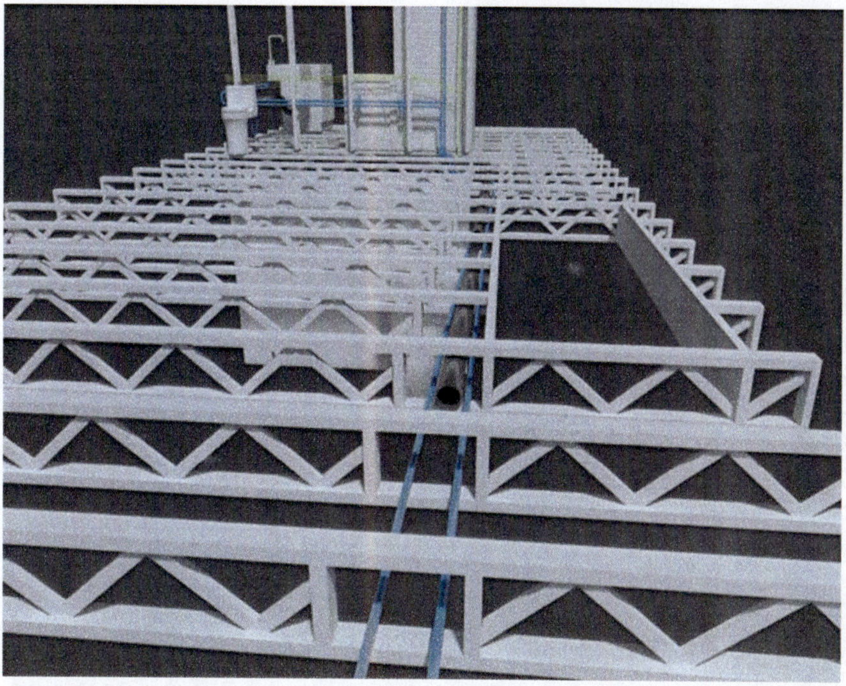

Fig. 7.16 Open web floor joists allow a single chase to run the length of the unit, for placement of all the home's utilities

This allows for their receptacles to be easily accessed for repair or modification without requiring walls to be cut and refinished.

The use of technology doesn't stop after the immediate occupancy of the home, an online platform functions as a data control center, giving users key information about the functioning of their house. The dashboard proves information related to energy and water consumption, security, and other relevant performance data. Additionally, it enables users to remotely control heating, internet, lighting, blinds, cameras, and appliances. Domus ex Machina streamlines the design of utilities within the home, using energy efficient systems and technologies to create high-performance dwellings and communities (Figs. 7.21 and 7.22). Its design demonstrates how a modern approach when it comes to systems and utilities can be used to improve efficiency and adaptability.

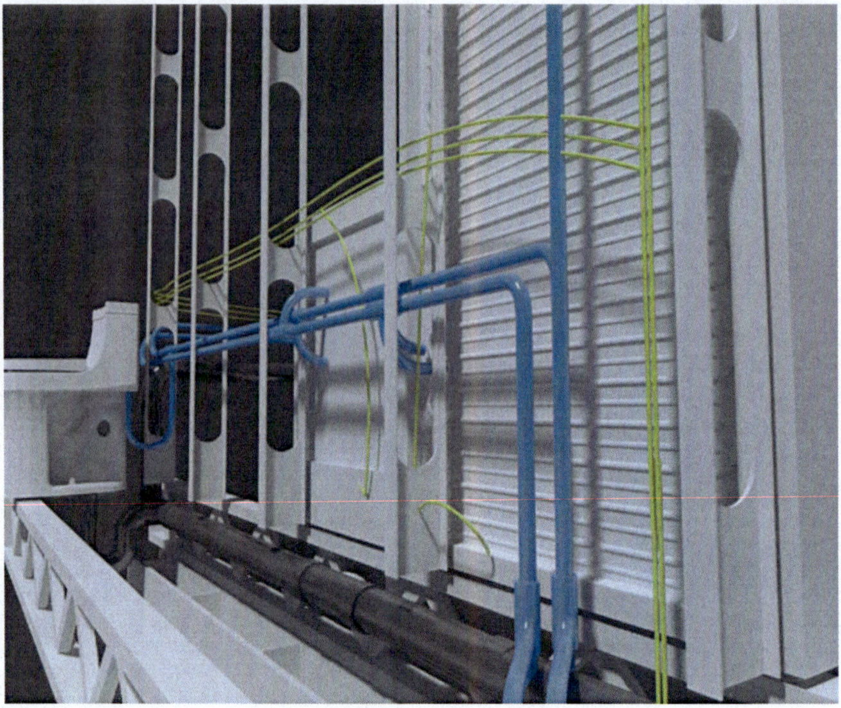

Fig. 7.17 The chase culminates at a wet wall, where the kitchen, bathroom, and laundry of the unit are located

7.8 Final Thoughts

Utility systems and their proper maintenance play a vital role in increasing or decreasing a dwelling's sustainable performance. Much like the engine of a car, the utility systems must not only function well, but also need to be accessed for replacement or repairs over time. The key goals that efficient and optimal utility systems will achieve are savings through labor, material, and time. When planning, keeping in mind structural strategies, water harvesting and recycling, and the centralization and mapping of electrical and plumbing systems will be advantageous to the overall fluid and working mechanism of a utility system. If a problem is not detected on time, it can escalate further. The selection of piping materials such as CIPP and safe electrical wiring systems such as baseboards would have taken repair and accessibility into account which would make the long-term maintenance of the household less stressful and disruptive to daily life. Reevaluating and thereby redesigning current utility systems will better achieve sustainable plumbing and electrical systems.

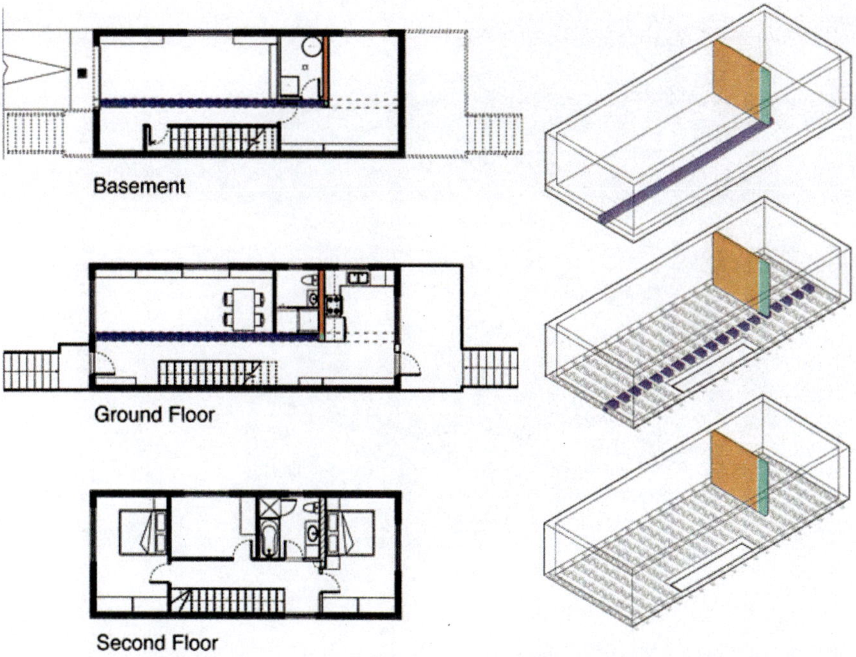

Fig. 7.18 The wet wall is placed in a consistent location on each floor of the unit, allowing for utilities to be stacked vertically, reducing material use

To better contribute to a circular economy, the proper treatment and reuse of grey and black water would profoundly reduce the freshwater demand. Various techniques such as xeriscaping and on-site water filtration systems will all be beneficial to reducing unnecessary water waste and preventing harmful bacteria and pathogens from being released into the surrounding environment through sewers. Keeping all types of water in a circular nature will promote and ensure the social, environmental, and economic wellbeing.

Questions for a Follow-Up Discussion

1. What are the key technologies used to achieve water-saving, harvesting, and recycling systems?
2. What are the criteria to affect the choice of efficient heating, ventilation, and air conditioning (HVAC) systems?
3. How did the Internet of Things (IoT) and Wi-Fi effected a modern lifestyle in a home?

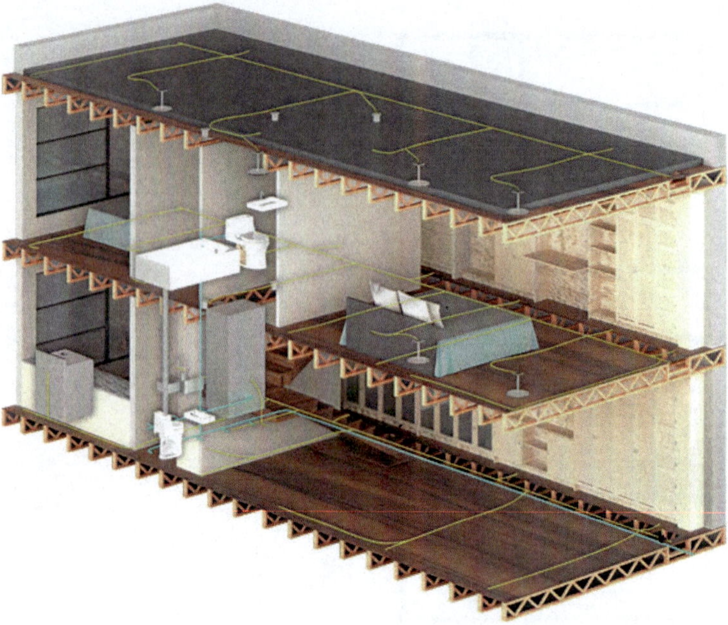

Fig. 7.19 In Domus ex Machina functions may be spread out across floors or shared on one wet wall depending on the size and layout of each unit

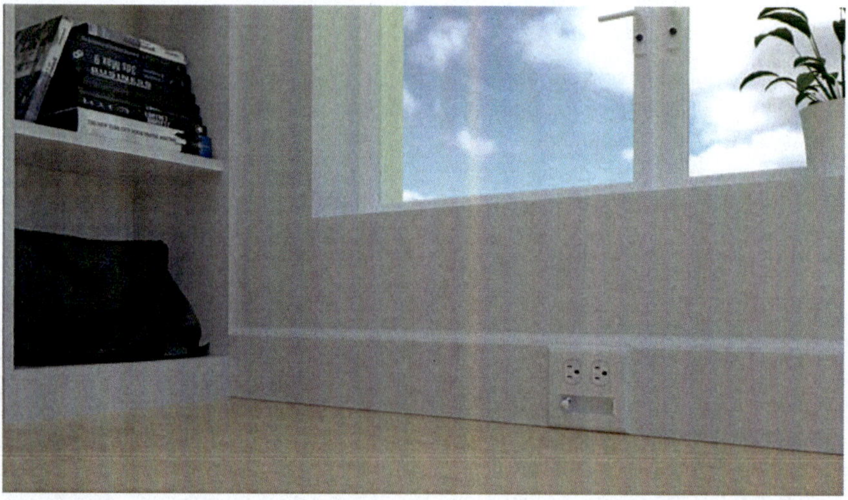

Fig. 7.20 Electrical wiring and internet connections are enclosed within base moldings rather than within the walls themselves

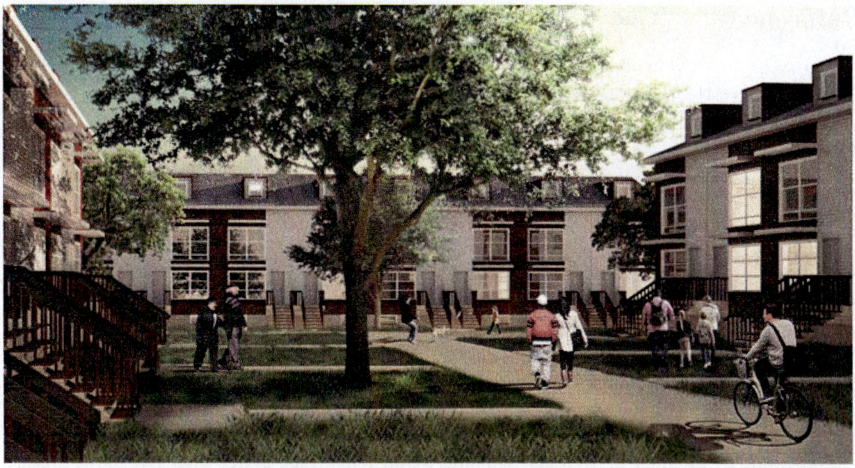

Fig. 7.21 A community made up of Domus ex Machina units

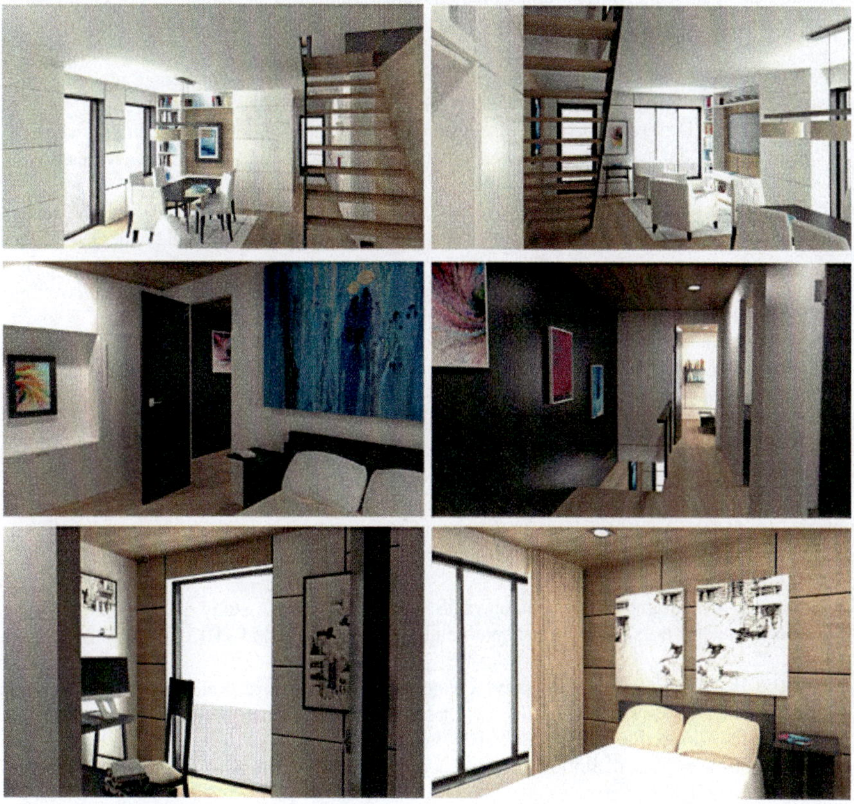

Fig. 7.22 Interior of a Domus ex Machina unit

References

AHAM (2021) AHAM history. In: The association of home appliance manufacturers. https://www.aham.org/AHAM/AuxAHAMHistory. Accessed 13 Oct 2022

Aquacell (2019) Inside salesforce tower's water recycling system. In: Aquacell. https://aquacell.com.au/v2/wp-content/uploads/2019/09/Saleforce-Tower.pdf. Accessed 13 Oct 2022

Boston University C of C (2018) WiFi and the future of connection | Center for Mobile Communication Studies. https://sites.bu.edu/cmcs/2018/09/14/wifi-and-the-future-of-connection/. Accessed 13 Oct 2022

Britnell G (2015) CIPP rehab of water mains running through backyards. https://ferpalinfrastructure.com/wp-content/uploads/2015/11/mstt-journal-2015.pdf. Accessed 13 Oct 2022

CAA (2022) Drains and foundations: prevention or repair? CAA Quebec. https://www.caaquebec.com/en/at-home/guides/drains-and-foundations-prevention-or-repair/. Accessed 13 Oct 2022

Canada NR (2014) Medium-density fibreboard. https://www.nrcan.gc.ca/our-natural-resources/forests-forestry/forest-industry-trade/forest-products-applications/taxonomy-wood-products/medium-density-fibreboard/15849. Accessed 13 Oct 2022

Clifton C (2020) Methods of venting plumbing fixtures and traps in the 2018 IPC. https://wabo.memberclicks.net/assets/pdfs/Plumbing_Venting_Brochure_2018.pdf. Accessed 13 Oct 2022

Delile H, Keenan-Jones D, Blichert-Toft J, Goiran J-P, Arnaud-Godet F, Albarède F (2017) Rome's urban history inferred from Pb-contaminated waters trapped in its ancient harbor basins. Proc Natl Acad Sci 114:10059–10064. https://doi.org/10.1073/pnas.1706334114. Accessed 13 Oct 2022

EPA SA (2022) Black and grey water management. https://www.epa.sa.gov.au/environmental_info/water_quality/programs/grey_and_black_water_discharge. Accessed 13 Oct 2022

Global WG (2022) Difference between blackwater and greywater—Global Water. https://www.globalwatergroup.com.au/our-blog/difference-between-blackwater-and-greywater. Accessed 13 Oct 2022

Hillel D (2005) Water harvesting. In: Encyclopedia of soils in the environment (ed), vol 4. Elsevier/Academic Press, pp 264–270. https://pubs.giss.nasa.gov/abs/hi09000a.html. Accessed 13 Oct 2022

HQ Editorial Group (2020) 6 tips on where to place your wireless router for the best signal/coverage | TP-Link. In: TP-Link. https://www.tp-link.com/us/blog/87/6-tips-on-where-to-place-your-wireless-router-for-the-best-signal-coverage/. Accessed 13 Oct 2022

Hunter RB (1941) Methods of estimating loads in plumbing systems. https://www.govinfo.gov/content/pkg/GOVPUB-C13-a13c5a2020dd0a7b1df8b91b165745ca/pdf/GOVPUB-C13-a13c5a2020dd0a7b1df8b91b165745ca.pdf. Accessed 13 Oct 2022

Ji HW, Koo DD, Kang J-H (2020) Short- and long-term structural characterization of cured-in-place pipe liner with reinforced glass fiber material. Int J Environ Res Public Health 17:2073. https://doi.org/10.3390/ijerph17062073.Accessed13October2022

Johnson R, Menzie K (1992) Here are seven of the basic rules for getting your ducts in a row—Baltimore Sun. https://www.baltimoresun.com/news/bs-xpm-1992-07-04-1992186077-story.html. Accessed 13 Oct 2022

Kantamneni S, Chandramouli DK, Chaitanya JSN, Sai AN (2021) A study on modular construction technique. 3. http://www.ijmtst.com/volume7/issue07/48.IJMTST0707087.pdf. Accessed 13 Oct 2022

Lasky J (2018, April 9) Two designers create a small but Luxe Retreat in the Catskills. Dwell. https://www.dwell.com/article/two-designers-create-a-small-but-luxe-retreat-in-the-catskills-953fcd77. Accessed 9 Feb 2019 and 13 Oct 2022

Law I (2022) House sewer definition. Law Insider. https://www.lawinsider.com/dictionary/house-sewer. Accessed 13 Oct 2022

Line C (2019) What is a plumbing stack? Clean Line Plumb. https://cleanline.ca/what-is-a-plumbing-stack/. Accessed 13 Oct 2022

Martín-Gómez C, Zuazua-Ros A, Del Valle de Lersundi K, Saiz-Ezquerre BS, Ibáñez-Puy M (2021) Integration development of a Ventilated Active Thermoelectric Envelope (VATE): constructive optimization and thermal performance. Energy Build 231:110593. https://doi.org/10.1016/j.enbuild.2020.110593. Accessed 13 Oct 2022

Murphy J, Harshaw J (2011) Understanding chilled beam systems 38–34. https://www.trane.com/content/dam/Trane/Commercial/global/products-systems/education-training/engineers-newsletters/airside-design/adm_apn034en_1209.pdf. Accessed 13 Oct 2022

National Geographic (2011) Xeriscaping. http://www.nationalgeographic.org/encyclopedia/xeriscaping/. Accessed 13 Oct 2022

Pradhan S, Al-Ghamdi SG, Mackey HR (2019) Greywater recycling in buildings using living walls and green roofs: a review of the applicability and challenges. Sci Total Environ 652:330–344. https://doi.org/10.1016/j.scitotenv.2018.10.226. Accessed 13 Oct 2022

Rubeiz C (2004) Benefits of PEX cross linked polyethylene for plumbing pipe. In: Green Building Solutions. https://www.greenbuildingsolutions.org/blog/flexing-pex-plumbing-possibilities-cross-linked-polyethylene-pipes/. Accessed 13 Oct 2022

Seyam S (2018) Types of HVAC systems. In: Kandelousi MS (ed) HVAC System. In Tech. http://www.intechopen.com/books/hvac-system/types-of-hvac-systems. Accessed 13 Oct 2022

States K, Mcquaid J (2022) Hand and portable power tools | Office of Environmental Health and Safety. In: Hand portable power tools. https://ehs.princeton.edu/workplace-construction/workplace-safety/hand-portable-power-tools. Accessed 13 Oct 2022

Triforce (2015) A little bit of Joist history…. In: TRIFORCE® Open Joist. https://www.openjoisttriforce.com/a-little-bit-of-joist-history/. Accessed 13 Oct 2022

Triforce (2020) A prefab home manufacturer's quest for the ultimate floor joist leads it to TRIFORCE®. In: Phoenix Build. https://phoenixbuilding.ca/wp-content/uploads/2020/06/triforce-case-study-maisons-bonneville.pdf. Accessed 13 Oct 2022

US Department of Energy (2010) Guide to home water efficiency. https://www.energy.gov/sites/prod/files/guide_to_home_water_efficiency.pdf. Accessed 13 Oct 2022

US Department of Energy (2022) Water-efficient technology opportunity: rainwater harvesting systems. In: Energy.gov. https://www.energy.gov/eere/femp/water-efficient-technology-opportunity-rainwater-harvesting-systems. Accessed 13 Oct 2022

Verizon (2022) What is Wi-Fi? | Definition, meaning and explanation | Verizon Fios. https://www.verizon.com/info/definitions/wifi/. Accessed 13 Oct 2022

Zhang J, Cui Y, Lu H, Lin Z (2021) Prediction of outflow parameters for a shower head. Water Res 202:117436. https://doi.org/10.1016/j.watres.2021.117436. Accessed 13 Oct 2022

Chapter 8
Green and Healthy Materials

Abstract The need of design and construction practices to align with contemporary environmental constraints has become urgent. Depletion of natural resources and their high costs have put pressure on key homebuilding actors to opt for "green" alternatives. The growing need for sustainable and renewable resources to follow an updated design methodology has ignited technological innovation and the creation of new design tools. By diversifying the industry's material palette, more appropriate cultural and economic design decisions can be made, suited to each project. This chapter presents insight into sustainable building products. It discusses a circular economy and life cycle approaches to design methods and choice of materials.

Keywords Big data · Biogeochemical cycle · Biosphere · Circular economy · Cradle to cradle · Design for assembly · Embodied energy · Green materials · Life cycle assessment · Linear economy · Material passport · Resource cycle · Systems thinking · 3-D printing

8.1 A Circular Approach to the Use of Materials

The depletion of finite natural resources has led to the rise of labor and material costs, thereby encouraging scientists, architects, and homebuilders to develop and source alternative building materials and introduce innovative designs. Conventional construction practices are often reliant on steel, aluminum, timber, zinc, and copper whose extensive harvesting is leading to eventual extinction. Those resources cannot be replaced at a pace rapid enough to sustain present consumption. At the current rate of extraction, the available supply of some common metals is said to run out within 50 years (Jowitt et al. 2020).

Furthermore, forests, especially tropical ones, are needed to absorb and store carbon from the atmosphere. When they are harvested, the stored carbon is released into the atmosphere, contributing to climate change. Deforestation accounts for 10% of total heat-trapping emissions, equivalent to annual emissions from 600 million cars (Union of Concerned Scientists 2016). One of the largest drivers of deforestation is timber used in construction. In Fact, deforestation for timber production results in

A. Friedman, *Fundamentals of Innovative Sustainable Homes Design and Construction*, The Urban Book Series, https://doi.org/10.1007/978-3-031-35368-0_8

380,000 hectares (939,000 acres) consumed annually (Union of Concerned Scientists 2016).

Commonly, architects and their clients often chose materials based on their availability, initial costs, and aesthetics. However, these choices are no longer sustainable and should not be the only defining selection parameters. Concerns over resource security, ethics, safety, and greenhouse gas reductions are shifting approaches to treating materials as assets to be preserved, rather than irresponsibly consumed and to a circular approach where products stay in use as long as possible (Stahel 2016).

A *Circular economy* is defined as a model where planning, resourcing, procurement, production, and reprocessing are designed and managed as both process and output, to maximize ecosystem functioning and human wellbeing (Murray et al. 2015). It aims to turn goods that are at the end of their service life into resources for other uses, closing loops in industrial ecosystems, therefore preventing waste (Fig. 8.1). Almost every biogeochemical cycle has been permuted by man—disrupting the homeostasis of the geo-biosphere. A system such as the circular economy seeks to restore *resource fluxes* back to their natural levels by reducing the exorbitant use of materials, and the disproportionate release of by-products into a cycle. The importance of transitioning from a linear to circular economy can be understood through the concept of *systems thinking*, a view that suggests that the world is made up of interconnected relationships rather than separate components that function individually (Arnold and Wade 2015). This view will be further explained below.

8.1.1 Systems Thinking

The *linear economy* is a one-way system, much like the flow of a river, driven by a "bigger-better-faster-safer" syndrome, exceptional at overcoming the challenge of scarcity, but prodigal at using resources from saturated markets. The production of waste leads to the deterioration of the environment by the removal of *natural capital* through unsustainable methods of extraction (Murray et al. 2015). Natural capital refers to the earth's reserves of natural assets which includes geology, soil, air, water, and living organisms.

Conversely, according to Stahel (2016), a circular economy could be compared to a lake. Its guiding principles are to eliminate waste and pollution, circulate products and high value materials, and regenerate natural systems (Ellen MacArthur Foundation 2022). This closed system should also be supported by the transition to renewable energy and materials to disable economic activities that fuels the consumption of finite resources (Stahel 2016). Circular-economy models can be categorized in two ways: one that promotes reuse and extends service life through repair, remanufacture, and retrofits; and another that recycles used material (Cantrell 2022).

As part of the design process, several tools need to be consulted. They include *life-cycle assessments* that look at a building's or a product's entire life and *material passports* that list products' vital attributes. This would minimize the *embodied*

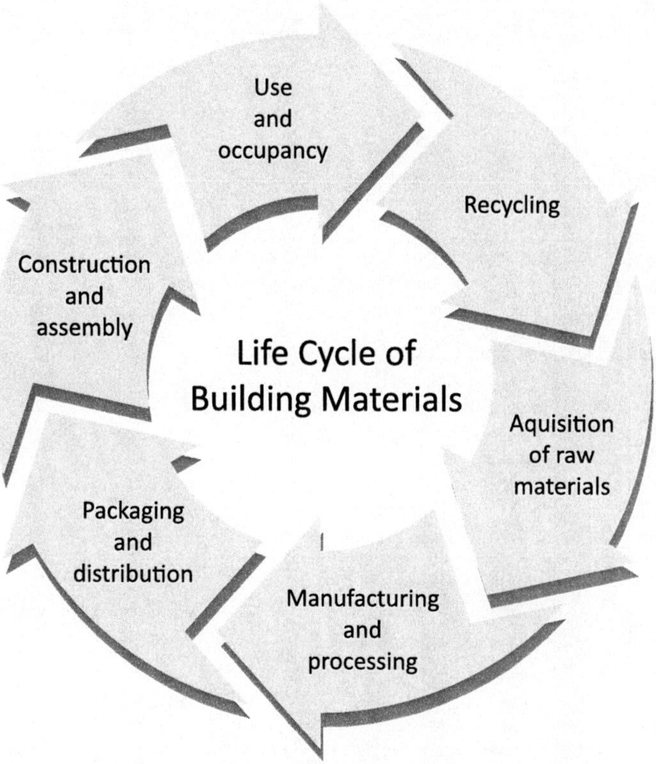

Fig. 8.1 The aim of circular economy is to turn goods at the end of their service life into resources for other uses, closing loops in industrial ecosystems thereby preventing waste

energy (i.e., the energy consumed during harvesting and processing) and the *operational energy* (i.e., the energy required to fulfill the building's function and maintenance) (Fig. 8.2) (Allan and Phillips 2021). More on these tools will be outlined below.

8.1.2 Life Cycle Assessment

The *life-cycle assessment* (LCA) method is a systematic tool that evaluates the embodied and operational energy of materials (Allan and Phillips 2021). This assessment provides information on how different materials and processes contribute to the overall environmental impact of a building. The process involves acquisition of raw materials, processing and manufacturing, packaging and distribution, construction, and assembly, use, and lastly, recycling or disposal.

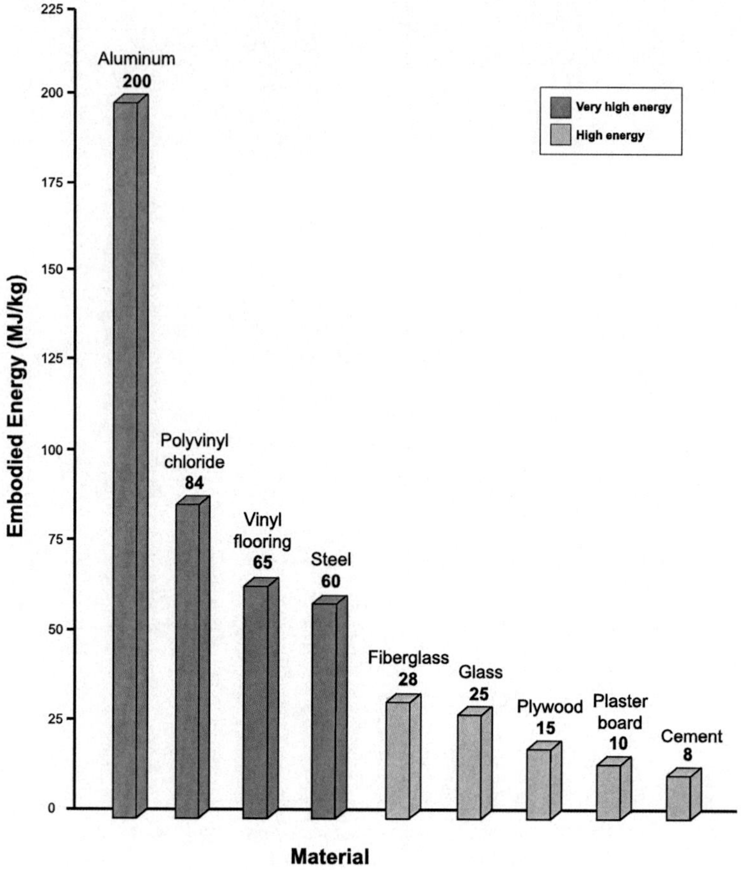

Fig. 8.2 Embodied energy used in the production of various products

Several terms are commonly used when life cycle cost materials are investigated (Fig. 8.3). Baseline date is the present, "today", when costs need not be discounted, as well as the point in time against which investment opportunities are measured when a householder takes possession of their unit, for example. Sunk costs are those costs which occur before the baseline years or before user takes possession. They are called "sunk" because they cannot be recovered. Usually, these costs are considered in the analysis only when the technique is being used to assess all the costs associated with a given decision—past, present, and future. Time horizon refers to the ending point of the *life cycle cost* analysis, the cut off, or last year of the analysis. This time, horizon may be wholly or at least partially a function of the householder's objectives, which may intentionally stay in the unit for only a limited, pre-determined number of years.

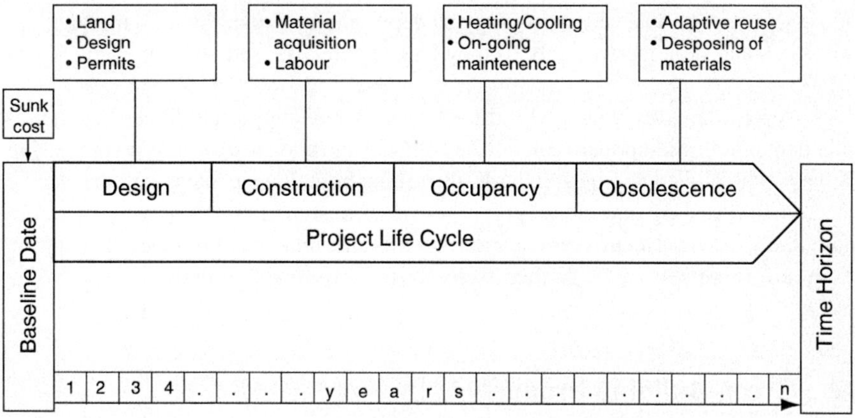

Fig. 8.3 The life-cycle assessment (LCA) method can be used to evaluate the embodied and operational energy of materials

8.1.3 Material Passports, Separation of Cycles, and Recycling

The circular economy is intertwined with resource cycling. This can be achieved through industrial symbiosis, where firms collaborate, and work in sync with one another to utilize each other's waste as resources. In other words, this could be seen as "*waste-as-food*", where unwanted outputs of one industrial process are used as raw materials in another industrial process (Murray et al. 2015). Increasing the longevity of products through streamlined manufacturing and high-quality maintenance would minimize the embodied energy of materials.

Material passports add value to the selection process by evaluating and listing the recycling potential of materials used in buildings. A building information modeling (BIM)-based material passport assesses the quantitative and qualitative properties of materials, such as their recycling properties and potential reduction of environmental impacts (Honic et al. 2019). The material passport serves as a design optimization tool that architects can refer to during their decision-making process about using a product.

Another component to efficient recycling is the separation of *biological* (e.g., Wood) and *technical* (e.g., Laminate) cycles (Velenturf et al. 2019). This separation is the way products can be deconstructed and reverted to their original state so they can be recycled (Velenturf et al. 2019). This would be through the removal of adhesives, coatings, and lacquers from biological products such as wood or metal.

According to Stahel (2016), exceptional metallurgical and chemical science analysis is necessary for a circular economy to be maintained. Yet, there is minimal research on disassembling material blends at the atomic level. More knowledge would be a great leap toward closing the circular-economy loop. The construction industry requires advanced technology to depolymerize, de-alloy, de-laminate, devulcanize, and de-coat materials so they can be safely reused and recycled (Stahel 2016).

However, the evolving social attitudes can also be supportive toward economically demanding yet innovative investments in construction research. With increasing global wealth and education standards, upcoming homeowners have the means to be selective in investing into more "green" and well-designed homes and are therefore more likely to select them over cheaper, and less sustainable alternatives. The notion of "green" products will be further explored in the following section.

8.2 Green Building Products

Green building is an overarching term that can be defined as the practice of creating structures and using processes that are environmentally responsible and resource-efficient throughout a building's life cycle from sitting to design, construction, operation, maintenance, renovation, and deconstruction (Vermont Journal of Environmental Law 2014). Green building design seeks to use energy, water, and other resources efficiently resulting in the protection of occupant health and reducing waste, pollution, and environmental degradation (Vermont Journal of Environmental Law 2014).

The following sections highlight "green" products that are advantageous by reducing the burden on the environment and preventing the exhaustion of natural resources all at a lowered cost. It is important to note that most of the modified materials used in the residential construction are still reliant on wood, metal, and concrete as key products. However, through incremental funding and research, other energy and material efficient methodologies are surfacing to maximize product value, thereby reinforcing the practice of reducing and reusing.

8.2.1 *Insulated Concrete Forms (ICFs)*

Insulated Concrete Forms (ICFs) are composed of expanded polystyrene sheets, a rigid cellular plastic with an expansion agent. It has been extensively tested and is deemed a suitable light material for wall construction (Latha and Arunraj 2014). The advantage of this material is that its grid wall system is robust against natural disasters, therefore suitable for seismically active regions. ICF walls provide energy efficient buildings due to the *thermal mass* of concrete that helps absorb and release heat slowly, the air tightness and continuous insulation of the system also minimize temperature fluctuations within the structure (Latha and Arunraj 2014). ICFs are energy saving, efficient to construct, soundproof, and require less maintenance

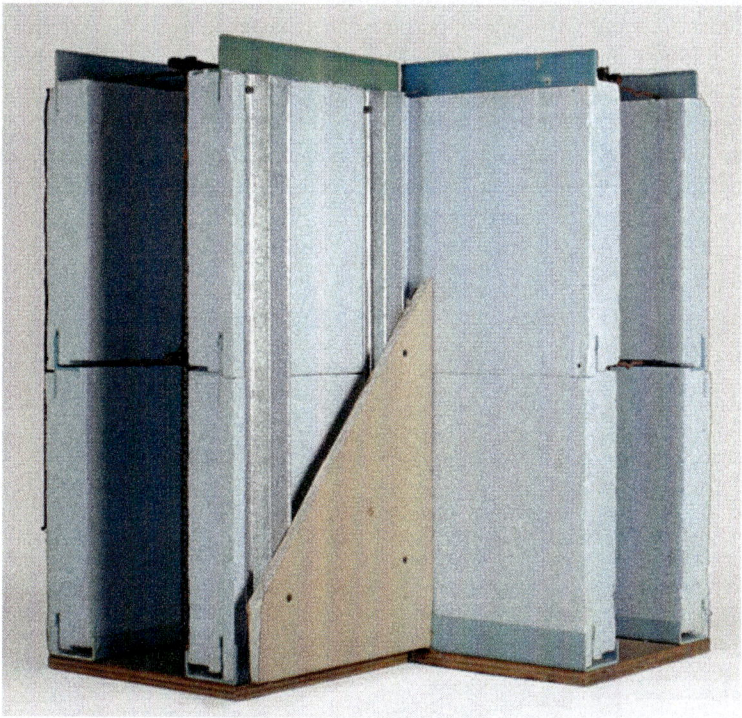

Fig. 8.4 Insulated Concrete Form (ICFs) walls improve energy efficient buildings due to the combination of insulation and thermal mass of concrete to minimize temperature fluctuations

compared to other forms of concrete use (Fig. 8.4). The underlying properties of ICF are benevolent to both environmental and social sustainability as research demonstrates; it is highly durable, resistant to natural disasters thus making it a safe material for mankind and the biosphere.

8.2.2 Structural Insulated Panels (SIPs)

One of the many steps toward creating more sustainable practices is the gradual transition from solid wood-reliant structures to modified structures in building practices. One alternative is incorporating *structural insulated panels* (SIP). They are composed of insulated foam with wood particles as the core layer, while having plywood, cement particle board, and fiber-cement boards as the surface layers (Thongchareon et al. 2021) (Fig. 8.5). The three kinds of commercial wood composite boards are adhered with polyurethane then placed within a clamping device and compressed until the polyurethane is cured. SIP is a multipurpose panel that is often used for exterior walls, roofing, flooring, and even foundation systems (Thongchareon et al. 2021). SIP is energy efficient as a lightweight material, therefore reducing the transportation cost from its production to construction site. Once it reaches the construction site, only a simple installation is necessary. The environmental benefit of integrating SIPs is that it requires less wood than conventional wood-frame constructed houses making it more sustainable since it incorporates other recyclable materials.

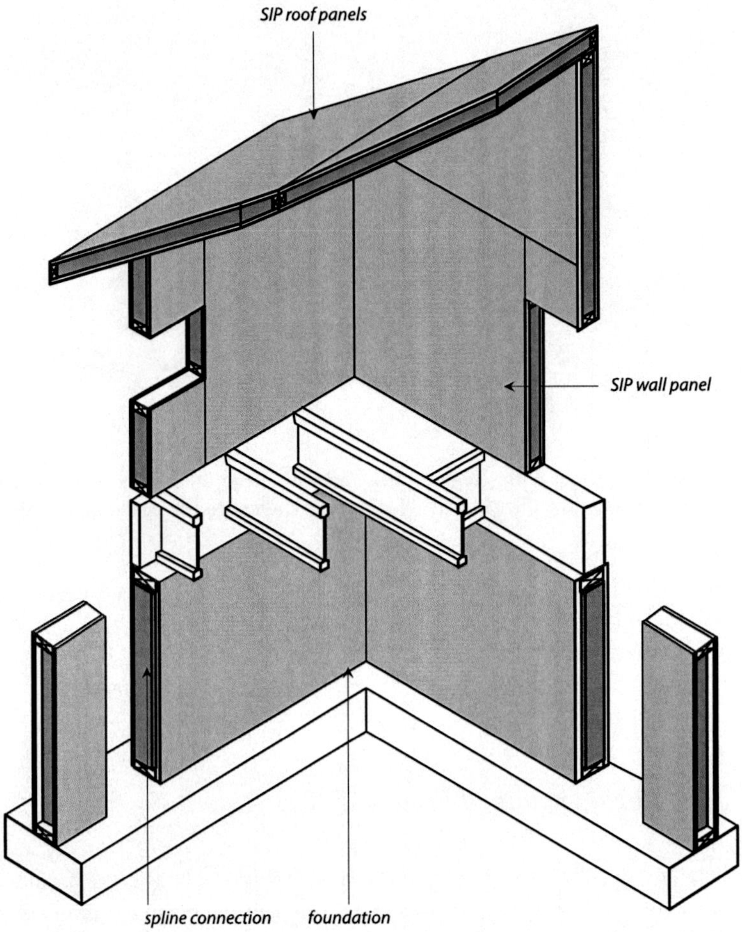

Fig. 8.5 Structural insulated panels (SIPs) are composed of insulated foam with wood particles as the core layer, while having plywood, cement particle board, and fiber-cement boards as the surface layers

8.2.3 Engineered wood and Laminated Timber

Engineered wood products (EWPs) maximize product performance by doing more with less wood than conventional wood-frame construction methods (Fig. 8.6). EWPs redistribute and reinforce natural defects that occur during tree growth such as knots, shakes, and spalting. This mechanism shapes the products into efficient structural molds that sawing is unable to achieve (O'Ceallaigh et al. 2021). Furthermore, EWPs are considered environmentally superior when the wood sourced comes from sustainably managed forests, and at the end of the building life, they can be deconstructed and reused (Khatib 2016).

Cross laminated timber (CLT) is a notable EWP to reach the market in increasing volumes to prevent the use of steel and concrete, mainly in multi storey construction projects (Khatib 2016). Cross laminated timber is composed of an uneven number of layers such as three, five, or seven each consisting of beams placed parallel to one another, which are crosswise arranged to each other at a 90-degree angle and connected by adhesives (Khatib 2016). Utilizing CLT offers the building better thermal performance and compared to concrete, requires less energy to heat and cool. The disadvantage in including adhesives are the health and environmental concerns associated with the release of toxic gasses such as formaldehyde. It demonstrates the

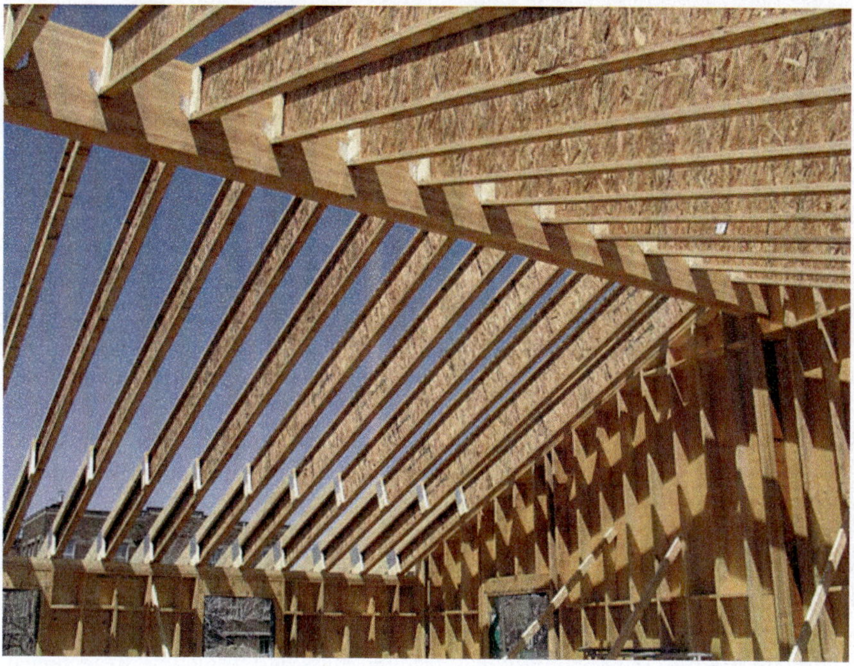

Fig. 8.6 Engineered wood products (EWPs) such as these joists maximize performance and sustainability by doing more with less wood than conventional construction methods

consequences of mixing biological and technical aspects which inhibits recyclability and reusability (Sotayo et al. 2019).

Another EWP product is *dowel laminated timber* (DLT). It is adhesive free which means it does not contribute to poor indoor air quality and have lower embodied carbon. However, these claims are based on qualitative research with the need for supplementary data on the life cycle impact assessment to quantify their environmental impact (Sotayo et al. 2019). The downside of EWPs is the additional energy needed to process them, and further research is still needed to make them reliable, affordable, fully non-toxic, reusable, and recyclable with well characterized mechanical properties and documented life cycle assessment (Sotayo et al. 2019).

8.3 Products Made from Recycled Materials

To further prevent demolition waste from ending up in landfills, *recycled concrete aggregate* (RCA) can be used as a substitute for *natural aggregates* (NA) in concrete products. RCA not only reduces the need to exploit natural resources but also subsides the expense of mining the components of concrete (Tang et al. 2020). RCA is generally a combination of crushed construction debris such as concrete, bricks, tiles, stone, timber, glass, plastics, and metal obtained from demolition sites.

The mining of raw materials for concrete is energy and resource intensive. Therefore, to offset these detrimental environmental impacts, using *fly ash*, a by-product of coal-fire plants, eliminates the practice of treating it as waste but rather, using it as a concrete mixture (Fig. 8.7). Fly ash has numerous advantages such as mechanical resistance, increased protection against chemical agent attacks, and resistance to freezing and thawing cycles in colder climates regions (Longarini et al. 2014). Due to the presence of lower *hydrolysis* content, the incorporation of fly ash leads to a more robust and impermeable cement paste to be used in construction (Longarini et al. 2014). However, with the demand to shift toward renewable energy, the dependence on fly ash is likely not a sustainable long-term practice.

Cellulose insulation is composed of recycled paper fibers and inorganic additives, its properties prevent mold growth and increase fire resistance. Cellulose fiber insulation has a lower embodied energy and therefore less resource intensive compared to other insulation materials (Pal et al. 2021). This natural insulator promotes a steady humidity value in indoor air which promotes better air quality. With a low mean thermal diffusivity, it can insulate buildings even under dynamic heat transfer conditions and increase indoor thermal comfort with less energy required, therefore is ideal compared to other insulation materials (Fig. 8.8). Furthermore, as a durable, biodegradable, and low-density renewable material cellulose reduces and reuses resources within the resource cycle (Pal et al. 2021).

Mineral wool is composed of remanufactured glass into insulation. Its technical commercial name is *Batt insulation*, a pink or yellow cotton-like fiber from pulled out

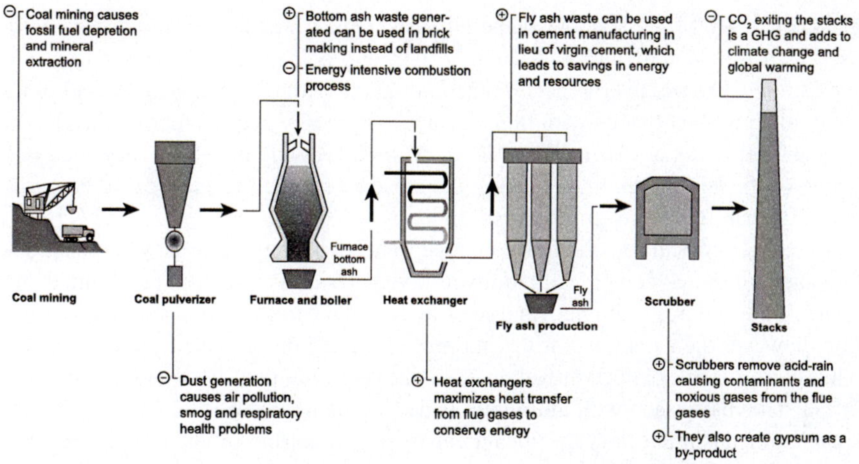

Fig. 8.7 Fly Ash, a by-product of coal-fired plants is used as a concrete mixture

Fig. 8.8 Cellulose insulation is composed of recycled paper fibers and inorganic additives, its properties prevent mold growth and increase fire resistance

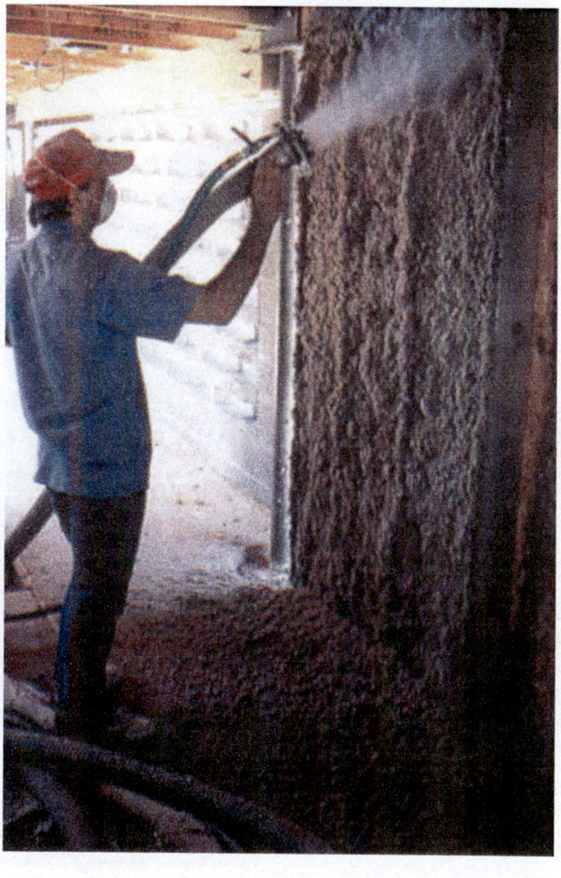

glass (Fig. 8.9). Mineral wool is generally used in mats and boards, but occasionally used as a filling material (Jelle 2016). Light and soft mineral wool products are applied in wood-frame houses and other structures with cavities. Heavier and more solid mineral wool boards with high densities are used when the thermal insulation is used for flooring or roofing (Jelle 2016). Mineral wool derived from recycled glass is one of the key drivers in achieving the standard requirements of passive houses and contributing to zero emission buildings.

Metals are critical construction components needed for multiple uses such as structure and plumbing yet their availability as raw materials is forecast to be limited due to their finite nature. Aluminum and steel are two of the most used metals in construction, however their manufacturing and processing are energy intensive (Fig. 8.10). Aluminum, despite being lightweight, embodies high compressive strength (Soo et al. 2017). The main issue with aluminum is despite being one of the most commonly recycled metals that offers significant energy saving during production, its benefits are offset by the lack of purity in its scrap sources. Perfect material separation during the end-of-life (EoL) phase is extremely challenging in the shredder-based recycling

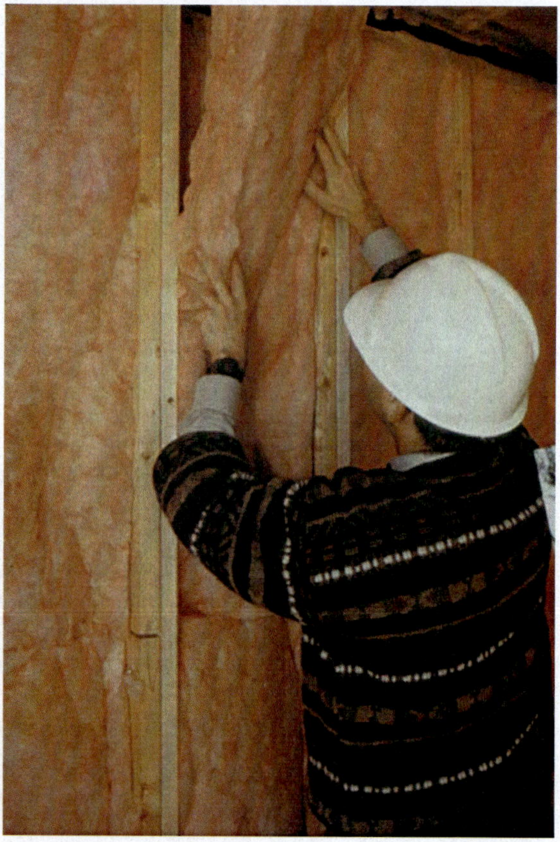

Fig. 8.9 Mineral wool, also known as Batt insulation, is composed of remanufactured glass

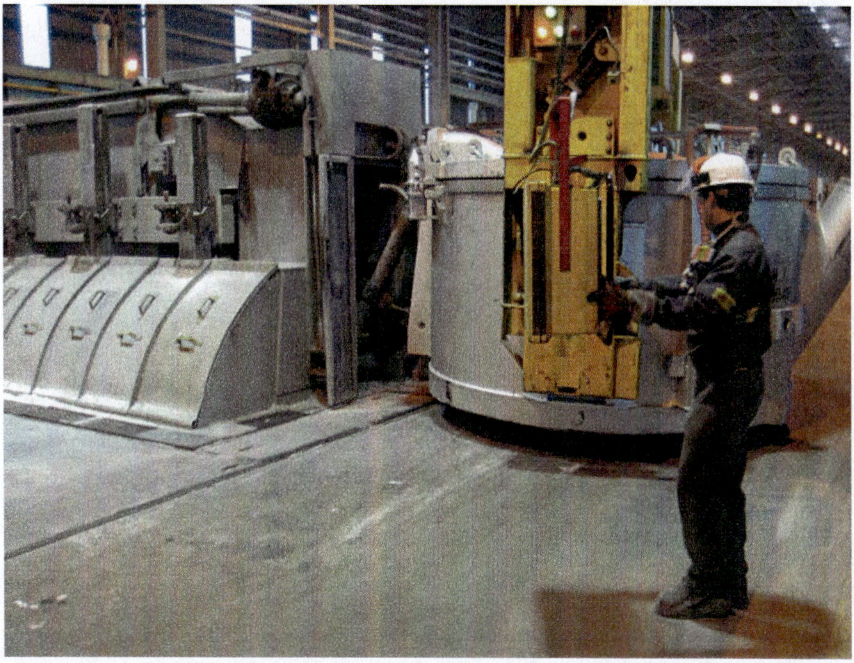

Fig. 8.10 Aluminum, seen here during production, is one of the most used metals in buildings and is energy intensive to manufacture and process

practices due to the complex product designs and the difficulty in separating different material types at the end of their life cycles (Soo et al. 2017).

Unlike aluminum, steel is slightly less environmentally damaging. One of its composites, iron ore, can be found globally, and therefore be distributed with a lower carbon footprint. Steel buildings have been found to have a 48–58% reduction in causing smog, acidification, and ozone depletion potential (Allan and Phillips 2021). As for its disposal and recycling, steel is one of the most recyclable materials available since it can be magnetically separated from the scrap source, and then processed into a high-quality alloy (Allan and Phillips 2021). Due to the combustion produced from manufacturing steel, going forward, recycled steel should be opted for rather than the mining of this precious finite natural resource.

8.4 Innovative Sustainable Materials

There is an immense potential for numerous alternative materials being at the forefront of those known as "green". However, many are still in the experimental stage, and therefore have failed to catch the attention of residential builders. Such innovative products as described below will unfortunately take time to reach mainstream construction practices but unlocking their full potential in the near future could radically transform the industry.

Hempcrete has shown to be a sustainable aggregate substitute in concrete that can reduce buildings' embodied energy while improving energy performance and indoor environmental quality (Abdellatef et al. 2020). Being plant based, it has non-toxic properties, and can be considered a local resource available worldwide. As a highly *porous* material it is easy to transport thereby reducing its carbon output throughout its material cycle.

Bamboo is a versatile material originating from tropical regions, it is commonly used as a laminate for its fiber, or as a dried woven material (Whitelaw 2014). As a fiber, bamboo is used as a composite conducive to decking, paneling, and veneers (Whitelaw 2014). The lightweight material makes it easy to lift and transport thus improving the material's life cycle. Furthermore, bamboo is a *carbon sequestering* agent that can thrive, even with minimal water, making it suitable for disaster relief housing, a crucial solution due to the increasing frequency and severity of extreme weather events.

The development and implementation of *earthen architecture* has been used in the past and is currently being used in certain parts of the world (Morel et al. 2021). *Ferrock* is currently a leading cement alternative due to its strength, accessibility, and cure time of four days rather than 28 days (Garcia et al. 2017). One of the key concepts that differentiates Ferrock from ordinary cement is the curing reaction, which requires carbon dioxide as an input. Not only does Ferrock absorb carbon dioxide in its hardening process, but it also does not require immense heat to catalyze the reaction or fire the ingredients in a kiln, so the main contributor to its carbon footprint is from its transportation.

Rammed earth calls for simple construction techniques, that do not require the labor force to have a specific skill set leading to more employment opportunities for unskilled workers. This earthen structure is an ancient building technique with intensive initial costs, however, requires little subsequent maintenance even in challenging climates (Easton and Easton 2014). Similarly, *Thatch* requires minimal training and specialized equipment, and it is widely available and lightweight, thereby reducing the carbon emissions associated with its transportation (Simpson 2021). Both rammed earth and thatch are affordable, waterproof, and entirely natural composites making them promising contributors to a circular economy through their sustainable material life cycles.

Linoleum is a flooring material made from biodegradable compound, such as linseed oil, resin, powdered wood, and limestone (Rosso et al. 2020). Therefore, it is almost entirely made of decomposable materials, which can be partially recycled and reused (Fig. 8.11). It fits the concept of *Design-for-Disassembly* (DfD) where products are designed intentionally for recovery, value retention, and meaningful future use (King 2020). Linoleum enables environmental stewardship throughout its material cycle. This is exemplified during its manufacturing and processing where discarded linoleum can be recovered and then reused. Made of bio composites, Linoleum is designed for installation as a sheet or rolls and only requires simple disassembly. Furthermore, the construction and assembly process of linoleum does not

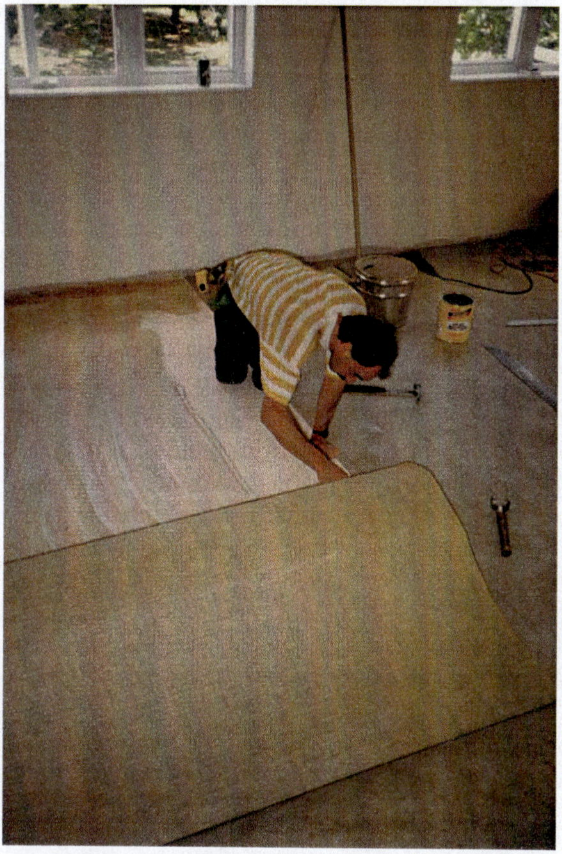

Fig. 8.11 Linoleum, seen here during application, is a flooring material made from biodegradable compounds, such as linseed oil, resin, powdered wood, and limestone

call for large amounts of energy to maintain its durability, greaseproof, waterproof, and fire-resistant properties.

Natural paints are derived from biodegradable composites such as plant resin, ethereal oils, mineral fillers, and pigments. This is a "greener" and more sustainable alternative in comparison to mainstream paint that produces several types of waste, many of which are hazardous. Therefore, new *eco-paints* that are waterborne with bio-based fatty acids incorporated into the mixture are being developed (Moreno et al. 2015). However, with other types of natural paint, most of the waste can be recovered and reworked back into marketable products during manufacturing and processing. But within the paint's material life cycle, its packaging and distribution pose a level of concern. To mitigate the environmental concerns associated with natural paints, recycled products should be used to package the paint, followed by local distribution. Leftover paint can be safely disposed of via municipal collection and recycling programs. At the demolition and renovation step, there is currently a lack of a post-consumer recovery. Therefore, it is "greener" to use biodegradable and water-soluble natural paints. Or perhaps the regular use of paint should be reconsidered in being a mainstream construction practice.

8.5 Materials for a Healthy Indoor

To a large measure healthy indoor environments are influenced by the types of materials used in construction. The COVID-19 pandemic which forced people to remain at home has also led them to pay attention to indoor air quality. A lack of daylight, poor ventilation, excessive noise, or toxic compounds emitted from some materials can have additional health consequences and may result in allergies and stress.

The advancement of technology in altering and fabricating new materials has been both a blessing and a curse. On one hand, it has led to efficient, sustainable, and cost-effective alternatives. On the other hand, the introduction of composite wood products, stain-resistant synthetic carpets, thermally efficient rigid insulation boards, cleaning agents, and furnishings, has also resulted in the emission of odors, and toxic compounds which harm occupants' health (Fig. 8.12).

"Green" buildings should offer indoor health benefits by emitting little to no pollutants. Building materials can be classified as naturally occurring or man made. *Naturally occurring* alludes to unprocessed biodegradable and organic materials such as

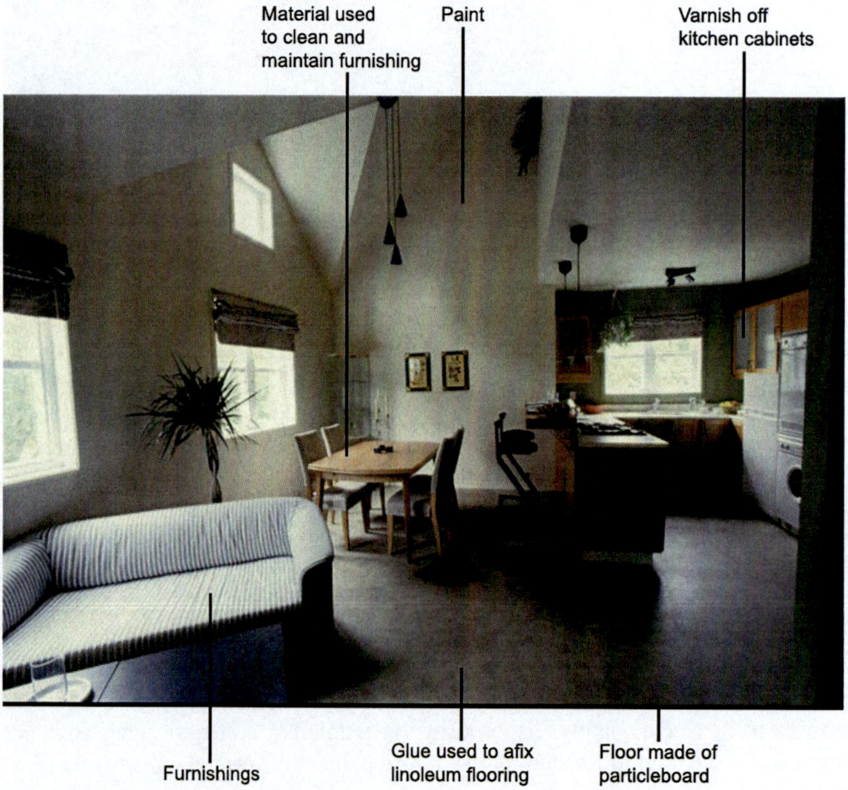

Fig. 8.12 Potential sources of emitted air pollutants in a dwelling's interior

wood, bamboo, and straw, as well as minerals and rocks. Whereas *man-made* products are derived from natural resources that undergo mechanical and chemical processing and include additional additives such as plastics and paints. These attributes will be outlined below.

8.5.1 Indoor Air Quality (IAQ)

Indoor air quality (IAQ) is determined among other factors by interior compounds and external air filtered through the building's ventilation system (Fig. 8.13). The dilution of indoor pollutants is predominantly achieved by increasing the rate at which outdoor air is supplied to the building. *Sick Building Syndrome* (SBS), which is manifested through respiratory symptoms is a major health concern and a result of poor indoor air quality (Suzuki et al. 2021). One of the biggest contributors to SBS is the use of adhesives. Multipurpose adhesives utilized in carpeting and plywood sheets, and tile flooring emit high concentrations of volatile organic compounds (VOCs). For example, adhesives used as binders in composite wood products such as particleboards are major sources of emitted harmful levels *formaldehyde* (Sterley et al. 2021). Soft plywood and waferboard are products that commonly contain *phenol formaldehyde*, and emit minimal amounts of formaldehyde, and are therefore preferable compared to *urea formaldehyde-based adhesives* (Sterley et al. 2021). Opting for insulation products such as fiberglass, mineral wool, or cellulose would be beneficial since they emit VOCs but at much lower levels. However, the accumulation of embedded adhesives within the various components indoors still poses a health risk and therefore construction practices should not be complacent with the current alternatives available (Fig. 8.14).

Type	Characteristics
Mechanical filters	• Flat or pleated; Effective at removing particles • Flat filters are effective at removing large particles • Pleated filters, such as high-efficiency particulate air filters (HEPA), collect the smaller, respirable particles
Electronic air cleaners	• Use electronic charge to remove airborne particles • May also produce ozone, a lung irritant
Ion generators	• Use electronic charge to remove airborne particles • May also produce ozone, a lung irritant

Fig. 8.13 Indoor air quality (IAQ) is determined by external air filtered through the building's ventilation system

Principle pollutants	Building product	Timing/duration of emission
Formaldehyde	Hardwood, plywood, particleboard and other glue-bonded wood products Some coatings and sealants may contain in lower concentrations	Highest emissions when new or recently applied; may continue for up to a year in lower amounts; emissions increase with temperatures and humidity
Toluene, Xylene hexane, aliphatic hydrocarbons 4-Phenylcyclo-hexane	Styrene Butadien Rubber (SB-R) sealants; Butyl Rubber (BR) or solvent based acrylic sealants	Time may vary from few to many hours depending on drying time
	Styrene Butadiene (SB) latex backed carpets	Low emissions, but may persist due to the high sink potential of carpeting
Volatile Organic Compounds (VOCs)	Vinyl tile and sheet flooring	Decrease substantially after 24 hours if well ventilated; lower emissions can continue for years
	Traditional multi-purpose carpet adhesives, both organic solvent and water based	Highest for the first 3-6 weeks; may persist depending on potential of floor material
	Latex (water or organic solvent - based) and acrylic paints; fiber glass and mineral wool insulation	During application and drying; lower emission may persist for months

Fig. 8.14 The accumulation of embedded adhesives within various components indoors poses a health risk. Construction practices should not be complacent with the current alternatives available

The two main aspects of IAQ are identifying the mechanism by which pollutants are released into the air and documenting the emission characteristics of the products made from these materials. Emissions from building materials are produced at all phases of a building's life cycle. Most emissions can be envisioned through careful consideration during design. Emissions can originate from three sources: the product itself, the surrounding air, and products used to install, clean, and maintain it. Therefore, to mitigate poor indoor air quality, a bottom-up approach should be taken by attempting to control the pollutant emissions directly at the source, when selecting materials for building.

Manufacturers have developed "low" and "zero-emission" versions of adhesives, which could reduce the concentration of toxic compounds released into the surrounding air (NPT 2019). Pollutants can also be significantly reduced by sealing a product to keep the volatile organic compounds from being released. Other strategies would be eliminating the need for adhesives completely by using methods that do not require gluing such as using solid sawn lumber or applying a tackless strip method.

8.6 Using Green Materials in Copenhagen, Denmark

TK-33 is a rural home located north of Copenhagen designed for a senior couple, by the Danish firm Tegnestuen LOKAL (Fig. 8.15). The project practices sustainability through its careful selection of materials, cutting down carbon footprint while providing a unique residential identity. The flexible space of the T-shaped TK-33 building includes a living space, open kitchen, and dining area all connected to a L-shaped, south-facing deck. Not having walls allows this space to be rearranged to accommodate future owners as well. Floor-to-ceiling glazing on the central wall takes full advantage of the southwest sun (Wang 2018).

The architects started the selection of materials by identifying those that make a large impact on carbon dioxide emissions. In a typical Danish home, concrete and brick are the primary materials. Concrete is a large cause of CO_2 emissions due to the production process of cement. Brick also makes a sizable impact during its production and because of its low reusability. In an environmental analysis of the construction

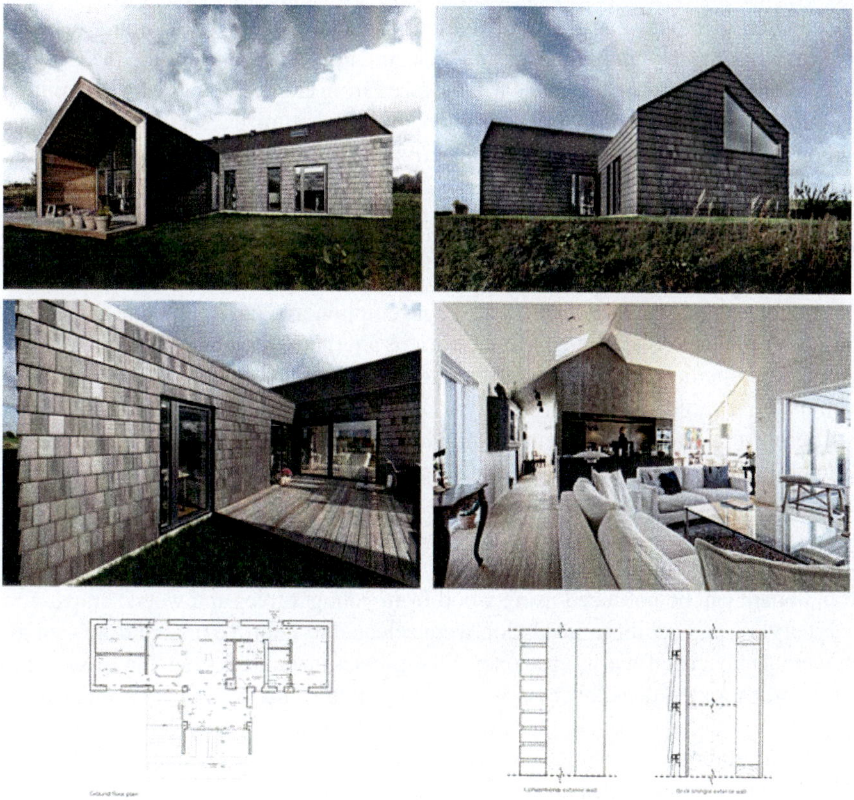

Fig. 8.15 The TK-33 home practices sustainability through its careful selection of materials and cutting down the carbon footprint while providing a unique identity for the house

of a typical house, outer walls make up nearly 30% of total emissions. In the design of TK-33, the load-bearing structure that is typically made of concrete was replaced by timber framing. It is a material that is both renewable and has lower emissions in the extraction and manufacturing stages. A thin layer of brick shingles envelope the facades and its reusability are ensured by its consumer-to-consumer (C2C) certification, in case replacements need to be made. The cladding brick is high in durability, increasing the lifespan of the wooden framing underneath and decreasing the emissions impact. The architects designed TK-33 with the environmental consequences in mind, not only in its construction stage but also considering the effects after years of use. Despite limitations in material choice, the project embodies a comfortable, ideal style of modern living that showcases the feasibility of sustainable design.

8.7 The Green Grow Home

The Green Grow Home is an elaboration of the Grow Home concept that was outlined in Chap. 3. The expansion focuses on integration and use of environmentally sustainable materials in the unit's interior and exterior (Fig. 8.16). During the design, the sustainability of building materials was assessed from cradle-to-cradle by examining the impact of all stages of their use. The goal was to optimize selection of materials by considering their environmental impact, quality, durability, and cost (Friedman and Cammalleri 1995).

A quantitative life cycle and circular analysis of materials was first completed to achieve an understanding of how components were acquired, processed, packaged, distributed, assembled, used, and disposed of; allowing designers and builders to make choices that would reduce energy consumption and environmental impact of the home. Tables were created outlining various alternatives for building components with their embodied energy for the proposed quantity, along with specifications for each material choice.

The most sustainable element of the Green Grow Home came from the use of structural wood because it is a renewable resource with low embodied energy, and it is easily recycled and reused (Fig. 8.17). Strand boards were selected in place of plywood for floors, walls, and roof sheathing, reducing the embodied energy for these components by 59%. Both materials have similar strength and durability, however strand board can be produced using wood from younger trees and wood scraps.

Fiberglass and cellulose insulation were selected to improve the building's environmental impact and energy efficiency. Fiberglass is made from sand, limestone, and borax; the mining process necessary to acquire these components can have negative consequences for biodiversity, air quality, and water quality. Supplementing fiberglass with cellulose insulation, a product derived from recycled newsprint, helped to offset the environmental impact of insulating materials while improving the thermal performance of the home. On the exterior, cedar boards and shingles were selected in place of vinyl siding and asphalt, saving 41,076 MJ of energy.

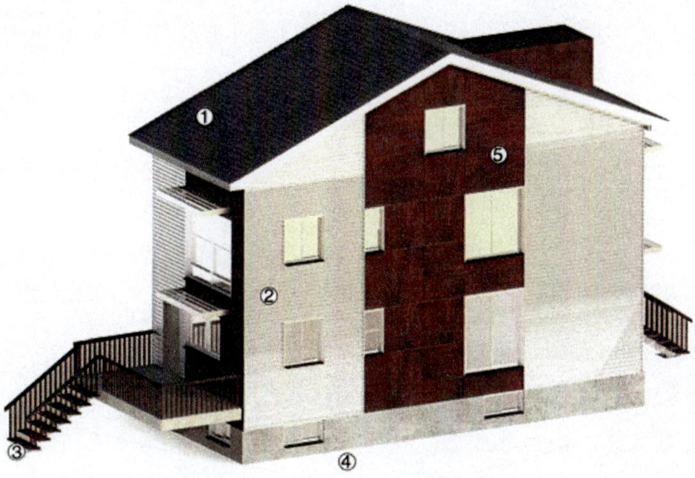

Fig. 8.16 During the design of the Green Grow home, the sustainability of building materials was assessed from "cradle-to-cradle" by examining their environmental impact, quality, durability, and cost

Materials were also assessed for their impact on indoor environmental quality, particularly for their potential to emit VOCs into the home (Fig. 8.18). Carpets can emit excessive quantities of VOCs; they are also more likely to trap dust and soil particles. Following the R-2000 standard, carpet was limited to a maximum of 50% of the interior floor area. Vinyl, often used in bathrooms and kitchens, is also responsible for VOC emissions, its production can be harmful to the environment, and it is high in embodied energy. By replacing carpet and vinyl flooring with parquetry and ceramic tile in the Green Grow Home, 12,342 MJ of embodied energy was saved.

These modifications to the materials used in the original Grow Home would result in 84,000 MJ of energy savings, equal to the energy used to heat the home for four years. The selection of materials for the Green Grow Home demonstrates how building components can be chosen by builders and designers to reduce the embodied energy and improve the sustainability of houses.

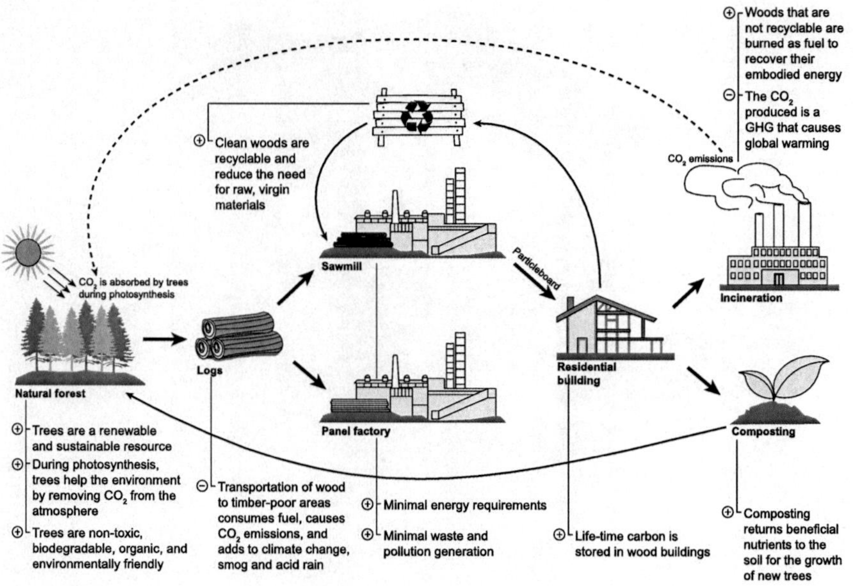

Fig. 8.17 The most sustainable element of the Green Grow Home came from the use of structural wood because it is a renewable resource with very low embodied energy

8.8 Final Thoughts

The commercialization of innovative materials highlighted above requires more funding and in-depth research. Although some materials exist, they are still rarely implemented due to minimal information about them in building codes. Therefore, the full integration of these relatively new materials requires changes to the regulatory level. These regulatory changes would also be underpinned by cultural influences that vary from region to region, which could slow down the integrative process of "green" materials into mainstream construction practices. The common theme in this chapter is the necessity to diversify our materials and practice concepts and use cradle-to-cradle assessment and use design-for-disassembly to enforce a full-fledged circular economy. None of the "green" materials and products should solely be relied upon, and if opted for, should be locally sourced and built within climates that are supportive to their maintenance and therefore maximize their useful life. It is a moral imperative to integrate more "green" materials and products into construction practices as soon as possible, to live sustainably in harmony with the environment.

Questions for a Follow-Up Discussion

1. Why is circular approach has become critical to homebuilding?
2. What is a material passport and how it can affect choice of products?
3. Name 3 products made out of recycled material and their use in home building?

Fig. 8.18 In the Green Grow Home materials were assessed for their impact on indoor environmental quality, particularly for their potential to emit volatile organic compounds (VOCs)

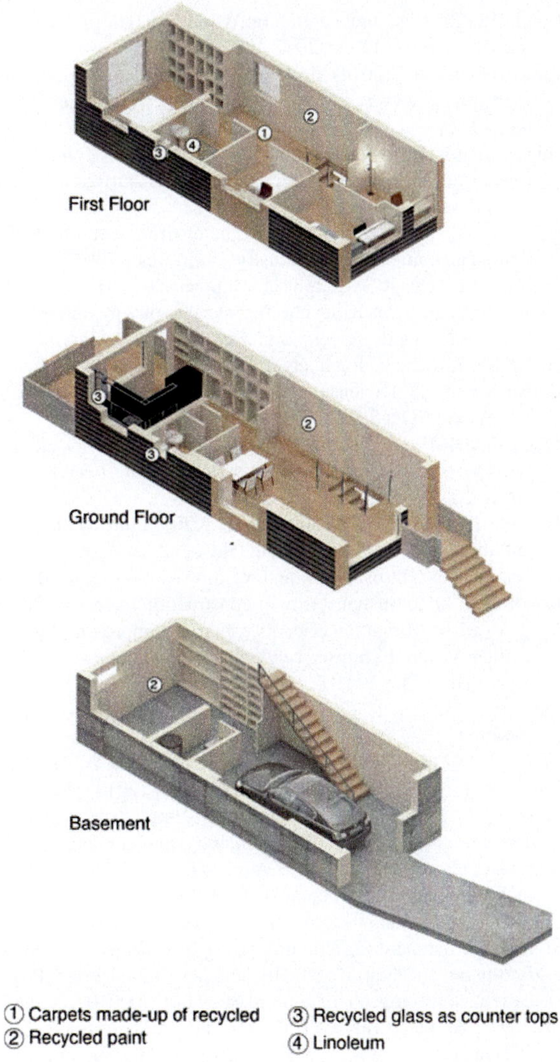

References

Abdellatef Y, Khan MA, Khan A, Alam MI, Kavgic M (2020) Mechanical, thermal, and moisture buffering properties of novel insulating hemp-lime composite building materials. In: MDPI. https://www.mdpi.com/1996-1944/13/21/5000. Accessed 11 Oct 2022

Allan K, Phillips AR (2021) Comparative cradle-to-grave life cycle assessment of low and mid-rise mass timber buildings with equivalent structural steel alternatives. In: MDPI. https://www.mdpi.com/2071-1050/13/6/3401. Accessed 11 Oct 2022

Arnold RD, Wade JP (2015) A definition of systems thinking: a systems approach. In: Procedia Computer Science. https://www.sciencedirect.com/science/article/pii/S1877050915002860. Accessed 11 Oct 2022

Cantrell J (2022) Natural capital. In: World Bank. https://www.worldbank.org/en/topic/natural-capital#1. Accessed 11 Oct 2022

Easton D, Easton T (2014) Modern rammed earth construction techniques. In: Modern Earth Buildings. https://www.sciencedirect.com/science/article/pii/B9780857090263500149. Accessed 11 Oct 2022

Ellen MacArthur Foundation (2022) New to circular economy overview. In: Ellen MacArthur Foundation. https://ellenmacarthurfoundation.org/topics/circular-economy-introduction/overview. Accessed 11 Oct 2022

Friedman A, Cammalleri V (1995) The environmental impact of building materials in the North American home building industry. Build Res Inf 23(3):162–166

Garcia AL, Achaiah AT, Bello J, Donovan T (2017) Ferrock final paper 4.24.17 Johnbello—IronKast. In: Iron Kast. http://ironkast.com/wp-content/uploads/2017/11/USC-Ferrock-Final-Paper-4.24.17.pdf. Accessed 11 Oct 2022

Honic M, Kovacic I, Rechberger H (2019) IOPscience. In: IOP Conference Series: Earth and Environmental Science. https://iopscience.iop.org/article/10.1088/1755-1315/225/1/012073. Accessed 11 Oct 2022

Jelle BP (2016) Nano-based thermal insulation for energy-efficient buildings. In: Start-Up Creation. https://www.sciencedirect.com/science/article/pii/B978008100546000008X. Accessed 11 Oct 2022

Jowitt SM, Mudd GM, Thompson JFH (2020) Future availability of non-renewable metal resources and the influence of environmental, social, and governance conflicts on metal production. In: Nature News. https://www.nature.com/articles/s43247-020-0011-0. Accessed 11 Oct 2022

Khatib JM (2016) Sustainability of construction materials, 2nd edn. Elsevier, Amsterdam

King A (2020) Mitigating criticality, part III: improving the stewardship of existing supplies. In: Critical Materials. https://www.sciencedirect.com/science/article/pii/B9780128187890000074. Accessed 11 Oct 2022

Latha H, Arunraj E (2014) Compressive strength of insulated concrete form blocks. In: Research Gate. https://www.researchgate.net/publication/261798817_COMPRESSIVE_STRENGTH_OF_INSULATED_CONCRETE_FORM_BLOCKS. Accessed 11 Oct 2022

Longarini N, Crespi P, Zucca M, Giordano N (2014) (PDF) The use of fly ash in high strength concrete mix design. In: Research Gate. https://www.researchgate.net/publication/282605450_The_use_of_fly_ash_in_high_strength_concrete_mix_design. Accessed 11 Oct 2022

Morel J-C, Charef R, Hamard E, Fabbri A, Beckett C, Bui Q-B (2021) Earth as construction material in the circular economy context: practitioner perspectives on barriers to overcome. In: The Royal Society Publishing. https://doi.org/10.1098/rstb.2020.0182. Accessed 11 Oct 2022

Moreno M, Lampard C, Williams N, Lago N, Emmett S, Goikoetxea M, Barandiaran MJ (2015) Eco-paints from bio-based fatty acid derivative latexes. Progress Org Coat. https://www.sciencedirect.com/science/article/pii/S0300944015000107. Accessed 11 Oct 2022

Murray A, Skene K, Haynes K (2015) The circular economy: an interdisciplinary ...—JSTOR. In: JSTOR. https://www.jstor.org/stable/44164300. Accessed 11 Oct 2022

New PT (NPT) (2019) NPT—New Polyurethane Technologies—sealing and bonding ... In: Adhesives and sealants for wood flooring and parquetry. https://www.nptsrl.com/wp-content/uploads/2020/06/CATALOGO-PARQUET-NPT-2019-ENG-mod.pdf. Accessed 11 Oct 2022

O'Ceallaigh C, Conway M, Mehra S, Harte AM (2021) Numerical investigation of reinforcement of timber elements in compression perpendicular to the grain using densified wood dowels. Constr Build Mater. https://www.sciencedirect.com/science/article/pii/S0950061821007509. Accessed 11 Oct 2022

Pal RK, Goyal P, Sehgal S (2021) Effect of cellulose fibre based insulation on thermal performance of buildings. Mater Today: Proceedings. https://www.sciencedirect.com/science/article/pii/S2214785321019027?via%3Dihub. Accessed 11 Oct 2022

Rosso F, Pisello AL, Pigliautile I, Cavalaglio G, Coccia V (2020) Natural, bio-based, colored linoleum: design, preparation, characteristics and preliminary life cycle assessment. J Cleaner

Prod. https://www.sciencedirect.com/science/article/abs/pii/S0959652620322496. Accessed 11 Oct 2022

Simpson A (2021) The effect of moisture on the thermal property of a reed thatch roof during the UK heating season. Energy Buildings. https://www.sciencedirect.com/science/article/abs/pii/S0378778821010616. Accessed 11 Oct 2022

Soo VK, Peeters J, Paraskevas D, Compston P, Doolan M, Dufleu JR (2017) Sustainable aluminium recycling of end-of-life products: a joining techniques perspective. J Cleaner Prod. https://www.sciencedirect.com/science/article/abs/pii/S0959652617332298. Accessed 11 Oct 2022

Sotayo A, Bradley D, Bather M, Sareh P, Oudjene M, El-Houjeyri I, Harte AM, Mehra S, O'Ceallaigh C, Haller P, Namari S, Makradi A, Belouettar S, Bouhala L, Deneufbourg F, Guan Z (2019) Review of State of the art of dowel laminated timber members and densified wood materials as sustainable engineered wood products for construction and building applications. Dev Built Environ. https://www.sciencedirect.com/science/article/pii/S2666165919300043. Accessed 11 Oct 2022

Stahel WR (2016) The circular economy. Nature. https://pubmed.ncbi.nlm.nih.gov/27008952/. Accessed 30 Jan 2022

Sterley M, Serrano E, Källander B (2021) Building and construction: timber engineering and wood-based products. In: Adhesive Bonding, 2nd edn. https://www.sciencedirect.com/science/article/pii/B9780128199541000216. Accessed 11 Oct 2022

Suzuki N, Nakayama Y, Nakaoka H, Takaguchi K, Tsumura K, Hanazato M, Hayashi T, Mori C (2021) Risk factors for the onset of sick building syndrome: a cross-sectional survey of housing and health in Japan. Build Environ. https://www.sciencedirect.com/science/article/pii/S0360132321003802. Accessed 11 Oct 2022

Tang W, Khavarian M, Yousefi A, Cui H (2020) Properties of self-compacting concrete with recycled concrete aggregates. Self-Compacting Concrete: Materials, Properties and Applications. https://www.sciencedirect.com/science/article/pii/B978012817369500009X. Accessed 11 Oct 2022

Thongchareon N, Khongtong S, Srivaro S, Wisadsatorn S, Chub-Uppakarn T, Chaowana P (2021) Development of structural insulated panels made from wood-composite boards and natural rubber foam. Polymers. https://pubmed.ncbi.nlm.nih.gov/34372099/. Accessed 11 Oct 2022

Union of Concerned Scientists (2016) What's driving deforestation? In: Union of Concerned Scientists. https://www.ucsusa.org/resources/whats-driving-deforestation. Accessed 11 Oct 2022

Velenturf APM, Archer SA, Gomes HI, Christgen B, Lag-Brotons AJ (2019) Circular economy and the matter of integrated resources. Sci Total Environ. https://www.sciencedirect.com/science/article/pii/S0048969719330220. Accessed 11 Oct 2022

Vermont Journal of Environmental Law (2014) Vol. 14, no. 4, Summer 2013 of Vermont Journal ...—jstor.org. Vermont J Environ Law. https://www.jstor.org/stable/vermjenvilaw.14.issue-4. Accessed 11 Oct 2022

Wang L (2018) Danish home champions wood over concrete for lower carbon emissions. Inhabitat. https://inhabitat.com/danish-home-champions-wood-over-concrete-for-lower-carbon-emissions/. Accessed 11 Oct 2022

Whitelaw F (2014) Bamboo build—jstor.org. In: JSTOR201. https://www.jstor.org/stable/sanctuary.28.72. Accessed 11 Oct 2022

Chapter 9
Energy Efficient Dwellings

Abstract Rising global energy demands in the residential sector are prompting design and construction of energy-efficient dwellings. Interest in such dwellings is also driven by climate change and higher costs. To account for these challenges, architects and homebuilders are focusing on designing and implementing innovative heating and cooling practices. This chapter outlines a selection of efficient systems, Net-Zero homes which produce on site power as much as they consume, renewable energy sources, landscaping for reduced consumption, and the monitoring of energy performance.

Keywords Energy retrofits · Energy efficient dwellings · Geothermal heat pumps · Heat pumps · High carbon housing · Low energy building · Low energy building · Micro hydropower · Micro-wing turbines · Net-zero homes · Passive solar design · Positive energy homes · Renewable energy · Solar heating system · Thermal mass

9.1 A Need for Energy Efficient Dwellings

Climate change is forcing governments and the construction sector to reevaluate the current methods of powering homes. Buildings are known to consume 30–40 percent of yearly primary energy produced in developed countries, and approximately 15–25 percent in developing countries (Wu and Skye 2021). Furthermore, rapid population growth and the expansion in home sizes have contributed to rising energy consumption (Stephan and Crawford 2016). These effects are exemplified by the construction industry accounting for 38 percent of all energy-related carbon dioxide emissions (Rukikaire and Collins 2020). Heating and cooling are energy intensive systems that consume most of a household's monthly energy. Therefore, selecting energy efficient heating means is critical in providing comfortable living conditions while also protecting the environment (Hydro Quebec, n.d.).

Energy efficient dwellings refer to those that consume less energy than conventionally built home while performing the same task (EESI, n.d.). Building energy efficient dwellings powered by renewable energy presents an opportunity for homeowners and communities to significantly reduce their *carbon footprint* which is defined as the

A. Friedman, *Fundamentals of Innovative Sustainable Homes Design and Construction*,
The Urban Book Series, https://doi.org/10.1007/978-3-031-35368-0_9

total amount of greenhouse gasses (GHG) that are generated by the occupants' actions (NC, n.d.). Taking advantage of a variety of renewable energy methods, landscaping for energy efficiency, and site planning considerations are the strategies and tools for environmental conservation which will be highlighted below.

Low carbon housing is also known to have a smaller *ecological footprint,* which is the amount of biologically productive land necessary to support a person's level of consumption and waste generation (Hopton and White 2011). At present, the burning of fossil fuels is one of the key causes of global warming. Initiatives to minimize the consumption of primary energy and fossil fuels will effectively reduce the size of carbon footprints, and subsequently, the ecological footprint as well (Hopton and White 2011).

Energy efficient buildings are well built and properly insulated. Therefore, an airtight house will minimize thermal losses (Ng et al. 2018). Heat can leak through the building envelope or through ventilation ducts (Fig. 9.1). The use of materials with high insulating values will therefore lead to lowered energy consumption (EPA 2009). Ventilation systems that incorporate heat recovery methods will also contribute to lower consumption and cost (Ng et al. 2018) (Fig. 9.2). A good design will focus on keeping heat in, making use of passive heat gains during winter months yet will lower summer solar gain to avoid overheating (Fig. 9.3). Furthermore, automated control systems for lighting and heating for example will increase the level of comfort in a home while also conserving energy.

Although in recent decades housing has become more energy efficient, household's energy use and related GHG emissions are not shrinking due to the expanding use of home appliances and information technologies among other factors (Goldstein et al. 2020; Lelieveld et al. 2019). Concurrent energy retrofits and reduced in home

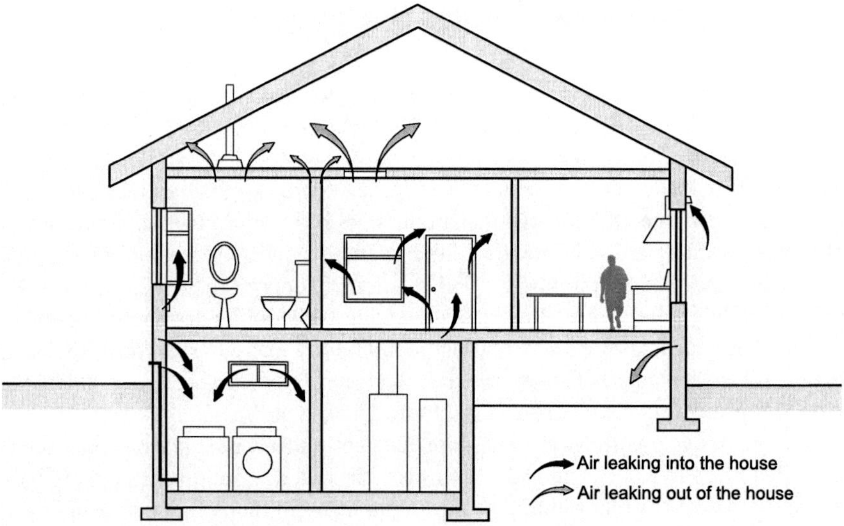

Fig. 9.1 Common points of air leakage into and out of a home

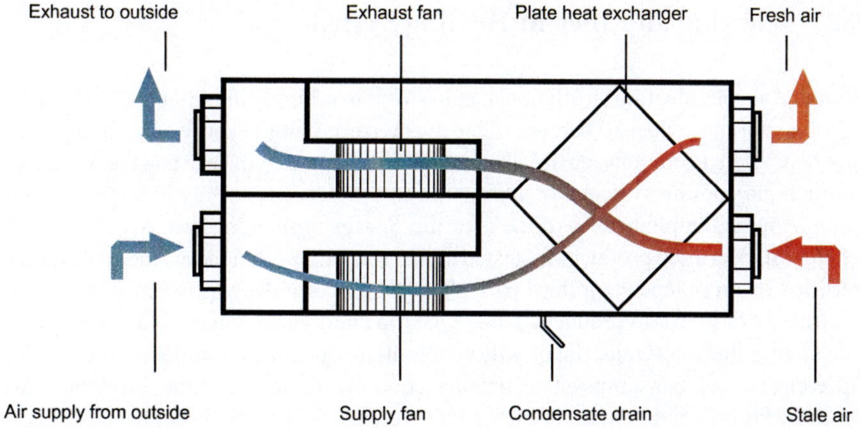

Fig. 9.2 Ventilation systems that incorporate heat recovery methods will contribute to lower energy consumption and cost

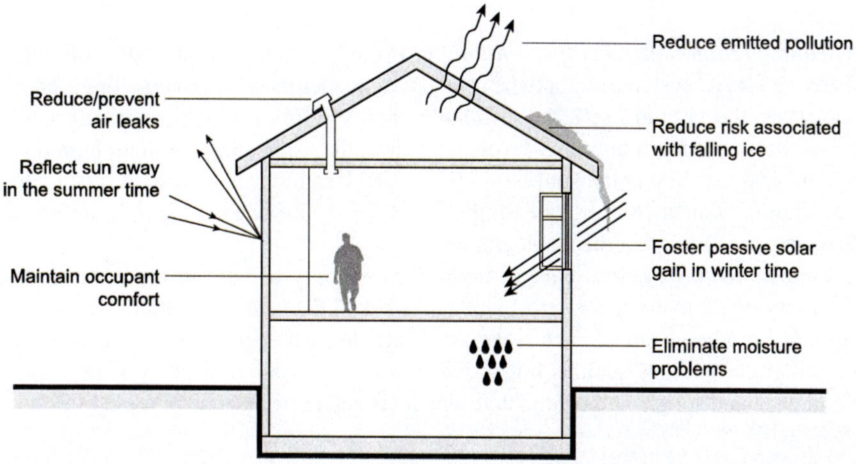

Fig. 9.3 A good design focuses on keeping heat in, making use of passive heat gains yet lowers summer solar gain to avoid overheating

fuel use through efficient heating systems are the recommended approach for energy conservation (Goldstein et al. 2020). Furthermore, the building of smaller homes through denser planning patterns, shape, orientation, and landscaping will play an important role in conserving energy. The details and characteristics of these factors will be further explored below.

9.2 Selecting an Efficient Heating System

As noted above selecting an efficient heating system is highly important, since heating accounts for more than 50 percent of the average household energy consumption (Li and Just 2018). Preferably, energy from renewable sources will be used to power those heating and cooling systems. In addition, a way to identify energy efficient heating equipment and appliances is to look for the *Energy Star* logo (NRCAN 2021). All Energy Star certified products are tested to meet strict US efficiency standards and are certified by an independent third party. The *Energy Star Most Efficient* designation is offered every year to products that are the top energy performers (NRCAN 2021). Consulting the EnerGuide label will enable homeowners and builders to compare different energy consumption of heating equipment and therefore select the best product. Further elaboration on the systems' choice will be made below.

9.2.1 *Commonly Available Heating Systems*

A *heating, ventilation, and air conditioning* (HVAC) system controls indoor climate. Different HVAC systems ought to be used based on factors such as a dwelling's characteristics, the system itself, and available energy sources to achieve comfortable temperature, optimum humidity levels, and clean air via heating, cooling, humidification, air purification, and ventilation (Fig. 9.4). Heating is a process of raising the temperature of an enclosed space for the primary purpose of ensuring the comfort of the occupants through a mechanical means.

Central heating systems have a single power supply which can heat a house by distributing heat to every room through a network of ducts or pipes. It generates heat by transforming a sourced chemical energy into thermal energy. The most common are radiation methods such as hot water, steam, or hot air furnace systems which circulate heat through walls or the floor via ducts and registers, and pumps (Martínez et al. 2019).

In some locations *hydronic radiant central heating* has been replaced with radiant floors which are gaining popularity. Rather than heating the air, the heat radiating from the floor warms people, objects, walls, and ceilings. They maintain optimal humidity levels by not drying the air, operate with minimal noise, and reduce dust and allergy risks. No maintenance is required, making it a preferred system for energy saving (Group 2022).

Forced air central heating utilizes a furnace or heat pump to heat the air and then disperses it through the house via ductwork and in-room vents (Fig. 9.5). Cold air from the home is pulled into the system where it passes through the air filter eliminating allergens such as pollen and dust (Rardin 2015). This system provides a constant and uniform comfort level and increases air flow throughout the house. The main energy-saving mechanisms in forced air central heating are through the installation of heat pumps. The permanently mounted *baseboard heater* is the most

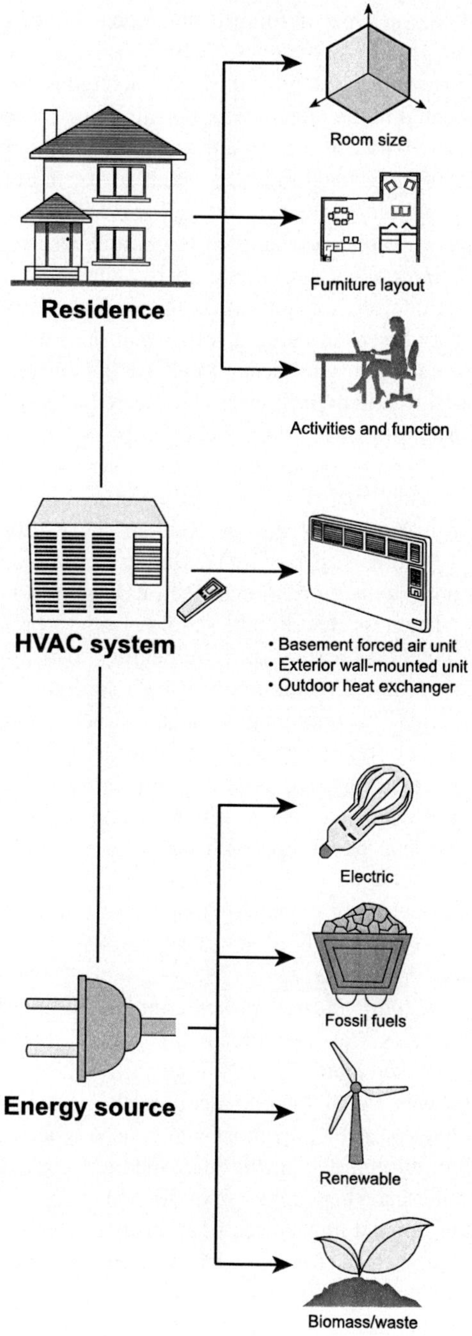

Fig. 9.4 Factors such as the dwelling's characteristics, the system itself, and the energy source will be included in the choice of a heating, ventilation and air conditioning (HVAC) system

common type of room heater. Ideally installed under windows, they have a 100 percent gross efficiency. Their effectiveness, however, is significantly compromised, since only a portion of the heat delivered is used to heat the air. Individual controls are located in the different rooms, allowing independent temperature control. The time delay between temperature readings using conventional thermostat and baseboard reaction is long and often leads to wasted energy and very warm temperatures in the vicinity of the heater. Electric baseboard systems require a substantial amount of power to operate, but their low initial cost explains their popularity.

Heat pumps are essentially two furnaces combined into a single unit. One furnace is a compressor that draws in external air to the building during the cold weather heating season using a refrigerant, and the other furnace is an electric resistance heater. Almost all heat pumps use forced warm air delivery systems to move heated air throughout the house. There are two common types: air-source and ground source heat pumps. Air source heat pumps use transfer coils to move heat from the outside into the interior, and therefore can deliver up to four times more heat energy than the electrical energy they consume during operation (Martínez et al. 2019). The design of heat pumps is energy efficient because they transfer heat rather than generate. Therefore, these passive heating systems require 60 percent less energy than standard homes with conventional electric resistance-based heating systems (NRCAN 2019).

Ductless Heat Pumps should also be a preferred heating system due to its energy efficient properties and save heating costs by 50 percent (NR Canada 2019). Ductless mini-split heat pumps use compressors and fans that can adjust speeds to save energy. They require minimal construction, are small and more efficient, as they have no ducts, reducing the potential for leaks and energy loss (Martínez et al. 2019). The ductless heat pump is a multipurpose system with the ability to customize the conditions within each room which are adjustable through wall consoles, remote controls, and smartphone apps (NRCAN 2019). Interior units are attached to the walls to connect to an exterior unit using narrow refrigerant lines. This heat pump is suitable for open-floor plans, distributes energy through refrigerant lines instead of water or air, and is widely used across various climates, as it performs well in both hot and cold environments.

A *solar thermal system* contains components that collect, store, and distribute the sun's heat. The three main types of solar collectors are unglazed, flat plate, and evacuated tube (Fig. 9.6). *Unglazed collectors* consist of black plastic or metal pipes through which the fluid medium is circulated (Lamnatou and Chemisana 2021). *Flat plate collectors*, the most widely used type of solar system, consist of a flat insulated box through which the fluid medium circulates. The flat plate systems consist of water-carrying copper pipes bonded to copper absorber plates. The fluid is heated by the sun and transported to an insulated water tank, from which is transferred to the domestic water through a heat exchanger (Lamnatou and Chemisana 2021).

Another sustainable heating system is a *heat pump water heater* (HPWH). It can produce hot water temperatures up to 90 °C (194 °F) without any operational problems, and the primary energy consumption can be reduced by more than 75 percent compared with electrical systems (Gong and Sumathy 2016). Studies have shown that the most efficient water heating system is the combination of the solar

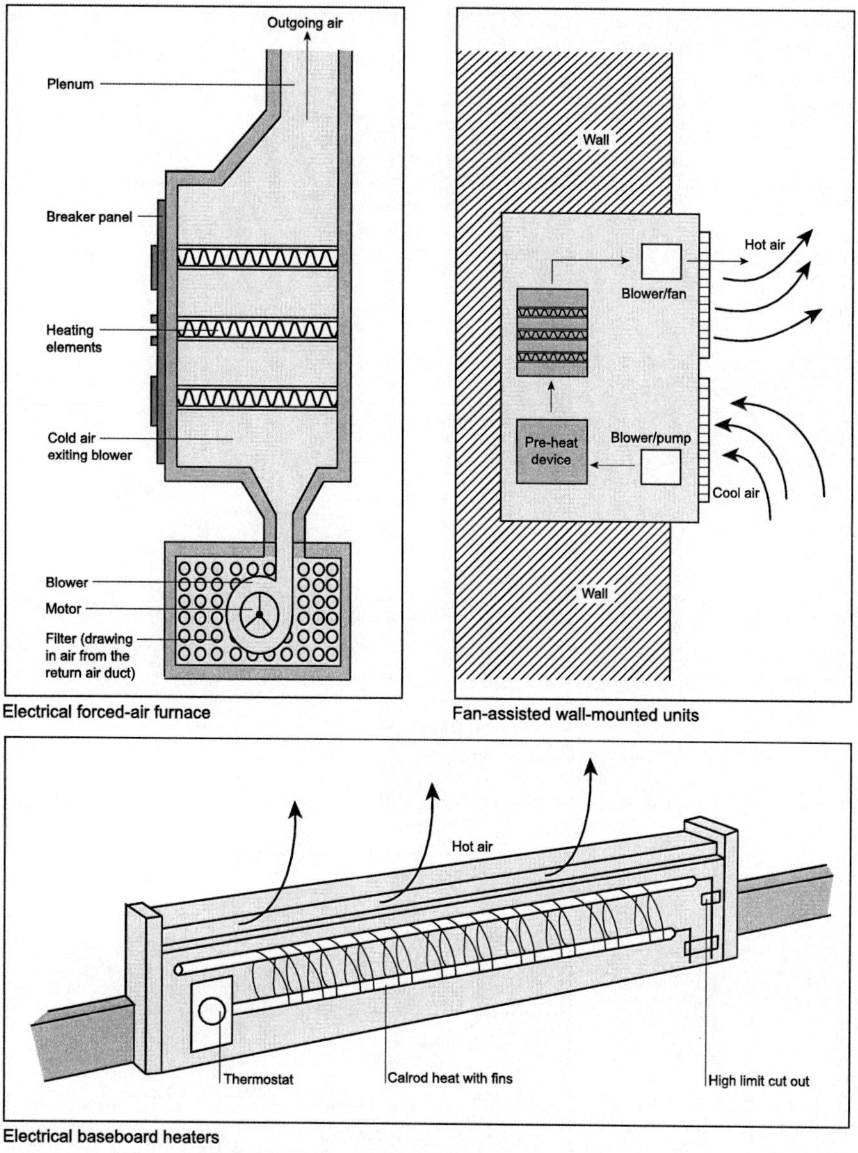

Fig. 9.5 Forced air central heating (top left and right) utilizes a furnace or heat pump to heat the air and then disperses it through the house via ductwork and in-room vents. The permanently mounted baseboard heater is the most common type of room heater. Ideally installed under windows, it has a 100 percent gross efficiency (bottom)

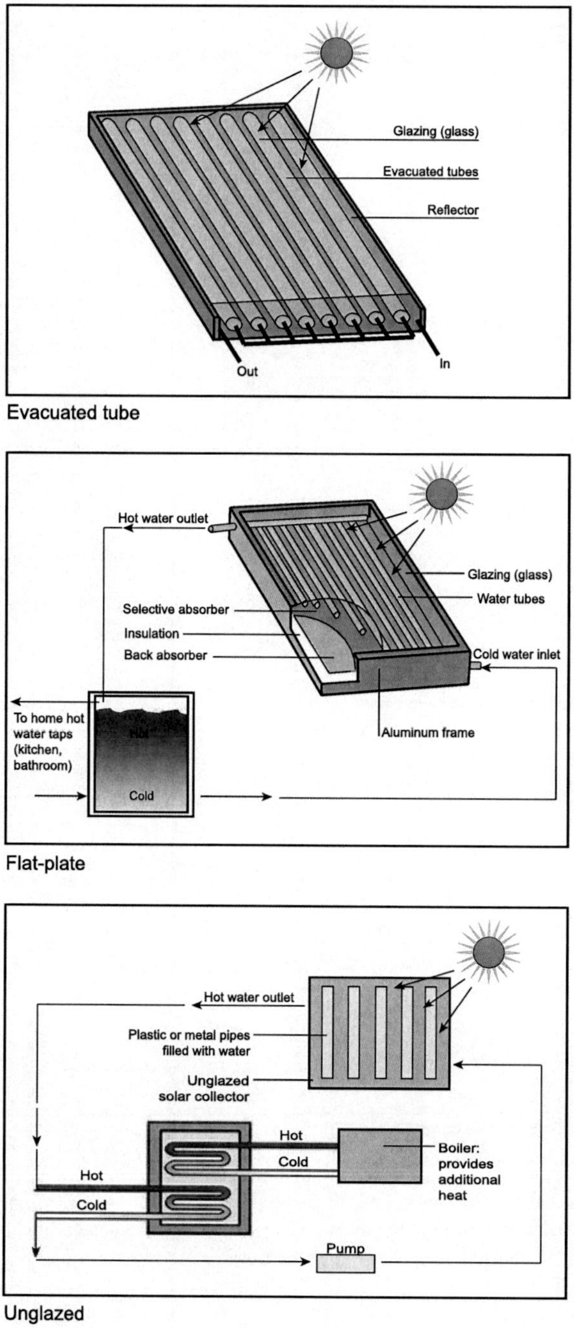

Fig. 9.6 The three main types of solar collectors are unglazed, flat plate, and evacuated tube

thermal preheat tank and a heat pump water heater (HPWH) (Balke et al. 2016). Modeling results reveal that this configuration is the most efficient of the systems, with the system's highest coefficient of performance (COPsys) being 2.87 (Balke et al. 2016). The efficiency of the solar thermal system with a heat pump as space conditioning equipment exceeded a standard resistance water heater by a factor of 2.4 and a standalone HPWH by a factor of 1.3 (Balke et al. 2016). Therefore, when selecting efficient heating systems, different combinations of mechanical systems should be installed, to optimize their function and therefore achieve a high degree of energy efficiency.

9.3 District Heating

District heating systems consist of infrastructures, where thermal energy is distributed to multiple buildings from a centralized plant through insulated underground piping networks (Fig. 9.7). The thermal energy is then transferred to the building's heating system without the need for boilers in each building or dwelling unit (IDEA 2019). This elaborate network can spread hot and cold water to heat and cool units with the water being returned to the main plant to be reheated (IDEA 2019).

District heating plays a key role in cost savings and energy security, since it uses local energy sources such as biomass and has a large potential to use available sources

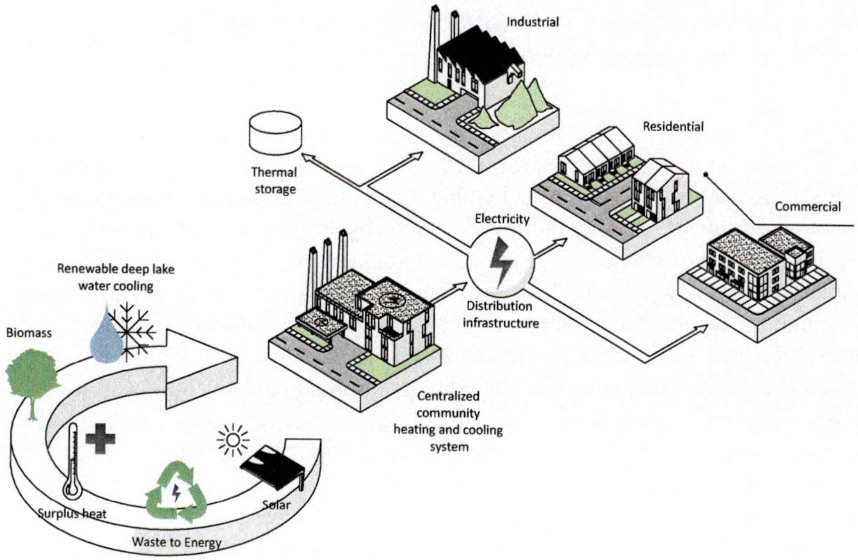

Fig. 9.7 In district heating systems thermal energy is distributed to multiple buildings from a centralized plant through insulated underground piping networks

of excess heat that would be unused otherwise (Falk 2021). A local energy system also prevents the need of importing and transferring energy from a further source to the district being powered, making it more efficient. The design of district heating is also favorable in integrating renewable energy sources into its grid system. In urban areas, district heating systems allow the wide use of combined heat and power and waste-to-energy sources. It also enables the integration of surplus heat sources such as industrial excess heat, and renewable heat sources such as geothermal and solar thermal heat (Falk 2021). The incorporation of heat pumps will increase the flexibility of district heating systems, by utilizing multiple sources of heat, including low temperature heat sources available naturally (Falk 2021). Therefore, the paired mechanism of district heating and heat pumps can harness geothermal energy and sewage or waste heat, by forcing heat flow from a lower heat source to a higher one using minimal amounts of electricity (Falk 2021).

The multi-faceted nature of district heating illustrates its advantages beyond that of lowering greenhouse gas emissions. Fundamentally, both district heating systems and Net-Zero-energy buildings operate using a similar grid system (IEA 2021). They harvest renewable energy in a centralized location, often a plant, and this heats or cools water which is then distributed to buildings within the neighborhood network (IEA 2021). Therefore, the concept of district heating systems is complementary when designing and constructing Net-Zero communities. However, prior to building Net-Zero communities, single Net-Zero dwellings must embody certain design principles to achieve energy efficient results which will be discussed below.

9.3.1 A Sustainable Community with District Heating in Stockholm, Sweden

Hammarby Sjöstad, where district heating is used, is recognized as one of the most sustainable communities in the world (Figs. 9.8 and 9.9). The project was initially planned as part of Stockholm's Olympic bid to host the most "environmentally conscious" games. Even after the city's rejection by the Olympic committee, officials recognized the importance of sustainable community design and so went forward with the 20,000-unit project. This was evidently a sound decision: Hammarby is now a community revered by environmentalists and planners worldwide—a true success in experimental sustainable development. The project is a refreshing approach to sustainable design. It is in fact an embodiment of principles such as transit-oriented development, mixed-use design, cyclist/pedestrian priority, carbon neutrality, and, of course, district heating net-zero energy consumption (White 2014).

An ENVAC waste collection system is used to dispose of the community's garbage and recycling. Pressurized air transports the waste to central collection points (Fig. 9.10). This eliminates the need for pick-up services and therefore heavily reduces the carbon produced in the process of waste collection. Only one-third of

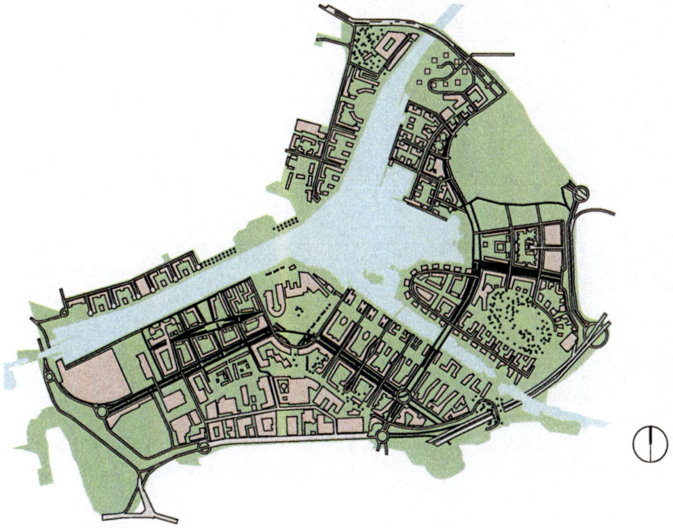

Fig. 9.8 Urban plan of Hammarby Sjöstad, near Stockholm, Sweden, where district heating is used, is recognized as one of the most sustainable communities in the world

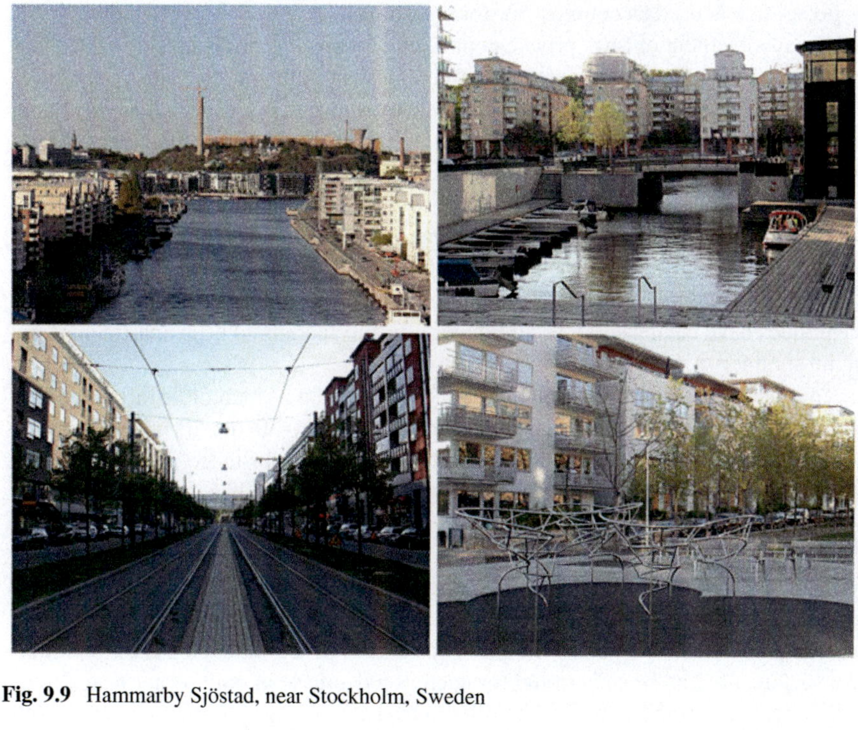

Fig. 9.9 Hammarby Sjöstad, near Stockholm, Sweden

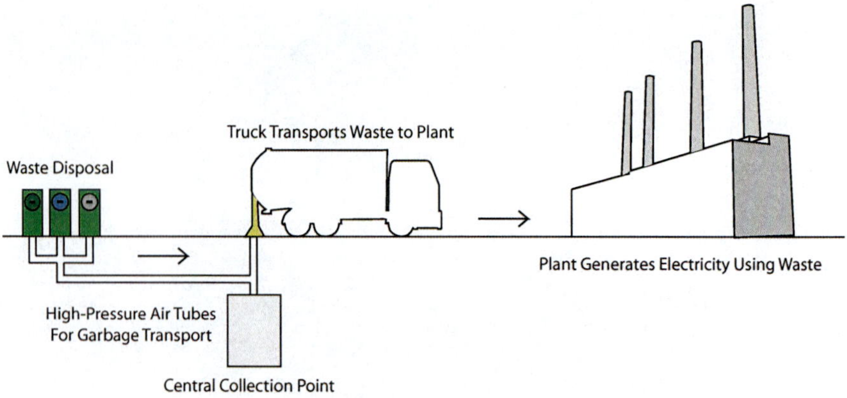

Fig. 9.10 In Hammarby Sjöstad an ENVAC waste collection system is used to dispose of the community's garbage and recycling (top). Tubes are used as part of the collection system (bottom)

the waste plant is devoted to incineration. The remaining two-thirds of are for emission management since 94 percent of emissions are released as water vapor. Liquid sewage is converted into heat and biogas, in turn used to power municipal buses. Solid sewage is used as compost for forested areas.

The involvement of forty private contracting companies drove up competition, as each wanted their contribution to be the "greenest" building on site. This competitive spirit gave rise to some of Hammarby's greatest achievements, especially in the realm of photovoltaic panels. Early on, contractors recognized the advantages conferred by installing solar panels on their structures. As a result, in the summer months, 50 percent of all energy used in Hammarby is sourced through photovoltaic panels. The solar energy harnessed, coupled with the energy produced by organic waste, together compose the vast majority of the district's supply—meaning that the community is run almost entirely on renewable energy. To further their achievements various techniques have been used to ensure maximum sun shading to keep apartments cool in the summer months.

The success of the Hammarby project lies not only in its incredible infrastructure, but also in the commitment of its citizens to sustainability. In the first place, Swedes, as a people, have widely embraced the green lifestyle. For instance, 79 percent of all Stockholmers walk, bike, or use public transport for their daily commute; and in the past ten years, bike usage has spiked 70 percent in the city. Seeking to improve on these already impressive commitments to environmental living, Hammarby's developer set up an education center to encourage pro-environmental behavior on a micro level. These initiatives have been highly successful in encouraging a variety of green behaviors such as water conservations. The success of the water conservation initiative specifically can be seen in figures such as a decrease in daily water usage from the citywide average of 200 L (53 gallons) to the Hammarby average of 150 L (40 gallons) per household.

9.4 Design and Construction Principles for Energy Efficiency Dwellings

Manipulating variables such as a unit's dimensions, configurations, size, shape, and the joining of units can save energy and as a result reduce the building's overall carbon footprint. It also stands to reduce heat loss and the amount of construction materials required, leading to cost savings without compromising the occupants' living comfort. These and other strategies will be outlined below.

9.4.1 Building Size and Shape

Building size has significant impact on energy consumption; downsizing to a smaller home will lower the carbon footprint, due to the reduced energy investment during construction and operation. Also, a result of downsizing will be cost savings on equipment since smaller volume will require less heating, cooling, and air conditioning (HVAC) systems. Dwelling size and type are the strongest predictors of residential energy consumption, making them focal points worth exploring, to reduce their environmental impacts (Huebner and Shipworth 2016). Given that the initial embodied energy of one square meter of floor area lies within 10–19 Gigajoules (GJ), each unit requires an additional 370–703 Gigajoules (GJ) for an increase in floor area as explained by Stephan and Crawford (2016). Furthermore, electricity use is strongly correlated with floor area and increases on average by 49 kWh (176.4 MJ) for every additional square meter of floor area. Therefore, the additional heating and cooling demands required for extra space offset a significant share of the energy and greenhouse gas emissions reductions that could have been achieved otherwise (Stephan and Crawford 2016).

The challenge in designing the interior of a small sized home efficiently is evidently given the minimal area available. To adapt to this limitation, architects should minimize the length of circulation paths and hallways, emphasize horizontal lines, use fewer partitions, slope ceilings, and select lighter colors, to make the living space feel larger through visual manipulation. An open concept is a type of floor plan where the walls and doors are removed, so the living area becomes one large space. An open-floor plan will also contribute to better interior heat management as there are fewer walls to block the flow of warm air from one area to the other.

Narrow and tall homes will enable warm air from lower levels to rise to the upper floors and homes with an open concept will let heat distributed evenly. On the other hand, a more complex shape with many corners will cause more wall and roof surfaces to be exposed, resulting in higher heat absorption during the day and greater heat loss at night as well as high labor and material costs (Cutland Consulting Ltd and Eco Design Consultants Ltd 2016).

A circle-shaped structure is the most energy efficient design because it encloses the greatest volume but has the least amount of surface area. Practically, however,

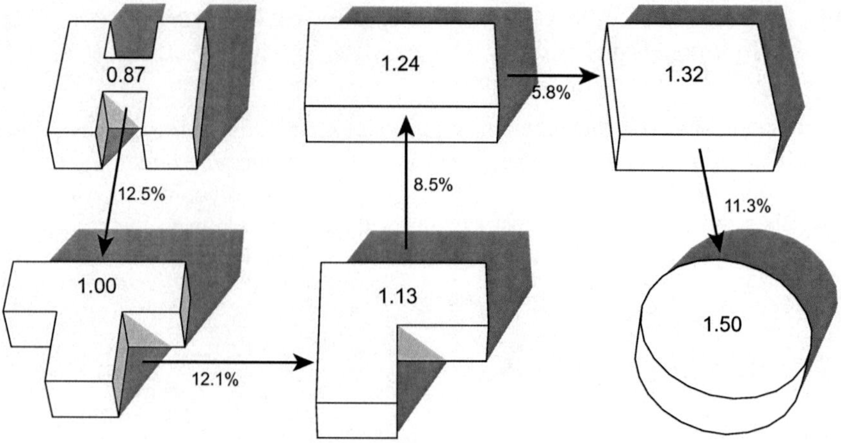

Fig. 9.11 A dwelling with a rectangular shape will have a good floor area to perimeter ratio (represented by the number in the shape) contributing to lower energy consumption and a smaller amount of construction material

rectangular and cubed homes are more efficient and easier to build (Fig. 9.11). In colder climates, cubed shapes reduce heat loss, whereas in more humid and warm climates, elongated shapes, such as rectangular designs can be opted for, so that cross ventilation for cooling is easier to achieve (Cutland Consulting Ltd and Eco Design Consultants Ltd 2016). Otherwise, making the floor plan less deep from north to south lets heat distribute from the north-facing rooms to other less sunny rooms (Donn and Thomas 2010). Furthermore, rectangular designs require less land, fit on narrower infill lots, and are easily grouped together to form neighborhoods, and therefore, will have an overall reduced carbon footprint (Donn and Thomas 2010).

A tall and more condensed home design has less surface area on its roof for a given volume and therefore slows down the loss of heat when warm air rises. A smaller roof would have less area available to absorb sunlight, which minimizes overheating and the need for cooling. Attached units are highly energy efficient because there are only two exterior walls, and the opportunity for cross ventilation is increased with front and rear openings (Cutland Consulting Ltd and Eco Design Consultants Ltd 2016) (Fig. 9.12).

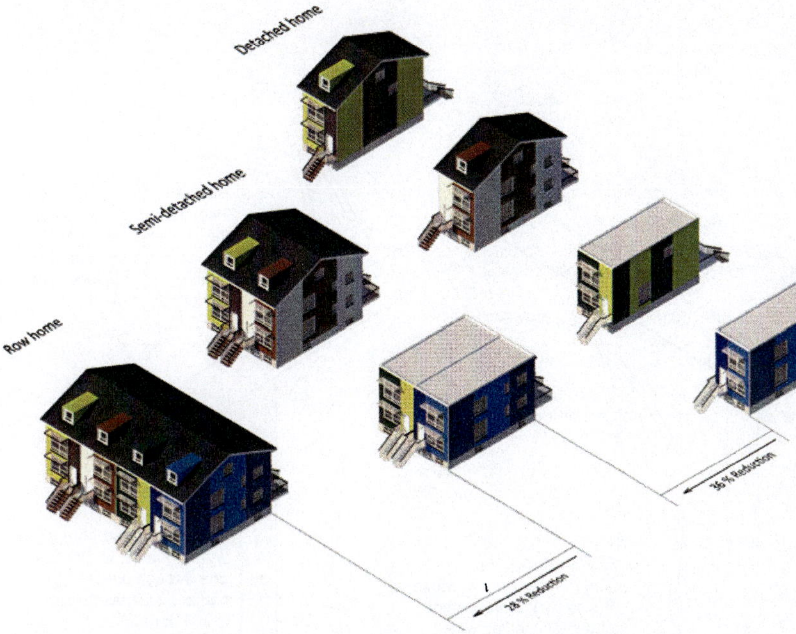

Fig. 9.12 Attached units are highly energy efficient since there are only two exterior walls, and the opportunity for cross ventilation is increased with front and rear openings

9.4.2 Thermal Insolation

Wall insulation is a critical contributing factor to a building's thermal performance and reduces its energy usage. Insulation's primary purpose is to slow down the transmission of heat via conduction, radiation, and convection through a building's walls, floor, ceiling, and roof (Martínez et al. 2019) (Fig. 9.13). An energy efficient dwelling should have robust insulation in the walls and its ceiling. Opting for good insulation with air sealing can deliver comfort and lower energy bills during the warmest and coldest times of the year. Insulation performance is measured by its R-value, used to determine its resistance to heat flow. R-value is physically represented by the ratio of a material's thickness divided by its thermal conductivity (Martínez et al. 2019). A higher R-value suggests more insulating capacity, and different R-values are recommended for walls, attics, basements, and crawl spaces, depending on the region (EPA 2009).

Since heat rises and escapes through the roof, investing in high-quality ceiling insulation will prevent heat loss. To be energy and cost efficient, the most convenient place to embed insulation is in the attic, where heat is frequently lost (EPA 2009). The recommended insulation level for most attics is R-38, however in cold climates, it is recommended to insulate up to R-49 (EPA 2009). Insulation works best when air is not moving through or around it, thereby sealing air leaks before installing insulation

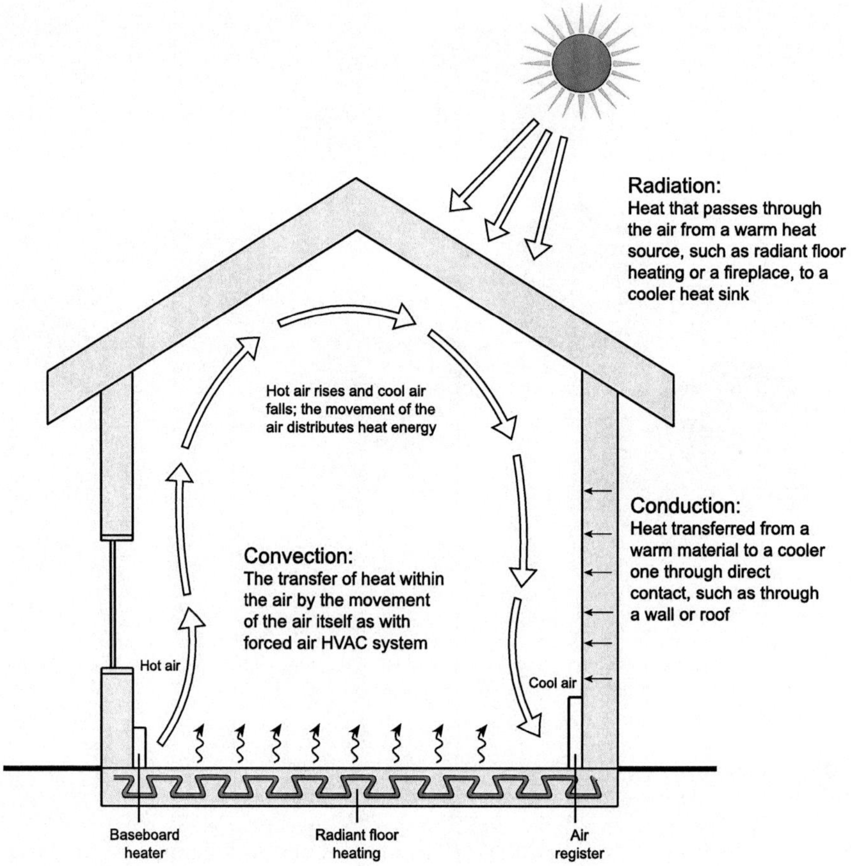

Fig. 9.13 Insulation's primary purpose is to slow down the transmission of heat via conduction, radiation, and convection through a building's walls, floors, ceilings, and roof

ensures that its heating effects are maximized (EPA 2009). Heat can leak through the building envelope or through ventilation ducts, therefore the use of materials with high insulating values will lead to lower energy consumption.

The more thermally insulating a material is, the slower heat transmission will occur, which will ultimately reduce heating and cooling requirements. To optimize insulation, one needs to ensure that the insulation fills the space completely and evenly since empty spots and corners will allow heat to bypass the insulation, reducing energy efficiency (Martínez et al. 2019). Thermal bridging which is any solid material that connects the warm side of the envelope to the cold side should also be minimized. When insulation is installed on one side of the thermal bridge, it acts like a barrier, therefore reducing heat loss. Loose-fill insulation must also be installed with the right thickness and density to be effective. The most used "green" insulating materials include fiberglass, mineral wool, and brown cellulose (Martínez et al. 2019).

Commonly, the thermal mass of a building envelope helps regulate temperature fluctuations within a given space (Donn and Thomas 2010). Thermal mass is the ability of a material to absorb, store, and release heat. When the sun's heat rays come in contact with a high thermal mass surface, the solar radiation is absorbed directly. Therefore, thermal mass is more effective when it receives direct sunlight and is dark in color, as darker surfaces absorb more heat compared to lighter surfaces. Simultaneously, thermal mass absorbs heat from the air within the home that is hotter than the thermal mass itself (Donn and Thomas 2010). The heat stored by the thermal mass is then released back into the room when the room temperature drops below that of the thermal mass, thus maintaining optimal thermal comfort. Thermal mass with 100 to150 mm (4 to 6 inch) concrete walls helps mitigate temperature fluctuations and also creates an acoustic barrier (Donn and Thomas 2010). Any dense and thicker material may take too long to heat, and those that are too thin would store insufficient heat. In most climates, a concrete slab insulated underneath floors when in direct contact with the ground is effective in increasing thermal mass (Donn and Thomas 2010).

9.4.3 Ground Cover and Passive Solar Design

The ground surrounding the home also plays a role in heat gain and loss. Dark colored paving such as asphalt can get very hot in the summertime, result in rising ambient heat causing overheating in the dwelling nearby. On another hand, having nearby grass and vegetation cover could absorb heat through evaporation and photosynthesis, resulting in the temperature surrounding the grass being considerably lower than bare earth or paving (Donn and Thomas 2010). Air flowing over grass, and into the home will therefore have a cooling effect for its surroundings in the warmer months.

Passive solar design is a strategy that takes advantage of the sun's heat when it is beneficial and avoids it when disadvantageous. A unit with passive solar design can achieve *passive solar gain*, which does not require mechanical means to gather and utilize the sun's energy. This design principle constitutes the most cost-effective approach to energy self-sufficient building and can reduce heating especially in cold climates. As a rule, for achieving the best passive solar design, architects working in the northern hemisphere can orient major openings toward the true south, limit windows on western and eastern elevations, and avoid using windows facing the north or extend the roof's overhung (Donn and Thomas 2010) (Fig. 9.14). Since little to no sunlight can reach the north facade, openings on that side will not contribute to solar gain. Therefore, the number and size of windows on the north facade should be minimized to prevent heat loss and will require greater insulation as it receives the least sun. On the other hand, south-facing windows are particularly beneficial to the dwelling, since the heat acquired through passive solar gain exceeds heat loss from the windows facing south. When considering shape and orientation, a rectangular home can benefit from passive solar gain. An elongated building sited parallel to an east–west axis will expose the longer south side to maximize heat gain, especially in the winter. The understanding of these dynamics can be achieved through analyzing the sun path diagram and applying it to the site of interest.

To further reduce heat loss, the floor plan of the dwelling can be arranged to create an insulating buffer and maximize solar input. Less commonly used areas, such as bedrooms, utility rooms, garages, or hallways, should be situated along the north side, while common spaces should face south (Donn and Thomas 2010).

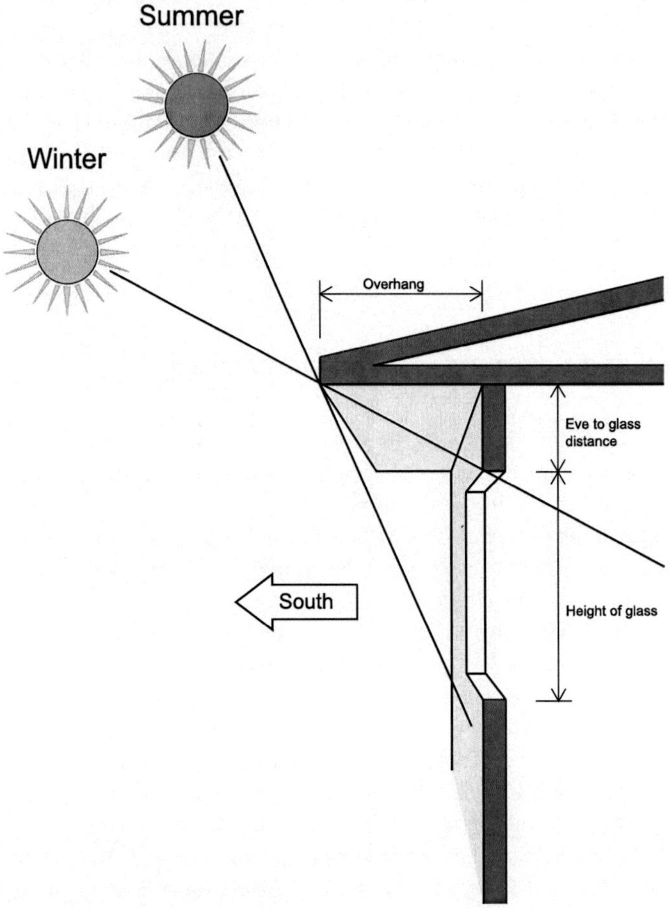

Fig. 9.14 For achieving the best passive solar design, architects can orient major openings toward the true south, limit windows on western and eastern elevations, and avoid using windows facing the north or extending the roof's overhung

9.5 Net-Zero Homes

The main feature of a *Net-Zero* home is that it does not rely on the common energy grid for power (Ng et al. 2018) (Fig. 9.15). The primary concept is that it generates its own power on site using renewable energy sources, which are then stored and used by the occupants (Wu and Skye 2021). Any excess energy created from the home's energy source can then be fed back into the power grid and used elsewhere. Net-Zero which are also known as *Off-the-Grid Housing* requires designers, builders, and occupants to be actively engaged in a culture of energy self-sufficiency. This shift in mindset would entail each unit being able to completely power itself through sustainable design and energy efficient heating systems.

Net-Zero homes have a very low carbon footprint and over time will emit no carbon dioxide. Therefore, to power a Net-Zero home is to use photovoltaic (PV) panels, wind turbines, and utilize heat production methods with geothermal pumps and solar heating systems (Fig. 9.16). The key characteristics of a Net-Zero home design are design for passive solar gain, proper insulation, innovative windows, low energy building principles, green roofs, and thermal mass. To achieve low energy consumption in Net-Zero homes, attention should be paid to high insulation levels, efficient heating and cooling systems, and the integration of renewable energy sources (Sarbu and Sebarchievici 2017).

Proper selection of windows is another key contributor in achieving a Net-Zero home. Windows are critical in preventing unwanted heat loss or gain, alongside insulation. Window systems should no longer be designed as mere aesthetics and should be regarded as a passive solar energy converter within the home. Implementing

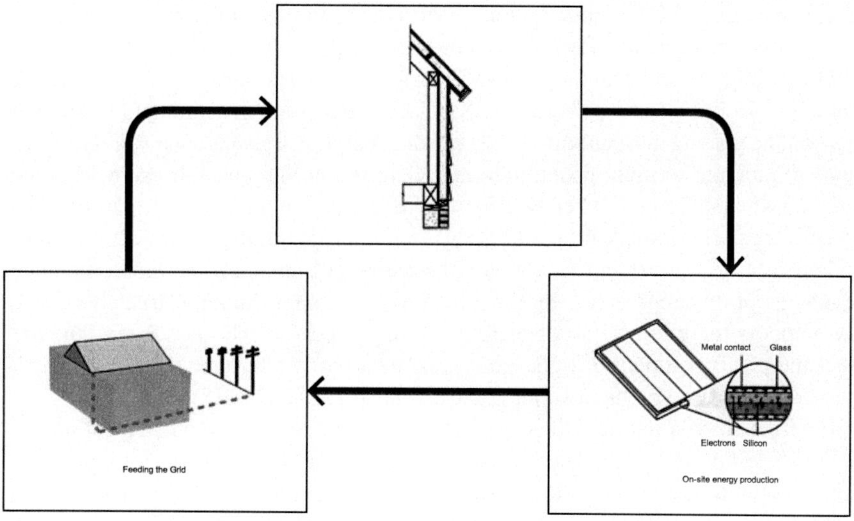

Fig. 9.15 The main features of a Net-Zero home are proper insulation, the use of renewable energy, and feeding the power grid with excess energy

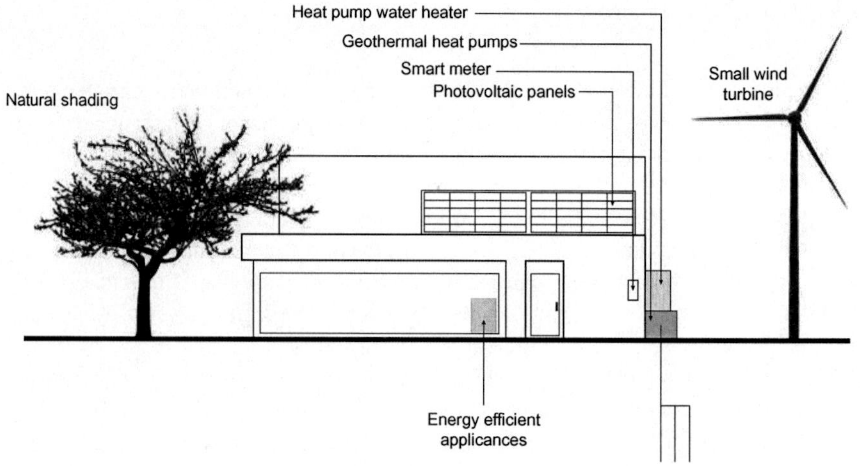

Fig. 9.16 A Net-Zero home can use photovoltaic (PV) panels, wind turbines, or geothermal pumps

multifunctional window systems would reduce the energy demand of buildings and achieve the goal of Net-Zero homes more easily. One of the more widely used types of glass in Net-Zero homes is krypton gas-filled low-emissivity glass, which can be available in double or triple pane windows and doors. Filling the gap between the glass panes with low-conductivity gas such as krypton is a viable alternative solution for thermally insulated glass windows (Ahmed et al. 2021). They are more efficient than traditional double pane windows due to their glazing systems that enhance thermal insulation (Fig. 9.17). The glazing technology includes the application of phase change materials (PCM), photovoltaic (PV), and vacuum glazing (VG) separately or in a hybrid structure, thus forming a multifunctional window (Ahmed et al. 2021).

Other innovative technologies and designs include *electrochromic dynamic windows* (Asdrubali and Desideri 2018). These are essentially dimmable windows that contain glazing that can shift the glass from opaque to transparent instantly, which helps to mitigate sunlight penetration and glare in interior space. It can be thought of as a switchable glass, where a burst of electricity can alter its transparency and reflectiveness. This small burst of energy can be programmed from wall switches, remote controls, movement sensors, light sensors, or timers. Furthermore, the main advantage of this approach is that, while giving occupants an unobstructed view to the outdoors there is no risk of glare discomfort because the glass is designed to meet the precise amount of light, glare, and heat passing through, or reflecting off the window desired by the dweller (Asdrubali and Desideri 2018).

External obstructions and the sun path must also be considered when locating and designing windows. For instance, an east–west oriented street will generally preclude sunlight admittance deeper into the street, and it is the top part of the façade facing south that will receive the most direct light (Donn and Thomas 2010). In general, direct sunlight on the south façade of a house is the easiest to manipulate since the sun has the highest altitude when it is due south. As a result, houses that make use

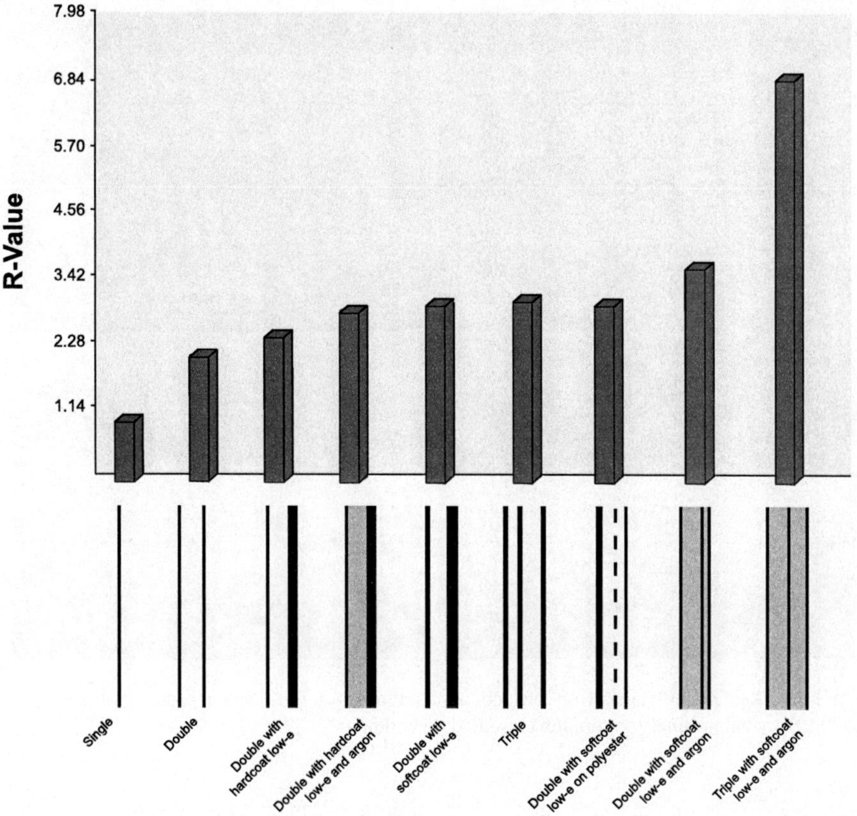

Fig. 9.17 Thermal performance of various glazing units

of natural daylight will be energy efficient, therefore reinforcing the importance of passive solar design. Natural lighting will reduce the need to power lights especially during daytime, while also offering health benefits to occupants to improve mood and productivity.

9.6 Renewable Energy Sources

There are a variety of renewable energy sources that in recent years found their ways to mainstream energy production either at the unit or the community levels (Fig. 9.18). Photovoltaic (PV) cells are renewable power generating technology suitable for an entire community or a single-family home (Sarbu and Sebarchievici 2017) (Fig. 9.18). The heating systems can operate through passive or active methods. Both harvest thermal energy from the sun and utilize the collected heat for space heating or to heat water. *Passive solar systems* are installed on a tilted roof to collect the sun's energy.

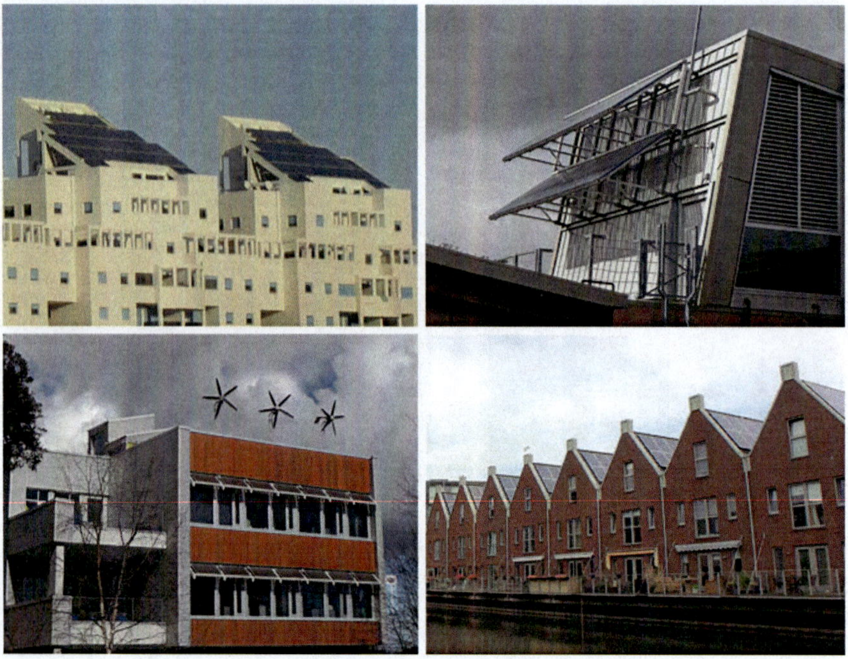

Fig. 9.18 Solar water heaters (top left), condescend tubs (top right), solar panels (bottom right), and micro-wind turbines (bottom left) installed on buildings

Active solar heating systems rely on heat pumps to transfer the collected heat from solar collectors to the building (Sarbu and Sebarchievici 2017).

Hydropower is one of the most efficient and "green" forms of electricity generating sources of energy. It is generated by converting the energy in flowing water with the aid of a turbine to produce power (DOE 2021). This power is then converted into electricity by a generator. Some micro-hydropower systems operate "run-of-river", so there is no need for dams or water storage reservoirs therefore having little environmental impact (DOE 2021). The majority of these systems only require a fraction of the available stream flow to generate power, which can generate up to 100 kilowatts (kW) of electricity, plenty to power a single dwelling (DOE 2021). Micro-hydropower systems are commercially available, and the associated turbines and generators can be stand-alone or grid-connected parts. Therefore, any excess energy produced from the hydropower system can be fed back into the grid system and be utilized elsewhere.

A windy building site can have advantages especially in the context of Net-Zero homes. Harnessing the wind's energy involves extracting power from the wind and converting it into useful energy (NRCAN 2003). Determining the wind speed at a proposed wind turbine site is critical to estimating the economic potential of the wind turbine. The preferred way is to measure the wind speed with accurate wind-recording equipment (NRCAN 2003). Wind speeds greater than 15 km/h (9.3 mph) are needed before a wind energy system can begin to generate electricity. This is known as the "cut-in" speed. The "cut-out" speed, usually around 70 km/h (43.5 mph), is where the system will pause to protect itself from damage (NRCAN 2003).

If a Net-Zero home is located on a site where there is sufficient wind speed between the "cut in" and "cut out" range, on a general basis, installing micro-wind turbines for each home would be suitable for residential energy production (Fig. 9.19). Micro-wind generation is a system that uses the flow of wind energy to produce electricity for a house. Depending on the amount of wind available for the turbine system to utilize, micro-turbine systems should be installed alongside other "micro" renewable energy systems such as solar and hydro (Ramenah and Tanougast 2016). If there is insufficient wind speed available at certain periods, the home would still be reliably powered by renewable energy, thereby maintaining its Net-Zero status. Another benefit of this energy source is that it can be connected to the grid system, so any excess power generated from one dwelling can be redistributed and used in homes that require more electricity (Ramenah and Tanougast 2016).

Geothermal Heat Pumps (GHP) take advantage of natural heat from the earth and are the most efficient and comfortable heating and cooling technology currently available (Fig. 9.19). GHP's gain heat from the ground, where temperatures are more constant year-round. In the wintertime, a GHP system extracts heat from the ground and distributes it to building spaces, whereas in the summertime, it removes heat from the building and transfers it to the ground for cooling (Zhang et al. 2020). The GHP's passive mechanism occurs via the following processes: by being connected to the earth, the heat pump circulates a working fluid, composed of water or an antifreeze solution to absorb heat or reject heat from the ground via a ground heat exchanger (GHE) loop. The heat pump then transfers heat between the building(s) and the earth connection, and lastly the heated or cooled air will be distributed throughout the building space (Zhang et al. 2020).

9.7 Landscaping for Energy Efficiency

Objects that buffer the dwelling from strong winds are commonly referred to as *windbreaks*. Windbreaks are most effective if they are not completely impermeable since a solid windbreak will produce a low-pressure area on the building's leeward side (Donn and Thomas 2010). Proper tree planting can capture summer breezes and mitigate harsh winter winds (Delgado et al. 2013). Trees can be a natural form of windbreak. When planted around the house they can act as a buffer from harsh winds and channel them away (Delgado et al. 2013) (Fig. 9.20). In comparison to hedges and walls, trees are semi-permeable. Therefore, they can simultaneously block wind and allow sunlight to pass through, so the home does not completely lose access to direct sunlight (Delgado et al. 2013). Trees are also more aesthetically pleasing carbon sequestering agents. They absorb carbon from the atmosphere thereby reducing the concentration of greenhouse gasses in the atmosphere (Eckley 2019).

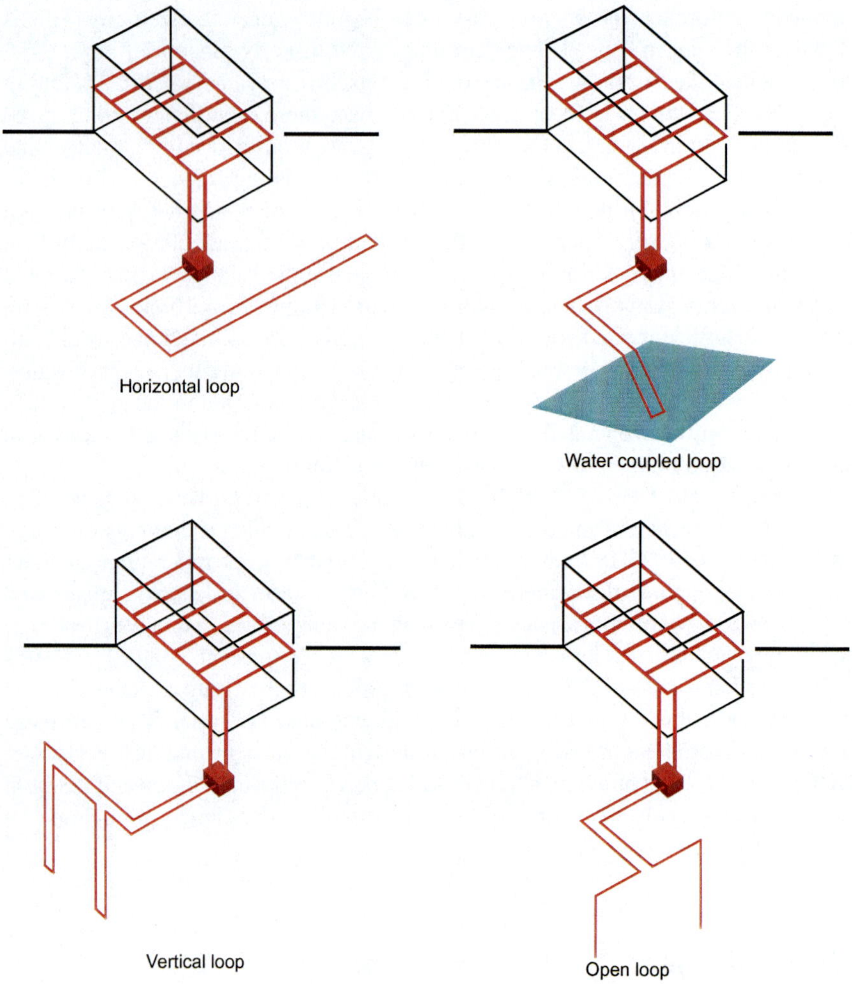

Fig. 9.19 Geothermal heat pumps (GHP) take advantage of the earth's natural heat

When considering the orientation of a home, it would be beneficial to plant trees along the south side, where the exposure is the greatest. Selecting an appropriate tree species is also critical, as it should be effective in its design purpose, but also indicative of the climatic conditions of its environment. Trees species that allow more sun rays to pass through when they are bare during winter are preferred. When selecting between coniferous and deciduous trees, the deciduous trees would make a more effective choice since they will drop their leaves seasonally, whereas coniferous retain their needle-like leaves year-round (Martin's Tree Service Inc. 2020). Deciduous trees will also provide shade for and absorb incoming radiation from nearby dwellings in the summer months acting as a natural cooling agent.

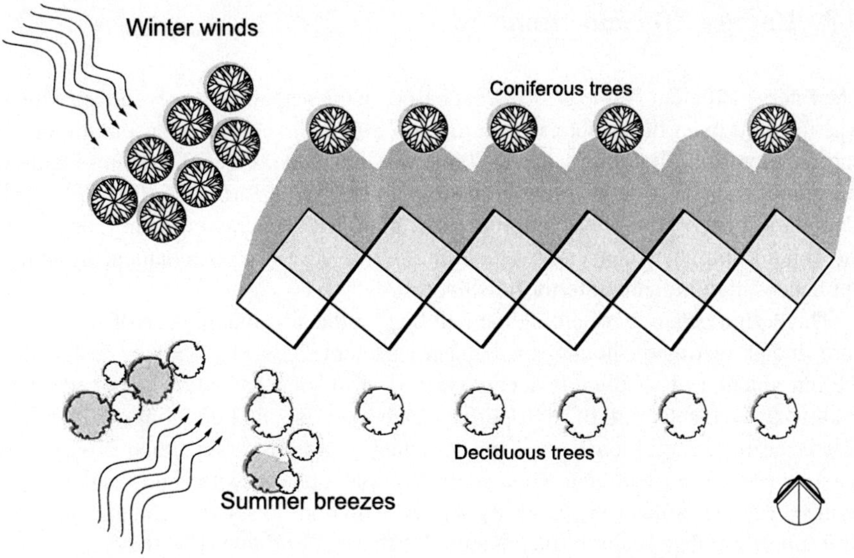

Fig. 9.20 Trees can be natural windbreaks. When planted around the house where incoming wind is the strongest, they can act as a buffer from harsh winds and channel them away

The thermal preservation of homes is particularly important for those with poor insulation, and in cold and windy climates. Wind velocity increases at the ridge of a hill and cold air subsides at the bottom of a valley, therefore it is best to locate dwellings halfway between the valley and the ridge to avoid both extremes (Donn and Thomas 2010). On the other hand, in hot northern climates, homes should be constructed at the bottom of the valley, on the south side of a landform (Donn and Thomas 2010). This is to increase the home's exposure to sun rays, which have a low angle in the winter. In warm or humid climates where cross ventilation is crucial, the dwelling should be sited on a hill's ridge to benefit from the ventilation that stronger winds provide. Sloping sites also receive different amounts of solar radiation, as north-facing slopes receive more solar gain than flat sites (Donn and Thomas 2010). This means that ground and air temperatures are higher and heat loss from a building is lower on a north-facing slope (Donn and Thomas 2010).

9.8 Energy Efficient Home

The Energy Efficient Home is a concept based on the narrow-front row house which was featured in earlier chapters, with the aim of reducing energy consumption and carbon footprints. It follows the R-2000 code, a standard established by Natural Resources Canada for construction quality and energy performance (Friedman and Cammalleri 1996). It takes a comprehensive approach to energy efficiency, detailing the relationship between energy, building form, envelope, mechanical systems, appliances, lighting, and interior finishings.

The design began by orienting the building on the site, making use of sun exposure, and protecting the home from harsher weather (Fig. 9.21). Window design and the introduction of shading devices used to control solar gains were planned next. Following this, selection of high-performance materials and detailing for thermal resistance makes the home as energy efficient as possible (Friedman 2012). For exterior walls, the traditional wood sheathing was replaced with rigid insulation to reduce heat loss. Studs were carefully aligned with floor joists, reducing the number of framing members to limit the potential for thermal bridging (Fig. 9.22).

High-performance double-glazed argon-filled windows with low emissivity (low-e) coatings improved the thermal performance of windows. Low-e coatings are used to reduce heat transfer through windows by reflecting radiation. Low-conductivity gas fills further prevented heat transmission through windows. Double glazed windows were recommended on the southern façade, with triple glazed windows on facades facing other directions (Friedman and Cammalleri 1997).

The design of the envelope helps to reduce the heating and cooling load by limiting air leakage and heat transmission. Therefore, the most important mechanical function is an adequate ventilation system, particularly in structures shared between multiple units. Individual mechanical systems were selected over shared ones, giving

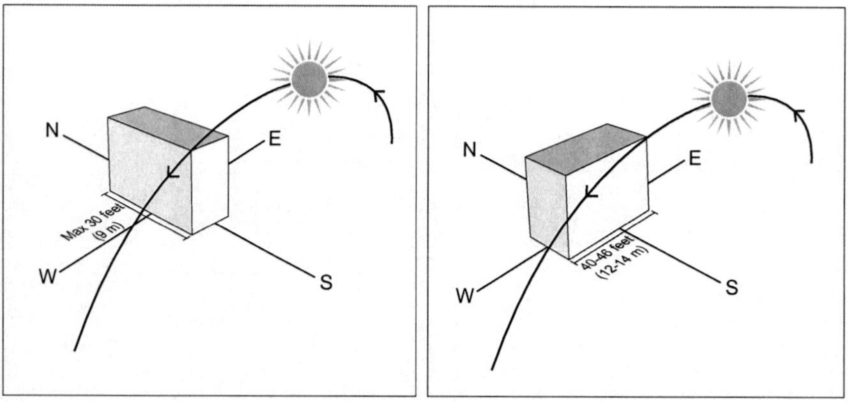

Fig. 9.21 The home energy performance was improved with proper orientation for passive solar gain

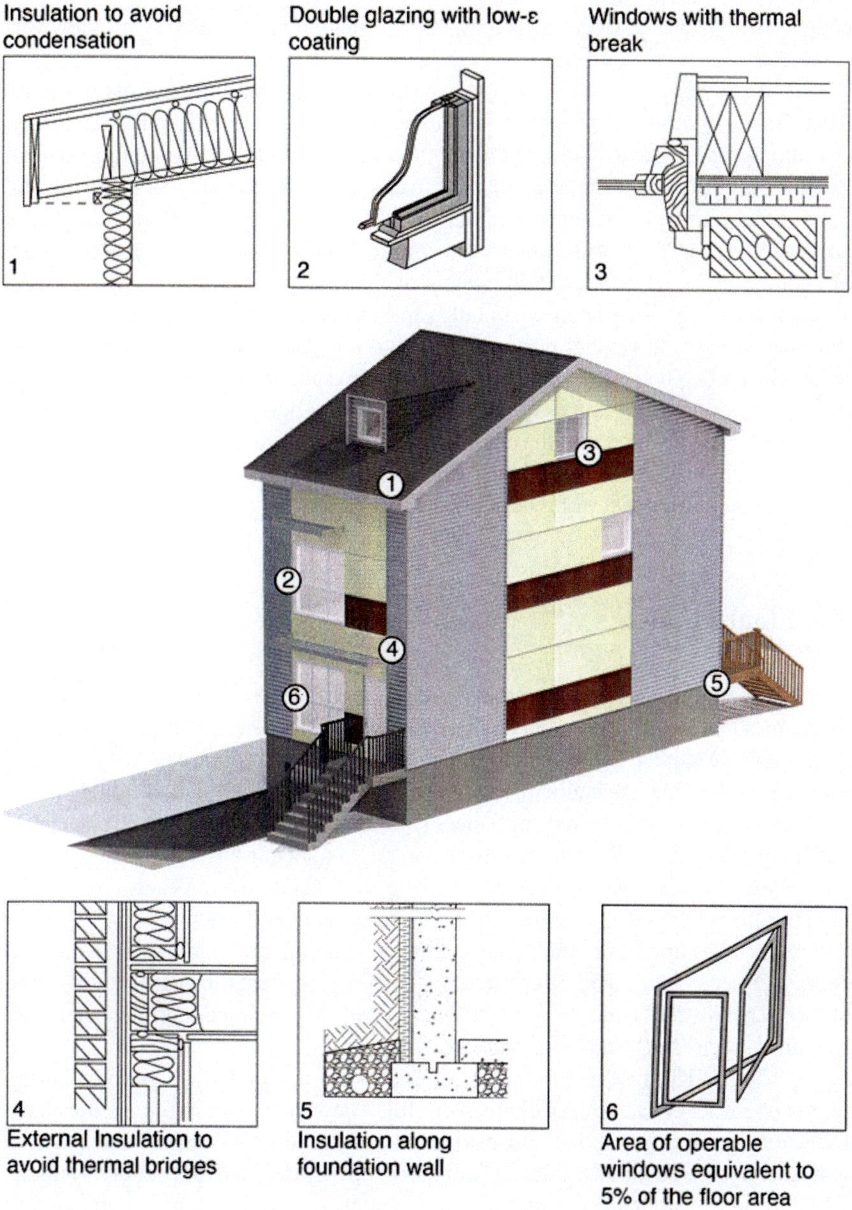

Fig. 9.22 Strategies to improve the energy performance of the home's exterior

occupants control over their own unit. Supplemental mechanical ventilation was introduced to provide controlled air exchange through the unit.

A heat recovery ventilator recovers heat from exhaust air and controls humidity levels. Exhaust vents are placed in wet areas including the bathroom and kitchen, with supply vents located elsewhere in the unit. Because of the smaller size of dwellings and thermal efficiency of the envelope, the requirement for mechanical heating was reduced. An integrated mechanical system that combines the water and space heating functions was used to improve energy use. In occasional, very cold conditions, radiant baseboard heaters or thermal storage heaters may be used to supplement the space heating. Finally, appliances and fixtures were selected to minimize energy and water consumption. Water efficient toilets, shower heads, and faucets are used. The selection of appliances follows the most up to date EnerGuide recommendations (Fig. 9.23). By implementing these design strategies, the Energy Efficient Home consumes half the energy of a similarly sized house constructed in the 1980s. Exemplifying the comprehensive approach which should be used to improve the environmental impact of modern houses.

9.9 Final Thoughts

Further refinement of the energy reducing strategies described above may enable Net-Zero homes to transition toward a Positive Energy Home (Energy Project 2019). The energy positive aspect suggests that they are efficient to the point that they produce more energy than they consume (Energy Project 2019). Therefore, additional energy could be used to power external appliances such as mobile devices, or even an electric car (Energy Project 2019). The transition can begin to occur through the installation of an energy monitoring system. Monitoring the energy use of each circuit will guide the changes needed to reduce the household's energy consumption. This could be done through upgrading appliances, better insulation, and equipment as needed (Energy Project 2019). The micro renewable energy systems in place will then be able to generate the same amount of energy with a guaranteed surplus therefore upgrading the home from Net-Zero to Positive.

According to Krangsås et al. (2021), the challenges associated with implementing the Positive Energy Home model are plentiful. They are rooted in the need to have innovative governance, social and environmental incentives, support and engagement, sustainable business models, a balance in energy demand and supply systems, as well as the consideration of regional and local differences. Incentives are contextual in the sense that they would need to adapt to the socio-political situation of the building location. Therefore, specific energy efficiency technology, costs, development level of the district, local markets, and consumer preferences need to be considered. Furthermore, promoting the use of renewable energy sources would require the offset of low-interest investment funding schemes, regulatory barriers, the absence of subsidies, and unstable policy frameworks in each region (Krangsås

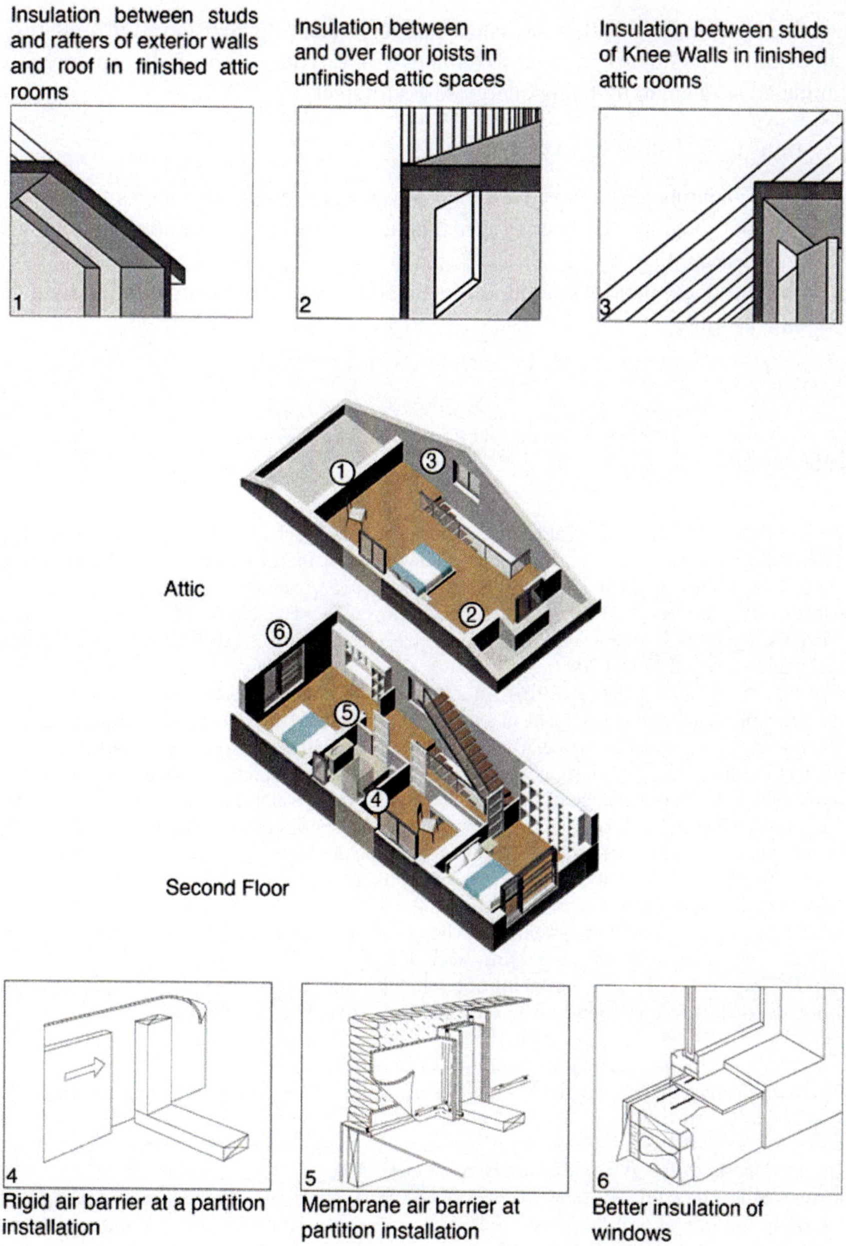

Fig. 9.23 Strategies to improve the energy performance of the home's interior

et al. 2021). Going forward, if the listed challenges are mitigated and addressed, residential building's impact on the environment will be reduced, energy bills potentially eliminated altogether, by being energy self-sufficient.

Questions for a Follow-Up Discussion

1. Name 3 heating systems and list their advantages and disadvantages?
2. Is in your view district heating offers an advantage in comparison to individual home heating?
3. Is in your view district heating offers an advantage in comparison to individual home heating?

References

Ahmed MMS, Radwan A, Serageldin AA, Abdeen A, Abo-Zahhad EM, Nagano K (2021) The thermal potential of a new multifunctional sliding window. In: Solar energy. https://www.sciencedirect.com/science/article/abs/pii/S0038092X21007015. Accessed 13 Oct 2022

Asdrubali F, Desideri U (2018) Building envelope. In: Handbook of energy efficiency in buildings. https://www.sciencedirect.com/science/article/pii/B9780128128176000395?via%3Dihub. Accessed 13 Oct 2022

Balke EC, Healy WM, Ullah T (2016) An assessment of efficient water heating options for an all-electric single-family residence in a mixed-humid climate. In: Energy and buildings. https://www.sciencedirect.com/science/article/pii/S0378778816309136. Accessed 13 Oct 2022

Cutland Consulting Ltd, Eco Design Consultants Ltd (2016) The challenge of shape and form: nhbc foundation. In: The challenge of shape and form. https://www.nhbcfoundation.org/wp-content/uploads/2016/10/NF-72-NHBC-Foundation_Shape-and-Form.pdf. Accessed 13 Oct 2022

Delgado JA, Nearing MA, Rice CW (2013) Conservation practices for climate change adaptation. In: Advances in agronomy. https://www.sciencedirect.com/science/article/pii/B9780124076853000025?via%3Dihub. Accessed 13 Oct 2022

DOE (2021) Microhydropower systems. In: Micro hydropower systems. https://www.energy.gov/energysaver/microhydropower-systems. Accessed 13 Oct 2022

Donn M, Thomas G (2010) In: Designing comfortable homes https://cdn.ymaws.com/concretenz.org.nz/resource/resmgr/docs/ccanz/ccanz_tm37.pdf. Accessed 13 Oct 2022

Eckley SN (2019) Carbon sequestration. In: Encyclopædia britannica. https://www.britannica.com/technology/carbon-sequestration. Accessed 13 Oct 2022

EESI (n.d.) Energy efficiency. In: EESI. https://www.eesi.org/topics/energy-efficiency/description. Accessed 13 Oct 2022

Energy Project Z (2019) Positive energy homes. In: Zero energy project. https://zeroenergyproject.org/buy/positive-energy-homes/. Accessed 13 Oct 2022

Environmental Protection Agency US (EPA) (2009) A guide to energy-efficient heating and cooling. In: Energy star. https://www.energystar.gov/sites/default/files/asset/document/HeatingCoolingGuide%20FINAL_9-4-09_0.pdf. Accessed 13 Oct 2022

Falk A (2021) Summary report on heat pumps in district heating systems. In: Celsius city. https://celsiuscity.eu/wp-content/uploads/2021/05/Heat_pumps_in_district_heating_systems_RISE_2021.pdf. Accessed 13 Oct 2022

Friedman A (2012) Fundamentals of sustainable dwellings. Island Press, Washington, D.C

Friedman A, Cammalleri V (1996) The impact of R-2000 building technology on Canadian housing. Build Res Inf 24(1):5–13

Friedman A, Cammalleri V (1997) "Selection criteria for energy-efficient windows." Build Res Inf 25(4):234–238

Goldstein B, Gounaridis D, Newell JP (2020) The carbon footprint of household energy use in the United States. In: PNAS. https://www.pnas.org/content/117/32/19122. Accessed 13 Oct 2022

Gong J, Sumathy K (2016) Active solar water heating systems. In: Penn state. https://pennstate.pure.elsevier.com/en/publications/active-solar-water-heating-systems. Accessed 13 Oct 2022

Group M (ed) (2022) Central heating: the master group: master group. In: The master group. https://www.master.ca/consumer/central-heating. Accessed 13 Oct 2022

Hopton ME, White D (2011) A simplified ecological footprint at a regional scale. In: J Environ Manag. https://www.sciencedirect.com/science/article/pii/S0301479711002556. Accessed 13 Oct 2022

Huebner GM, Shipworth D (2016) All about size?—The potential of downsizing in reducing energy demand. In: Applied energy. https://www.sciencedirect.com/science/article/pii/S0306261916302033. Accessed 13 Oct 2022

Hydro Quebec (n.d.) What is the breakdown of household electricity use? In: Hydro Quebec. https://www.hydroquebec.com/residential/customer-space/electricity-use/electricity-consumption-by-use.html. Accessed 13 Oct 2022

IDEA (ed) (2019) District heating. In: District heating—International district energy association. https://www.districtenergy.org/topics/district-heating. Accessed 13 Oct 2022

IEA (2021) District heating—Analysis. In: IEA. https://www.iea.org/reports/district-heating. Accessed 13 Oct 2022

Krangsås SG, Steemers K, Konstantinou T, Soutullo S, Liu M, Gieancola E, Prebreza B, Ashrafian T, Murauskaitė L, Maas N (2021) Positive energy districts: identifying challenges and interdependencies. In: MDPI. https://www.mdpi.com/2071-1050/13/19/10551/htm. Accessed 13 Oct 2022

Lamnatou C, Chemisana D (2021) Solar thermal systems for sustainable buildings and climate change mitigation: recycling, storage and avoided environmental impacts based on different electricity mixes. In: Solar energy. https://www.sciencedirect.com/science/article/pii/S0038092X21009816?via%3Dihub. Accessed 13 Oct 2022

Lelieveld J, Klingmuller K, Pozzer A, Burnett RT, Haines A, Ramanathan V (2019) Effects of fossil fuel and total anthropogenic emission removal on public health and climate. In: Proceedings of the National Academy of Sciences of the United States of America. https://pubmed.ncbi.nlm.nih.gov/30910976/. Accessed 13 Oct 2022

Li J, Just RE (2018) Modeling household energy consumption and adoption of energy efficient technology. In: Energy economics. https://www.sciencedirect.com/science/article/pii/S0140988318301440?via%3Dihub. Accessed 13 Oct 2022

Martin's Tree Service Inc. (2020) The difference between coniferous and deciduous trees. In: MARTIN'S TREE SERVICE INC. https://www.martinstree.com/blog/difference-coniferous-and-deciduous-trees/. Accessed 13 Oct 2022

Martínez DM, Ebenhack BW, Wagner TP (2019) Chapter 8-Residential and commercial sector energy efficiency. In: Energy efficiency: concepts and calculations. https://www.sciencedirect.com/science/article/pii/B9780128121115000081?via%3Dihub. Accessed 13 Oct 2022

Natural Resources Canada (NRCAN) (2003) Stand-alone wind energy systems - NRCAN. In: Wind energy systems. https://www.nrcan.gc.ca/sites/www.nrcan.gc.ca/files/canmetenergy/files/pubs/WindEnergy_buyersguide_ENG.pdf. Accessed 13 Oct 2022

Natural Resources Canada (NRCAN) (2019) Government of Canada. In: Natural resources Canada. https://www.nrcan.gc.ca/energy/products/categories/heating/heat-pumps/air-source/ductless/21316. Accessed 13 Oct 2022

Natural Resources Canada (NRCAN) (2021) Government of Canada. In: Natural resources Canada. https://www.nrcan.gc.ca/energy-efficiency/products/product-information/heating-equipment-for-residential-use/13740. Accessed 13 Oct 2022

NC (n.d.) What is your carbon footprint? In: The nature conservancy. https://www.nature.org/en-us/get-involved/how-to-help/carbon-footprint-calculator/#:~:text=A%20carbon%20footprint%20is%20the,is%20closer%20to%204%20tons. Accessed 13 Oct 2022

Ng L, Poppendieck D, Dols WS, Dougherty BP, Emmerich SJ (2018) Balancing energy and IAQ: NIST net-zero energy residential test facility. In: ASHRAE J. https://www.ncbi.nlm.nih.gov/pmc/articles/PMC6508590/. Accessed 13 Oct 2022

Ramenah H, Tanougast C (2016) Reliably model of microwind power energy output under real conditions in France suburban area. In: Renewable energy. https://www.sciencedirect.com/science/article/pii/S0960148115304377. Accessed 13 Oct 2022

Rardin N (2015) Forced air heating system guide. In: HVAC solutions. https://www.hvac.com/resources/what-are-forced-air-systems-here-are-3-things-you-should-know/. Accessed 13 Oct 2022

Rukikaire K, Collins T (2020) Building sector emissions hit record high, but low-carbon pandemic recovery can help transform sector—UN report. In: UN environment. https://www.unep.org/news-and-stories/press-release/building-sector-emissions-hit-record-high-low-carbon-pandemic. Accessed 13 Oct 2022

Sarbu I, Sebarchievici C (2017) Solar heating and cooling systems || Solar water and space-heating systems. In: Solar heating and cooling systems || Solar water and space-heating systems. https://www.sciencedirect.com/science/article/pii/B9780128116623000050?via%3Dihub. Accessed 13 Oct 2022

Stephan A, Crawford RH (2016) The relationship between house size and life cycle energy demand: implications for energy efficiency regulations for buildings. In: Energy. https://www.sciencedirect.com/science/article/pii/S0360544216314724?via%3Dihub. Accessed 13 Oct 2022

White (2014) "Hammarby Sjostad." http://www.white.se/en/project/15-hammarby-sjostad. Accessed 13 Oct 2022

Wu W, Skye HM (2021) Residential net-zero energy buildings: review and perspective. In: Renewable and sustainable energy reviews. https://www.sciencedirect.com/science/article/pii/S1364032121001532?via%3Dihub. Accessed 13 Oct 2022

Chapter 10
Home Automation

Abstract Technological innovation and social trends have led to an increase in the use of digital means. The prevalence of devices such as smartphones, computers, and smart household appliances is becoming common place with future growth projections. One can argue that the Internet of Things (IoT) to a large degree has improved the efficiency and quality of domestic life and contributes to the home's sustainable performance. By investigating the larger context of smart technology, it can also notice how it became deeply woven into city planning and architectural design. When successfully adopted, these technologies provide benefits that touch upon the four pillars of sustainability which were outlined in Chap. 1 and other social spheres such as health, mobility, and home security. This chapter discusses home automation by providing examples of how smart technology and the IoT can effectively accommodate residents and lead to the creation of a more sustainable environment.

Keywords Home automation internet of things · Privacy and security · Smart city · Smart homes · Smart mobility · Telehealth

10.1 Digital Advancement and Sustainability

It is no cliché to suggest that the use of the internet and digital devices are more common than they ever have been. The COVID-19 pandemic has led to a sharp increase in internet usage due to lifestyle and work habit changes as discussed in Chap. 1. The International Telecommunication Union (ITU) reported that in 2021 internet usage had reached 4.9 billion people, equivalent to 63% of the world's population (ITU, n.d.). With this increase, new strides are being made to incorporate the Internet of Things (IoT) into more aspects of people's daily lives. It works as a dynamic worldwide network of billions of smart objects that have the capability to sense, acquire, share, and exchange loads of information through interactions with one another. Smart objects include electronics like computers and smartphones as well as home appliances such as fridges, microwaves, and other devices. IoT and Wi-Fi access have been incorporated into an array of social aspects such as healthcare,

A. Friedman, *Fundamentals of Innovative Sustainable Homes Design and Construction*, The Urban Book Series, https://doi.org/10.1007/978-3-031-35368-0_10

education, and the supply chain management operations (Fig. 10.1). The expected worth of the global IoT market is set out to be between \$3.9 and \$11.1 trillion by 2025 (Shuhaiber and Mashal 2019). An exploratory study conducted in the UK on the adoption of the IoT found that the perceived usefulness, ease of use, privacy and security, and knowledge of the technology were significant predictors of the adoption of IoT in a given home (Shuhaiber and Mashal 2019). The overall acceptance of such technologies means that consumers will have many smart products to choose from. In turn, these technologies are being incorporated into consumers' living spaces slowly converting them into smart homes one device at a time.

When investigating smart technology in the urban context, one can notice how it is being weaved into city planning and architecture. In Chap. 1, smart cities were described as urban environments that utilize communication, information technologies, and other means to improve quality of life, the efficiency of urban services, operations, and competitiveness while ensuring the needs of future generations in a sustainable manner (Fig. 10.2). These new digital development principles aim to be data driven and built for sustainability, privacy, and security. Importantly, smart

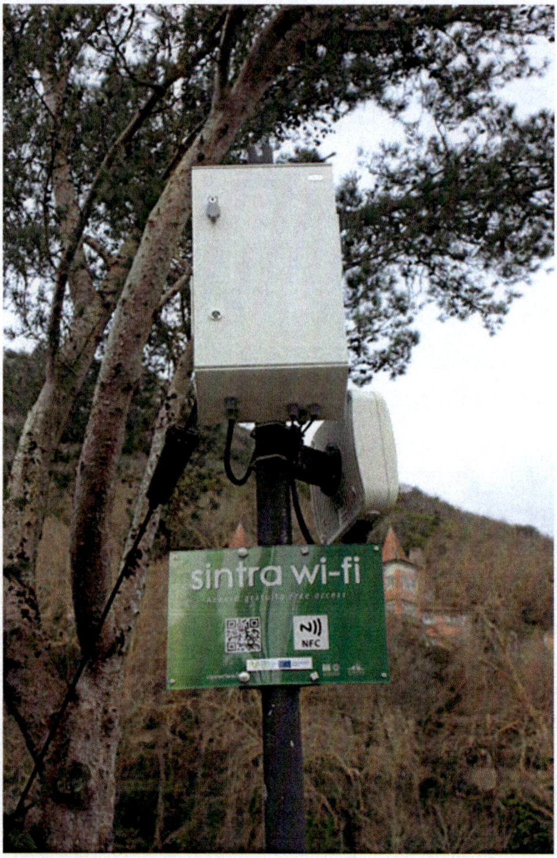

Fig. 10.1 The internet of things (IoT) and Wi-Fi access was incorporated into an array of aspects of human life such as healthcare and cities' public spaces like this one in Lisbon, Portugal

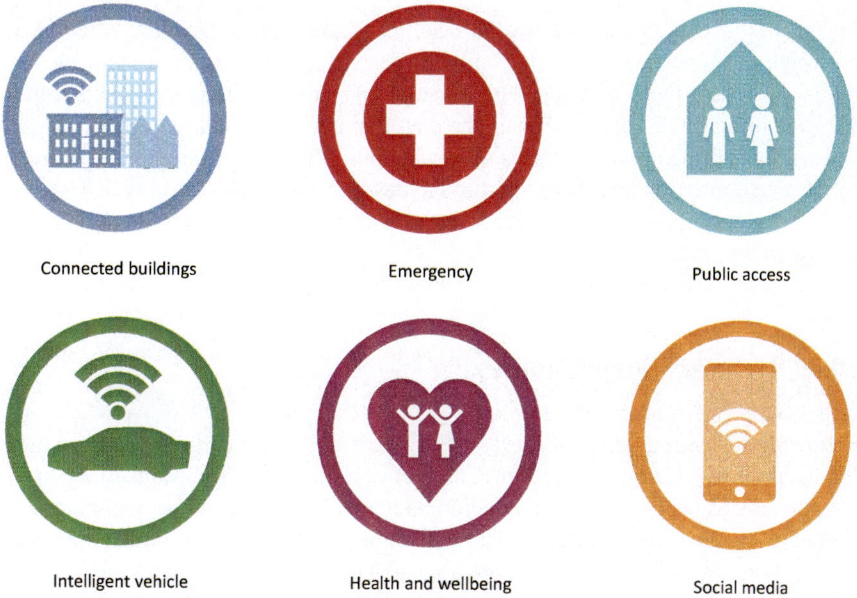

Fig. 10.2 Smart cities are described as urban environments that utilize communication and information technologies to improve quality of life, the efficiency of urban services, operations, and competitiveness while ensuring the needs of future generations in a sustainable manner

cities which also include smart homes, aim to work with the existing ecosystem to facilitate human development objectives established by the global community. In such, the IoT becomes a key element to the ecosystem of smart cities.

Smart cities and their complexity can vary according to what part of the world one is living in. Though not all smart city transitions are successful and like any other form of technology, have their downsides, certain parts of the world are utilizing the IoT to foster human development. In South Africa for example, the IoT is slowly being integrated in various dimensions of the urban environment to make cities more efficient. In this context, the vision of a smart city involves making sure that all municipal buildings and libraries have internet, clinic health records are made electronic, traffic management systems are smart and that meter-ready systems for water and electricity are automated.

Regardless of how different or basic smart city transitions between places may be, their efforts should be progressive when the best interest of the population is in mind. Using technology to accelerate development is challenging to say the least and comes with a myriad of valid concerns about privacy, equity, politics, government investment priorities, and environmental sustainability. With time, the constant use of smart installations causes a reduction of their efficiency leading to consumption of more power than they originally were planned for. Greenhouse gas emissions from the heavy usage of electronic devices and appliances is also of concern (Tetteh and Amponsah 2020).

Yet, in the same token, the overall benefits of smart cities have the potential to outweigh the consequences by becoming urban spaces that are more inclusive and in tune with meeting people's needs. All solutions have consequences whether they are good or bad. The key is choosing the ones that have the smallest negative impact and will help bring about the forms of sustainability. There is no single magic bullet solution to improving quality of life through development while working in harmony with the environment. However, smart cities and dwellings can be seen as a step in the right direction.

10.2 Defining Smart Homes

Following the introduction of the IoT and smart cities, we can now address their implications using a micro level lens and focus on the home. Today, dwellings have the potential to become spaces that combine human relationships and technological connectivity under the same roof. These dwellings that have superior technological connectivity are referred to as smart homes. They can be understood as a cyber physical system where homes are equipped with sensors and smart appliances which are controlled by and connected to the Internet (Shuhaiber and Mashal 2019). The sensors in question detect things like temperature, heat, motion, and light and are intelligent enough to make decisions about the home's environment. Smart appliances and sensors can regulate themselves or can be programmed with the help of devices like computers, mobile devices, tablets, and remote controls (Javed et al. 2021) (Fig. 10.3).

Perhaps the true meaning of a smart home lies in its ability to be "context aware". A digital home system has the capacity to obtain and manage contextual information and adapts its functions automatically. As the home seeks to provide useful information or services to the resident, it can recognize these dynamics change and becomes adaptable to the user's particular needs (Moreno et al. 2017).

There are two ways of having a smart home. It can be designed and built anew or alternatively, an existing structure, not a smart home, can be modified into one (Tetteh and Amponsah 2020). Smart home devices include anything from fridges and coffee makers to front doors and fire alarms. These wireless connections are possible through telecommunication networks and Wi-Fi. Smart home providers can help consumers make the transition by offering their services to facilitate and centralize the control of smart devices installed in the home through interactive digital platforms known as a *hub*. They collect and analyze the data produced by mobile devices and smart home appliances to provide interactive services for their customers (Shuhaiber and Mashal 2019).

In 2021, global consumer spending on smart home-related devices reached $62 billion US. This trend is expected to reach $88 billion US by 2025 (Larrichia 2022). As widespread adoption of such technology is set to provide massive financial gains, many tech giants such as Amazon, Google, Samsung, and Comcast are embarking on the journey to develop smart home products of their own. Voice controlled speakers

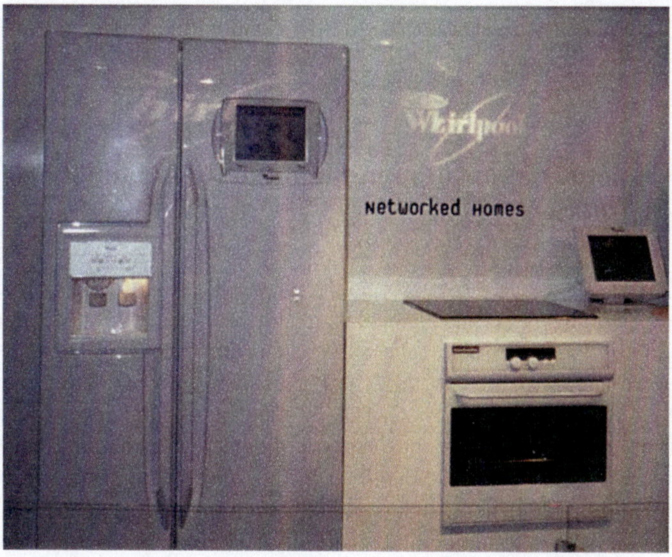

Fig. 10.3 Smart appliances and sensors like the ones shown here can regulate themselves or can be regulated remotely with the help of computers and mobile devices

Fig. 10.4 Voice controlled devices such as Amazon's Eco Dot turn words into actions by allowing users to verbally ask their speakers to perform tasks or to operate other IoT devices

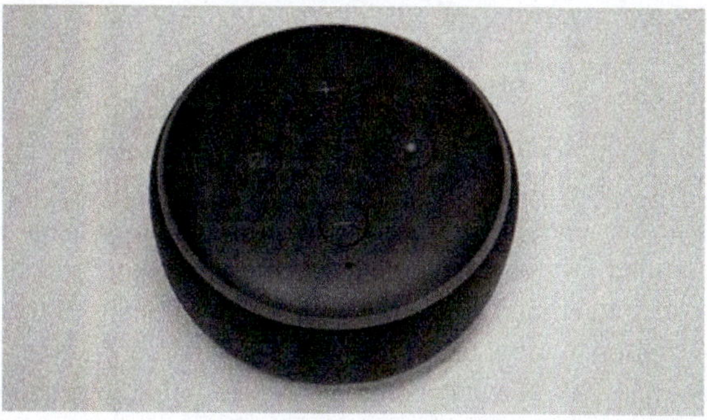

such as Google Home and Amazon's Eco Dot equipped with "Alexa" technology are among some of the popular products currently on the market. These gadgets turn words into actions by allowing users to verbally ask their speakers to perform tasks or to operate other devices connected to the IoT. These examples are indicators of what is possible with smart home devices (Fig. 10.4).

Data collection practices of these technologies make people rightfully weary about privacy concerns. In addition, high price tags remain one of the main reasons why people are reluctant to embrace smart systems. These installations are also associated with long-term financial commitments which raise important questions about affordability and what economic classes may be included or excluded from using them. A lack of consumer awareness and understanding is also a barrier to smart technologies adoption (Tetteh and Amponsah 2020). Below is an outline of the benefits of smart technologies when adopted.

10.3 Technology and Automation in the Residential Environment

With their diverse network of sensors and connectivity, smart homes offer support and benefits for many aspects of life: social, healthcare, security, independent living, and sustainability (Javed et al. 2021). Some smart home designs can also promote inclusiveness for people with disabilities and save time while completing certain tasks (Tetteh and Amponsah 2020). This wireless effect implicitly means that whether someone is at home or away on a business trip, home devices and residential safety can be managed from a distance. Not only does this simplify domestic life, but it also provides environmental benefits that could outweigh the above-mentioned negative consequences. Home automation and smart city technology also touch upon other aspects of residential environments such as mobility which will be discussed below.

10.3.1 Energy Conservation

Through certification standards, products are manufactured to consume less energy. For example, the Energy Star certification touched upon in Chap. 9 was drafted by the US Environmental Protection Agency (EPA) in 1992 to deliver products that have the least negative impact on the environment. In 2019 alone, Energy Star's certified products have helped Americans save $39 billion US in energy costs and reduce electricity consumption by approximately 500 billion kilowatt-hours. Since its inception, more than 7 billion Energy Star certified products have been sold (Energy Star, n.d. a). The certification can be applied to products worldwide and expands to buildings. What makes home distinctly smart is when these types of certified products and household appliances are linked to the IoT. For example, smart thermostats can be installed to regulate the home's temperature automatically or from a distance with the use of cellular or desktop devices (Fig. 10.5). Heating and cooling consume more energy than any other appliance, making smart thermostats one of the best ways to reduce energy consumption (Energy Star, n.d. b). Studies have found that the average household could save approximately 15% on cooling and 10–12% on heating costs with a smart thermostat (Enercare 2019).

In addition, other energy-saving devices include smart light bulbs which have longer lifespans and consume less energy than traditional non-LED light bulbs. Smart bulbs can be controlled from a smartphone or turned on or off from a distance (Fig. 10.6). There are many different types with a varying prices range depending on what additional features they have. Some other examples of their unique features

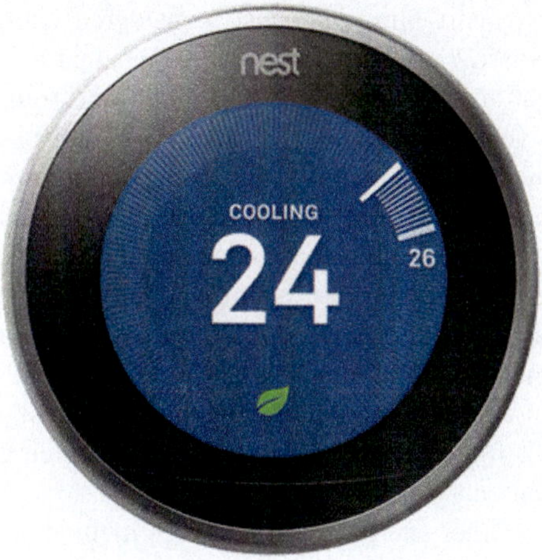

Fig. 10.5 Smart thermostats can be installed to regulate the home's temperature automatically or from a distance with the use of cellular or desktop devices

include location-based controls called geo-fencing that can be used to automatically turn off lights in a room based on the location of one's smartphone. Some also have built-in sensors which turn lights on or off based on if a person is present in the room. Certain bulbs can also be paired with smart security systems and thermostats to activate energy-saving modes which minimize energy use while away for extended periods of time. This mode also operates lights at a minimal brightness as a security measure (Energy Star, n.d. b). Continuing along with affordable options of smart technology, smart plugs are also a popular low-cost option. They can be plugged into an electrical outlet to manage energy consumption. Using mobile apps, the user can control times of use of plugged devices and turn them on or off remotely as well (Enercare 2019). What these two technologies have in common is that they allow the user to reduce energy consumption by setting timers to turn off devices or lighting in the room.

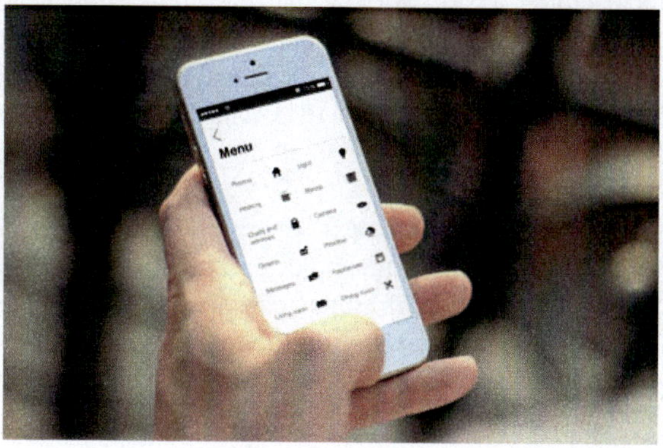

Fig. 10.6 Smart light bulbs and other domestic gadgets can be controlled via smartphone from a distance

Smart blinds are also considered energy-saving devices. They are motorized window coverings that can be opened and closed through a remote-control application installed on a smartphone or through voice command and powered through solar, battery, and electrical cords. Of course, each source of power has its own benefits and inconveniences depending on the type of home in which the device is being integrated (Bradford 2022). When used in combination with adequate insulation, smart blinds can reduce the need to heat and cool a home. Smart blinds can be programmed to close automatically during the sunniest part of the day in summertime. It may seem counterintuitive to close the blinds on a sunny day, but this can mean avoiding unnecessary use of air conditioners when temperatures get too warm indoors. Beyond energy savings, cordless motorized blinds may also provide safer alternatives than traditional blinds as their cords are hazardous to small children and animals (Glennon 2018).

Other energy-saving devices include larger appliances like smart fridges, ovens, washers, and dryers. Smart dryers can automatically adjust their cycle time with the help of sensors to reduce the appliance's energy use by shutting off when clothes are dry (Energy Star, n.d. b). Many of these options come with a much higher price tag and may not be the best alternative for everyone. Yet, despite this barrier, their popularity has been steadily growing. In essence, by automating tasks or sending reminders directly to a phone, smart devices are not only more energy efficient, but can also help break away from habits that lead to higher energy consumption (Fig. 10.7).

10.3.2 Water Conservation

The need for energy and water conservation goes hand in hand when considering the characteristics of a sustainable home. Water is indispensable to human life and development and the price of its use varies in many cities around the world. In 2021, for example, Oslo, Norway had the highest tap water price among many global cities at $6.69 US per 1.02 square meter (11 square feet) followed by San Francisco, California where 832.8 L (220 gallons) of water cost $6.07 US (Tiseo 2021). In fact, North Americans are known to lead the chart in water consumption (Fig. 10.8). Water conservation is therefore crucial not only for saving on utility bills but also for the environment.

In the residential context, Energy Star and other certified appliances like dishwashers and washing machines can greatly reduce water usage when performing chores at home. Smart homes can go a step further by providing useful information about water consumption to the homeowner through digital platforms such as smartphone mobile apps. Certain tasks such as watering the garden can also be automated, saving both time and money for the user and reducing environmental waste. According to the United States Environmental Protection Agency (EPA), as much as 50% of water used outdoors is wasted due to wind, evaporation, and runoff from inefficient irrigation systems and methods (Kral 2021). Smart sprinklers and

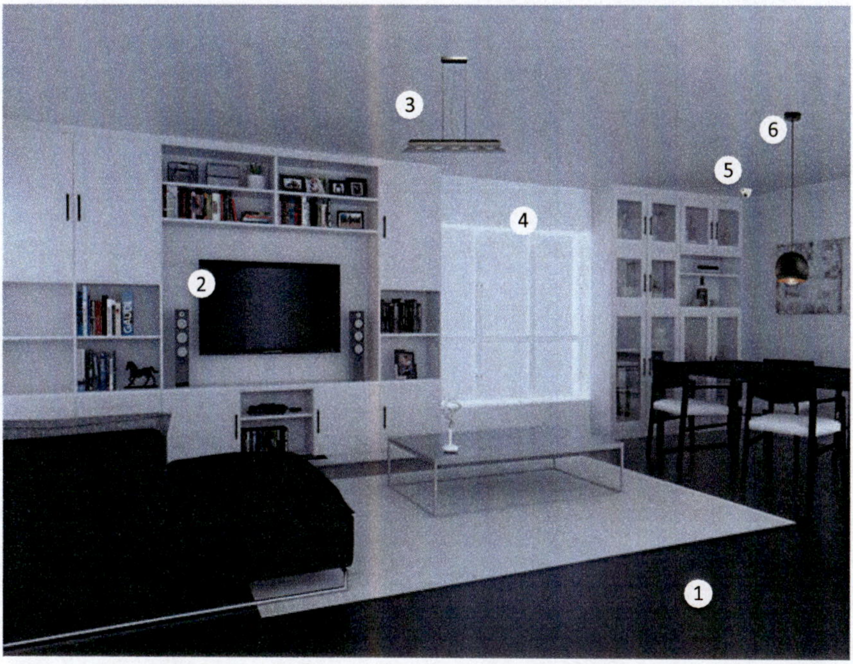

1	Heating	4	Blinds
2	TV / Wi-fi	5	Camera
3	Living room lights	6	Dining room lights

Fig. 10.7 Automating tasks or sending reminders directly to a smart device to control items shown here help break away from bad habits that lead to higher energy consumption

irrigation methods can offer a solution. Replacing standard sprinklers to ones with clock-based controllers can save an average American household up to 47,280 L (12,490 gallons) of water annually (EPA 2016). Due to global warming, drought is a real threat to many communities resulting in municipalities taking action against outdoor irrigation. Depending on the location of one's home, some municipalities provide water fines to homeowners who water their lawns on days where it is not permitted. Using smart irrigation systems allows a homeowner to program the time of day and frequency of watering their lawn. This way, water fines (where applicable) are avoided, and less water needs to be used during the hottest parts of the day when evaporation is high (Kral 2021).

Other water-saving devices include smart water leak detectors which can be installed under sinks, hot water tanks, and any other water source to warn a homeowner of any potential water problems before they happen. These devices can send a homeowner a text or an email note about a potential leak or pipe freeze. It can help avoid plumbing damage to the home and wasted water in the event of a leak (Enercare 2019).

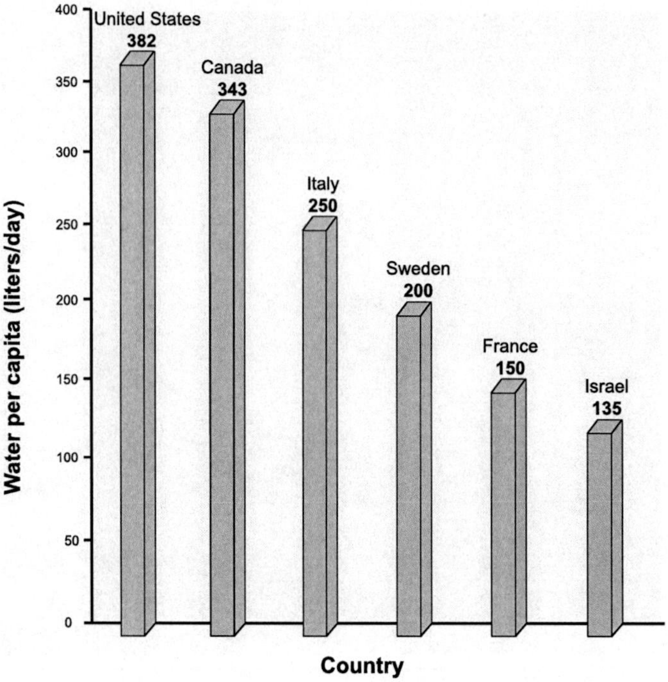

Fig. 10.8 North Americans are known to lead the chart in water consumption

Smart shower timers or water-saving shower heads can provide another alternative to water conservation. In the US, showering accounts for approximately 17% of indoor water use (EPA 2017). Unlike traditional water-saving shower heads that reduce flow and the amount used, smart timers can help reduce the amount of time spent in the shower. Some companies have thought of savvy programmable devices that keep shower usage to a minimum. A user can program their timer to the desired amount of time spent in the shower. As the time on the device approaches the end, cold pulses of water are sent out from the shower head to warn the user that their time is almost up. After the timer hits zero, the hot water is completely turned off, discouraging the user from extending shower time. This may be a slightly more unpleasant yet very effective way of reducing shower times, therefore reducing the cost of utility bills, and reducing water waste once again. These different energy and water-saving devices can be easily combined and incorporated into one's home to provide environmental and economic benefits for the homeowner. The advantage is that even homeowners on a low budget can opt to automate their homes with more affordable technologies such as smart plugs and lightbulbs as well as smart sprinkler systems without spending large sums of money on smart appliances.

10.3.3 Smart Mobility

The COVID-19 pandemic has created a sense of caution about the use of public transit due to overcrowding. As a result, the importance of creating mobility that is inclusive and promotes safety and wellbeing in the urban environment has been highlighted. Different mediums are available to help fulfill these needs by making it easier for people to plan and book their journeys in a seamless way (Tabbitt 2021). Smart mobility involves leveraging the IoT to manage multiple forms of transportation in more efficient and sustainable ways. Here, the IoT has an important role in collecting and aggregating data for analytical purposes. This data can then be transmitted into digestible formats for planners and road users who include pedestrians, commuters, cyclists, and drivers alike (Tabbitt 2021). For example, mobile transit apps can be used on the go to help road users make better informed decisions about travel information such as traffic congestion, how to reach a desired destination through various means of transportation, as well as real-time public transit schedules. These applications are easily downloadable and often free. The IoT is, therefore, also a powerful tool used in increasing mobility in the urban environment. It ensures that people feel safe about, and in control of, their travel. Further development of user-friendly transit apps may also be the key to incentivizing more people to adopt sustainable active mobility modes to include bicycles, electric scooters, shared cars, and commuting by public transit (Fig. 10.9).

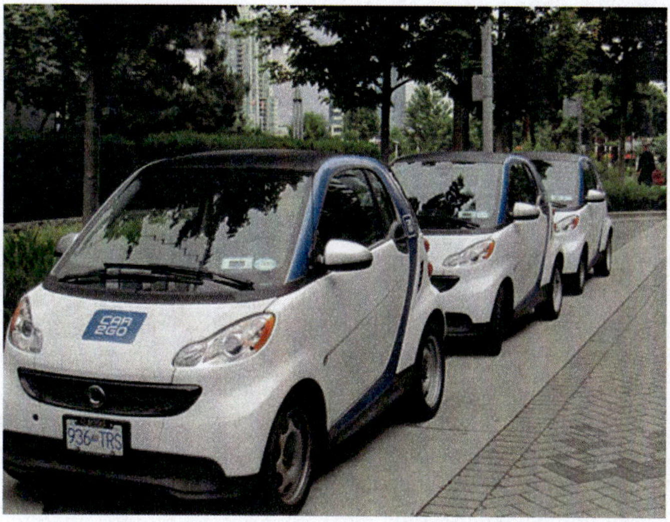

Fig. 10.9 Mobile devices are instrumental to the use of shared cars

10.3.4 Innovations in Telehealth

In the last decades, the medical field underwent a digital revolution. With a raging pandemic, telehealth is now common in the public health sector of many nations. It involves the ability to deliver remote healthcare-related services for patients at home, saving both monetary and human resources as well as reducing response times for emergency situations. Large numbers of patients can be dealt with through communication technologies from a centralized location without having to send medical personnel to peoples' homes (Li 2012). There are three kinds of home monitoring for patients in a telehealth scenario: reactive alert detections which involves sending alerts to telehealth centers in the event where devices like fire and smoke detectors are set off, medical monitoring which is used to measure patient's health status, and activity monitoring used to track lifestyle changes and track patterns of possible pathologies (Moreno et al. 2017). The concept of telehealth has two main components: telecare and telemedicine.

Telecare aims to help the elderly as well as physical and cognitively disabled individuals live at home independently for as long as possible. Through various monitoring systems, patient's families can have peace of mind and know that their loved one is safe. Telecare is especially beneficial for patients who live alone and are concerned about being isolated and those who suffer repeated risk situations such as falling (Moreno et al. 2017). The second component is *telemedicine* which uses information and communications to provide health services to patients remotely through video consultations, medical evaluations, diagnostics, and treatments. It is a valuable tool for professionals such as radiologists and neurosurgeons who can analyze and deliver results to patients without a need to meet in person (Moreno et al. 2017).

At its most basic level, telemedicine can be practiced anywhere and in any home no matter if it is smart or not since all the patient usually needs is a phone or internet connection to speak with a care practitioner. However, certain smart homes can excel in providing medical assistance to homeowners and allowing medical conditions to be monitored from a distance through the practice of telecare. To do so, home must be equipped with certain features to support medical needs. Communication management components need to be integrated into the home that communicates with a health care facility. For example, one such system gathers information from a person's defibrillator to help cardiologists monitor and diagnose someone's health status (Fig. 10.10). Internal communication such as system interfaces and sensors as well as external communication components that will advise medical staff about a possible emergency both connected to the cloud and necessary (Moreno et al. 2017).

Devices used to track temperature, magnetic switches on doors to monitor residence entries and exits, motion sensors used to detect presence and mobility, and microphones used to detect call for help and detect abnormal noises are among some of the other technologies one may find in a home designed for telecare (Li 2012). However, as the concept of telehealth remains relatively new and its adoption is still slow, questions about privacy and data security can again be seen as a potential

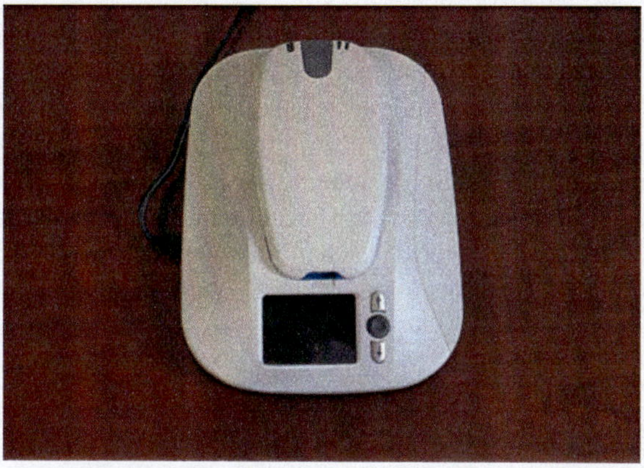

Fig. 10.10 Communication devises can link a home with a health care facility. For example, one such system gathers information from a person's defibrillator to help cardiologists monitor and diagnose someone's health status

barrier to telehealth adoption in many contexts. Infrastructure limitations at home and in some health facilities also means that these services may not be available to everyone and any place. But, as society keeps emerging in a post-pandemic world, there is no doubt that telehealth, telecare, and telemedicine will eventually find a permanent role in public health.

10.3.5 Smart Home Security

A more common and widely available method of increasing safety at home is through a security system. It is a key component to many of today's modern homes. There exists a wide range of options but consumer preference for smart security and fire alarm systems is on the rise. In 2020, the global revenue from the smart security market reached $2.49 billion US and is forecast to grow even further reaching $5 billion US in 2025 (von See 2021).

Smart home security systems have added many features that traditional systems do not have. Doorbells can be wireless and equipped with a camera to monitor visitors from anywhere while sophisticated alarms can be connected directly to emergency services. Diagnostic equipment can warn the homeowner about hazardous situations such as broken water pipes and fire as well (von See 2021). Other features include the option of live streaming one's property and instantly being communicated via a smartphone notification when there is an intruder. Smart systems are also less likely to trigger false alarms than traditional ones because homeowners can see for themselves if an emergency is real thanks to live stream features (Enercare 2019). Additionally, forgetting to set the alarm before heading out is a thing of the past. A smartphone app let the homeowner program the security system from a distance while doors can also be locked and unlocked using the same method with smart locks. Again, this is all possible because of the IoT.

When considering the installation of smart security devices, the costs of the devices themselves as well as their installation fees need to be kept in mind. Other expenses include monthly subscription fees billed by the company providing the security monitoring service in the consumer's home. Such costs can be dispersed through a monthly or annual rate based on a multiyear plan like the ones available for home internet or TV cable. In essence, smart home security systems are an effective way to have peace of mind while away and have more added features than traditional security systems do.

10.3.6 Lifestyle and Technology

The symbiotic relationship between the IoT and people's lifestyles is an important aspect needed to fully illustrate the implications of tech and automation of the built environment. Technology allows people to build resilient ways of staying connected despite potentially difficult situations. It is shaping a new way of life and consumer behavior around the world while simultaneously providing benefits to the young, adults, and the elderly. Today, the speed at which services can be delivered to consumers is unprecedented. In some cases, whether one orders groceries online or a phone case from a foreign country on Amazon, deliveries can be expected within the next or at times, within the same day. In 2020, around 5% of Canadian and 5% of French consumers had bought items online for the first time while an even smaller percentage of 2% was recorded for Japanese and German consumers (Tighe 2020). These small numbers of first-time online shoppers during the start of the pandemic are indicative that methods of spending and consumer habits are increasingly moving to the digital world, including the remaining "slow-moving" adopters. Therefore, people are adopting new habits without having to leave the comfort of their home. It can be argued that coupling telehealth and other facets of technology such as online shopping can provide many benefits to those in need; namely the elderly, individuals with reduced mobility, and those with other impairments.

When looking at our social interactions, more of them are being made possible through online messaging apps, and video calls to name a few platforms. From a productive point of view, remote learning and work also allow people to move past the constraints of mandatory mobility and physical attendance to acquire new skills or to earn a living. We can also consider that the IoT has been integrated into many wearable devices that have become our daily companions. For example, smart watches can be useful in helping the user make important daily decisions. Monitoring health status and activity such as heart rate, blood pressure, and steps can make the user more conscious about the choices that will influence their health. However, the very same technology that does so many great things for people's daily lives is also a source of major distraction, taking away many opportunities for quality in-person interactions. With the fast-paced environment we live in, it is important to recognize that we are distinct social beings after all, not futuristic cyborgs.

10.4 A Connected Home in Puglia, Italy

BS House was designed by Italian firm Reisarchitettura for a couple in the countryside of Puglia, Italy (Fig. 10.11). The house functions both as a working studio and a relaxing retreat for the clients. In the design, the architects drew inspiration from the traditional homes of the region while combining smart home technology for modern convenience.

The elevated site offers a generous view of the surrounding landscape. The house is organized around a central courtyard facing the north. As the massing of the house blocks direct sunlight, this space becomes passively cooled for enjoyable outdoor living. This spatial arrangement has been used since ancient times as protection from the harsh sun. Indoor, the living room and workspace are to the east, while private quarters are to the west. The kitchen and dining area are in the middle, with direct access to the courtyard. A separate annex to the north provides relaxing amenities. The chosen materials reflect the traditions of Puglia. The architects used a combination of lime plaster for the walls and oak wood for openings in a contemporary fashion.

The clients wanted to have control of their home while they are away on business trips, so a KNX automation system was integrated into the house. Not only are lighting and air conditioning remotely adjustable, but the system also features control of the

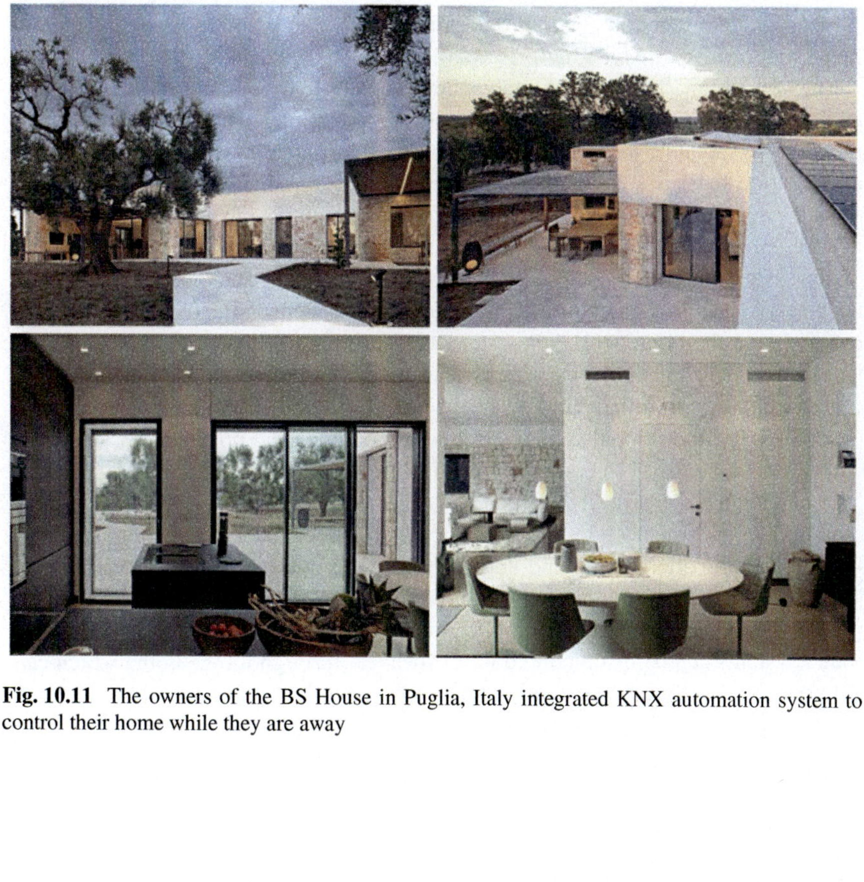

Fig. 10.11 The owners of the BS House in Puglia, Italy integrated KNX automation system to control their home while they are away

house's security system, including the entry buzzer, door lock, and alarms. Along with these elements, energy consumption management can be monitored through a smart device app. The house generates power from photovoltaic panels, which serve the home as well as a charging port for an electric car. Reisarchitettura designed the home with convenience and comfort in mind, combining traditional methods and modern technology to make the most of the site. BS House fully embraces home automation as a tool for sustainability.

10.5 The Smart Home

The Smart Home is a conceptualized design by the author that centralizes control of various systems to facilitate the use of advanced technologies and home automation strategies. These technologies help elevate a dwelling's sustainable performance by reducing energy consumption and facilitating daily activities such as controlling domestic appliances and maintaining security.

The home contains a series of sensors, appliances, and devices that can be monitored by the user through a control platform. Sensors have the capability to detect air quality, temperature, motion, light, and sound. This information is transmitted to servers to facilitate monitoring and control of devices. There are four different areas of focus of smart technology: control of appliances, indoor environment and energy regulation, security, and media (Fig. 10.12).

The home's smart appliances enable the occupants to monitor and control kitchen appliances, laundry machines allowing chores to be completed more efficiently while tracking energy and water usage. For example, smart laundry machines use weight and moisture sensors to adjust the cycle time, improving their energy consumption. Interior environment and temperature regulation monitoring include the management of temperature, air quality, and light to improve energy efficiency and internal comfort. The smart thermostats automatically regulate internal temperature, and they can be set to a timer and controlled remotely. This system limits the energy used to heat and cool the home while unoccupied and overnight. Smart light bulbs used in conjunction with light sensors, which determine when there is insufficient natural light, can also be controlled remotely by the timer to reduce their energy consumption.

Motorized blinds can also be controlled remotely, based on timers, or automatically by light sensors to reduce overheating from sun exposure. Garden sprinkler systems can also be timer controlled to prevent water overconsumption. Security systems use motion detectors, door and window sensors, and cameras to allow users to monitor their home and activate alarm systems with their security code. Various settings can be configured for different scenarios such as away, stay, or night modes, which use different alerting systems and measures based on users' choices. Sound systems can be installed to control speakers throughout the home which can additionally be coupled with entertainment systems in the living room.

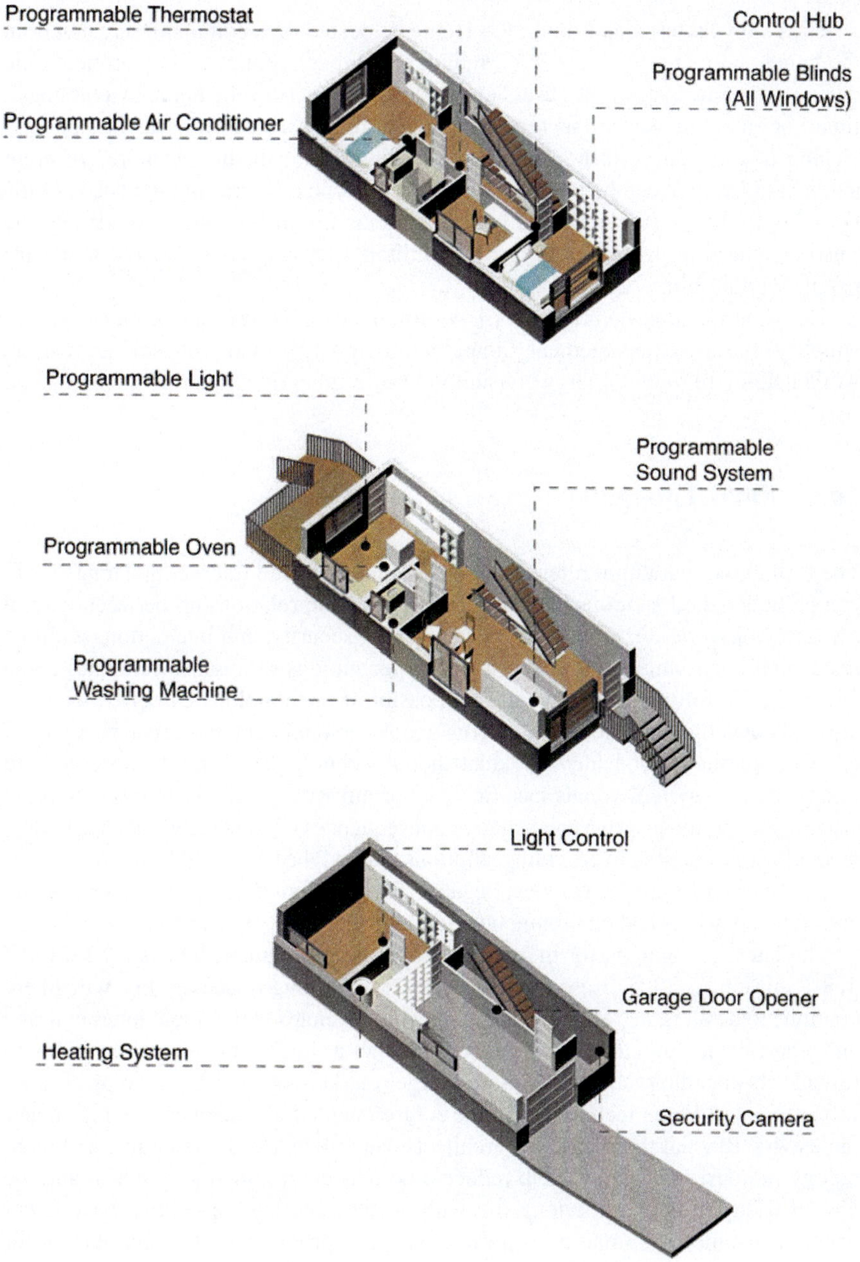

Fig. 10.12 Elements of home automation in the Smart Home

The control interface of the Smart Home is located on a command screen within the home and can also be accessed remotely from cellphones and other electronic devices. The interface allows users to manage all aspects of the home by command, timer, or automatic sensor such as arming/disarming the security system, opening/closing the garage door, turning on/off appliances, playing media, and more. Different automated features can be included based on the type of home and the needs of the household. Moreover, centralized control systems for all functions provided by the Smart Home improve accessibility by facilitating easier control of the home for people with mobility or vision issues.

The Smart Home demonstrates how automated features can be integrated to simplify household management. Smart features aim to provide convenience to users while helping to reduce energy consumption by efficiently regulating the home.

10.6 Final Thoughts

The technology paradigm referred to in this chapter as the Internet of Things (IoT) can be understood as a mechanism that solidifies the relationship between natural resource conservation, mobility, people's health, security, and interactions within a residential environment. However, adopting technology at a wide scale does come with risks. To fully grasp the wonderful possibilities technology can offer, we must also acknowledge its potential concerns about personal data and privacy. Again, it is also important to consider that smart home technology will not be accessible to everyone as many individuals face financial or infrastructural constraints. Thinking critically about the positive and negative consequences of smart city and smart home technology is essential to providing solutions that will be beneficial for human development while leaving the smallest negative footprint possible on social, economic, cultural, and ecological environments. So then where are we headed next?

The latest predictions for the global adoption rate of smart homes are set to be 478.2 million in 2025 (Lasquety-Reyes 2021). Although accessibility will likely continue to be an issue for many, the increasing revenues from global internet usage and smart device purchases could potentially act as an incentive for tech giants to provide cheaper alternatives to capture a larger market segment. Developing cheaper forms of smart home technology in the future may act as a mechanism for greater inclusivity. Beyond these barriers, architects can utilize and incorporate smart technology in housing design to help reduce a home's environmental footprint and the cost of utility bills over the long run with devices and appliances that have lower overall consumption. Smart home designs can also promote a safer environment for a homeowner who belongs to vulnerable communities or is in need of medical care. Conversely, one need not forget multiple facets of connectivity within a smart home have the potential to make life easier overall through task automation and home diagnostic information provided directly to the user.

Questions for a Follow-Up Discussion

1. How can home automation contribute to achieving sustainability in dwellings?
2. What are Smart Homes and what is their main features?
3. Name 3 home automation technologies that can lead to cost saving and sustainability?

References

Bradford A (2022) 8 best smart blinds. https://www.familyhandyman.com/list/best-smart-blinds/#:~:text=Smart%20blinds%20are%20window%20coverings,honeycomb%2C%20roller%20and%20light%20filtering.text=Blinds%20can%20be%20hardwired%20or,solar%2C%20battery%20or%20electrical%20cord. Accessed 13 Oct 2022

Enercare (2019) 8 ways your smart home can save energy. https://www.enercare.ca/blog/smarter-home/8-ways-your-smart-home-can-save-energy. Accessed 13 Oct 2022

Energy Star (n.d. a). What is Energy Star. https://www.energystar.gov/about. Accessed 13 Oct 2022

Energy Star (n.d. b). Smart home tips for saving energy. https://www.energystar.gov/products/Smart_home_tips. Accessed 13 Oct 2022

Glennon M (2018) What are the benefits of motorized blinds? https://www.somfysystems.com/en-us/blog/post/what-are-the-benefits-of-motorized-blinds. Accessed 13 Oct 2022

International Telecommunication Network (ITU) (n.d.) Internet use. https://www.itu.int/itu-d/reports/statistics/2021/11/15/internet-use/. Accessed 13 Oct 2022

Javed AR, Fahad LG, Farhan AA, Abbas S, Srivastava G, Parizi RM, Khan MS (2021) Automated cognitive health assessment in smart homes using machine learning. Sustain Cities Soc 65:102572. https://doi.org/10.1016/j.scs.2020.102572. Accessed 13 Oct 2022

Kral H (2021) 5 main benefits of a smart sprinkler system. https://www.familyhandyman.com/list/benefits-smart-sprinkler-system/. Accessed 13 Oct 2022

Larrichia F (2022) Consumer spending on smart home related devices worldwide from 2019 to 2025.https://www.statista.com/statistics/873607/worldwide-smart-home-annual-device-sales/. Accessed 13 Oct 2022

Lasquety-Reyes J (2021) Number of Smart Homes forecast in the World until 2025. https://www.statista.com/forecasts/887613/number-of-smart-homes-in-the-smart-home-market-in-the-world. Accessed 13 Oct 2022

Li KF (2012) Smart home technology for telemedicine and emergency management. J Ambient Intell Humanized Comput 4(5):535–546. https://doi.org/10.1007/s12652-012-0129-8. Accessed 13 Oct 2022

Moreno VL, Martín Ruiz ML, Malagón Hernández J, Valero Duboy MÁ, Lindén M (2017) Chapter 14—the role of smart homes in intelligent homecare and healthcare environments. In: Dobre C, Mavromoustakis C, Garcia N, Goleva R, Mastorakis G (eds.) Ambient assisted living and enhanced living environments. Butterworth-Heinemann, pp 345–394. https://doi.org/10.1016/B978-0-12-805195-5.00014-4. Accessed 13 Oct 2022

Shuhaiber A, Mashal I (2019) Understanding users' acceptance of smart homes. Technol Soc 58:101110. https://doi.org/10.1016/j.techsoc.2019.01.003. Accessed 13 Oct 2022

Tabbitt S (2021) The technology that is driving the future of smart mobility. https://www.smartcitiesworld.net/special-reports/special-reports/the-technology-that-is-driving-the-future-of-smart-mobility. Accessed 13 Oct 2022

Tetteh N, Amponsah O (2020) Sustainable adoption of smart homes from the Sub-Saharan African perspective. Sustain Cities Soc 63:102434. https://doi.org/10.1016/j.scs.2020.102434

Tighe D (2020) Share of consumers that shopped online for the first time since the coronavirus pandemic was declared in 2020, by country. https://www.statista.com/statistics/1192388/first-time-online-shoppers-since-covid-19/. Accessed 13 Oct 2022

Tiseo I (2021) Most expensive prices of tap water in selected cities worldwide in 2021. https://www.statista.com/statistics/478870/leading-cities-by-highest-freshwater-prices/. Accessed 13 Oct 2022

United States Environmental Protection Agency (EPA) (2016) WaterSense labeled controllers. https://www.epa.gov/watersense/watersense-labeled-controllers. Accessed 13 Oct 2022

United States Environmental Protection Agency (EPA) (2017) Shower better. https://www.epa.gov/watersense/shower-better. Accessed 13 Oct 2022

von See A (2021) Smart home security market value worldwide 2018–2025. https://www.statista.com/statistics/1056057/worldwide-smart-home-security-market-value/. Accessed 13 Oct 2022

Chapter 11
Cooking and Dining at Home

Abstract The development of the kitchen has been greatly influenced by economic, technological, political, and historic circumstances. Parallel to the evolution of cooking and dining, their social importance in a dwelling also transformed. Green minded tendencies have simultaneously introduced design and construction aspects along sustainable principles through building materials, energy saving, recycling, composting, and growing food. This chapter discusses the latter as well as the rise of the social importance of the kitchen and key design concepts such as layout, modularity, storage, and adaptability for individuals with special needs.

Keywords Adaptable kitchens · Kitchen design · Kitchen layout · Social kitchen · Sustainable kitchens

11.1 The Evolution of Cooking and Dining Spaces

Over centuries, various economic and political events have influenced the design and functionality of kitchens and dining spaces (John Desmond Limited 2016). This section looks at kitchens of the past to fully grasp how the contemporary kitchen has evolved to become the home's showpiece. Historically, in many dwellings, the sole purpose of the kitchen was to satisfy basic cooking needs. It was commonly located at the back of a home with limited storage capacity (Fig. 11.1). It was also a place for women to do laundry and sew making sewing machines and washing boards staple kitchen items (Monark 2017). Families stored perishable food items in cellars to prevent spoiling. In the early 1900s, using a hearth which hung over the fireplace was a common cooking staple. It required paying close attention to the fireplace while simultaneously preparing meals. A few decades later, nearly every kitchen was equipped with a cast-iron stove, an icebox, and a sink once they were invented (Monark 2017).

A defining moment in the history of the kitchen began in the 1920s when it took on a higher sense of importance within the home. The first refrigerator with shelves known as the *Shelvadore* was invented and widely adopted by higher-income families while the middle and lower classes opted for a more affordable version (Monark

A. Friedman, *Fundamentals of Innovative Sustainable Homes Design and Construction*, The Urban Book Series, https://doi.org/10.1007/978-3-031-35368-0_11

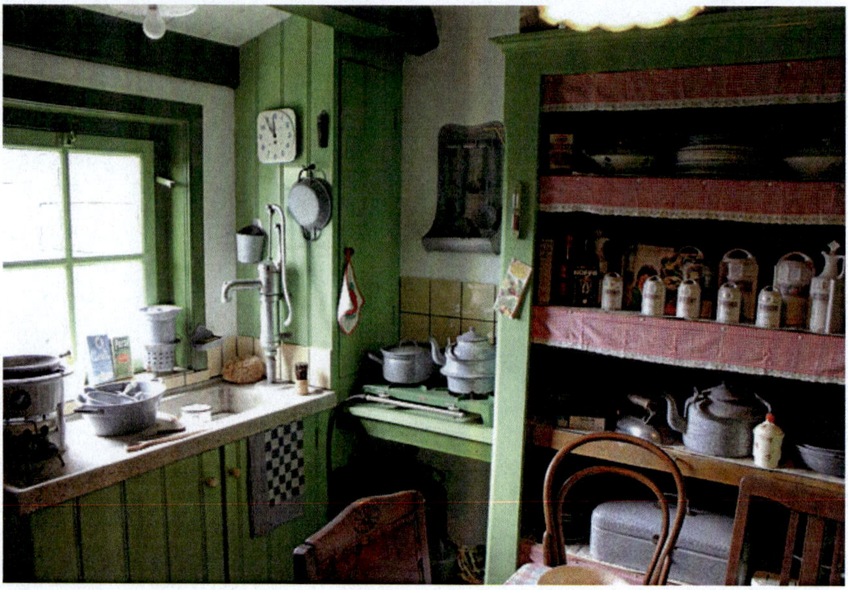

Fig. 11.1 Historically, in old dwellings, the sole purpose of the kitchen was to satisfy basic cooking needs. It was often hidden away in the back of the home with a limited storage capacity

2017). As technology and design further evolved, families began to recognize the social importance of this space. The kitchen was no longer the dark workroom at the rear. The use of wood or coal burning ovens were replaced with electric ones that offered a more efficient way of cooking (John Desmond Limited 2016). Following World War II, when smaller dwellings were built, the kitchen took on a more inviting feel with colorful walls and cabinets, decorative dinnerware, and curtains (Fig. 11.2). Other typical features included the infamous black and white checkered floor tiles and improved storage capacity using built-in cabinets (Monark 2017).

In the 1950s, due to the home's small space, the walls between the living room, dining area, and the kitchen were taken down to create an open-floor plan. The introduction of advanced electric stoves and refrigeration technology as well as the emergence of appliances such as hand-held mixers and toaster ovens kept making cooking more convenient. A decade later with more women joining the workforce, additional time-saving appliances for cooking and cleaning were introduced. Appliances such as microwave, dishwashers, freezers, and garbage disposals in the sink become common. The 1970s saw the birth and popularization of the kitchen island (Monark 2017; Fig. 11.3). This era also saw the phrase "the heart of the home" coined during the mid-1980s when the kitchen became a hub of social activities as we know today.

The open-floor plan grew common and trendy interior design saw the matching of flooring, cabinets, and countertops. Designers began to include wine racks, pot racks, and cookbook shelves when celebrity television cooking shows became popular. The

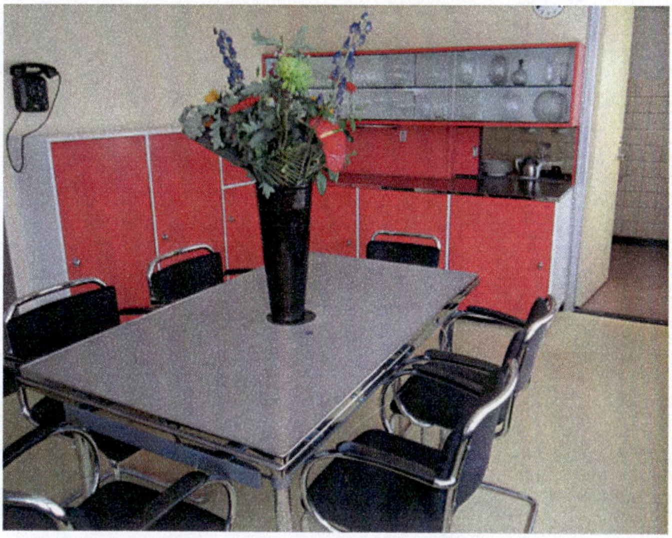

Fig. 11.2 Following World War II, when smaller dwellings were built and kitchen, dining, and living areas combined, the kitchen took on a more inviting feel with colorful walls and cabinets, decorative dinnerware, and curtains

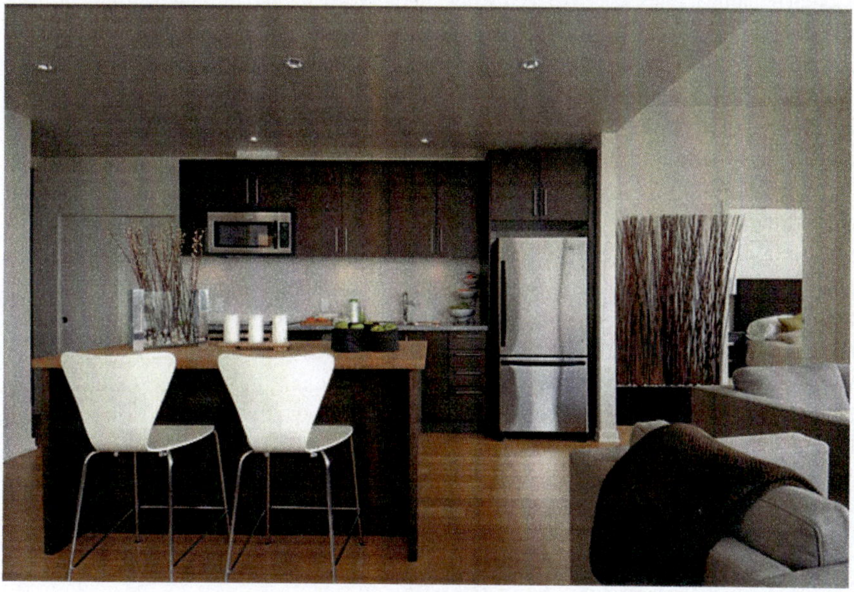

Fig. 11.3 The 1970s saw the birth of the kitchen island

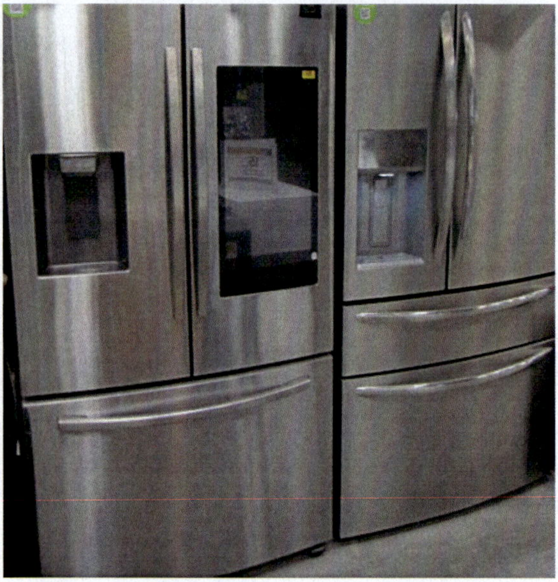

Fig. 11.4 Smart appliances such as programmable fridges with monitors are increasingly being introduced to the kitchen environment

kitchen became a place for displaying designer cookware and demonstrating culinary skills (John Desmond Limited 2016). The end of the century made way for black and stainless-steel appliances which continue to be of popular choice to this day (Monark 2017). While the designs and styles of kitchens vary greatly, generous lighting, efficient storage space, and uncluttered decor are common practices among interior designers and architects (Monark 2017). Contemporary kitchens are now status symbols and main selling points of real estate agents and builders (Friedman and Krawitz 2002).

Automation, smart energy, and water-saving appliances—as discussed in Chap. 10—are becoming increasingly common in kitchens (Fig. 11.4). Other staples of modern kitchens are convenient and portioned meal kits that can be ordered online, similarly frozen meals and precut fruits and vegetables can be found in supermarket isles. Media advancements have also contributed to changing kitchens and cooking habits by enabling people to expand their culinary knowledge and skills. Ultimately, decades of evolution in kitchen design and appliance technology have profoundly changed the way people perceive cooking and dining experiences at home.

11.2 The Kitchen as a Social Hub

The kitchen's design ultimately influences the types of interactions and social activities households will have (Paay et al. 2015). A study conducted about people's social interactions while cooking demonstrated that certain types of kitchens can generate patterns such as working individually, helping one another, overlooking, and learning. For example, kitchens with long counter tops and those placed along a wall will foster side-by-side interactions while kitchens with freestanding countertops will favor "L-shaped" interactions. Interestingly, in both cases, the collaborative nature of the tasks that take place will be influenced by the layout (Paay et al. 2015). Therefore, promoting design practices that foster collaborative relationships in the kitchen will increase the homeowner's sense of belonging to the home and to others living in it.

The contemporary kitchen has become a multipurpose space where social interactions, work, cooking, and entertainment coexist. Since the 1950s when the dining room and the kitchen were combined, many more activities have been introduced (Fig. 11.5). Nowadays, with Wi-Fi, people study and work from their dining tables or kitchen counters as more hybrid modes of learning and working are becoming common. Watching television while cooking or sharing a meal has also become a common staple of the kitchen environment. All to say, modern kitchens have the power to bring people together, facilitate conversations and create a sense of place. Spanning beyond the scope of the kitchen layout itself, its aesthetic appeal is also an important proxy for promoting social interactions. Presentability and an eye-catching design will likely also encourage homeowners to invite guests to their homes to spend time over a meal or drinks.

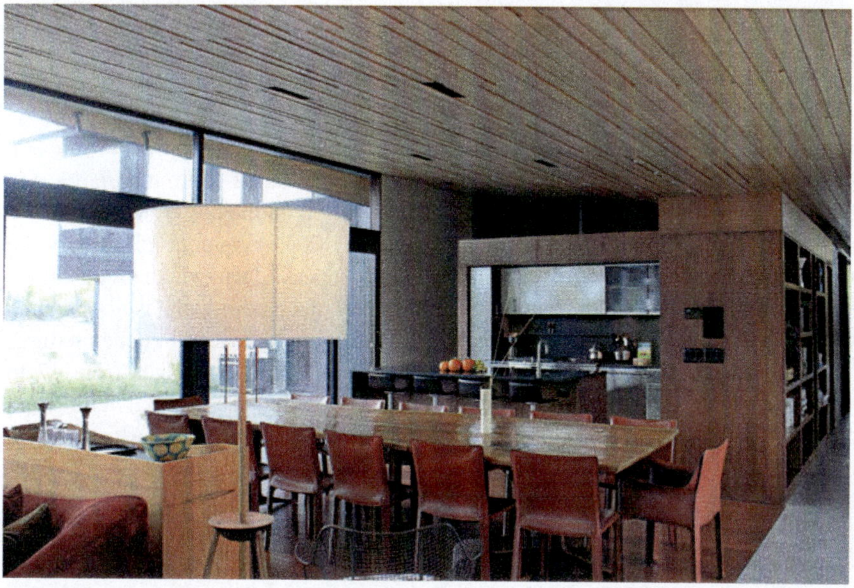

Fig. 11.5 Since the 1950s when the dining room and the kitchen were combined, many more uses have been allocated to this space

11.3 The Sustainable Kitchen

Cooking and dining areas are regarded as a space for resource consumption from high water and electricity use to food consumption and waste generation. However, it is possible to transform this naturally high consumption space into one that is designed and constructed to function alongside sustainable principles through building materials, energy savings, recycling, composting, growing food, and being the home's social hub.

The first step in designing a sustainable kitchen is its structural base. Ensuring that the kitchen is adequately insulated will regulate its temperature. It is possible to use green options such as cellulose insulation, one of the most sustainable options on the market that consists of 80% recycled newspaper (Grundig, n.d. c).

Beyond the macro design scale and component choices, sustainable appliances and gadgets can be used to make cooking and dining areas more eco-friendly. For example, opting for non-plastic cooking utensils made of recycled materials or wood (SpaceWise 2021). Substituting plastic wrap and single-use plastic bags for reusable containers, beeswax wrap, biodegradable, or cloth bags can also make an impact. Additional options to make the kitchen's environment more eco-friendly will be discussed below.

11.3.1 Modular Customizable Components

Due to their modular nature, contemporary kitchens can be customized to meet a homeowner's needs thereby also saving space and materials. Different prefabricated kitchen components can be individually selected based on the preference and needs of the occupants. On average, the kitchen space will take up about 10–15% of the overall home's area. A 6.5 square meter (70 square foot) kitchen, which is considered small is commonly found in apartments and smaller-sized homes and is often joined to smaller dining and living areas (Fig. 11.6). They include essential kitchen fixtures such as a stove, fridge, and dishwasher. The average kitchen ranges from 17 to 41 square meters (180–440 square feet) depending on the size of the dwelling and allowing more space for food preparation and storage as necessary. Alternately, a large kitchen may measure 68 square meters (720 square feet) (Yascaribay 2019).

Before considering the common types of kitchen layouts, it is important to understand the general design of a kitchen since each has unique purposes and flows. First, the food appliance, cookware, and utensil storage will consist of pantries, a refrigerator, drawers, and cabinets. The cooking area consists of the stove and oven while the preparation space consists of countertops. Lastly, the sink area is dedicated to cleaning food items and kitchenware. The sink, pantry, preparation, and cooking spaces are usually permanently combined to allow the meal preparation process to be efficient. Additionally, the preparation, cooking, and sink areas can produce a narrow

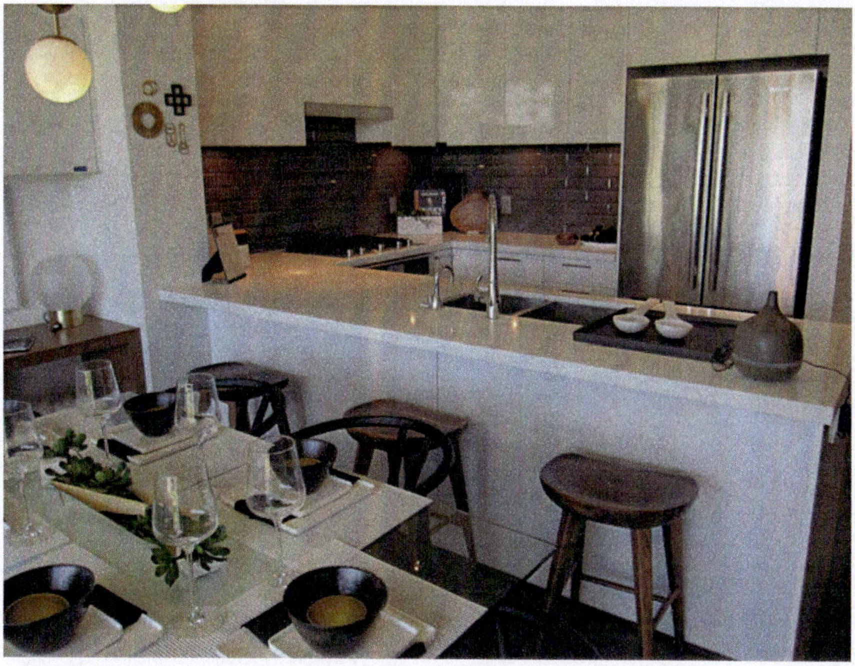

Fig. 11.6 A 6.5 square meter (70 square foot) kitchen can commonly be found in apartments and smaller homes and are often join smaller dining and living areas

triangular workflow based on the different types of kitchens' layouts (ArchDaily 2016a).

As for design, there are four common types of kitchen arrangements in today's homes (Fig. 11.7). These configurations include the efficient wall kitchen, the linear or galley layout, the "rush hour" U-shaped, and the traditional L-shaped layouts. The one wall kitchen layout is typically found in smaller kitchens with a simple and space efficient arrangement. It consists of cabinets and countertops being installed along a single wall. This layout can have upper and lower cabinets and shelving to create a minimalistic aesthetic. The fridge can be installed on one end, the oven and counter space in the middle and the kitchen sink on the other. Additional storage space can be created above cabinets if they do not reach the ceiling (Grundig, n.d. d).

The *linear kitchen* layout also known as the *galley kitchen* is designed with two rows of cabinets facing one another. This design eliminates the need for corner cupboards saving space and offers a simple design with flexible storage places. Work areas should be along one wall only to help avoid traffic throughout the work triangle flow and eliminate the risk of injury while cooking (Grundig, n.d. d). The third common type is the *L-shaped kitchen*. In this configuration, cabinets are installed along two perpendicular walls creating an open plan design that allows flexibility in the placement of appliances and design of the preparation areas.

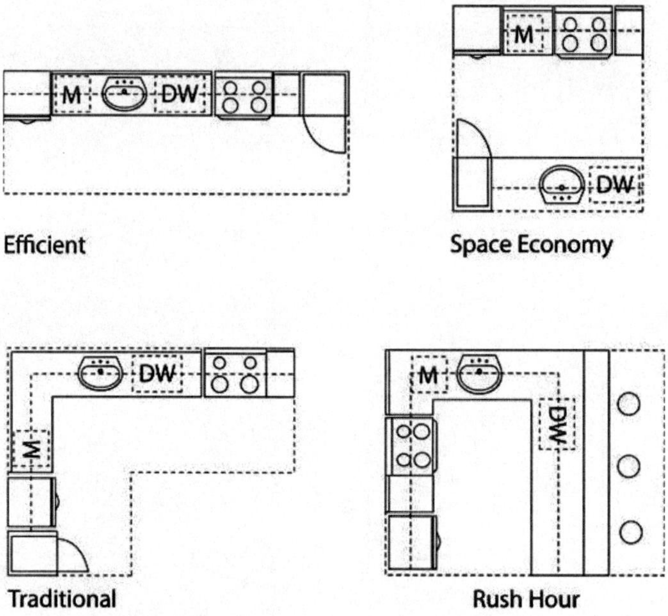

Fig. 11.7 Common types of kitchen arrangements in today's homes include the linear wall kitchen, the galley layout, the U-shaped, and the L-shaped layouts

For larger kitchens, the *U-shaped* design is optimal as it consists of cabinetry along three adjacent walls which allows for multiple users to work at the same time and efficient workflow (Grundig, n.d. d). Lastly, the *kitchen island* design remains one of the most popular choices today as it allows large central work and storage areas while simplifying traffic flow. The island can be turned into a multipurpose area for preparation, a bar or to save counter space, it is also a great storage area for other kitchen gadgets (Stanley 2021).

Considering the amount of storage space in the kitchen is essential. In other words, carefully designing kitchen storage space is about finding a balance between function and appearance. There are many design solutions to fulfill this need. For example, kitchen cabinet rollouts can allow a homeowner to store appliances and food more efficiently (Fig. 11.8). Rollout pantries are useful for inventory purposes as they provide a quick way to visually organize food items by category and date thus reducing the chances of wasting food items or buying excess quantities of the same food item. Beyond visual appeal and organized cabinets, the rollout pantry can provide important benefits for those with physical difficulties and the elderly. These types of pantries put less physical strain on the back and knees when reaching for desired items (Ferris 2021). Other simpler design practices include hanging up kitchen utensils on the wall as decoration rather than using additional cabinets and drawers to store them (Stanley 2021).

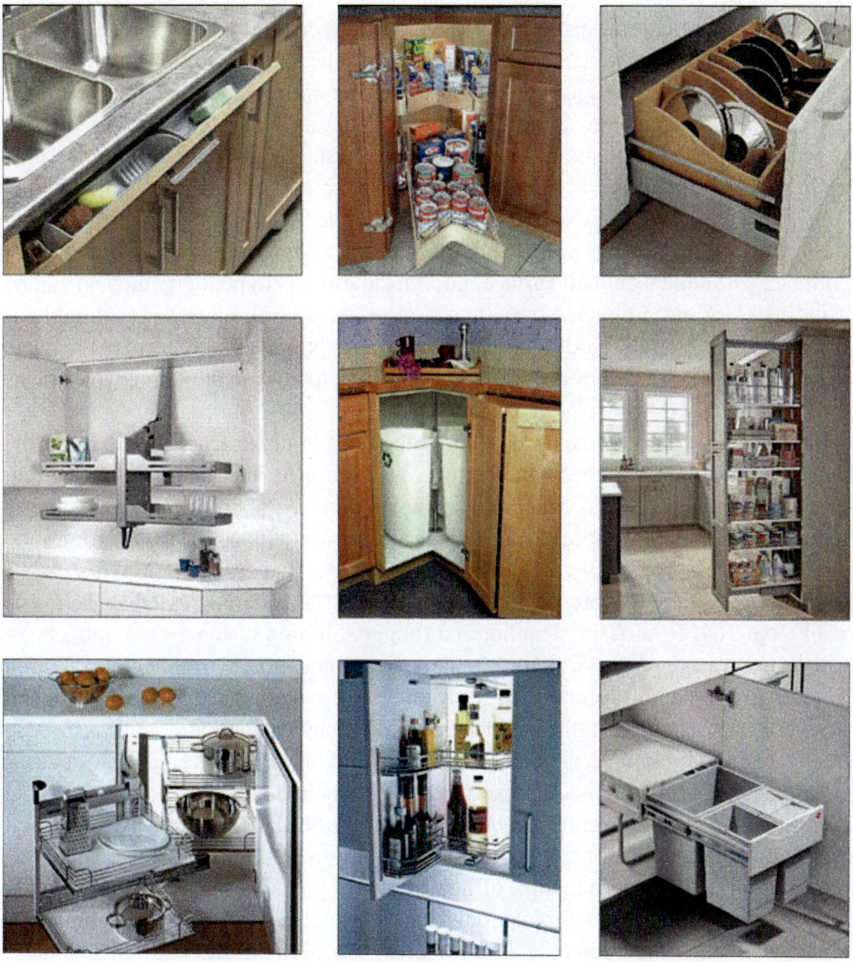

Fig. 11.8 Designing kitchen storage space is about finding a balance between function and appearance. Many innovative items have been recently introduced

11.3.2 Sustainable Building Materials

Making a kitchen truly sustainable begins with the basic design of the entire home and the room in question. A number of kitchen components can be built using renewable sources or with materials that are produced in eco-responsible ways to minimize their impact on the environment. Such materials include bamboo, cork, recycled glass, and ethically sourced timber. Other lesser-known materials even include milk-based paint, fiberboard made from potato starch, bioplastics, and bricks made from plastic bags (Grundig, n.d. c).

Kitchen floors can be made from recycled tiles, natural stone as well as some of the materials mentioned above making them cheaper than traditional flooring (SpaceWise 2021). Although some of these flooring options such as cork or bamboo have been around for quite some time, their popularity is being renewed thanks to the recent green movement. Cork is sourced from the bark of the cork oak tree, is completely natural and can be replenished. To make cork floor tiles, the material is ground up, compressed, and formed into sheets bonded with layers of resins (Lewitin 2021). Cork flooring has a warm and natural look adding uniqueness to the room. It offers a soft and cushioned surface, good insulation, is hypoallergenic and can be easily installed and refinished over time. Cork tiles are not, however, as durable as other flooring materials. Additionally, they are susceptible to many forms of damage over time such as discoloration from sunlight or scratches if the homeowner owns pets (Lewitin 2021).

Bamboo is a more damage resistant material. It is made from the fibers of the bamboo grass plant which continues to grow even after the stalks have been cut making it a valuable commercial renewable resource. Bamboo flooring also gives a unique and natural look appeal to the kitchen as does cork flooring. However, bamboo flooring does not provide thermal properties in the way that cork will (Lewitin 2021). It is also important to note that while bamboo is considered replenishable, it is mainly exported from Asia meaning that longer shipping distances are applied for those using it in the West which ultimately contributes to environmental pollution (Roberts 2020). Therefore, choosing the right type of sustainable flooring will depend on several factors such as distance to the market, environmental conditions inside and outside of the home, aesthetic preference, durability, and price.

When considering the modular nature of the kitchen, many choices of sustainable countertops are also currently available. They can be made using salvaged wood from old counters to encourage a circular economy which reduces waste and over-consumption from the production of new countertops. When considering recycled wood, it is important to look for materials that have not been treated with chemicals or painted with lead-contaminated paint (Roberts 2020). If a different design and material is desired, opting for recycled glass or metal countertops can be other favorable options (Fig. 11.9). Recycled glass materials are non-porous and considered maintenance free as they do not need to be sanded or varnished with toxic sealants. They are also crack and chip resistant making them a durable option and can ultimately be recycled after their useful life (SpaceWise 2021). Quartz countertops can also be an option as they offer the same benefits as recycled glass countertops. Some niche companies produce countertops primarily sourced from crushed wasted quartz left over in mining quarries making them less environmentally damaging than mining for additional resources (Roberts 2020).

Kitchen cabinets will largely influence the room's aesthetic appeal and storage capacity. The most sustainable material for the construction of kitchen cabinets is wood. The best options are certified high-quality hardwood and formaldehyde-free medium-density fiberboard or plywood. According to a study conducted by the World Wildlife Fund, the global demand for low-cost timber causes illegal logging activity in some of the world's endangered tropical forests. The study expresses that up to

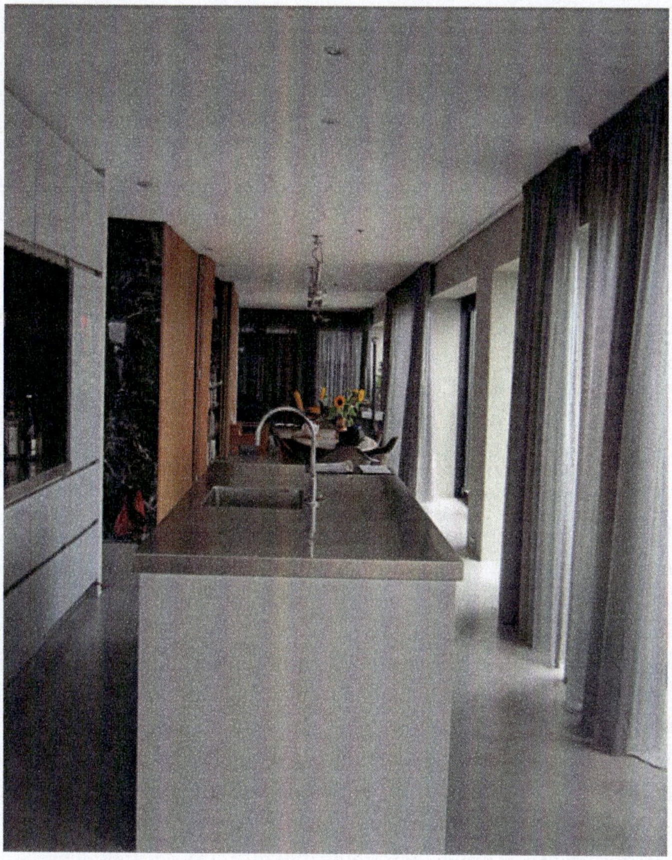

Fig. 11.9 If a different design and sustainable material is desired, opting for recycled metal countertops can be an option

50% of harvested tropical hardwood is produced through illegal logging. As such, it is important to keep the source of the product in mind while it is always better to opt for Forest Stewardship Council (FSC) certified products that are sustainably managed. In return, these responsibly managed forests can offer a sustained source of income to the rural populations of an area (Roberts 2020). On the other hand, opting for plywood cabinets can provide a truly more sustainable alternative as it can be locally sourced. Plywood is used for its strength, durability, and versatility. It consists of thin sheets of wood veneer and adhesive which are compressed, layered, and bonded together (Grundig, n.d. a). What makes this material so sustainable is that its production waste is far less significant than with timber, it can be locally sourced and is more affordable. Plywood has many other uses in the kitchen such as for the construction of walls, floors, and countertops. This material comes in different grades and their type will vary according to their use (Grundig, n.d. a).

When decorating and thinking about interior design, paint becomes one of the most important tools in giving life and character to a particular room through color choices and combinations. Eco-friendly paint can be a great option to make decorating the kitchen more sustainable. It is composed of natural and non-toxic ingredients made from solvent-free materials that have no volatile organic compounds (VOCs). Some certified paints also help reduce indoor air pollution during the application and drying and lower the risk of chemical exposure which can occur with traditional paint. Natural paint is composed of materials like clay, milk-proteins as previously mentioned, and even citrus (SpaceWise 2021). There are countless combinations that can be made when considering the choice of materials used to make a kitchen more sustainable.

11.3.3 Natural Light and Energy Efficient Appliances

Another defining characteristic of a sustainable kitchen is its ability to decrease energy consumed by appliances and artificial light. Designing the cooking and dining space with large windows that face in the desired orientation will allow natural light to illuminate the space (Fig. 11.10). To save on cooling costs it is essential to have fresh air flow and cross winds for ventilation purposes (Grundig, n.d. c). Additionally, choosing high-quality windows can help further reduce utility bills by approximately 12% and ultimately reduce greenhouse gas emissions (SpaceWise 2021).

As a rule of thumb, the selected appliances should be energy efficient and when possible, reduce overconsumption of non-essential items. For example, opting for manual can openers and stove-top kettles rather than electric ones will save electricity and get the task done just as efficiently. Using human-powered appliances is also good for one's wallet as these kitchen tools and appliances are sold at a more affordable price and consume less energy (Roberts 2020). Applying this rule of thumb also forces one to think about their spending habits and question whether they really need the newest electronic kitchen gadgets or if they can opt for items they already have at home or that have less of a carbon footprint when in use.

When considering larger appliances, energy star certified products are the best choices when considering energy efficiency as mentioned in Chap. 10. Dishwashers, fridges, ovens, and even microwaves can be designed to be energy efficient. It is common to assume that washing dishes by hand saves more energy than using a dishwasher. However, some studies suggest that Energy Star dishwashers can be a lot more efficient (Roberts 2020). These dishwashers will save the average household approximately 14,650 L (3.870 gallons) of water over its lifetime compared to washing dishes by hand (SpaceWise 2021).

The addition of smart appliances such as smart fridges can also be a great addition to a modern home if the homeowner's budget permits. Smart fridges will help reduce food wastage by detecting when food is about to expire. Fridges with smaller freezer sections will also help avoid energy spikes and reduce the cost of electricity as well (Grundig, n.d. c). Having proper kitchen ventilation will also help reduce heating

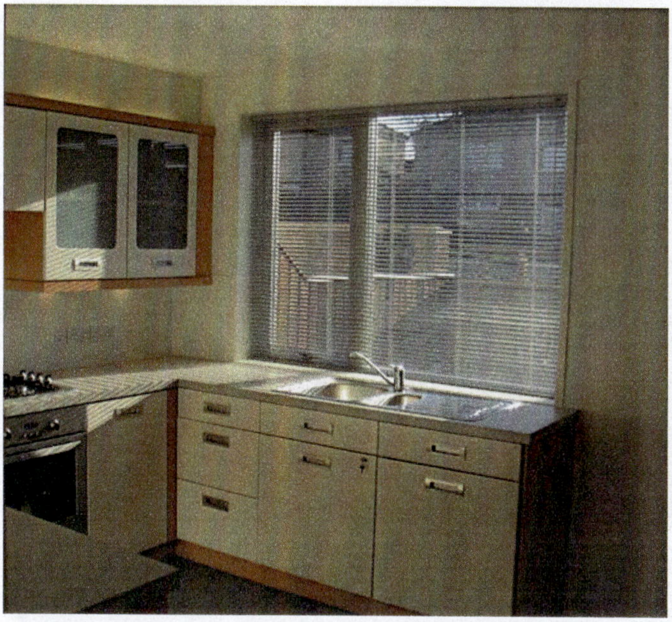

Fig. 11.10 Designing cooking and dining spaces with large windows that face in the optimal orientation will allow natural light to shine through and save energy

within the home while cooking, eliminating particle, heat, and odor buildup (Roberts 2020). Lastly, opting for induction cooktops over conventional ovens will reduce cooking times and save energy. With an induction cooktop, heat is concentrated onto the pan reducing the amount of heat released into the kitchen whereas a convection oven will use approximately 20% less energy than a standard oven as heated air continuously circulates reducing cook times as well (SpaceWise 2021). As now evident, designing kitchen spaces with the right appliances and ample natural light will make all the difference when aiming for a sustainable home.

11.3.4 Growing Food in the Kitchen

The kitchen is no longer a place to only prepare and cook food. Under the right conditions, it can become a place of food cultivation as well. One of the simplest ways to enhance the kitchen's sustainability is to use it as a growing space for vegetables and herbs from table scraps or purchased seeds. Building shelves or using windowsills as plant homes are affordable ways for making the kitchen an eco-friendly environment. Growing produce without having to rely on the outdoor environment can reduce food waste and help save money by having to purchase fresh produce less frequently. Growing plants, herbs, and vegetables also adds color and life to the

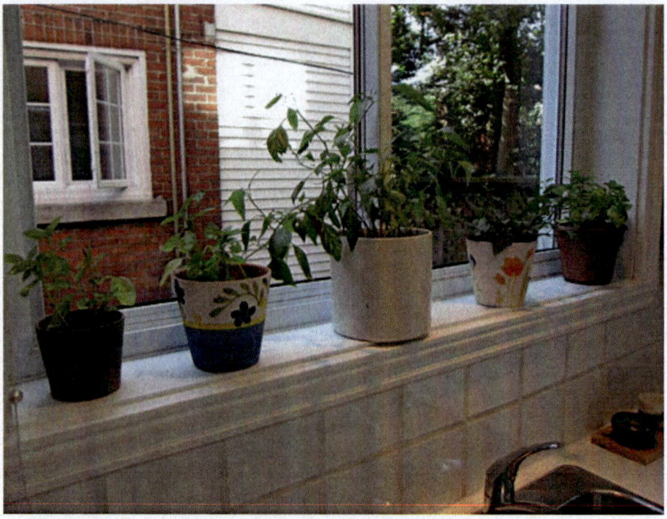

Fig. 11.11 Growing plants, herbs, and vegetables adds color and life to the kitchen environment and saves money

kitchen's environment to serve as a practical and affordable way to decorate the kitchen and save money (Fig. 11.11).

Currently, new technology is also making it easier to grow to produce indoors 365 days/year regardless of the environmental conditions within and outside of the home. An *urban cultivator* is a medium-sized appliance with dimensions like that of a dishwasher which allows a homeowner to grow food indoors. The urban cultivator provides the preferred environmental conditions to grow microgreens (Fig. 11.12). Seeds are placed on the cultivator's shelves and the appliance self-regulates the conditions necessary for optimal growth such as lighting, humidity, and water. The process requires little overall maintenance on the part of the owner (Urban Cultivator, n.d.). The cost of an urban cultivator varies depending on the brand and is not as budget friendly as a small fridge or wooden shelves. It remains a great innovative option for indoor cultivation, nonetheless and its cost is expected to decrease as it becomes common.

11.3.5 Composting, Recycling, and Reusing

As people's conscientiousness about protecting the environment keeps growing, daily actions such as recycling have become a way of life. Compositing is also gaining in popularity as many municipalities are encouraging citizens to compost food scraps rather than throwing them away to reduce landfill waste and give these leftover food items a second life. Compost provides an affordable way to boost soil and plant

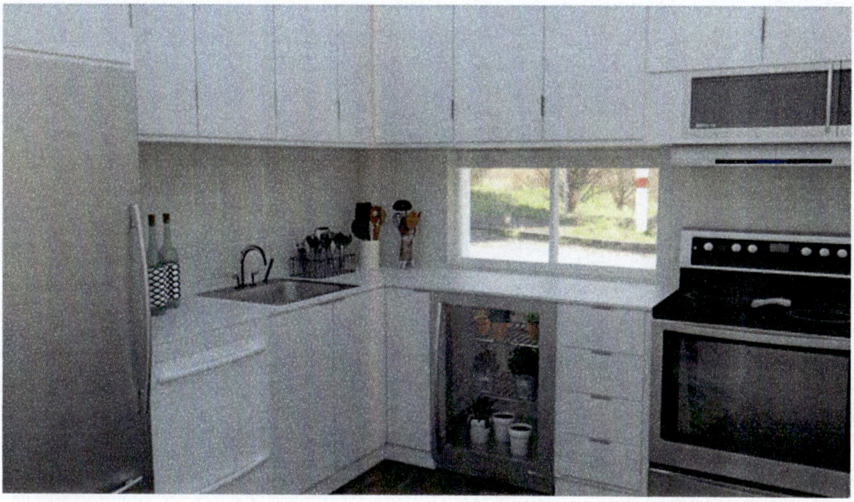

Fig. 11.12 The Urban Cultivator provides good environmental conditions to grow microgreens at home

health in one's garden in a sustainable fashion. Organizing the kitchen waste system can be done in a way to maximize and encourage composting and recycling at home. In terms of design, the recommended place to store composting and recycling bins is in the cleaning area or near the exit to make taking out the bins easier and away from fresh food (Grundig, n.d. b). These bins can be hidden away to preserve the visual integrity of the kitchen. They are an inexpensive way to manage kitchen waste and can be fitted into a cabinet door to slide, pivot or swing as the door opens. Other options include designing countertops with a lidded waste shoot storing the bins in the cabinet below. Recycling bins can also be separated into compartments like paper, plastic, and glass as they are all treated differently once they get to a sorting facility (Grundig, n.d. b; Fig. 11.13). Sorting recycled materials in the right way at home makes the entire system more efficient and sustainable.

Just as it is important to compost and recycle, reusing is also a key part to a sustainable waste management system. Reducing consumption by reusing items in the kitchen is one of the best ways for it to be sustainable. For example, old file organizers from offices can be used in a cabinet to store cutting boards, sheet pans, and muffin tins in an organized fashion rather than getting thrown away (Stanley 2021). Reusing plastic containers and glass jars rather than recycling them after emptying their contents is also a great way to avoid buying additional storage containers. Beyond storing leftovers, jars can be used to make decorative candles, soap dispensers, pots for growing plants and herbs, and much more (Nystul 2022). There are numerous ways items can be reused in the kitchen environment since it is such a multipurpose space. An organized waste management system as mentioned above can encourage creative thinking about the second life of an item to reduce overall consumption and waste.

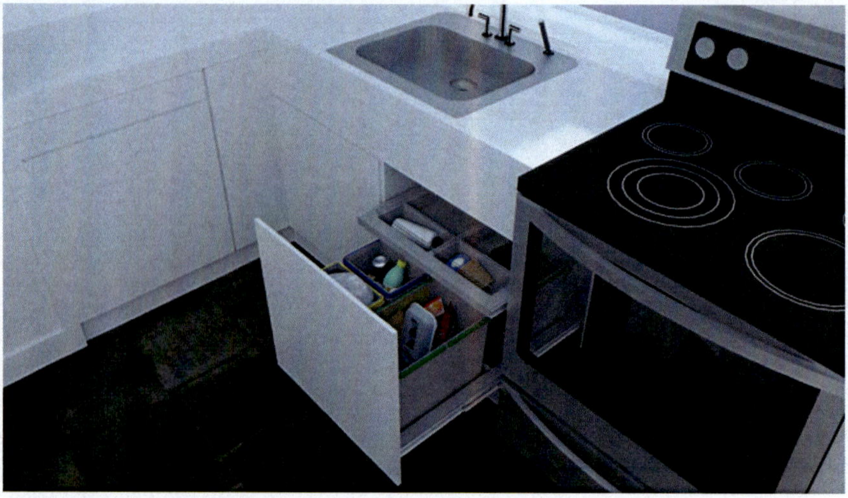

Fig. 11.13 Kitchen bins for items that can be recycled can be separated into compartments for paper, plastic, and glass products

11.3.6 Place-Making for Social Interactions

As expressed above, architecture will play an important role in influencing the types of interactions and social activities that take place in the kitchen. Now that the dining room and cooking area are closer, designing a “social kitchen” is much easier. A social kitchen is designed to encourage gatherings, turning it into a meeting place for the homeowners and their guests.

To achieve this, a kitchen island or rolling cart can be a good option as it creates a focal point in the room while providing often much-needed extra storage space as previously mentioned. In terms of layout, the open-floor plan is the top choice for social place-making as it makes traffic flow much easier (Giaquinto 2019). Additionally, creating a workspace in the kitchen can be a simple way to bring people together and generate productivity. This could mean adding stools around one side of the kitchen island or a folding table (Fig. 11.14). Incorporating bookshelves and a chalkboard wall can also help boost productivity and favor a mixed-use environment where parents can cook while supervising their children’s homework time for example. Designing banquet seating next to windows and adding other sit-down furniture such as stools and benches in and around the kitchen are all ways to highlight social zones through decor and design. Importantly, to make sure that the kitchen is up to the task and can support such activities, material choice for countertops becomes salient to avoid long-term wear and tear (Giaquinto 2019).

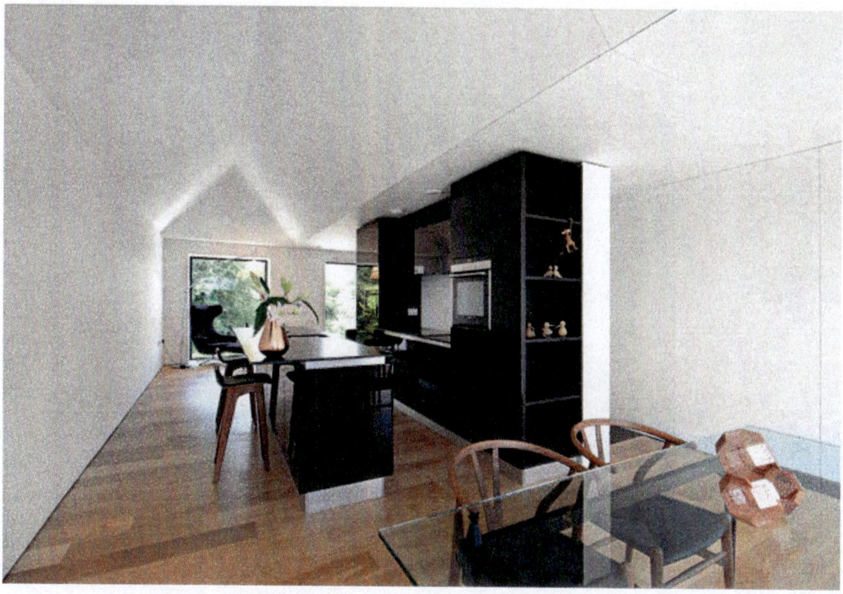

Fig. 11.14 Adding stools around one side of a kitchen island can create a gathering space

11.3.7 Accommodating People with Special Needs

When considering what a sustainable kitchen can be, adapting design to suit the needs of those who require specific accommodations is important. With adequate design, occupants with reduced mobility, those with sensory and cognitive impairments and the elderly can be given more independence (Fig. 11.15). Many guides are offered by the government of various countries to be used as references for adaptive kitchens. For example, The Americans with Disabilities Act (ADA) provides design guidelines for adaptable kitchens. The ADA was signed into legislation in 1990 as the most comprehensive piece of civil rights legislation at the time. It prohibited discrimination and guaranteed that individuals with disabilities have the same opportunities as any able-bodied individual to participate in mainstream life (ADA.gov, n.d.).

As mentioned above, pull-out pantries can be a great option to store food items in an organized way as well as heavier appliances such as a blender, slow cooker, or toaster as these storage methods will reduce the physical strain on a person's body. It is also possible to label these pull-out pantries making it easier to visually see its content (Ferris 2021). An important design principle for adaptive kitchens is the concept of rounded edges. Avoiding 90-degree angles on countertops will reduce the likelihood of injury and bruising (Ferris 2021). Another equally important principle is making sure that the sink is close to the stove. With a small-medium workspace separating the two, the one-wall kitchen layout is optimal. If this layout is not possible, the galley kitchen with the sink facing the stove is the best option so that a user must only turn 180 degrees to reach the sink if mishaps happen while cooking (Ferris 2021).

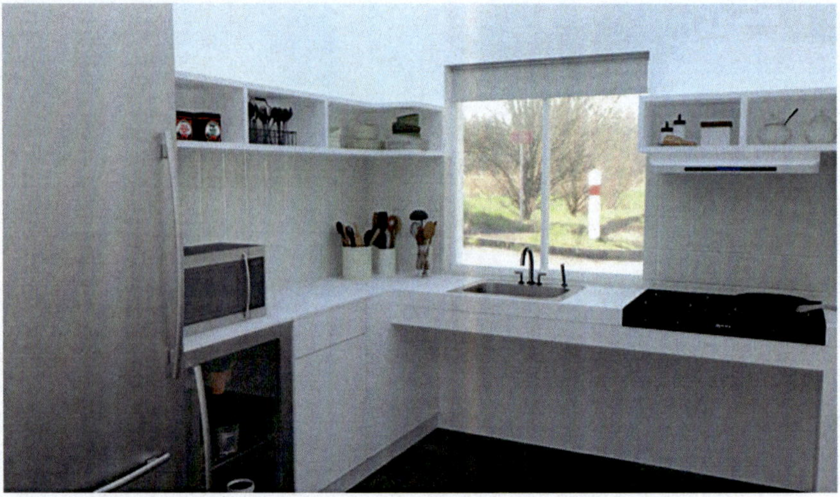

Fig. 11.15 With adequate design, occupants with reduced mobility, those with sensory and cognitive impairments, and the elderly can have more independence in using kitchens

For the elderly or someone with reduced mobility, opting for shallow sinks approximately 15–20 cm (6–8 inches) deep is the best option for reachability. Designing a kitchen storage with more drawers instead of pantry doors also makes it easier to sort out kitchenware, dishes, utensils, and pots and pans which will not require a person to bend down as often. There are many creative cabinet and shelving styles that can be both accessible and efficient for storage (Fig. 11.16). Corner drawers and corner cabinets with hip-level sliding shelves are great examples. Installing slip-resistant floors is also a good way to reduce the risk of injury in the kitchen (Ferris 2021). For wheelchair accessible kitchens, spacing and height become a key design principle. For example, accessible countertops should be installed at 86 cm (34 inch) height rather than the typical 91 cm (36 inch) height. Of course, measuring the homeowner's comfort will ultimately determine the appropriate height of the countertop when designing. Wider passages and doorways raised bottom cabinets and knee clearance for the sink and cooking area are also essential (Easterseals, n.d.).

Many kitchen tools can also be paired to make the cooking and dining environments more adaptable to special needs and to increase safety while cooking. Things like safe cutting boards and easy pour kettles are among some of the vast possibilities for adaptable kitchens (Price, n.d.). The combination of adaptive design practices with sustainable ones mentioned throughout the chapter can help people maintain autonomy while using a sustainable kitchen.

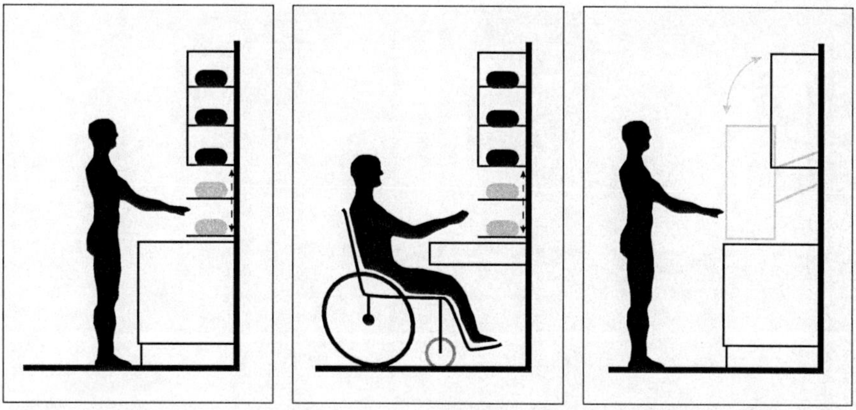

Fig. 11.16 There are many creative cabinet and shelving styles that can be both efficient for storage and accessibility

11.4 A Family Kitchen in a Tokyo Apartment

Tenhachi House is a renovation project of an apartment in Tokyo, designed by husband-and-wife architects for their family (Archdaily 2016b). The apartment was originally divided into several rooms, yet for the renovation, the architects stripped down all interior walls and ceiling, revealing the concrete structure to facilitate a more spacious, open plan home (Fig. 11.17).

Private areas of the apartment are defined not by walls, apart from the washroom, but rather by open boxes. These cubes act as connectors between them and the more public living area. The bedroom box divides the space vertically in two. Below is the main bed space, while above is a lofted kids' space, accessed by a ladder. The bathroom box is left open on one side, the only divider being a white curtain. This provides enough privacy while maintaining the spacious feeling of an open plan.

A 4.5 m (14.8 feet) long table made of Japanese cedar acts as the centerpiece of the living space, serving functional and decorative purposes. The table contains the kitchen, dining, and workspace for both adults and children. There are no physical divisions between each area allowing for the space to be used flexibly. The functional layout of the table may change throughout the day, from breakfast, to after school, to dinner gathering. A sink and stovetop occupy one half of the table. The kitchen is completed by storage in floor-to-ceiling cabinetry on the nearby wall as well as overhanging shelves above the table. This open storage above becomes decor, artfully displaying the family's kitchenware. The opposite side of the space is for working, where a computer can be placed. Storage is also available below, supported by a row of bookshelves, easily accessible for children. The table naturally becomes the dining area during mealtime. For large parties with friends, the table can accommodate seating up to twenty people.

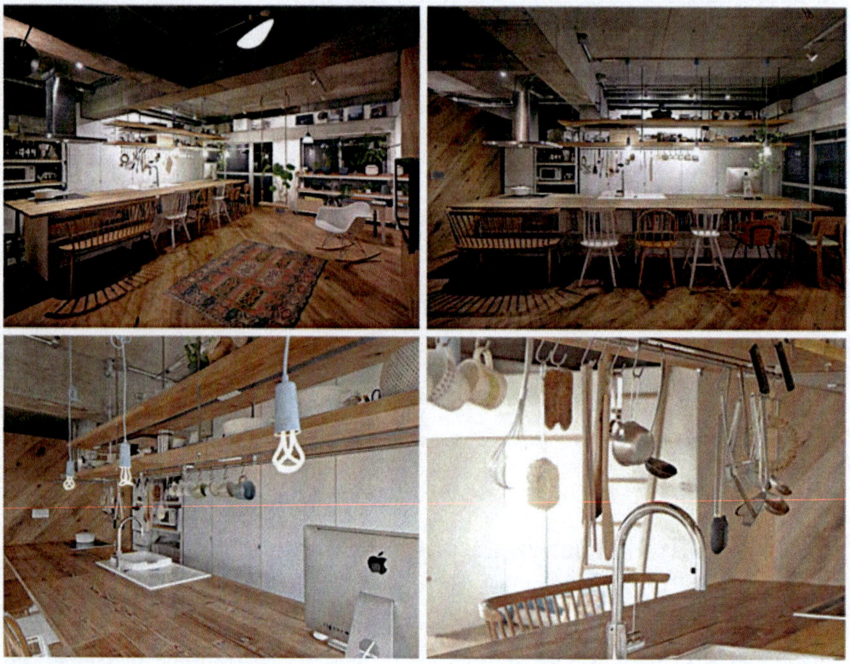

Fig. 11.17 In the Tenhachi House a 4.5 m (14.8 feet) long table made of Japanese cedar acts as the centerpiece. The table contains the kitchen, dining, and workspace for both adults and children

Tenhachi Architects personalized their apartment in this thorough renovation. The multifunctional table enlivens the home and brings all members of the family together.

11.5 A Home with a Sustainable Kitchen

A Sustainable Kitchen was designed by the author to accommodate the needs and lifestyles of modern households. Recognizing the kitchen as a family space for various activities beyond food preparation, the Sustainable Kitchen provides a highly adaptable, flexible space for the household. It was designed to be sustainable by using recycled materials and energy efficient building practices and appliances, while promoting a healthier and more environmentally sustainable lifestyle for its occupants (Fig. 11.18).

Homebuyers were presented with a selection of potential kitchen layouts; the design itself was comprised a variety of modular designed components which were arranged into several configurations (Fig. 11.19). In general, two different layouts were offered, the L-shape and the line kitchen; six different configurations were created from these two arrangements, which offered different choices of appliances,

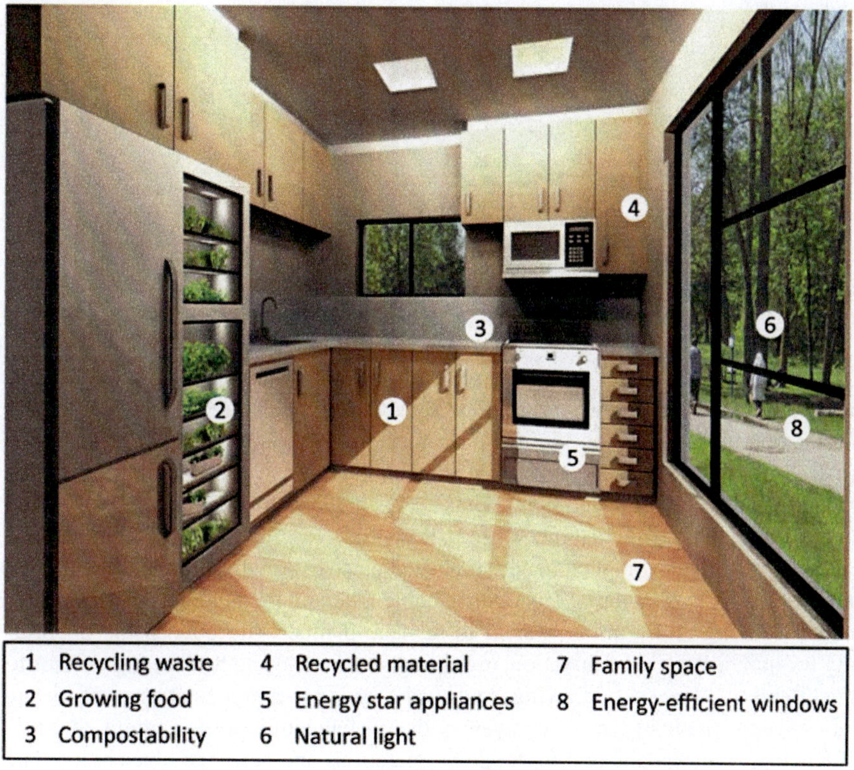

Fig. 11.18 Features that contribute to the kitchen's sustainability

fixtures, and furniture locations. The Line Efficient kitchen offered a smaller, more condensed layout, where all the functions were aligned to one wall. This option provided limited countertop space yet made effective use of a small area to provide occupants with more open-floor space in their unit. The Line Efficient kitchen layout was best suited for smaller households, or occupants for whom having a large cooking space was not a priority. Whereas the L-Shaped kitchen was designed to provide a more flexible space for families. A variety of different L-shaped kitchen options were offered, fitted with different appliances, locations of fixtures, and special features (Fig. 11.20). A kitchen for people with reduced mobility was also offered, which provided an open-floor space and lower cabinetry and shelving units, for easier access and circulation.

Kitchen options provided different arrangements of standardized, modular elements. A selection of cabinets, countertops, shelving units, pantries, and drawers was offered, alongside energy efficient kitchen appliances and fixtures (Fig. 11.21). All the different layout options included an additional wall of cabinet and pantry space with a small countertop for a variety of uses (Fig. 11.22). Some options for

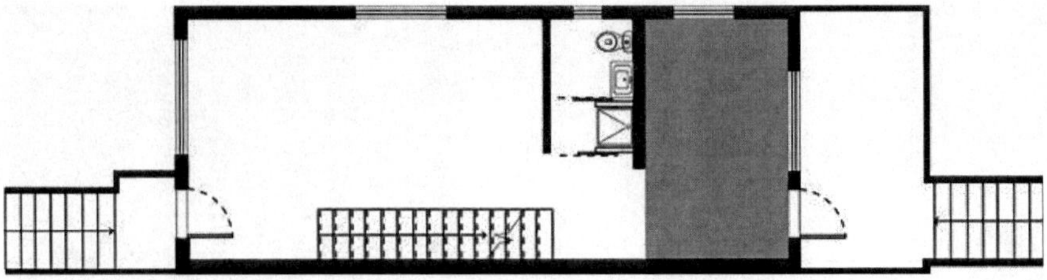

Fig. 11.19 Ground floor of the dwelling. The shaded area indicates the place in which occupants fit their selected kitchen's layout

countertops and kitchen islands featured built-in fold out tabletops, to provide additional seating for occupants. These tables provided a flexible seating area, allowing the kitchen to be used as a family space for conversation, chores, work, or mealtimes. The kitchen included several sustainable features. Alongside high-efficiency appliances, waste sorting compartments were included to facilitate recycling and composting, and a grow cabinet would allow users to grow their own ingredients. The Urban Cultivator, a specialized cabinet dedicated to growing greens as mentioned in Sect. 11.3.4, was included in the Sustainable Kitchen; it contained a built-in automated watering and lighting system to manage plants. It introduced the opportunity for individual food cultivation, providing sustainable and healthy ingredients to inhabitants at a low cost. Drawings and renderings of the various arrangements and choices were created to provide buyers with a visualization of their future kitchen.

Kitchens are fundamental to the daily operations and habits of households. The Sustainable Kitchen illustrates how the kitchen can be designed to accommodate the lifestyles of modern families, while promoting sustainable, healthy habits.

11.6 Final Thoughts

The use and design of kitchens have evolved significantly over the centuries and have now become an inclusive mixed-use environment for food preparation, socializing, and other domestic activities. The ability to customize the kitchen is now far beyond what could have ever been imagined in the 1950s and 60s. Recently, as people have spent much more time at home due to lifestyle changes resulting from the COVID-19 pandemic, even greater importance has been given to the kitchen's environment. Notably, people have rediscovered the importance of social interactions in their everyday lives. Therefore, designing for social activities in the kitchen will be one of the benchmarks for a desirable sustainable home. For it to be right for consumers, designing kitchens that are suitable, efficient, and flexible are among some of the key principles to keep in mind (Friedman and Krawitz 2002). Additionally, with increasing food prices, designing a kitchen environment that allows people to grow produce at home can be a great way to help homeowners save money and contribute to the reduce, reuse, and recycle principles. Efficient and organized waste

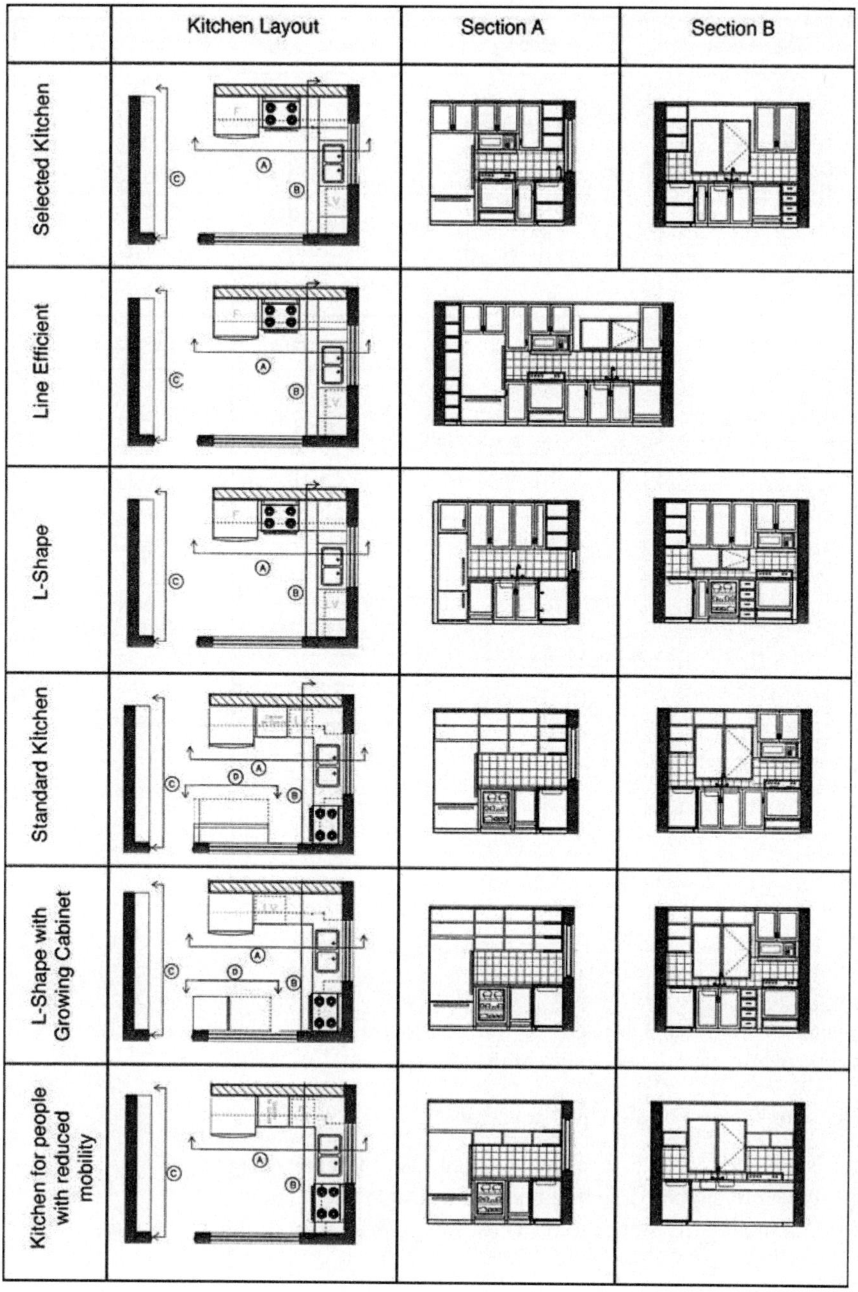

Fig. 11.20 Optional kitchen layouts for choice by occupants according to their lifestyle and budget

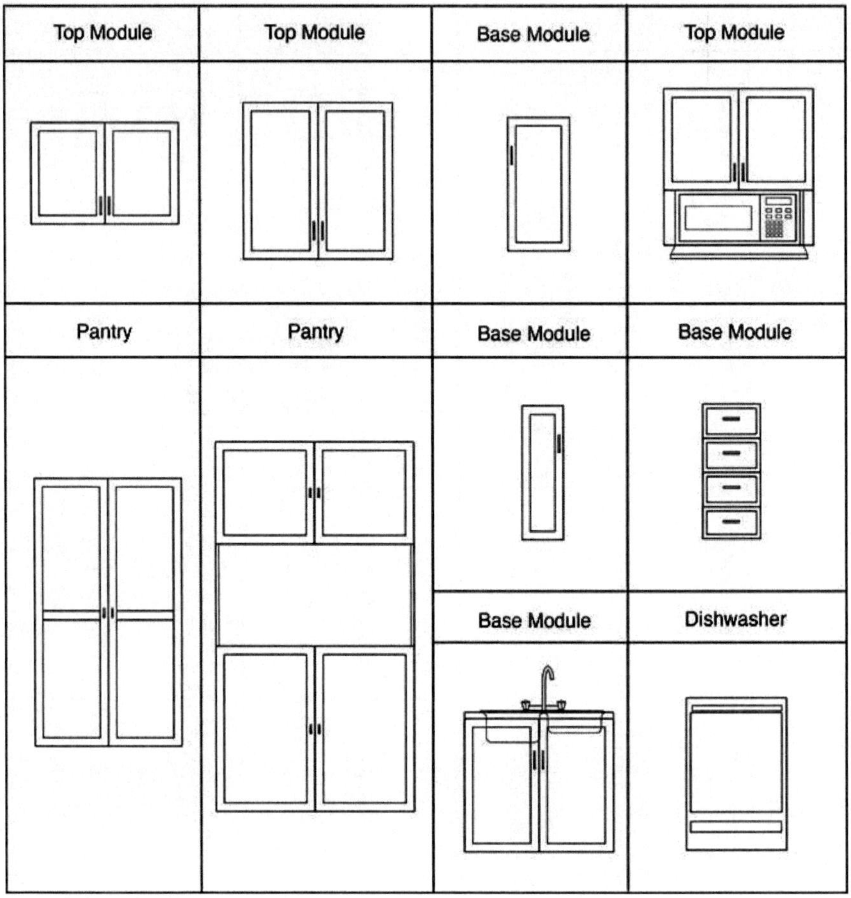

Fig. 11.21 A menu of kitchen cabinets for the occupants' choice

management systems are also key for considering sustainable practices and should not be overlooked in the design of the kitchen.

Questions for a Follow-Up Discussion

1. Name the key milestones in the evolution of kitchen design?
2. What are the key design features of a sustainable kitchen?
3. Can modularity stands to be a cost saving in kitchen design?

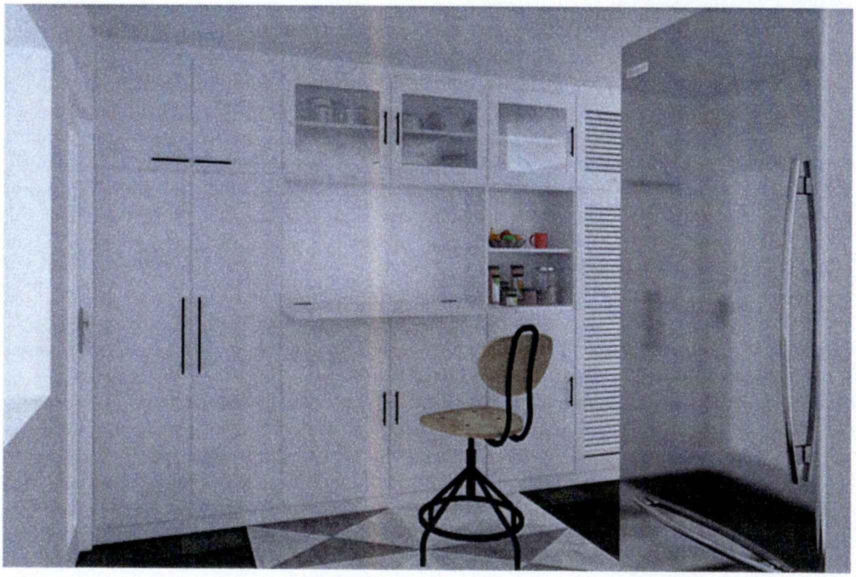

Fig. 11.22 Additional storage and workspace is offered

References

ADA.gov (n.d.) Introduction to the ADA. https://www.ada.gov/ada_intro.htm. Accessed 16 Oct 2022

ArchDaily (2016a) How to correctly design and build a kitchen. https://www.archdaily.com/789894/how-to-correctly-design-and-build-a-kitchen. Accessed 16 Oct 2022

Archdaily (2016b) Tenhachi house/Tenhachi architect & interior design. https://www.archdaily.com/789221/tenhachi-house-8-tenhachi-architect-and-interior-design. Accessed 16 Oct 2022

Easterseals (n.d.) 16 ways to make your kitchen more accessible. https://blog.easterseals.com/16-ways-to-make-your-kitchen-more-accessible/. Accessed 16 Oct 2022

Ferris S (2021) 10 ways to design a kitchen for aging in place. https://www.houzz.com/magazine/10-ways-to-design-a-kitchen-for-aging-in-place-stsetivw-vs~56039514. Accessed 16 Oct 2022

Friedman A, Krawitz D (2002) Peeking through the keyhole. McGill-Queen's University Press

Giaquinto G (2019) Design ideas for the social kitchen. https://blog.kitchenmagic.com/blog/bid/174536/design-ideas-for-the-social-kitchen. Accessed 16 Oct 2022

Grundig (n.d. a) How sustainable plywood is changing the way we design kitchens. https://www.grundig.com/ktchnmag/blog/how-sustainable-plywood-is-changing-the-way-we-design-kitchens/. Accessed 16 Oct 2022

Grundig (n.d. b) How to organize your kitchen waste system. https://www.grundig.com/ktchnmag/blog/how-to-organise-your-kitchen-waste-system/. Accessed 16 Oct 2022

Grundig (n.d. c) Sustainable homes that generate more electricity than they consume. https://www.grundig.com/ktchnmag/blog/sustainable-homes-that-generate-more-electricity-than-they-consume/. Accessed 16 Oct 2022

Grundig (n.d. d) The complete guide to kitchen layouts. https://www.grundig.com/ktchnmag/blog/the-complete-guide-to-kitchen-layouts/. Accessed 16 Oct 2022

John Desmond Limited (2016) A brief history of the kitchen. https://www.johndesmond.com/blog/design/a-brief-history-of-the-kitchen/#:~:text=The%20origins%20of%20the%20kitchen,were%20hanging%20above%20the%20fire. Accessed 16 Oct 2022

Lewitin J (2021) Cork flooring review: pros and cons. https://www.thespruce.com/cork-flooring-pros-and-cons-1314688. Accessed 16 Oct 2022

Monark (2017) The heart of your home: a history of the kitchen. https://monarkhome.com/history-of-the-kitchen/. Accessed 16 Oct 2022

Nystul J (2022) 13 brilliant things you can do with a glass jar. https://www.onegoodthingbyjillee.com/13-ways-to-use-glass-jars/. Accessed 16 Oct 2022

Paay J, Kjeldskov J, Skov MB (2015) Connecting in the kitchen: an empirical study of physical interactions while cooking together at home. Proceedings of the 18th ACM conference on computer supported cooperative work & social computing, pp 276–287. https://doi.org/10.1145/2675133.2675194. Accessed 16 Oct 2022

Price M (n.d.) Aging in place? Here's 8 kitchen tools to make cooking easy. https://www.rehabmart.com/post/aging-in-place-heres-8-kitchen-tools-to-make-cooking-easy. Accessed 16 Oct 2022

Roberts T (2020) Designing a sustainable kitchen. https://www.buildwithrise.com/stories/sustainable-kitchen. Accessed 16 Oct 2022

SpaceWise (2021) Upgrade your kitchen with these 17 sustainable ideas. https://www.extraspace.com/blog/home-organization/diy-projects/how-to-design-an-eco-friendly-kitchen/. Accessed 16 Oct 2022

Stanley J (2021) 41 genius small kitchen storage and organization ideas. https://www.familyhandyman.com/list/40-kitchen-organizing-ideas-that-will-save-your-sanity/. Accessed 16 Oct 2022

Urban Cultivator (n.d.) Eat better food at home. https://www.urbancultivator.net/kitchen-cultivator/. Accessed 16 Oct 2022

Yascaribay C (2019) Average kitchen sizes: what are the standard sizes for kitchens? https://marble.com/articles/average-kitchen-sizes. Accessed 16 Oct 2022

Chapter 12
Storing Stuff and Furnishing a Home

Abstract Increasing housing costs are putting a greater strain on an average family's budget. To cut costs in recent years the provision of smaller, affordable, and adaptable living spaces has become a priority. Ownership of "stuff" and the ability to store it is an important factor in choosing a small dwelling for one's needs. As a result, designing units with efficient storage spaces is essential. This chapter discusses how a combination of multifunctional furnishing, greener, and more innovative storage methods can be introduced. Lower costs, a smaller ecological footprint, and a more visually appealing home are among the factors highlighted in this chapter.

Keywords Adaptability · Consumerism · Furnishing · Multifunctional design · Small home storage

12.1 A Consumer-Oriented Society

An influx of visual information, sales, and marketing strategies have been luring people into rampant consumption supported by buy-and-throw-away habits. It is said that the average American home contains approximately 300,000 items including anything from pots and pans to the number of owned pairs of shoes (Becker n.d.). Endless store "deals" and periodic shopping events like Black Friday in America and the international Boxing Day holiday provide an incentive to perpetuate consumerism.

Consumerism is supported by the idea that the increased consumption of goods and services is always a desirable outcome and that an individual's wellbeing and happiness fundamentally depend on obtaining material possessions (Hayes 2021). Importantly, consumerism creates an unsustainable culture where overconsumption is normalized and encouraged. Although consumer-generated revenue is unarguably essential to the health of the global economy, the world is currently in a position where reducing harm done to the environment is a dire necessity to fight climate change. Overconsumption explicitly generates more waste, pollution, resource depletion, and land degradation (Hayes 2021). Moreover, cluttered homes resulting from overconsumption have been shown to have adverse psychological effects. In 2011,

A. Friedman, *Fundamentals of Innovative Sustainable Homes Design and Construction*, The Urban Book Series, https://doi.org/10.1007/978-3-031-35368-0_12

the Princeton Neuroscience Institute conducted a study which found that too much visual stimuli (objects) make it difficult for the brain to focus on tasks and process information efficiently (Kell n.d.). Over the years, purchasing and collecting "stuff" also results in requiring additional space. This self-inducing cycle can easily become overwhelming for the dwellers.

There is hope, however, that design can be an effective method in combating excessive consumption. To reduce consumerism, dwellings need to be smaller, and their interior spaces reimagined to eliminate useless features making them more usable at the same time. Efficient storage space therefore becomes an important factor by which homes can be more sustainable (Friedman and Krawitz 2002).

12.2 Household Demographics and Storage Needs

Household sizes have gradually decreased in many developed countries. The need for smaller homes that are adaptable to the smaller size of contemporary families is a key feature of future sustainable housing. Single headed households are among the more recent family types whose number increased. However, larger families are not to be overshadowed as they have specific housing needs as well. Additionally, in times where high rent and mortgage costs take up a larger portion of the annual income, having children live at home into their later adult years has become the new normal. Partially exacerbated by the COVID-19 pandemic, the global housing crisis characterized by a shortage of homes and rising prices will likely lead to the reduction of household size. For first-time homebuyers, purchasing a house has also become increasingly difficult making apartments and condos more attractive choices. Regardless of family type and size, an important determinant in the appeal when buying a home is its storage capacity. Therefore, efficiently utilizing space to accommodate a family's belongings becomes a necessity.

Schmidt and Austin's book *Adaptable Architecture Theory and Practice* (2016) outlines a select number of contemporary adaptable architectural approaches. They suggested that future adaptive housing designs should aim to be movable, adjustable, versatile, convertible, and scalable. These approaches consider how the intricate multifunctionality of a building on its space can be leveraged to satisfy dwellers' unique needs and desires (Zhou 2019). The adjustability approach can be particularly considered in the context where maximizing the use of small space is of utmost priority (Fig. 12.1). Adjustable strategies are associated with the elements within the home such as furniture and walls. Their strong flexibility supports the end user in accomplishing dynamic tasks without the need to upgrade features constantly. The degree of control the user has over the products and their multifunctional properties are key to adjustability (Zhou 2019). Moveable components like modular storage furniture and other accessories are essential part of the adjustable strategy.

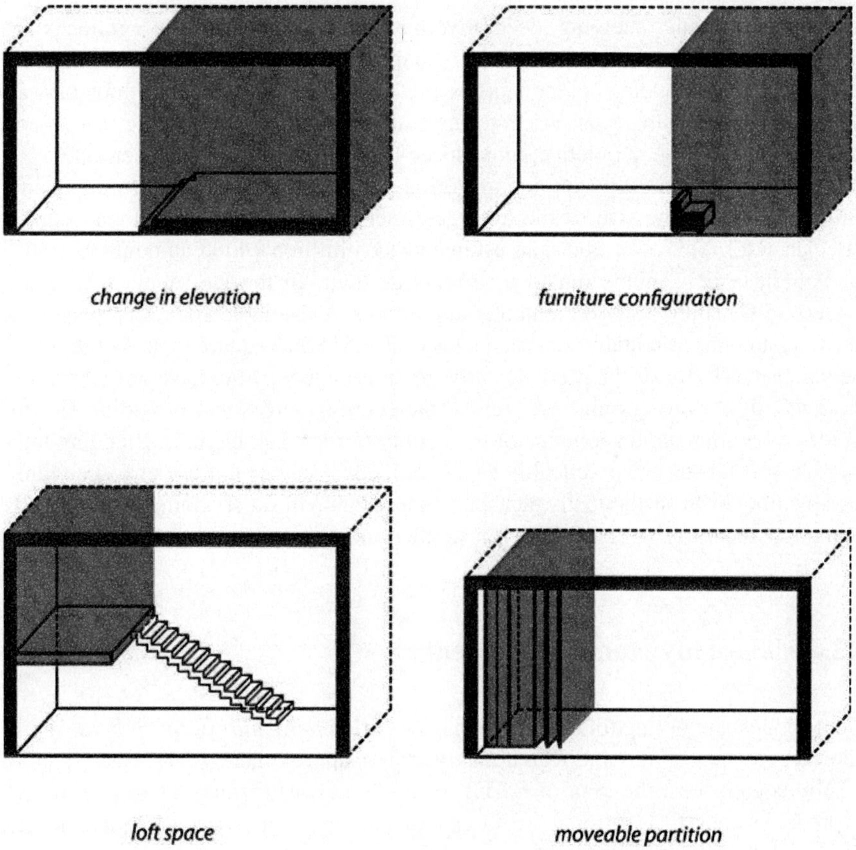

Fig. 12.1 Strategies for maximizing use of small space

12.3 Organizational Space Practices

Once occupied, the responsibility of finding space for and properly storing things is the homeowner's responsibilities (Friedman 2020). In many dwellings the available storage space is commonly limited to a few small closets and pantries sparsely located in rooms such as the kitchen, bedroom, and bathroom. In very small apartments known as micro units or studio apartments, storing items can be even more challenging.

To resolve the storage challenge, many products have been conveniently manufactured by retailers using a targeted approach. Currently, home design consultation services and certain large retailers enable homeowners to visually image and furnish their homes using 3D modeling software prior to occupancy according to their specific needs. Redesigning closets, kitchens, and bedrooms are among some of the popular services provided to customers. For those looking to redesign their organizational spaces in a more autonomous way, purchasing stand-alone storage cupboards, modular shelving, decorative baskets, and industrial storage racks are among the common choices homeowners make. However, the main issue with many of these larger modular components is that they consume valuable floor or wall space and serve a single purpose: storing stuff. In addition, as working from home has become a common practice, more space is now required for such activities.

Another contemporary solution to storing items, is to rent external storage units. Since the 1950s, the size of the average American home has more than doubled (Becker n.d.). Yet, as of 2021, an estimated 13.5 million which amounts to 10.6% of American households rented a self-storage unit. In the US, about 0.5 square meters (5.9 square feet) of rentable self-storage is available for every person as the total amount of rentable space equates to 176,515,776 square meters (1.9 billion square feet) (Harris 2021). More recently, green storage facilities have started making headway by powering units with renewable energy sources and providing electric vehicle charging stations outside of the facility (Green Storage n.d). Blending interior design and storage practicality can confidently create better use of space within the dwelling in an aesthetically pleasing manner that will satisfy homeowners' needs and allow them to live comfortably in smaller, more sustainable spaces (Fig. 12.2).

12.4 The Tiny Home Movement

Some people are opting for a minimalist lifestyle in *micro units* or *tiny homes*. While the average area of an American home measures approximately 241 square meters (2,600 square feet) the area of a Tiny Home is between 9 and 37 square meters (100–400 square feet). Micro units' area typically does not exceed 46 square meters (500 square feet) as well. Apartment buildings with micro units tend to offer shared amenities for entertaining, laundry, fitness, and storage (Livingston 2022).

Tiny homes are free standing structures of different shapes that may either be owned or rented. They can be constructed on a set foundation or designed to be mobile with wheels (The Tiny Life n.d.). Some are fabricated from old metal shipping containers while others can be built by manufacturers and sold as prefabricated units. Additionally, some homeowners build their own tiny homes as they are highly customizable, require minimal labor when compared to a typical non tiny home and create a sense of pride and ownership (Livingston 2022).

What sets tiny homes apart from conventional dwellings is that approximately 68% of owners do not pay a mortgage on them, only for the land on which they reside (The Tiny Life n.d.). Therefore, tiny homes are significantly more financially sustainable than owning a typical home as a smaller portion of income must be dedicated to housing-related expenses (Livingston 2022).

In addition to greater financial freedom, tiny homes offer other benefits and sustainable practices. Foremost, they consume less electricity, water, and generate less waste than traditional homes. Those situated off the grid can even heat and power their homes using solar panels for energy self-sufficiency. Some tiny homes have composting toilets to break down waste thereby avoiding a need to be hooked

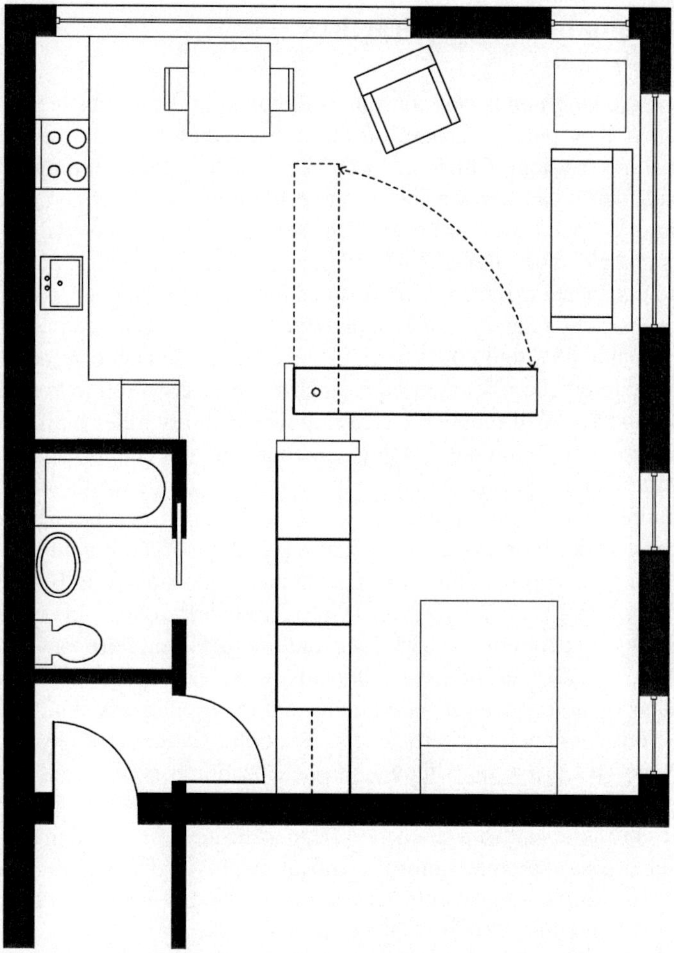

Fig. 12.2 Blending interior design and movable storage items can lead to better use of space as is the case in this micro apartment

up to a sewage system (Livingston 2022). For reference of popularity, the majority of tiny homeowners are under the age of 30 or retirees over the age of 50 (Saxton 2019). However, certain zoning laws, obtaining loans for construction, and shortage of sites are among the common barriers to being part of the tiny house movement (Livingston 2022).

12.5 Sustainable Storing Practices

Currently, a growing number of consumers are making conscious decisions about what products they purchase based on the impacts they have on the human and environmental ecosystem (Christian 2022). As a result, greener furniture and home accessorizing options are being made available. Furniture, modular storage units, and other forms of decor can be made from more sustainable materials. As designers and consumers make choices about how to furnish and decorate homes, paying greater attention to associated manufacturing certifications rather than unknowingly falling for attractive packaging and *greenwash marketing* is a better habit to practice. There are many sustainability and ethics certifications that consumers can look for when deciding what to purchase for their home such as the Energy Star logo as discussed in Chaps. 9 and 10. As discussed in Chap. 1, sustainability relies on the symbiotic relationships between economy, society, environment, and culture. As such, being aware of certifications that encompass some of these pillars are worthy of discussion.

For example, the B corporation (B Corp) certification can be applied to any industry and certifies businesses who pursue a social mission throughout their business activities. This certification is recognized in 37 U.S. states and B Lab partners globally (Christian 2022). B Lab is a non-profit network existing in regions of the world such as North America, Africa, East and South Africa, Europe, Oceania, and parts of Asia. They create policies, standards, tools, and programs that shift the culture, behavior, and structural underpinnings of the capitalist system. They mobilize the B Corp community toward collective action to address society's most critical challenges (B Corp n.d.). Although this certification is not without its potential faults, it is considered a gold standard among sustainable brands. Other certifications include the Forest Stewardship Council (FSC) certification discussed in the previous chapter, and the Fairtrade International certification provided by a multi-stakeholder non-profit that promotes fair working conditions and wages in a variety of industries including textile production (Christian n.d.).

Although certifications are not a perfect solution to sustainable production, they can be extremely useful to decipher greenwashed products from those that are genuine. However, an important caveat with any type of certification is that not all small businesses can afford to get certified as this process can cost upward of $25,000 US annually. Therefore, when in doubt, it is always best to reach out to the business personally when unsure about whether their product is truly sustainable or ethically sourced (Christian 2022).

In combination with certified production, some storage furniture and decor can be purchased as refurbished items encouraging a circular economy. Some online marketplaces work with sellers who upcycle and reclaim all types of unique furniture that may not be found at typical retailers (The Good Trade n.d.). Other sustainable storage options include products fabricated using natural and renewable materials. It is no secret that plastic has become a problematic material for the health of all natural ecosystems. In the UK alone, for example, an estimated 4.9 million metric tons of plastic enter the market annually with 75% of it becoming waste (Tiseo 2021). Willow, rattan, seagrass, and jute are among many natural, sustainable, and biodegradable materials that are currently being used as an alternative to creating storage baskets, boxes, and more to replace existing plastic options (Colville 2022). Though simple, this example along with product certifications illustrates some of the greener choices consumers have at their disposal when purchasing storage modules and decor.

12.6 Residual Spaces for Added Storage

Since the purchase of storage modules can only go so far in solving the occupant storage needs, designing better spaces and places for items within the home becomes essential. One best practice to maximize that space is to use residual space. Residual spaces include unusable spaces such as awkward corners that are a great opportunity for storing innovation (Campbell 2013). Below are recent and innovative architectural practices and products that can inspire homeowners and designers alike.

12.6.1 Wall-Mounted Seating and Bedding

If space for seating is hard to find, wall-mounted chairs and bench-shelves can be a great option as they do not clutter floor space (Fig. 12.3). Floating bench-shelves can be installed along a wall and can be made using a variety of materials depending on budget and desired look. Inexpensive materials like plywood make for a budget friendly option (Sims and Ulloa 2020). These types of floating bench-shelves also make for relatively easy building projects for do-it-yourself (DIY) enthusiasts. Similarly, integrated wall storage cabinets that act as seating can be installed under a window (Fig. 12.4). They can be designed to be folded to maximize space and can be topped with cushions. Small chairs and a round table can also be added to it to make it feel like a legitimate part of the cooking and dining area, eliminating the need for a larger kitchen table and more chairs (Sims and Ulloa 2020).

For homes with small bedrooms, beds can be integrated into the wall unit saving additional space for other purposes. Unlike a pull-down Murphy bed or hidden bed, the bed unit can be placed into the wall by carving out a space for an integrated platform or shelving for the mattress to lay flat on. The platform can have drawers at the bottom for additional storage space. Although they are often less comfortable than traditional beds, Murphy beds and hidden wall beds can also provide solutions to restricted spaces and are commonly used in micro units or guest bedrooms (Fig. 12.5).

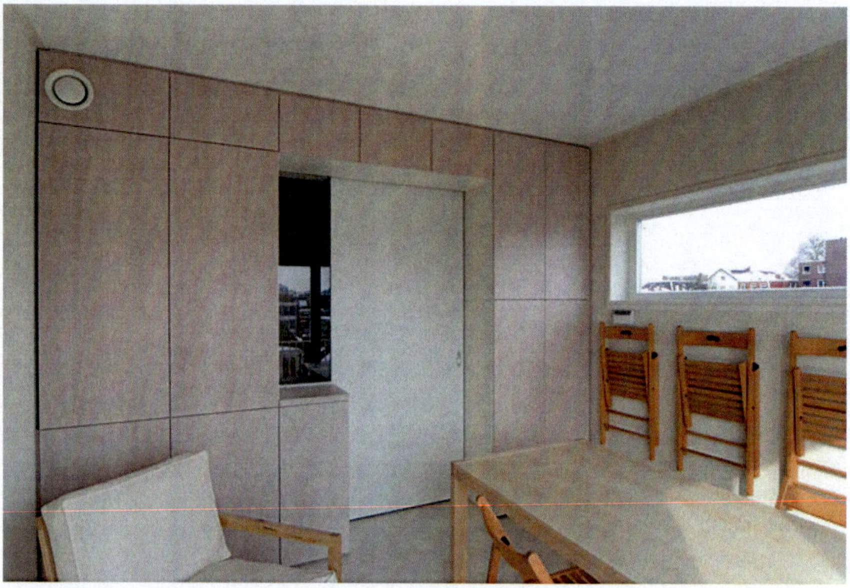

Fig. 12.3 If a space for seating is hard to find, wall-mounted chairs can be a great option as they do not clutter the floor

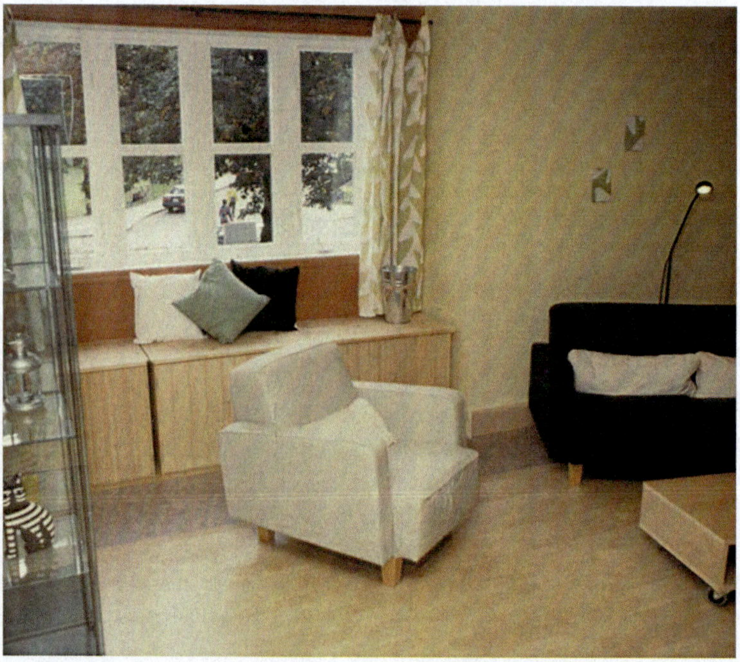

Fig. 12.4 Storage cabinets that act as seating can be installed under windows

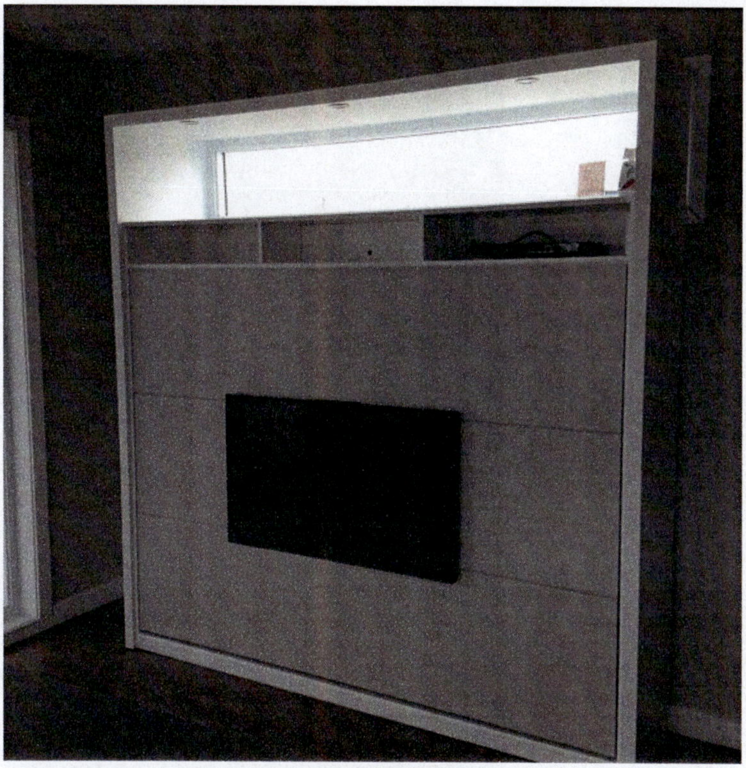

Fig. 12.5 Murphy beds shown here in an up-right closed position can provide a solution to restricted spaces and be used in micro units or guest bedrooms. Note the TV which serves the space when used as a living room

12.6.2 Storing Bicycles

For those living in dwellings without a dedicated storage room or garage, storing a large item like a bicycle can be frustrating. Wall-mounted hooks can provide a solution to maximize storage space. The bicycle can be hung from the hooks, eliminating floor clutter (Fig. 12.6). There are also *pulley systems* that draw bicycles out of view and into the ceiling (Sims and Ulloa 2020). Some homeowners may even decide to display their bicycles in the living room or kitchen as decorative wall art.

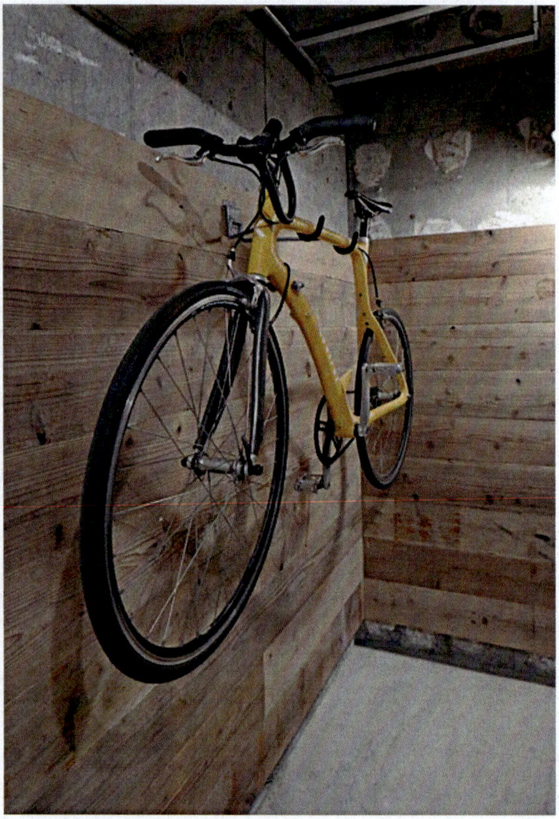

Fig. 12.6 A bicycle can be placed along a wall using hooks to eliminate floor clutter

12.6.3 *Floor Cabinets*

If the ceiling height in a home is tall enough, high storage can be a preferred option for making use of a room's space. In addition, floor cabinets can be constructed by building a raised platform. The key is to have varying floor heights where the raised section can either be opened from the top or drawers can be integrated into them (Sims and Ulloa 2020). The raised section of the floor can be placed below closets and dressers. Alternatively, the raised platform can be easily integrated into the living area. Steps leading to the raised section of the living room can create a storage space for an extra mattress and suitcases, for example (Sims and Ulloa 2020).

12.6.4 Staircase Storage

A way to take advantage of residual space is to utilize a wide staircase (Fig. 12.7). Under the stairs nooks can have multiple functions from being a mini to creating storage space (Fig. 12.8). The space can be turned into a cozy reading corner by installing a few cabinets, shelves, and a small chair. Alternately, a cushioned bench can also be installed rather than chairs (SpaceWise 2022). The space under the stairs can also be turned into a children's playroom requiring no additional floor space within the home while neatly tucking away any toys, books, and games. The area can also integrate sliding drawers instead since it is a great place to store items. This can be an appealing option for homes with little closet space (SpaceWise 2022).

A door can be added instead of drawers to allow more flexibility in the size of the items stored under the stairs. This method will also allow the space to act as a food pantry or space dedicated to storing excess dishes when shelves are added. Other uses of the space can be dedicated to storing dishes, wine glasses, and silverware. Additionally, if the workspace is lacking in the home, the area under the stairs can

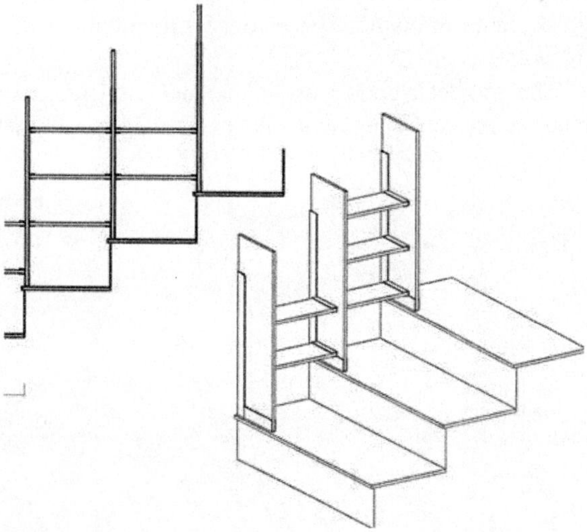

Fig. 12.7 Residual space can be found on a portion of a wide staircase

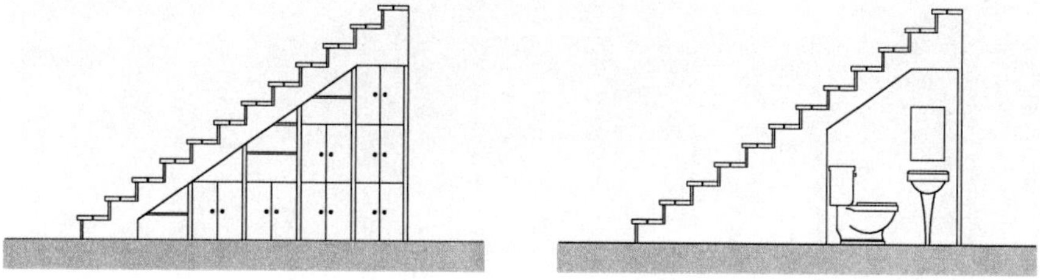

Fig. 12.8 Under the stair nooks can have multiple functions such as storage spaces or a small bathroom

also become an office by adding shelving along the length of the wall or a small desk and a chair (SpaceWise 2022). Overall, the staircase offers residual space as a versatile way of adding functionality to an otherwise obsolete area within the home.

12.7 A Home with Creative Storage

Designing creative storage was part imperative to the design process for the 3500 mm House in South Jakarta, Indonesia, by AGo Architects. As the name suggests, the three-story house has a width of only 3.5 m (11.5 feet), and naturally, finding storage space was an issue (Fig. 12.9). Through clever storage solutions, the architects were able to design a comfortable-sized home for a family of three (Archdaily 2019).

Although the house has distinct floors, there are additional levels between them that blur the divisions, creating one cohesive space throughout. The staggered levels allow for a double height living area, creating a more spacious atmosphere. The façade is a screen of perforated steel and polycarbonate, filtering the sunlight that reaches in. Behind the façade is a space that feels neither indoor nor outdoor. Plants grown in this transition space connect it with the exterior while bringing nature into the home.

The architects chose sustainable and cost-efficient materials, using wood for all interior fixtures. The architects also strived to eliminate walls, choosing instead to

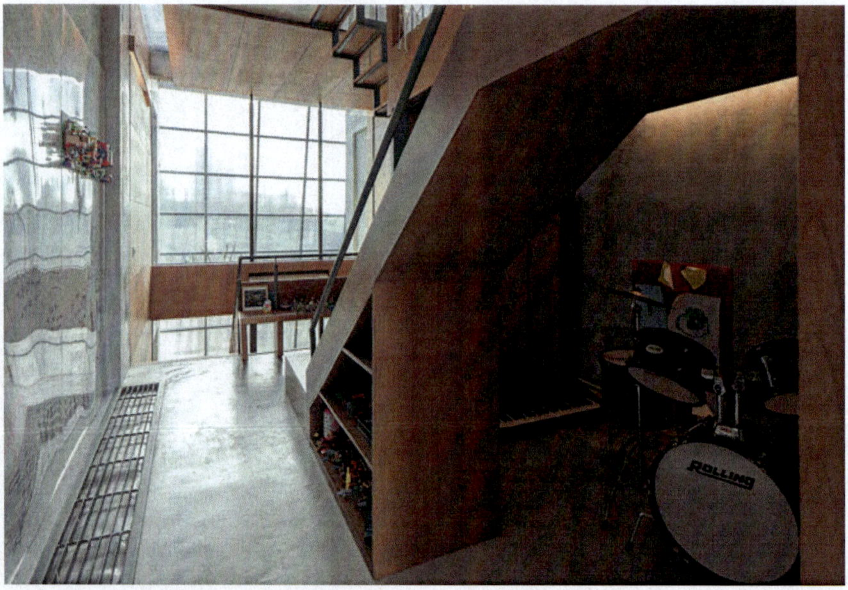

Fig. 12.9 The 3500 mm House in South Jakarta, Indonesia, has a width of only 3.5 m (11.5 feet). The architects were able to design innovative storage methods for a family of three

integrate storage in dividers and stairs. The kitchen island contains cabinets on two sides, as well as a set of cupboards above it. The living area has built-in storage beneath the wooden "floor" surrounding the sofa, a TV console and additional storage are nestled into the stairs leading up from the living area. The child's bed is raised above a study space; and more storage shelves are tucked behind stairs. These storage solutions make each element of the house multifunctional, reducing clutter. The plug-in storage system, level changes and an abundance of natural light transformed living in a narrow home into a spacious experience. The 3500 mm House is an example of urban living that is dense, yet comfortable with smart storage solutions.

12.8 Multifunctional Furnishing

When it is not possible to redesign a dwelling's interior, multifunctional furnishing can provide additional storage and maximize floor space for a clutter-free living environment by combining style, comfort, and practicality. Discussed below are a few examples of how furniture can fulfill these criteria in unique ways.

12.8.1 Bedroom Furniture

Bed Frames can provide ample storage space if drawers are integrated within them (Fig. 12.10). Some sleeping arrangements also provide raised drawer components in an L-shaped arrangement as well, doubling the storage capacity of the bed unit. This may eliminate the need for additional dressers and nightstands since the L-shaped configuration provides surface area for lamps and decoration while the drawers can be used to store clothing and other items. Other unique bed frame designs are raised bunk-bed style arrangement with desk and storage space under the bed frame platform. This type of bedding arrangement can be convenient for families with young children as it saves a large amount of space in the bedroom. Headboards can also become storage spaces as some are fitted with integrated drawers located at each extremity while others have small shelves used to store books and other small items. Though these furnishing items have been around for quite some time, making greater use of them in smaller living arrangements can be key to maximizing the usable floor space and a have clutter-free bedroom.

12.8.2 Fold-Down and Convertible Furniture

Dedicating space for a home office is important as improvised locations such as a kitchen or dining table may not be suitable. For example, a fold-down work surface can be fixed to a wall and conveniently used or tucked away as needed (Sims and

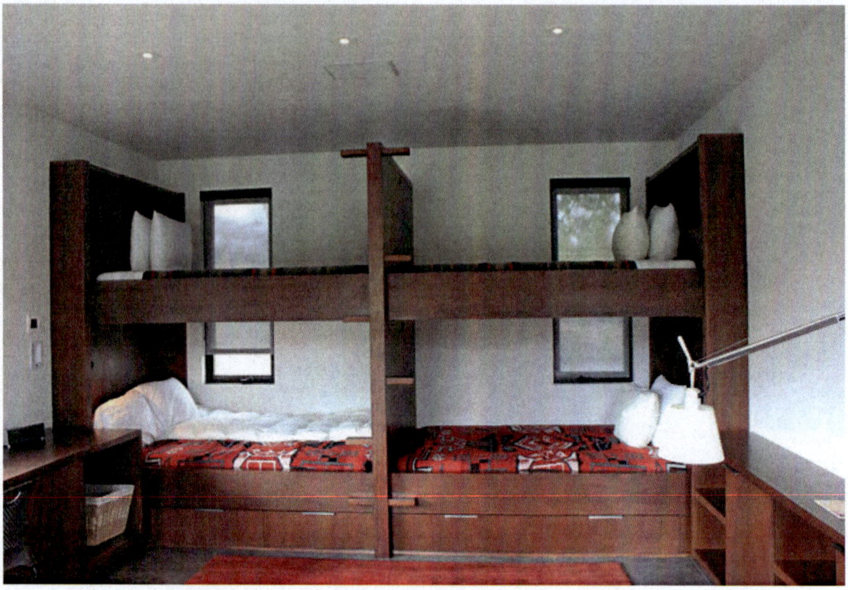

Fig. 12.10 Beds can provide ample storage space when they house drawers

Ulloa 2020). Convertible furniture is an efficient way of using them for more than one purpose. Some companies have found a way to turn wall panels into additional seating (Hays and Mason 2020). Other convertible options include drop leaf tables as they can be turned into a desk, a console, or even a dining table depending on the setting (Sims and Ulloa 2020) (Fig. 12.11).

12.8.3 Space Dividers

In micro units with a single open-floor plan, the need to divide space to accommodate different functions such as dining, and living may be important to the dweller. When there are restrictions on adding permanent partitions, tall bookcases can become great space dividers eliminating the need for walls and creating storage space at the same time (Mendelsohn 2021). Other innovative space dividing methods include MOLO Design's Softwall and Softblock modular system. The product is a freestanding interior partitioning design system made from 100% recycled Kraft paper that can be stacked, folded, and shaped into any form desired offering different heights to create dynamic wall surfaces and seating. This product can be particularly convenient for families who expect to grow and need additional space within the home. It's also easily foldable and requires little storage space making it a viable option for smaller homes and apartment units (Zhou 2019).

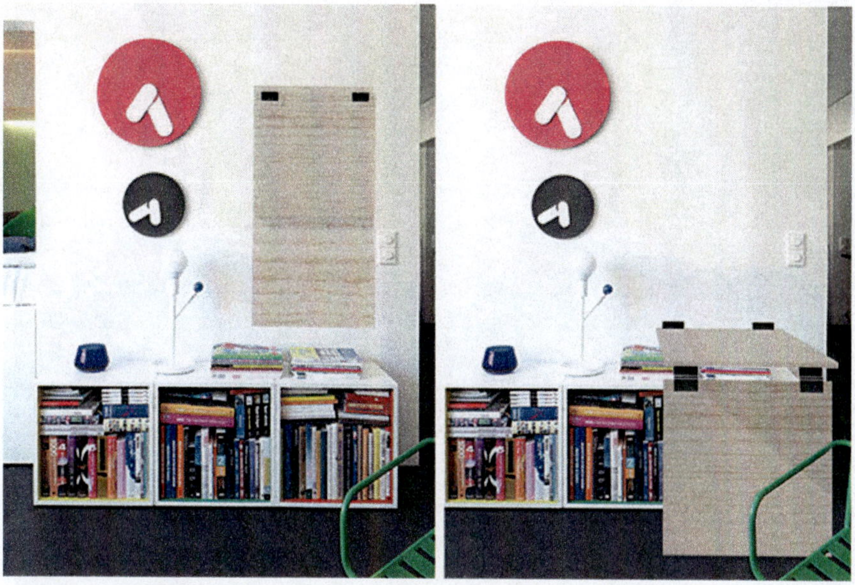

Fig. 12.11 Convertible furniture options include drop leaf tables which can be turned into a desk, a console or even a dining table

12.8.4 Modules for Micro Units

Characterized by living spaces under 46 square meters (500 square feet), micro units need extreme spatial organization to adequately store belongings in a clutter-free manner by having modules that integrate storage within the unit (Fig. 12.12). Because it is not possible to occupy all the rooms at the same time, unoccupied space can arguably be seen as a waste (Zhou 2019). To respond to this challenge, many innovative prototypes have been suggested. To maximize space perception and space augmentation, in 2014 the Massachusetts Institute of Technology (MIT) media lab announced a preliminary version of a living system called City Home. The prototype was tested in an 18.5 square meter (200 square foot) micro apartment in Boston, Massachusetts with the goal of redefining people's perceptions about the use of space within the living environment.

The robotic motion and voice-controlled City Home system integrated several spaces into one piece of furniture that is set on a track at the center of the room. On one side of the unit was a full-sized table, a single bed, and a folding bench chair that could all be pulled out and tucked away individually within the unit's compartments depending on the desired use. On the other side of the main living space, a more private section functioned as the shower and toilet area along with a storage closet. The unit could slide back and forth for convenience (Zhou 2019). Ultimately, these forms of indoor living system designs are beginning to provide efficient and flexible solutions to maximize the usefulness of confined living spaces while enabling

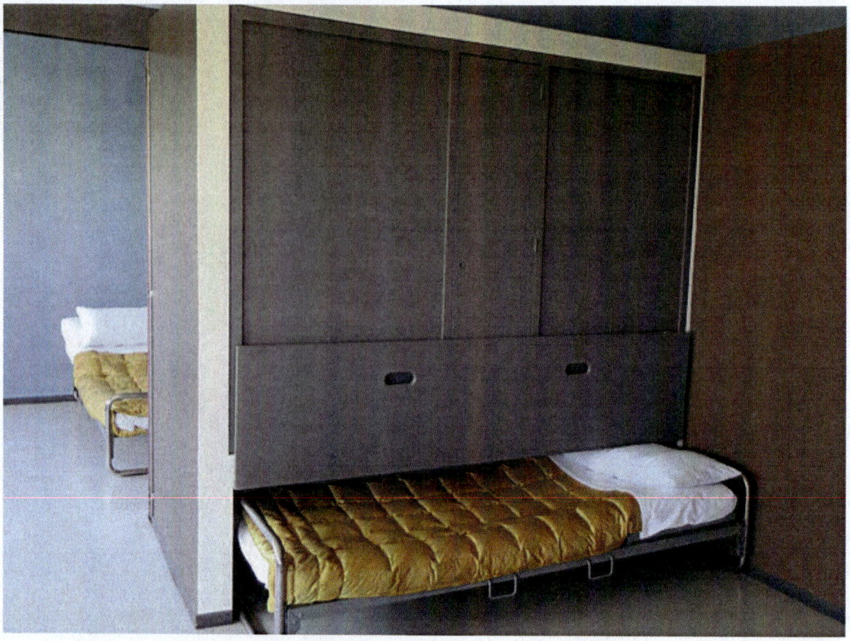

Fig. 12.12 Micro units need extreme spatial organization to adequately store belongings in a clutter-free manner by having modules that integrate storage and the bed

maximal comfort. Tiny homes and micro unit living systems demonstrate that it is possible to live in small homes that promote a simple lifestyle without sacrificing comfort.

12.9 The Max Storage Home

The Max Storage Home was designed to make efficient use of space with a specialized modular storage system. The goal was to maximize the available storage space in a smaller dwelling while providing an adaptable system to suit different occupants' needs and lifestyles (Fig. 12.13). The system consisted of a selection of specially designed storage modules made with sustainable materials. When choosing their unit, buyers could customize the combination of storage modules placed throughout the home, along with the finishings, types of doors, and fittings they wished to have (Fig. 12.14). There were fourteen different modules available for selection: two measuring 61 cm (24 inches), eight measuring 91 cm (36 inches), three measuring 120 cm (48 inches), and one measuring 180 cm (72 inches) in width. The standardized dimensions allowed modules to be easily mixed and matched by buyers.

Modules featured different options of open shelving, cabinetry, and closet space. One module included a built-in television console, and another featured a built-in

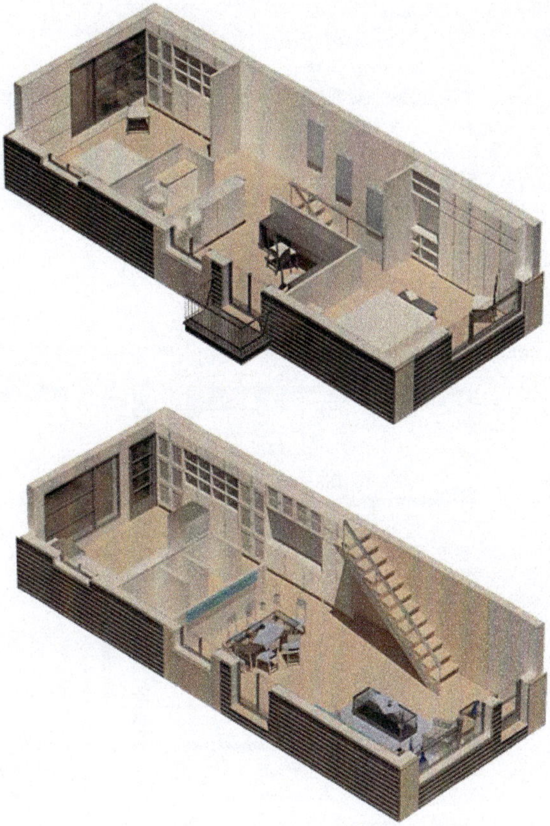

Fig. 12.13 Lower floor (bottom) and upper floor (top) of the Max Storage Home. The home was designed to make efficient use of space with a specialized modular storage system

desk space. Buyers could choose from solid, glass, and louvered doors, with a variety of different colors of finishing (Fig. 12.15). Within modules, specialized fittings allowed for storage spaces to be customized and made more efficient. In bedroom closets: drawers, hooks, shoe racks, jewelry organizers, laundry compartments, and specialized clothing hangers could be included. In the kitchen, garage, or laundry room: pull out garbage and recycling compartments, roll out shelving, and drawer organizers could be added.

The home itself was designed to facilitate personalized combinations of storage modules in each room. In the garage, any combination of modules could be placed along open walls. When the unit's staircase is in the middle of the home, rather than along the longitudinal wall, 11 m (36 feet) of uninterrupted wall space could be used for storage along one wall of the garage. On the main floor, a vestibule area at the front entrance included 120 cm (48 inches) of space available for one to two modules. The living space of the home provided another 3.4 m (11 feet) of wall space for storage modules. Depending on the internal plan and location of the staircase, corridors, and areas adjacent to the kitchen or bathroom could be fitted with additional storage modules. On the upper floor, corridors offer the option to include storage modules,

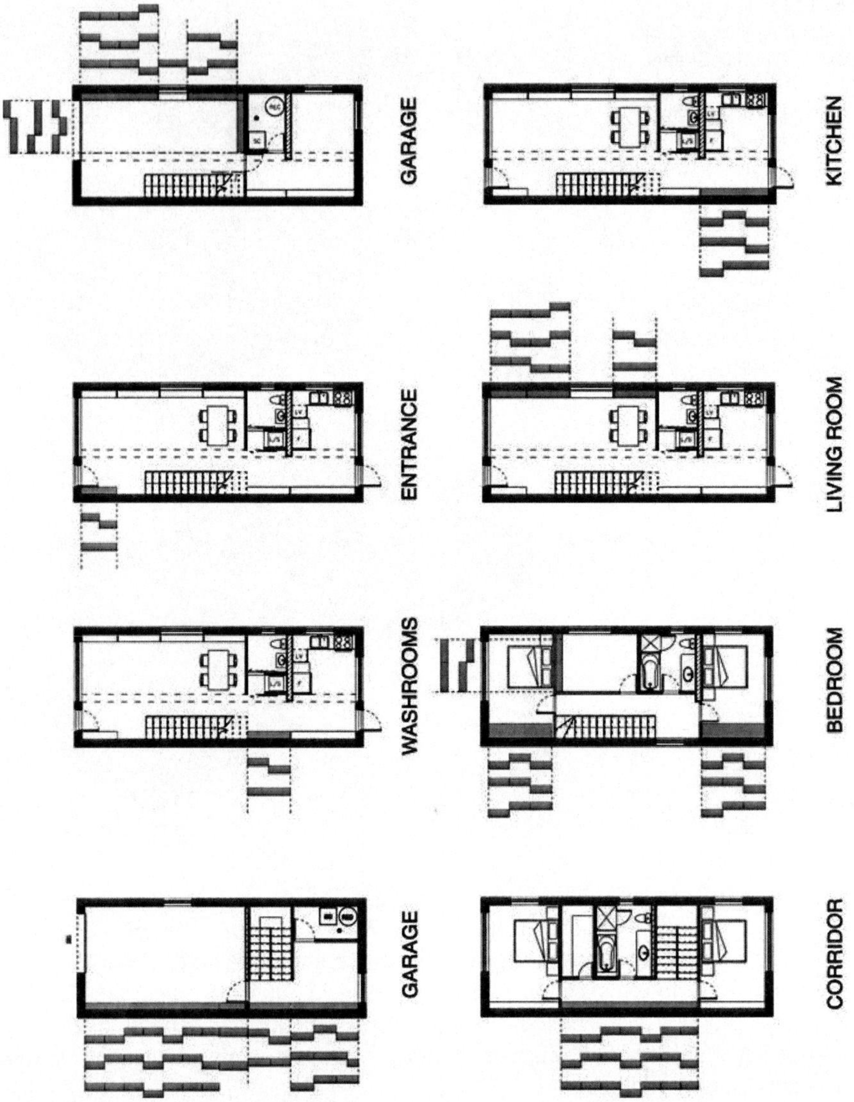

Fig. 12.14 When choosing their unit, buyers could customize the combination of storage modules placed throughout the home

and each bedroom had 2.4 m (9 feet) of storage space (Fig. 12.16). The placement and combination of modules were highly flexible and adaptable for the specific demands of each room and the priorities of buyers. Three-dimensional renderings of units and optional storage modules gave buyers the opportunity to visualize their choices. Computer software allowed buyers to make selections and see how their unit would look prior to its construction. Overall, the Max Storage Home demonstrates how

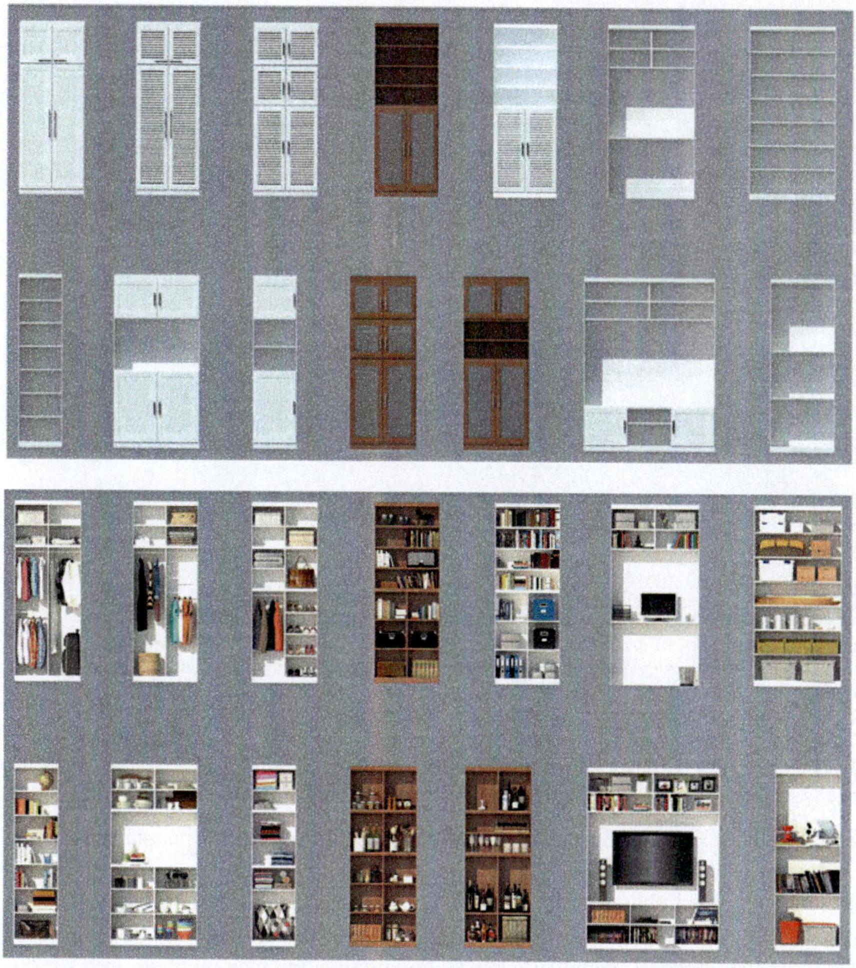

Fig. 12.15 The storage modules featured different options of open shelving, cabinetry, and closet space. Buyers could choose from solid, glass, and louvered doors, with a variety of different colors of finishing

a highly adaptable, functional storage system can be included in smaller dwellings using a modular design. Modules allow buyers to customize their units to their needs and lifestyles, making the most efficient use of small spaces.

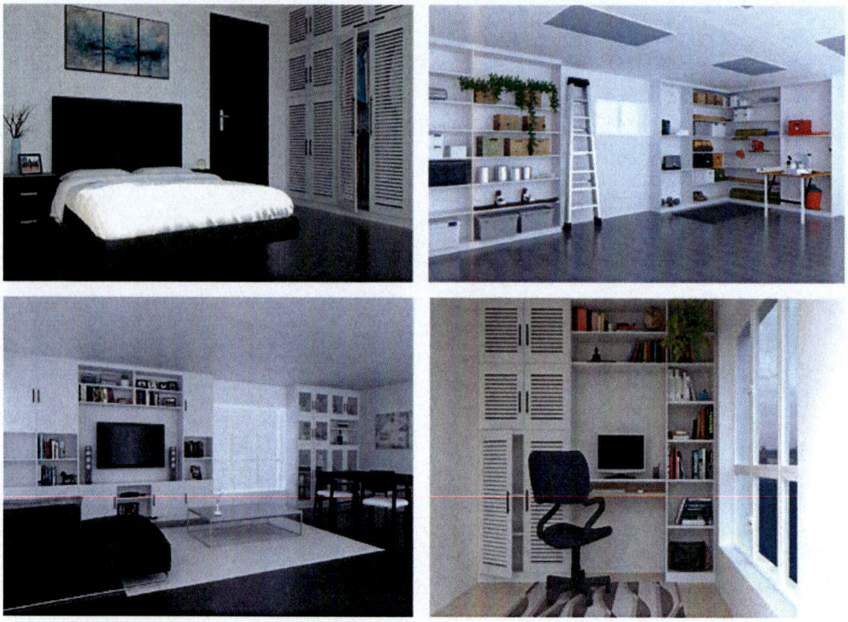

Fig. 12.16 The Max Storage Home was designed to facilitate personalized combinations of storage modules in each room

12.10 Final Thoughts

It seems that avoiding the buy-and-throw-away culture will ultimately have to be driven by one's own values. Though housing design practices can help leverage the change toward a more simplistic lifestyle through smaller homes with sufficient space and storage capacity. In combination with multifunctional furnishing, greener and more innovative storage methods can potentially convince more individuals that bigger is not always better if the design and quality of a smaller home is just as good as a large one. The benefits of owning or renting a smaller home are clear. Lower costs, a smaller ecological footprint, and higher chances of achieving financial freedom are among some of the top factors highlighted in this chapter.

Additionally, no matter whether dwellers rent or buy, the increasing cost of living will most likely provide another incentive for people to progressively become more conscious about spending on items that are of little use and will eventually collect dust in a storage box, at a thrift shop or simply end up getting tossed away. Rather, practicing living a more modest lifestyle can teach about consumerism and what it means to live happier with less. This process starts within oneself and the home environment.

Questions for a Follow-Up Discussion

1. Draw a link between household demographics and storage needs?

2. What is a Tiny Home and how storage is considered in these dwellings?
3. What is the advantage of multifunctional furnishing?

References

Archdaily (2019) 3500 Millimetre house/ago architects. https://www.archdaily.com/909456/3500-millimetre-house-ago-architects. Accessed 15 Oct 2022

B Corp (n.d.) About B Lab. https://www.bcorporation.net/en-us/movement/about-b-lab. Accessed 15 Oct 2022

Becker J (n.d.) 21 surprising statistics that reveal how much stuff we actually own. https://www.becomingminimalist.com/clutter-stats/. Accessed 15 Oct 2022

Campbell C (2013) Visualizing residual spaces in a new light. https://www.sjsu.edu/urbanplanning/docs/honors-reports/URBP%20298%20%20Honors%20Report%20-%20Campbell.pdf. Accessed 15 Oct 2022

Christian K (2022) Sustainability and ethics certifications: what do they actually mean? https://www.thegoodtrade.com/features/sustainable-certifications-and-standards. Accessed 15 Oct 2022

Colville C (2022) Sustainable storage solutions. https://www.countryandtownhouse.com/interiors/sustainable-storage-inspiration/. Accessed 15 Oct 2022

Friedman A (2020) Prefab living. Thames & Hudson Ltd., London, UK.

Friedman A, Krawitz D (2002) Peeking through the keyhole. McGill-Queen's University Press.

Green Storage (n.d.) What makes us green. https://www.greenstorage.ca/what-makes-us-green. Accessed 15 Oct 2022

Harris A (2021) U.S. self-storage industry statistics. https://www.sparefoot.com/self-storage/news/1432-self-storage-industry-statistics/. Accessed 15 Oct 2022

Hayes A (2021) Consumerism. https://www.investopedia.com/terms/c/consumerism.asp. Accessed 15 Oct 2022

Hays J, Mason B (2020) 8 transforming furniture solutions for small spaces. https://www.thespruce.com/transforming-furniture-for-small-spaces-4058276. Accessed 15 Oct 2022

Kell S (n.d.) 13 clutter statistics that will shock you. https://www.suzykell.com/blog/clutter-stats. Accessed 15 Oct 2022

Livingston A (2022) What is the tiny house movement – plans, resources, pros & cons. https://www.moneycrashers.com/living-tiny-house-movement-plans/. Accessed 15 Oct 2022

Mendelsohn H (2021) 27 clever room divider ideas to get the most out of any space. https://www.housebeautiful.com/room-decorating/g27348734/room-divider-ideas/?slide=15. Accessed 15 Oct 2022

Saxton M (2019) Who is buying tiny homes? https://www.buildwithrise.com/stories/who-is-actually-buying-tiny-homes. Accessed 15 Oct 2022

Schmidt R III, Austin S (2016) Adaptable architecture theory and practice (1st ed). Routledge. https://doi.org/10.4324/9781315722931. Accessed 15 Oct 2022

Sims A, Ulloa G (2020) 20 genius storage ideas for small spaces. https://www.architecturaldigest.com/story/storage-ideas-for-small-spaces. Accessed 15 Oct 2022

SpaceWise (2022) 17 creative under stairs storage ideas you need to try. https://www.extraspace.com/blog/home-organization/diy-projects/creative-under-stairs-storage-ideas/. Accessed 15 Oct 2022

The Good Trade (n.d.) 9 sustainable furniture brands for a stylish home in 2022. https://www.thegoodtrade.com/features/eco-friendly-furniture-brands-for-a-stylish-and-conscious-home. Accessed 15 Oct 2022

The Tiny Life (n.d.) What is the tiny house movement? https://thetinylife.com/what-is-the-tiny-house-movement/. Accessed 15 Oct 2022

Tiseo I (2021) Plastic waste in the UK – statistics & facts. https://www.statista.com/topics/4918/plastic-waste-in-the-united-kingdom-uk/#topicHeader__wrapper. Accessed 15 Oct 2022
Zhou Y (2019) Adaptable prefabricated interiors in urban dwellings. Post-Professional Master of Architecture: Urban Design and Housing McGill University, Peter Guo-hua Fu School of Architecture.

Chapter 13
Getting Old at Home

Abstract With the growth in the sizes of elderly population, countries and cities are facing mounting challenges. The need to rationalize investment in social and health care services forces them to think innovatively. Aging also poses a challenge to seniors since trivial daily tasks become increasingly difficult as people get older. Adequate aging is also an important pillar of sustainability since poor management can drain a nation's finance and contribute to deteriorating personal wellbeing. Several residential design strategies such as aging in place and multigenerational living are known to alleviate those challenges. Therefore, the need to design inclusive environments and dwellings that support these principles and are adaptable to people's needs will support aging in the community and at home. This chapter investigates these approaches in the design of the kitchen, bedroom, living room, bathroom, and the outdoors. The potential benefits of home automation are also explored as well as the merits of multigenerational living.

Keywords Aging in place · Adaptable design · Universal design · Multigenerational homes · Retirement communities · Senior care · Universal design

13.1 The Growing Aging Population

Over the last century, enhanced healthcare systems have been helping people live longer. Many nations are experiencing a growth in the size of their elderly population as most people are expected to live well into old age as is the case in Canada (Fig. 13.1). According to the world health organization (WHO), by the year 2030, one in six people will be aged 60 years or older and by 2050, the world's aging population of 60 years and older is expected to reach 2.1 billion people (WHO 2021). Although health care systems have improved, events such as the COVID-19 pandemic have exposed important system flaws that had been ignored by many governments over the years. Most noticeably, the beginning of the COVID-19 pandemic disproportionately affected seniors in long-term care homes to claim many lives.

A. Friedman, *Fundamentals of Innovative Sustainable Homes Design and Construction*, The Urban Book Series, https://doi.org/10.1007/978-3-031-35368-0_13

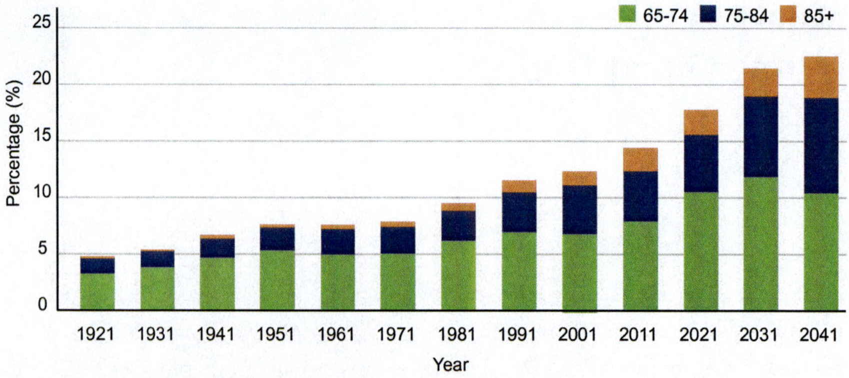

Fig. 13.1 Many countries are experiencing a growth in the size of their elderly populations as most people are expected to live well into old age as is the case in Canada

For both the seniors and their families, aging at home rather than in a long-term care facility provides an option to help alleviate concerns about potential neglect. A study conducted in New Zealand surveyed 121 seniors aged 56–92 years of age and found that older adults want choices about where and how they age. Aging at home was seen as being advantageous in terms of connection and sense of attachment as well as feelings of familiarity and security in their homes and communities. Considering that the size of the aging population is growing, the demand for housing options that allow people to live independently well into their older years will likely follow the same trend.

Aging is a non-linear process influenced by a range of lifelong factors such as one's physical and social environments which can provide incentives or barriers to good health. Aging is associated with other important life transitions such as retirement, relocation, and the loss of friends and partners which can all have impacts on health opportunities and behaviors (WHO 2021). The loss of friends and loved ones can create sentiments of isolation, loneliness, and depression. Biological changes associated with aging also inevitably lead to medical conditions such as fragility, cataracts, hearing loss, diabetes, neck and back pain, dementia, memory loss, and depression (WHO 2021). Such conditions often lead to an eventual reduction or loss of mobility and independence requiring additional care and assistance (Fig. 13.2). Importantly, functions in the community become difficult for the elderly as using services such as public transit, healthcare, buying groceries and other necessities can be difficult as they require a person to be active and mobile. Activities within the home can be equally as difficult when considering tasks such as climbing stairs, lifting items, cooking, operating digital appliances, and personal care.

Several factors are known to influence healthy aging such as a supportive built environment favoring aging in place. That involves the provision of an adequate physical environment that will allow an individual to live independently at home for as long as possible. On a macro level, when planning for aging in place one needs to

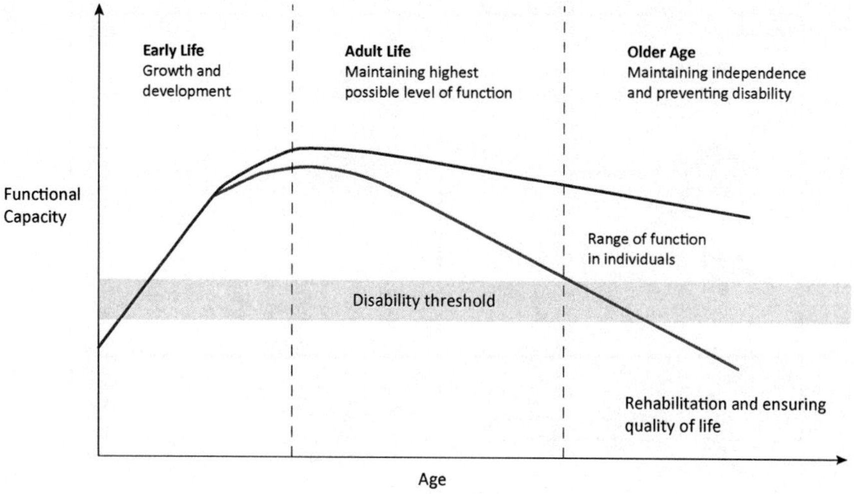

Fig. 13.2 Biological changes often lead to a reduction or loss of mobility and independence requiring additional care and assistance

consider the needs of the elderly as an age-friendly community setting that encourages seniors to stay active and engaged through adequately maintained benches and sidewalks, and accessible buildings (Government of Canada 2016; Fig. 13.3). Additionally, affordable, and reliable public transit becomes important as many may lose their ability to drive or walk. Having community support services is another valuable factor as people get older. Some may require assistance with daily tasks such as washing, and food preparation to continue to reside at home (Government of Canada 2016). Lastly, the home becomes a critical component as it needs to be adapted to an individual using a variety of measures that will be outlined below.

13.2 Aging in Place

Aging in place is a complex process that goes beyond the home's physical characteristics. From a sociological perspective, aging in place considers the emotional attachment to a home and how an occupant act in the face of dynamic landscapes of social, political, cultural, and personal shifts (Wiles et al. 2012). Similarly, the feeling of a home as a *place* is a process involving changing ideals and meanings about the dwelling and its external environment (Wiles et al. 2012). In this regard, a range of products and services can be offered to help seniors adapt to living at home independently and integrate them in the larger social spheres mentioned above. They can be offered assistance with anything from household chores to financial management to make them feel empowered about remaining at home (NIH 2017).

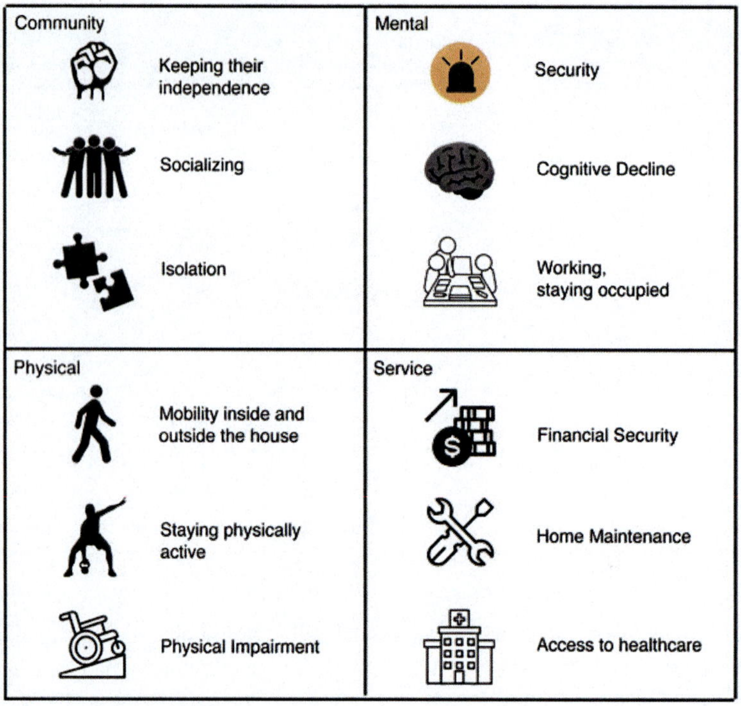

Fig. 13.3 When planning for aging in place one needs to consider the needs of the elderly as an age-friendly community. This should include encouraging seniors to stay active and engaged through adequately maintained benches and sidewalks, and accessible buildings

Physically the concept of adaptable elderly housing is partially about space management and making it safe. By prioritizing safety, a higher level of personal independence can be achieved. It is unlikely that most people will have the financial means to hire a personal caretaker or to be moved to a senior care institution as they age. Therefore, by designing homes that are readily and universally adaptable, better aging conditions can be provided.

In general, new housing should include universal design features that favor an environment that is usable for every member of society. The principles of universal housing design include equitable and flexible use, accessibility, and low physical effort while conducting basic tasks. It incorporates features that are appropriate for every life stage and are also convenient for the homeowners changing needs (Edmonton Senior Council n.d.). More specifically, universal design aims to create a living environment that can be used by people of all abilities, ages, and mobility levels without needing specialized designs or adaptations. It focuses on attractive aesthetic appeal, comfortability, and safety (Edmonton Senior Council n.d.). One of its main features is *visitability*, a design strategy that integrates three main features: wider doorways, a zero-step entrance, and a bathroom on the home's main floor (Fig. 13.4). This strategy allows seniors and people with reduced mobility to access the main floor without restrictions. For example, a zero-step entrance reduces the chances of falling and is equally convenient for seniors and wheelchair-dependent individuals as it is for parents with a baby stroller and young children. Zero step entrances also make it easier for moving large items in and out of the home (Edmonton Senior Council n.d.). Another feature of a universal design approach is accessibility. It is intended to make necessities accessible for all ages and mobility levels throughout the home particularly in the kitchen, bedroom, and bathroom (Edmonton Senior Council n.d.). An additional component could be the strategic placement and installation of large windows to allow ample sunlight to shine through, helping increase both luminosity and improve wellbeing. Especially for seniors, having access to large windows can help reduce the feeling of isolation and improve mental health.

Considering lighting levels and brightness is also an important element of a well-designed home especially for seniors. Adequate lighting that is neither too bright nor too dim can help with completing tasks more easily and alleviate depression. Therefore, a light switch should be placed in arm's reach near the main entrance of any room for convenience and safety. In terms of lighting types, recessed light fixtures make for added illumination to any room which can be a better option than

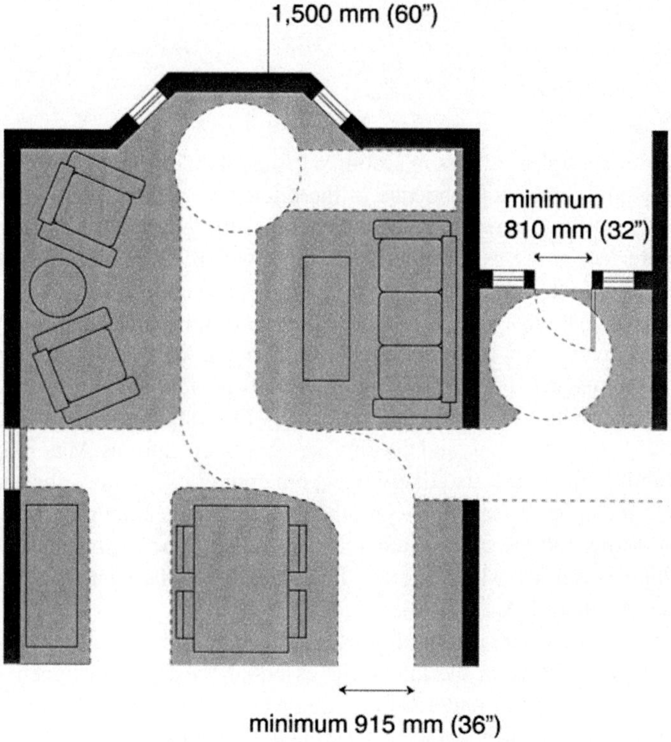

Fig. 13.4 One of the main features of designing for aging in place is *visitability*, a design strategy that integrates three main features: wide doorways, a zero-step entrance, and a bathroom on the home's main floor

traditional light fixtures that provide less light. Modern LED lights can be added to recessed fixtures to reduce energy consumption as well (Aging in Place 2018b).

In addition to universal design, the adaptable approach includes features that allow homeowners to adapt their living environments to meet their changing needs (Edmonton Senior Council n.d.). This can include aspects such as ensuring adequate flow and circulation specifically for wheelchair accessibility, having lower light fixtures, installing a stair lift, and changing round doorknobs to handle types. The adaptable approach also involves the inclusion of modular components that can be gradually added as special needs arise throughout one's life course. For example, the purchase of additional home furniture such as chairs and benches can be an added safety measure, creating adequate resting areas throughout the home for those with reduced mobility. For those who want to keep their existing doors but require additional space for wheelchair or walker accessibility, special hinges can be installed to move the edge of a door out of the passageway to provide a wider doorway (Aging in Place 2018b). The examples presented above are some of the possibilities that can be used throughout the home using the adaptable design approach. Discussed below are both universal and adaptive housing design concepts applied to the different areas in the home environment.

13.3 The Kitchen

The kitchen is central to a person's ability to maintain autonomy. As mentioned in Chap. 11, safety and independence in the kitchen is reliant upon its layout and accompanying utilities and appliances. Generally, the one wall or galley kitchen layout is preferred for use by the elderly and those with reduced mobility. Designing homes that are suitable for seniors also involves reachability. As such, installing shallow sinks or adapting cabinets to reduce physical strain can be helpful. Attention should also be given to the design of kitchens for people with Dementia where many of the components will be visible (Fig. 13.5). To prevent falls, floors should be clear of rugs and mats and have non-slip flooring. Flooring materials like non-glare coverings, cork, linoleum, or wood flooring are convenient options. Materials such as hardwood and linoleum are also likely to become preferable options for wheelchair dependent as it can be easier to move around than tiles and carpeting. The benefit of installing non-slip flooring in the kitchen is that spills can be easily cleaned without leaving a slippery surface which is often the case with regular stone or ceramic tiles for example (Aging in Place 2018b).

Similar to any other area in the home, an adequate mix of natural and accompanying artificial lighting in the kitchen is essential for a well-designed place. In addition to ceiling fixtures, under-cabinet lights can be installed to help illuminate countertops to make food preparation easier and reduce the risk of injury. In terms of countertops, designing them to be multilevel increases the occupant's maneuverability and choice so that anyone can reach them conveniently. For example, a

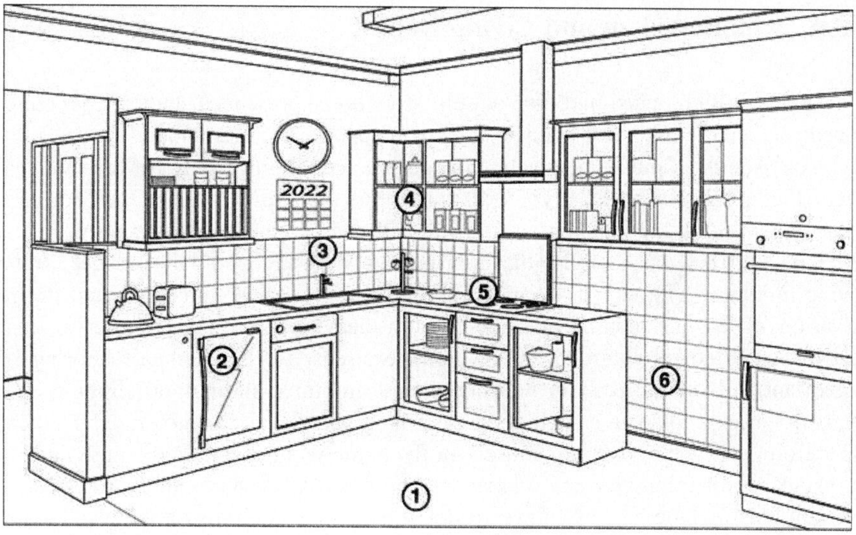

(1) Open and accessible floor plans for orientaiton

(2) Glass fronted fridge to show contents. If it is not possible a clear label is acceptable

(3) Clocks and calendars for temporal orientation the rooms and a board for leaving

(4) Open shelving which allows residents to easily see and use the items

(5) Kitchen's cooked top is equipped with a "crash barrier"

(6) Pastel colors in interior since reflections and repetitive pattern can cause fear

Fig. 13.5 Attention should also be given to the design of kitchens for people with dementia where many of the stored components are made visible

standard 91 cm (36-inch) countertop and a section where the height is 76 cm (30 inches) will provide variety for those who are using a wheelchair or would rather sit than stand. For seniors, an emergency grab bar can also be installed along the edge of a countertop to minimize potential falls (Aging in Place 2018b).

When it comes to heating and cooling, working in the kitchen can quickly increase the room's temperature level. If the kitchen layout is designed in a more confined manner, it can be preferable to have an easily accessible thermostat to adjust temperatures and avoid instances of lightheadedness or fainting from overheating and exhaustion. If the home does not have a thermostat or central heating, a conveniently placed window can help regulate temperature and air flow (Aging in Place 2018b). Other elements that can be added to the kitchen are hands-free faucets, placing the microwave oven at eye level for easier access, and purchasing automatic shut-off appliances to provide solutions for aging in place which are useful for all household members (Aging in Place 2018b).

13.4 The Bedroom and Living Room

For seniors with reduced mobility, a multi-story home presents a challenge. Having a bedroom on the main floor rather than on an upper one is recommended for aging in place. Attention should be given to the design and the furnishings of the room as well (Fig. 13.6). If a home's type or layout limits the ability to place the bedroom on the main floor, the installation of a stair lift can be considered (Aging in Place 2018a). In 2018, it was recorded that approximately 30 percent of falls in older adults living in homes without stairs happen in the bedroom. These falls can result from a variety of reasons such as side effects from medication or stiffness after getting out of bed (Medical Guardian 2018). For these reasons, the choice of flooring is important. Adequate bedroom flooring should minimize sliding and slipping and make it easier to maneuver with a cane, walker, wheelchair, or motorized scooter. Therefore, the type of flooring chosen for the home will depend on the individual's tastes, capability, and be easy to clean (Aging in Place 2018a). Other strategies to prevent falls include having a lower bed and installing side rails on the wall or the bed unit (Medical Guardian 2018).

For the bedroom's closet space, its design should be spacious and well-lit with low shelving for maximum reachability. For walk-in closets, the doorway should be wide enough to accommodate mobility assistance devices and wheelchairs. Regular closets can also be modified to suit the needs of an older adult by replacing the swing with sliding doors for example. Sliding doors are easier to use by users of wheelchairs or walkers (Aging in Place 2018a).

Akin to the bedroom, the living room accounts for about 31 percent of falls in the home. Here, they are likely to occur because of throw rugs, unsecured cables, and awkwardly placed furniture (Medical Guardian 2018). As such, it is important to keep cords and cables away from high-traffic areas which constitute tripping hazards. It is possible to make cables less visible through planning the locations of electrical outlets. They can also be placed at a more convenient height to eliminate the need to bend over to plug in electrical devices. This is especially useful for people with mobility issues and can be made part of a universal housing design (Aging in place 2018a).

13.5 The Bathroom

The bathroom is a highly used room and one of the most likely places for injuries from slipping or falling. In the US, over 230,000 people fall in the bathroom every year (Medical Guardian 2018). Especially for seniors, personal hygiene activities can be difficult to perform in a typical bathroom layout that lacks adaptability. More specifically, bathroom components such as the tub, shower, sink, and toilet areas that are not adapted to individuals with reduced mobility can lead to the risk of injury

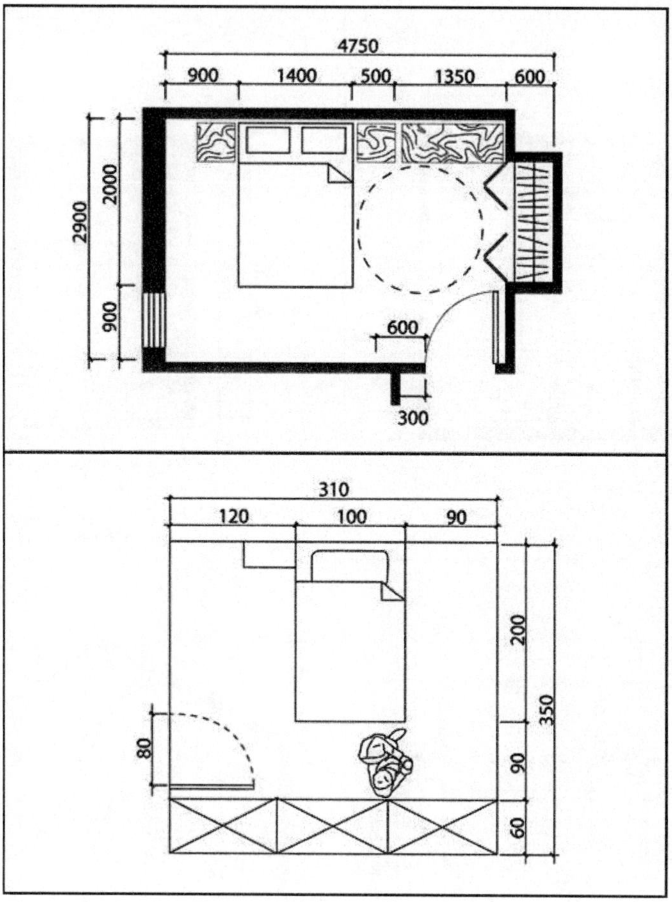

Fig. 13.6 To accommodate use, attention should be given to the bedroom's design and furnishings

and need to be adapted (Fig. 13.7). To make bathrooms safer for seniors, several interventions are recommended.

Traditional tubs require lifting one's feet over the edge of the tub which can be difficult as it requires a certain amount of balance to get in and out easily. Therefore, a stand-alone curb less shower that does not have an attached tub can be a desirable option. A curb less shower has no lip on the edge making it convenient for a walker or wheelchair and reduces the risk of falling (Aging in Place 2018c). Benches and chairs can be placed into a walk-in shower more easily as well as there are no curvatures on the shower floor (Fig. 13.8). However, some people would still rather have a tub in their home. Options such as a walk-in tub can be an alternative to a traditional one as it does not have a high step to climb over making getting in and out much safer. They also include a bench to sit on making it a convenient choice for those with limited mobility (Aging in Place 2018c). Other practical components to make a bathroom

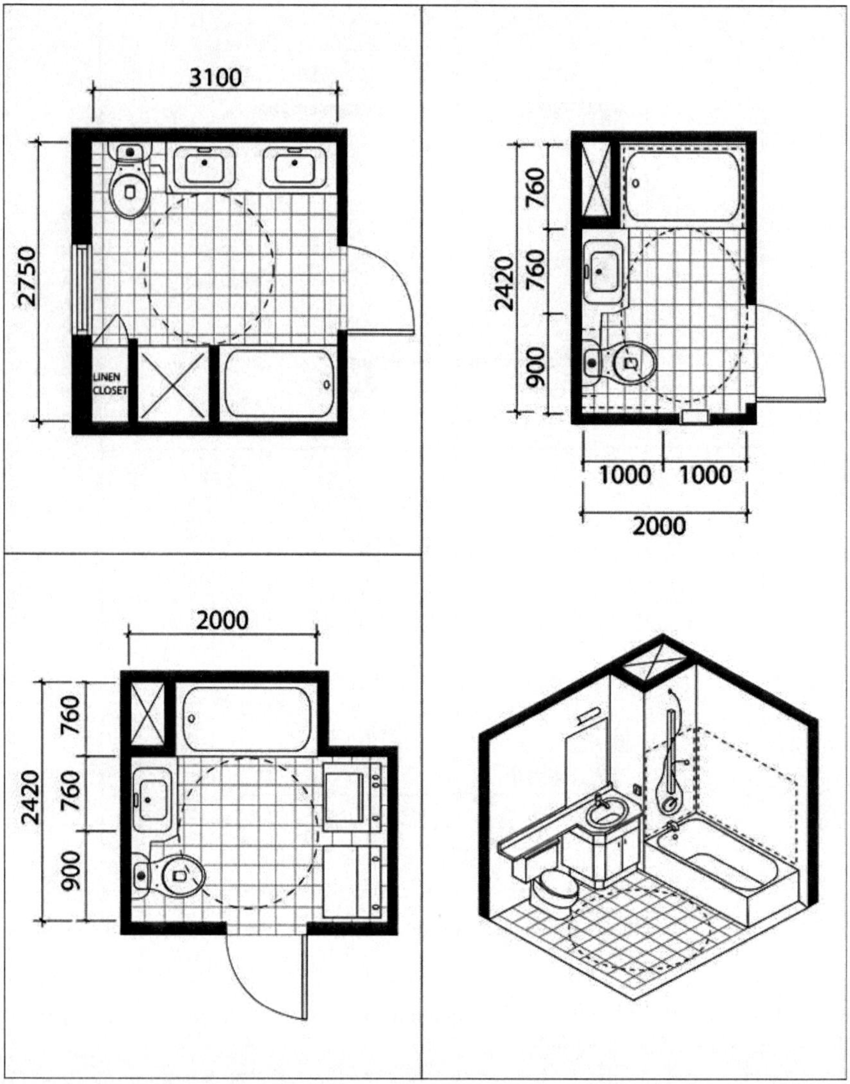

Fig. 13.7 Bathroom components such as the bathtub, shower, sink, and toilet areas that are not adapted for people with reduced mobility can lead to risk of injury

accessible include installing an adjustable shower head, grab bars, an anti-slip floor, thin taped-down bathroom mats, lower shelving units, and countertop vanities with under-sink clearance for wheelchairs (Aging in Place 2018c; Fig. 13.9).

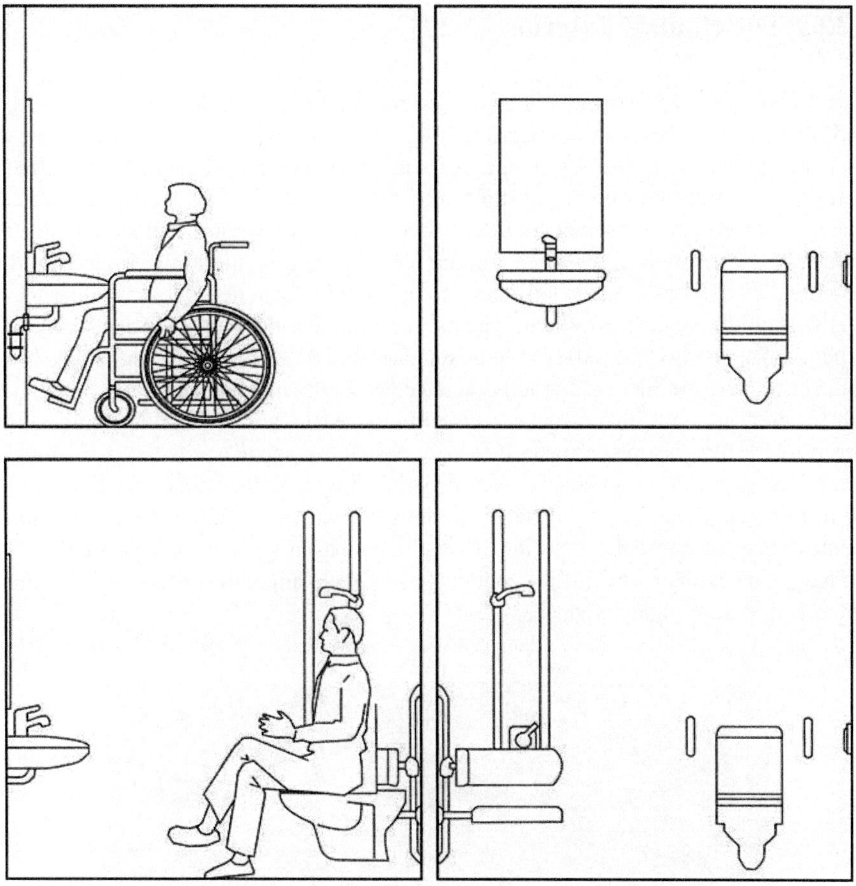

Fig. 13.8 Benches and chairs can be placed into a walk-in shower and curvatures on the shower floor should be avoided

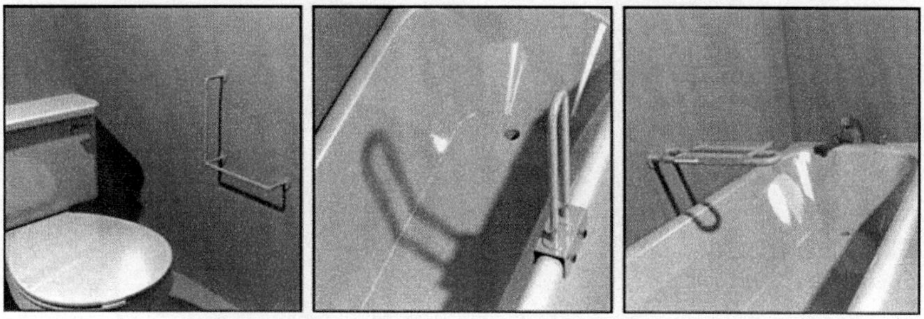

Fig. 13.9 To make a bathroom accessible, changes may include grab bars of different kinds

13.6 The Home's Exterior

The home's outdoor environment is just as important as the indoor. For people with reduced mobility a lift can be installed (Fig. 13.10). For seniors, hard surface such as concrete can be especially dangerous when falls occur. For these reasons, the driveway and any pathways around the home should be adequately leveled. Handrails can be installed in regions that are affected by frequent snow, rain, and ice for added safety. To increase visibility when walking outside at night, motion-activated lights could be installed as conveniently they do not require switches (Medical Guardian 2018). Additionally, outdoor maintenance such as clearing walkways especially in winter is key to ensuring safety (Medical Guardian 2018). For older adults who are still active, performing outdoor activities like gardening no matter how big or small, can promote good health through exercise. For those who are less mobile, companies are available to perform such tasks making it possible to age in a properly maintained home. Outdoor chairs and benches can also make for an aesthetically appealing yard or front porch while increasing safety in the same way that adding them inside the home creates more frequent resting areas. Such measures foster increased physical activity and healthy aging as they encourage seniors to enjoy the outdoors providing them with desired independence.

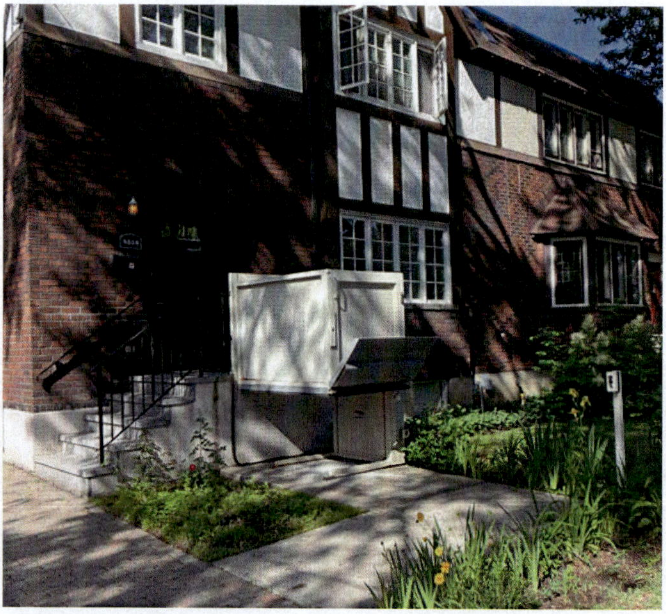

Fig. 13.10 A lift can be installed for people with reduced mobility

13.7 The Potential Benefits of Digital Technologies

One can assume that the elderly might be resistant to change and reluctant to embrace new technology. Indeed, barriers to using technologies among the aged arises with a lack of proper knowledge about their function. A study published by the *Healthcare Journal* suggests that frustration with technology use made older adults unsure about their ability to use it, leaving them feeling unmotivated to try and learn (Jefferson 2019). However, as many more seniors begin to open to the potential benefits of technology, incorporating them into the home environment can be particularly useful for aging in place. Examples such as smart home security systems, motorized blinds, motion-censored, and smart light bulbs were discussed in Chap. 10 to illustrate how these can assist people with reduced mobility. Being able to control devices from a distance offers a unique advantage by increasing independence regardless of someone's physical mobility status. With advancements in telehealth and telemonitoring, the aging population will likely be able to receive care from a distance more often in the future allowing them to remain at home rather than in a senior care home or a hospital. Additionally, electronic devices and the internet can help seniors feel more socially connected as they can video call friends and loved ones year-round without having to leave their home. Importantly, having internet access can help seniors order products and services such as groceries when these essential services are located at a distance and cannot be reached easily or when the weather poses a challenge. Moving forward, it will become increasingly important to educate seniors about using the internet and digital technologies so that they can make informed decisions about whether they choose to incorporate it into their daily lives.

13.8 Multigenerational Living Arrangements

Multigenerational homes are made up of extended family members of at least two generations who are living in the same structure (Richardson 2022). According to data collected from the Pew Research Center, the trend of living in multigenerational homes has been steadily increasing since the 1980s. In 2016, 64 million Americans—equivalent to 20 percent of the US population—lived in multigenerational households (Cohn and Passel 2018). With the COVID-19 pandemic, the demand for multigenerational homes in the US saw a 15 percent increase between April and June of 2020 (Richardson 2022).

The Global Conference on Aging has concluded that there are two major emerging themes with multigenerational housing to include encouragement of harmony and the reduction of conflicts. Alongside these themes are five key aspects that designers must consider. These include the level of privacy, personal independence, crowding, territorial issues, and personal space. Through careful consideration of public and private space, these five aspects can support the creation of a high-quality dwelling (Friedman 2013).

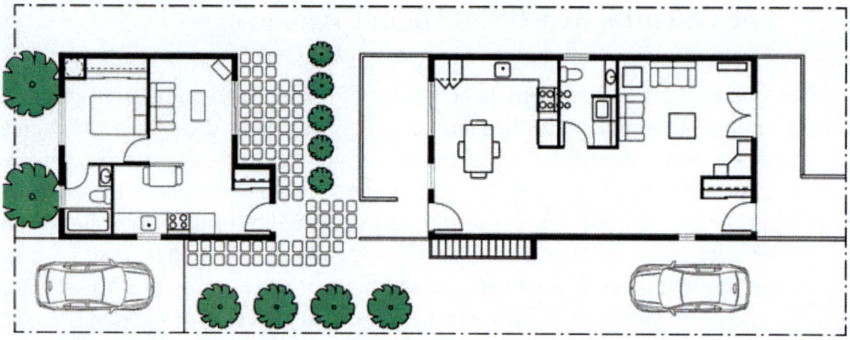

Fig. 13.11 A garden suite is a smaller, separate housing unit at the rear or the side of a main home

There are four main types of multigenerational housing: the side-by-side or bi-family household, plex units, the accessory apartment, and garden suite. The side-by-side living arrangement shares the same facade and sidewalls but has no internal connections. Plex units on the other hand are often stacked on top of one another to form two (duplex) or even three (triplex) living units each with their own entrances (Friedman 2013). The accessory apartment arrangement entails having two separate living arrangements under the same roof, usually with the second one located in the dwelling's basement with preferably separate entrances. Lastly, the garden suite is a smaller, separate housing unit in the rear or the side of a main home (Fig. 13.11).

In the context of aging in place, multigenerational housing can offer solutions for assisted-independent living. They can provide older adults with the privacy and space they desire while maintaining immediate contact with family members. This form of housing can also provide financial benefits through shared costs of housing-related expenses. What is important to keep in mind with these types of homes is accessibility to upper levels. For seniors, accessible single-level units with few stairs are a preferable option. If stairs are needed to access the unit, sturdy handrails are recommended (Medical Guardian 2018).

In addition to social and financial benefits, multigenerational homes can also be more sustainable than traditional single-family homes. A study conducted in Wollongong, Australia found that multigenerational family living presents opportunities to save energy, water, building materials, and land (Klocker et al. 2017). More specifically, the results from in-depth interviews with 10 multigenerational families found that they inadvertently reduced material consumption by sharing space and everyday objects such as cooking equipment, furniture, electronics, clothing, books, and food. Even though their sharing practices were not intentionally sustainable, they eliminated the need to purchase additional objects nonetheless (Klocker et al. 2017). Therefore, multigenerational homes are not only beneficial for aging in place, but they can become sustainable environments in various manners.

13.9 A Home for Aging in Place

Situated in an attractive setting in the Netherlands, Villa Deys, designed by architect Paul de Ruiter, blends in nicely with the landscape. Conceived for a couple in their sixties who wished to continue residing in the same house at an older age, the one-story 344 square meters (3,703 square foot) home is hardly perceptible with its overall simple design and clever use of materials. The structure merges into the grassland and façade to resemble the design of traditional Dutch farmhouses (Saieh 2009).

The interior responds to the functional needs of the occupants. With a love of fitness and swimming, the integration of an indoor pool was one of the client's wishes. The centrally located pool is safe for both the occupants and their visiting grandchildren as it is protected by glass panels which can be locked up. Along with vast amounts of light that infiltrates from both ends of the house, the water and its reflections can be seen in the living rooms, contributing to a relaxing atmosphere (Figs. 13.12 and 13.13).

The living spaces are located around the pool and the open plan design allows them to nicely flow into one another. The living room, kitchen, and study are oriented to the south, with adjustable blinds to control the intensity of the penetrating light. The southern façade consists of sliding doors which can be opened or closed according to weather conditions. The blinds, made of horizontal wooden slats, form a porch above the outdoor terrace.

Most of the house's appliances and fixtures are programmable and electronically operated to respond to some of the challenges faced by the occupants. To avoid obstacles and create an aesthetically pleasing effect, the control hubs are in the basement walls and ceilings. Sliding doors, lighting, and curtains have automated controls to avoid risky physical activity.

As the occupants age, it is assumed that the risk of accidents may rise. Therefore, light and color differentiation was considered for possible vision deterioration. The use of non-slip flooring around the bathroom and the kitchen was introduced to prevent accidents. In addition, the garage was designed to be converted into a nursing room to house live-in help in case it is needed.

13.10 An Adaptable Home

The Adaptable Home was designed to be accessible to occupants with a range of mobility and health concerns. It was designed by the author and his team as an entry to the Canada Mortgages and Housing Corporation (CMHC) Flex Housing National Competition (Fig. 13.14). The primary objective of the project was planning affordable accommodation for ongoing modifications to facilitate aging in place and independent living for individuals with disabilities and chronic health conditions (CMHC 2000).

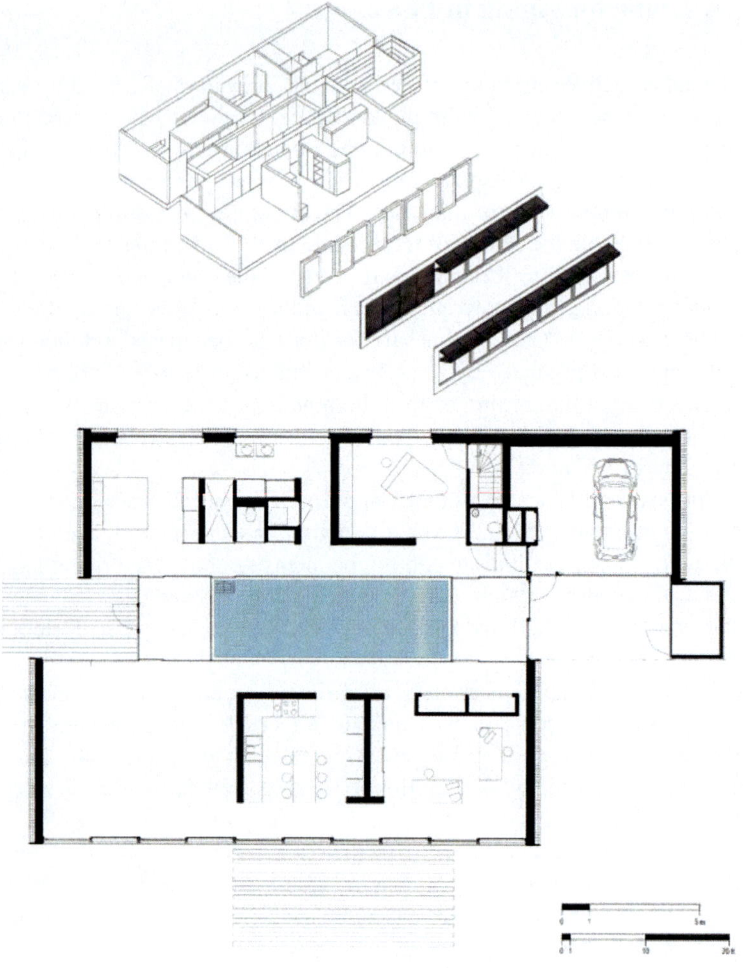

Fig. 13.12 Axonometric (top) and floor plan (bottom) of Villa Days

The original design consisted of semi-detached two-story homes within a large tract housing development. Buyers were offered to purchase the number of levels that they wished to inhabit and were given the opportunity to customize the features of their home. A menu of selections for interior and exterior components was offered, in consultation with an expert on design for individuals with reduced mobility (Fig. 13.15). Buyers were able to select the internal configuration and fittings of their unit to suit their needs and lifestyles. Additionally, units could be structurally prepared for the future installation of accessibility features, simplifying their later modification, unlike a typical housing development.

To illustrate the design process and ongoing adaptation for different households' needs, model units with different cases were presented. One was designed for a young

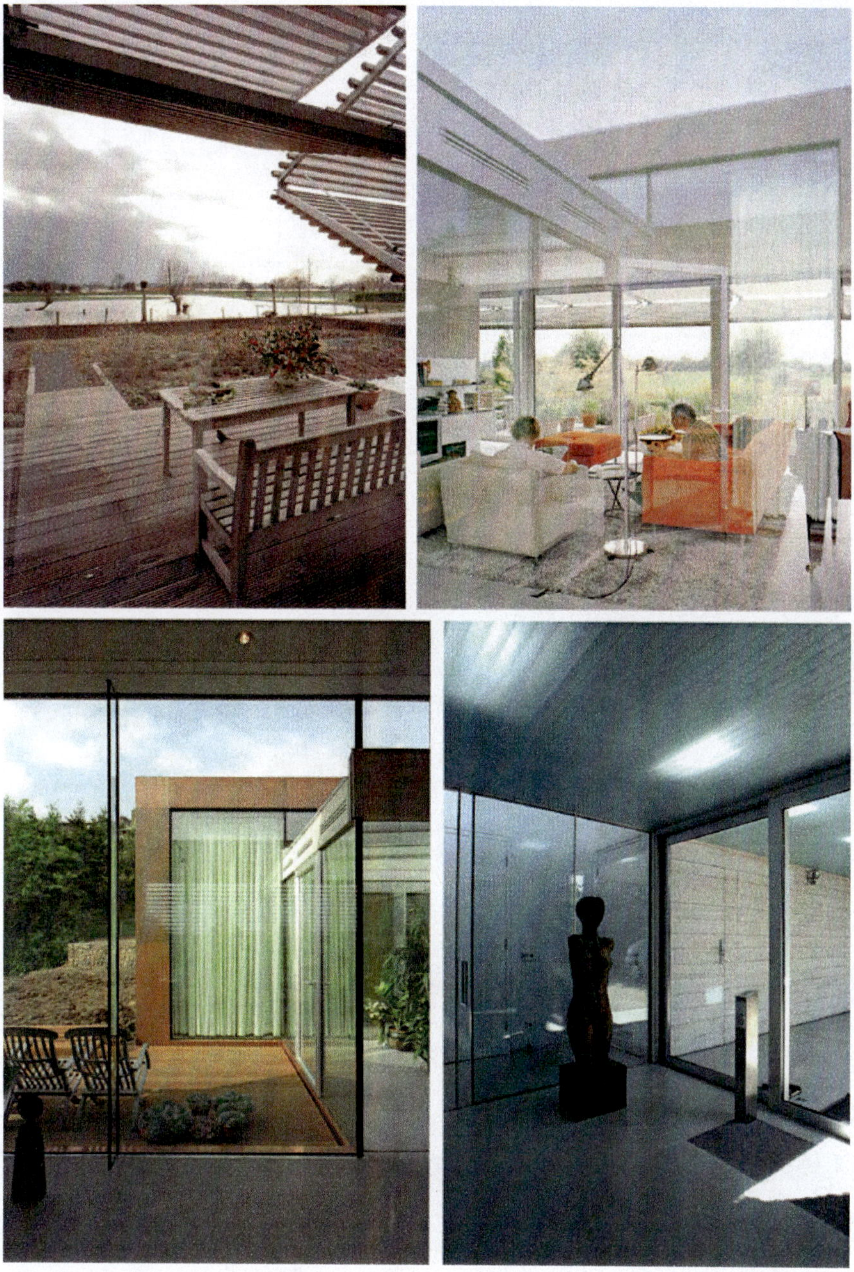

Fig. 13.13 Interior views of Villa Days

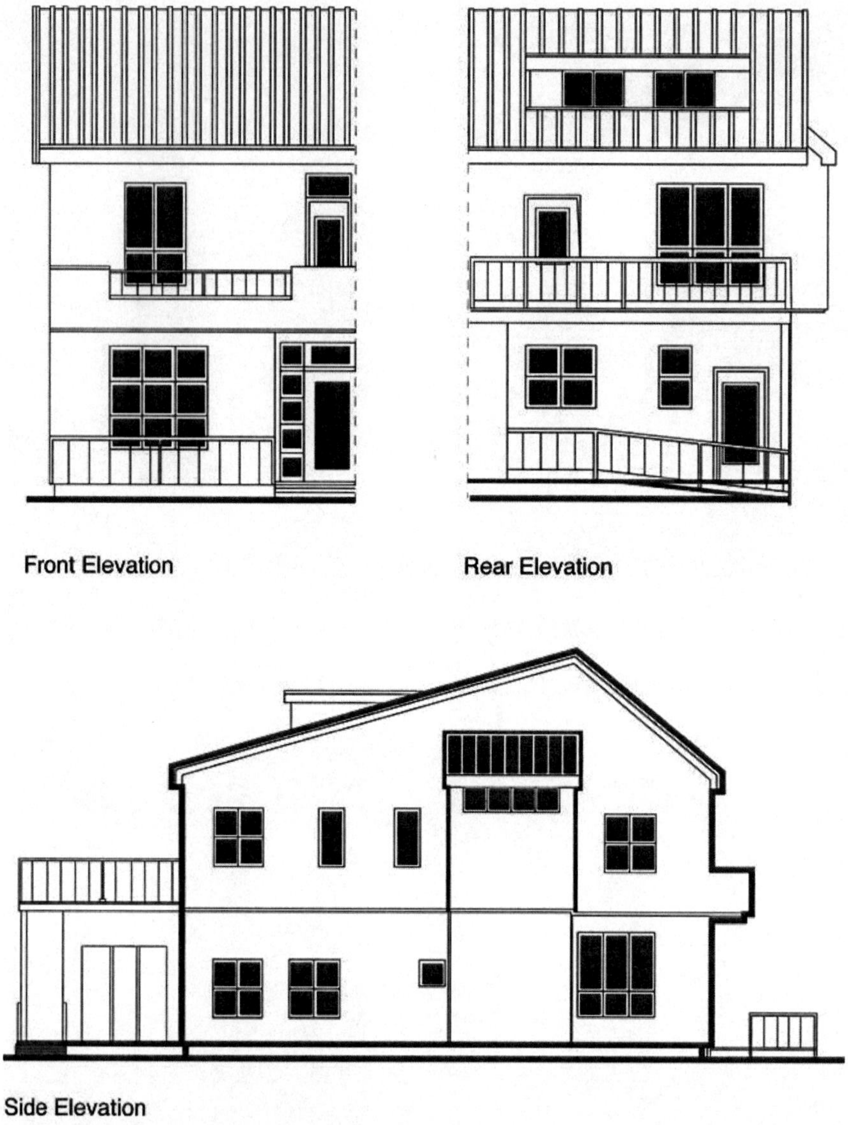

Fig. 13.14 Façade of the Adaptable Home

couple with a toddler-aged daughter. The mother, Ann, was diagnosed with Multiple Sclerosis (MS), therefore, the family wanted to ensure that their home could accommodate her independent living if her condition worsened. MS is a chronic disease of the central nervous system, it is known to cause pain, mobility issues, neurological issues, weakness, numbness, and vision problems among other symptoms. It is characterized by periods of relapse and remission of symptoms that can be progressively

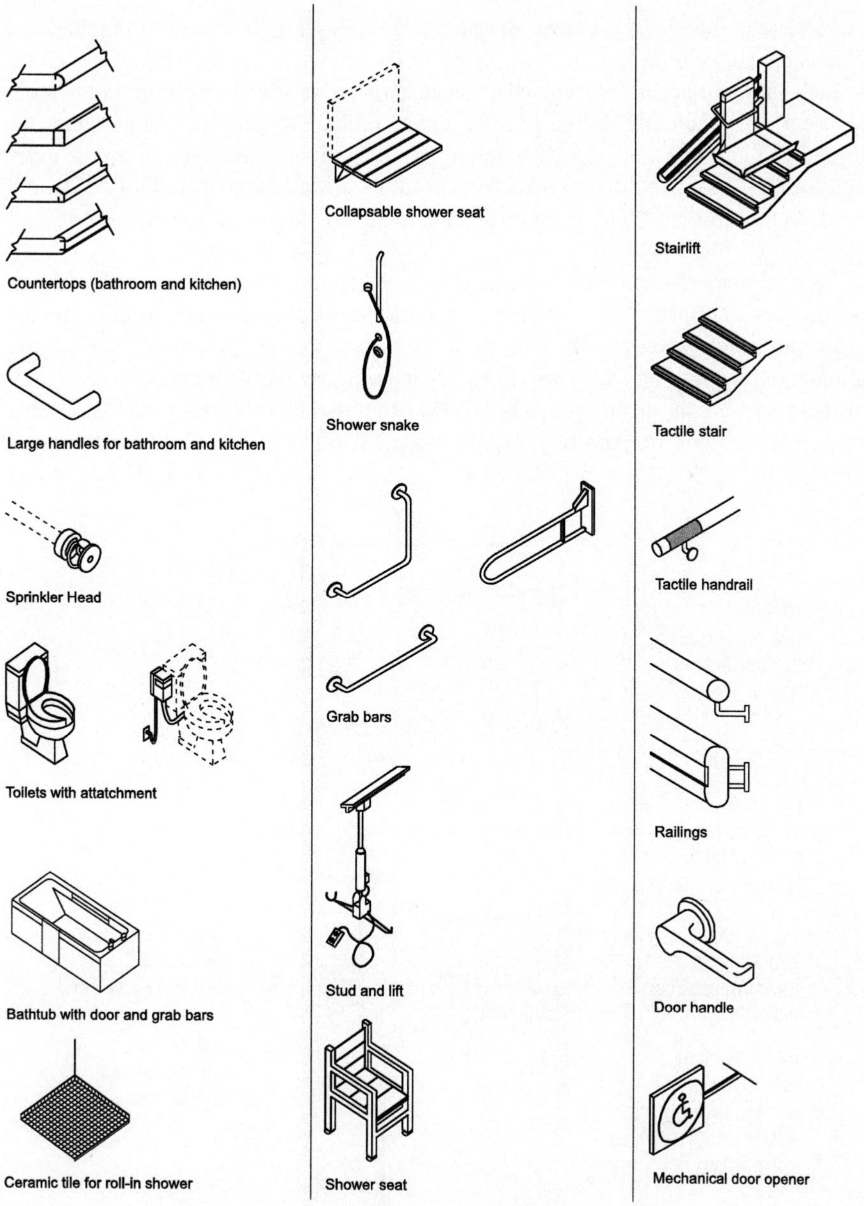

Fig. 13.15 A menu of items was offered to the Adaptable Home's buyers to help adapt their home for aging in place

more severe, though it is often unpredictable. Ann's family chose to purchase an Adaptable Home unit because it would allow them to customize the home according to their ongoing accessibility needs. Because of the progressive and unpredictable nature of MS, the main design priority was to prepare the unit for future additions that would aid Ann to continue living at home and managing her symptoms independently. The unit was designed for the future installation of mobility aids and modification to the arrangement of objects to ensure accessibility (Figs. 13.16 and 13.17).

Several years after moving in, Ann began to experience vision impairment and loss of balance. To limit physical exertion, a handrail was installed along the hallway and grab bars were installed in the bathroom. Storage places were modified to limit the amount of reaching up and down necessary to complete daily tasks such as cooking. To help with visual acuity, objects such as light switches were made black. Five years after moving in, Ann was required to use a wheelchair due to weakness and inability to walk. A lift was installed on the stairs to reach the upper floors and an

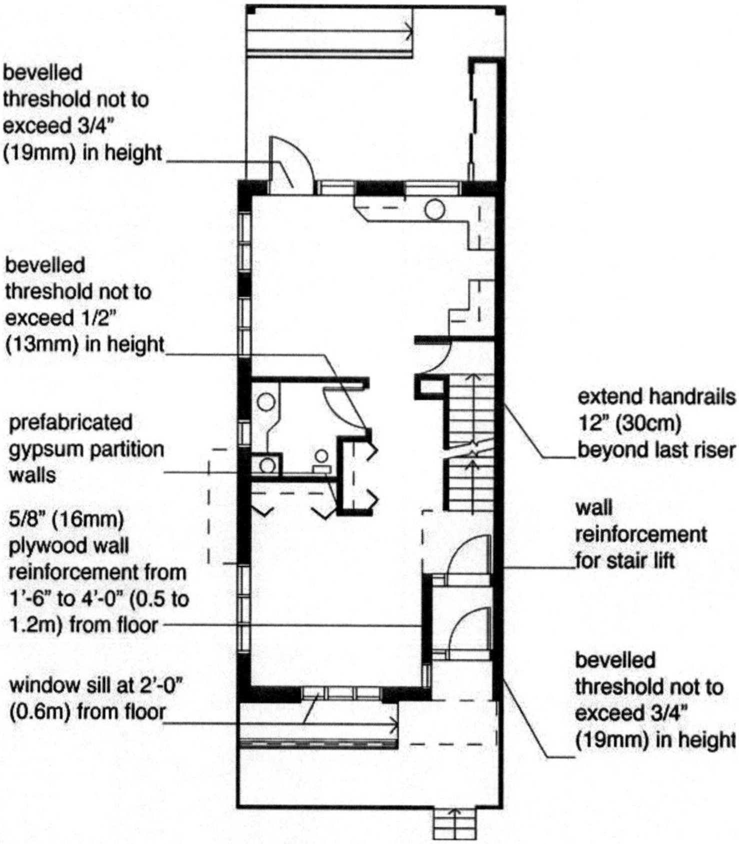

Fig. 13.16 The modified first floor of Ann's unit

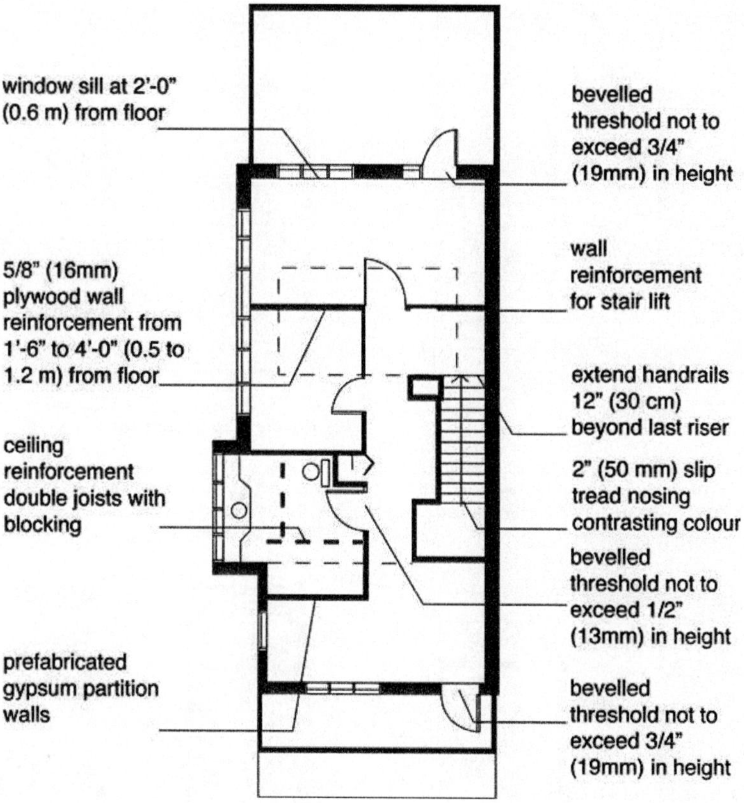

Fig. 13.17 The modified second floor of Ann's unit

outdoor ramp was installed at the front. The initial design of the home ensured that the width of doors and access ways would be sufficient to accommodate wheelchair access and the kitchen and bathroom will be adjusted to the need of a person with reduced mobility. After another five years, Ann experienced additional limitations in mobility and cognitive function (Fig. 13.18). A shower with a commode chair was installed allowing her to bathe independently and similar modifications were made in the kitchen (Fig. 13.19). Each of these additions was made possible by the initial design of the home which ensured the simplicity of future modification. Ann's unit of the Adaptable Home exemplifies how designers can help in the planning stage to ensure accessibility and aging in place. By providing buyers with a selection of accessible options and preparing units for later modification, homes can be made more accommodating to individuals with special needs.

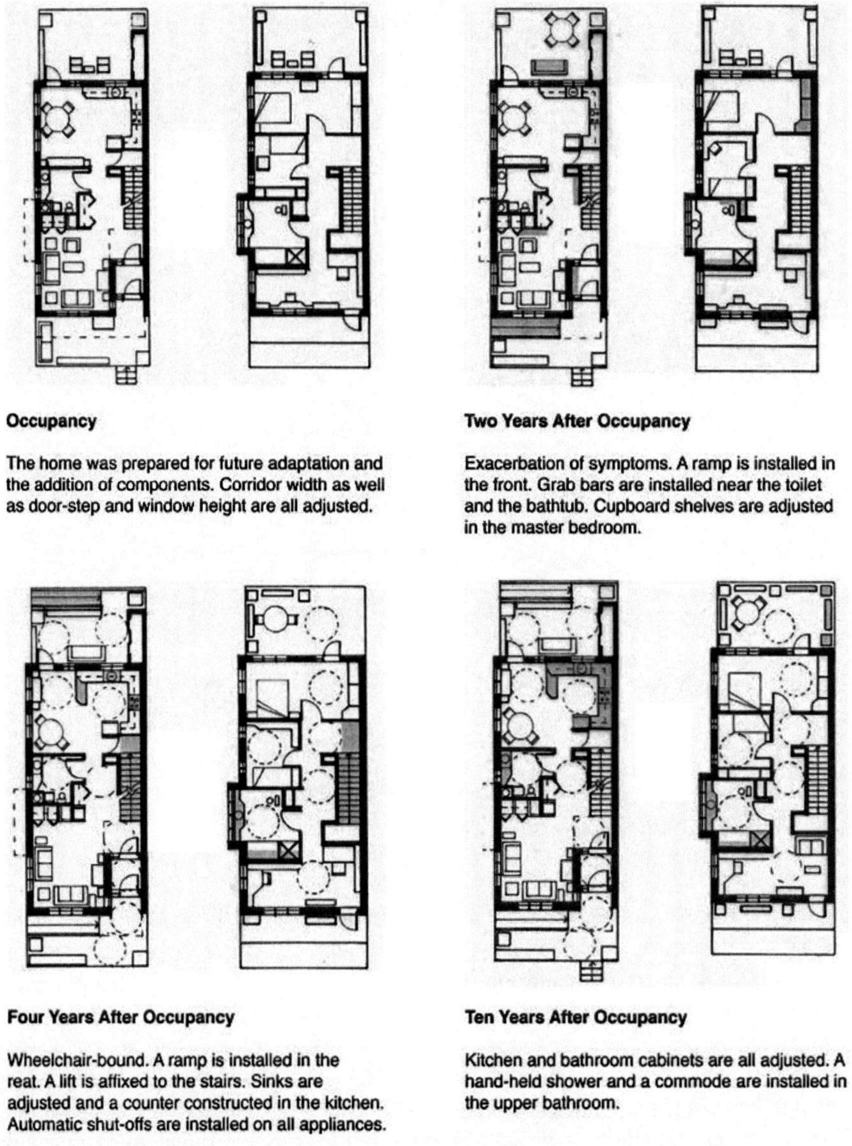

Fig. 13.18 Modifications that were made to Ann's home as her illness progressed

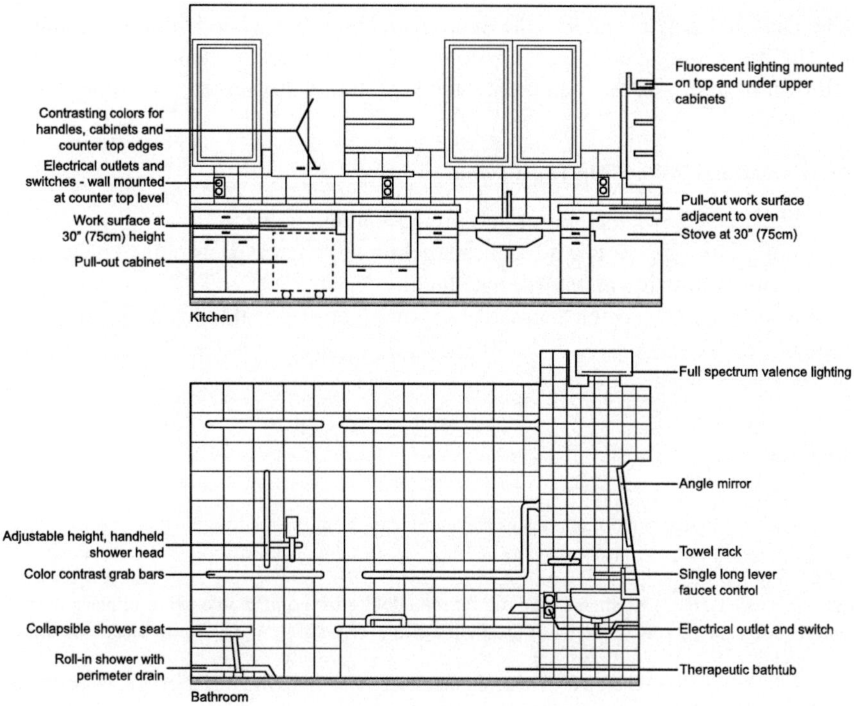

Fig. 13.19 Elevations of adapted kitchen (top) and bathroom (bottom)

13.11 Final Thoughts

As medical practices continue to advance, it is likely that people will live longer. The COVID-19 pandemic has created a paradigm shift in health service, health education, and the way authorities view caring for the aging population. As people continue to rely on technology for daily activities, there is no doubt that it will become more prominent in the home environment. Technology could be a key to allowing more individuals to age independently at home if a greater effort is dedicated to assisting seniors with understanding and utilizing devices.

Additionally, relocating to retirement communities has been on an upward trend among older adults hinting at a persistent increase of their future popularity. Retirement communities are unique living environments specifically designed for older adults and their needs. These communities foster active mobility, social connectivity, and functionality. In recent years, green retirement communities where people can age in place have become more conscious about their environmental footprint. Some of the homes within these micro ecosystems are built to include environmentally certified buildings, communal gardens, recycling programs, larger amounts of greenspace, and smaller energy efficient homes (Baker and SCAN 2018). As such, lessons can be learned about these retirement communities and applied to cities to make them more age friendly. The aging population has value within the community and designing a better built environment that will allow them to contribute in their own meaningful way can unlock the potential to strengthen the overall fabric of society.

Questions for a Follow-Up Discussion

1. What are the key strategies of designing for aging in place?
2. In designing for people with reduced mobility to what aspects of the kitchen and the bathroom you will pay special attention?
3. What are the key design alternatives of multigenerational living arrangements?

References

Aging in Place (2018a) Important updates and modifications for the bedroom. https://aginginplace.org/aging-in-place-important-updates-and-modifications-for-the-bedroom/. Accessed 15 Oct 2022

Aging in Place (2018b) Kitchen of the future: remodeling for comfortable aging in place. https://aginginplace.org/kitchen-of-the-future-remodeling-for-comfortable-aging-in-place/. Accessed 15 Oct 2022

Aging in Place (2018c) What to do when you redo bathroom. https://aginginplace.org/what-to-do-when-you-redo-your-bathroom/. Accessed 15 Oct 2022

Baker B, SCAN Foundation (2018) Green retirement communities are sprouting. https://www.nextavenue.org/green-retirement-communities-are-sprouting/. Accessed 15 Oct 2022

Canada Mortgage and Housing Corporation (CMHC) (2000) Flex housing: the professional guide. Publish by Canada Mortgage and Housing Corporation, Ottawa

Cohn D, Passel J (2018) A record 64 million Americans live in multigenerational households. https://www.pewresearch.org/fact-tank/2018/04/05/a-record-64-million-americans-live-in-multigenerational-households/#:~:text=Record%2064%20million%20Americans%20live%20in%20multigenerational%20households%20%7C%20Pew%20Research%20Center. Accessed 15 Oct 2022

Edmonton Senior Council (n.d.) Making our houses lifelong homes. https://www.seniorscouncil.net/uploads/files/AccessibleHousing_LORES_Feb171.pdf. Accessed 15 Oct 2022

Friedman A (2013) Innovative houses; concepts for sustainable living. Laurence King Publishing, London, UK

Government of Canada (2016) Thinking about aging in place. https://www.canada.ca/en/employment-social-development/corporate/seniors/forum/aging.html. Accessed 15 Oct 2022

Jefferson R (2019) More seniors are embracing technology. But can they use it? UCSD researchers suggest asking them. https://www.forbes.com/sites/robinseatonjefferson/2019/06/28/more-seniors-are-embracing-technology-but-can-they-use-it-ucsd-researchers-suggest-asking-them/. Accessed 15 Oct 2022

Klocker N, Gibson C, Borger E (2017) The environmental implications of multi-generational living: are larger households also greener households? https://ro.uow.edu.au/sspapers/3182. Accessed 15 Oct 2022

Medical Guardian (2018) The most common places you're most likely to fall around the home. https://www.medicalguardian.com/medical-alert-blog/senior-safety-/the-most-common-places-youre-most-likely-to-fall-around-the-home. Accessed 15 Oct 2022

National Institute on Aging (NIH) (2017) Aging in place: growing older at home. https://www.nia.nih.gov/health/aging-place-growing-older-home#support. Accessed 15 Oct 2022

Richardson S (2022) Multigenerational homes: a buyer's guide. https://www.rockethomes.com/blog/home-buying/multigenerational-homes. Accessed 15 Oct 2022

Saieh N (2009) Villa Deys/Paul de Ruiter Architects. https://www.archdaily.com/17627/villa-deys-paul-de-ruiter. Accessed 15 Oct 2022

World Health Organization (WHO) (2021) Aging and health. https://www.who.int/news-room/fact-sheets/detail/ageing-and-health. Accessed 15 Oct 2022

Chapter 14
Working from Home and in Common

Abstract The recent surge of working from home (WFH) can be attributed to accelerated technological developments and the forced isolation during the COVID-19 pandemic. Working from home offers a sense of freedom and flexibility of time management to the worker. Furthermore, it also eliminates the cost, stress, and loss of time associated with commuting, leaving more time to pursue leisure activities thereby contributing to sustainability in many ways. However, certain barriers and challenges remain. The key question that this chapter explores is how a live-work residence can be designed to enhance both productivity and family life. The chapter will explore residential designs that include spaces for work.

Keywords Coworking · Home office · Remote work · Telecommuting · Workspace design

14.1 History and Evolution of Live-Work Arrangements

It is estimated that by 2028, approximately 73 percent of all employers will have some remote workers in their ranks (Wrike n.d.-b). Although remote work is seemingly a recent phenomenon, it has been embedded in human history since ancient times. Ever since people congregated in settlements up to the Industrial Revolution, working from home was common. The innovative technological advancements of the Industrial Revolution including the emergence of large-scale factory production led to a separation between work and domestic realms. In recent decades with the development of digital appliances, a shift in favor of working remotely saw resurgence and was accelerated during the COVID-19 pandemic as the majority of workforce involuntarily was obligated to work from home (WFH). To get a more holistic understanding of the transitions that occurred in the way people work, a closer look into the history of work is highlighted below.

During medieval times, craftsman set up shops in their homes which were planned to support work/life arrangements. Houses were often built with the kitchen, dining, and living areas at the rear. The work function which included craftsmanship such metal work and tannery was placed at the front of the building facing a street to

A. Friedman, *Fundamentals of Innovative Sustainable Homes Design and Construction*, The Urban Book Series, https://doi.org/10.1007/978-3-031-35368-0_14

welcome patrons (Reynolds and Bibby n.d.). In later times, craftsmen and merchants separated live-work activities by residing above the shop (Reynolds and Bibby n.d.) (Fig. 14.1). During the Renaissance, the growing interest in record keeping, administering state business, and historical archives simultaneously led to centralization of activities and the beginning of the physical separation between work and domestic activities. The Uffizi Gallery built by the Medici family in Florence, Italy in 1580 is a prime example of early architectural success and one of the first buildings designed for administrative work (Reynolds and Bibby n.d.) (Fig. 14.2).

The Industrial Revolution brought about profound changes in production, the economy, and housing. With the expansion of factory work, families flocked to growing cities to seek employment. Despite this new way of working, some continued to work from home in what became known as *cottage industries,* well into the nineteenth and twentieth century (Reynolds and Bibby n.d.).

The typewriter, telephone, telegraph, and electricity revolutionized office work toward the end of the nineteenth century. The development of public transportation also enabled more workers to travel from home to work inexpensively (Reynolds and Bibby n.d.). During World War II, women were called upon to work in factories and take the places of men who had gone to war. As many women returned home at the war's end, some found new business opportunities with their newly acquired skills and knowledge gained while working outside the home. Perhaps the most notable remote-work business venture at the time was the sale of Tupperware products and

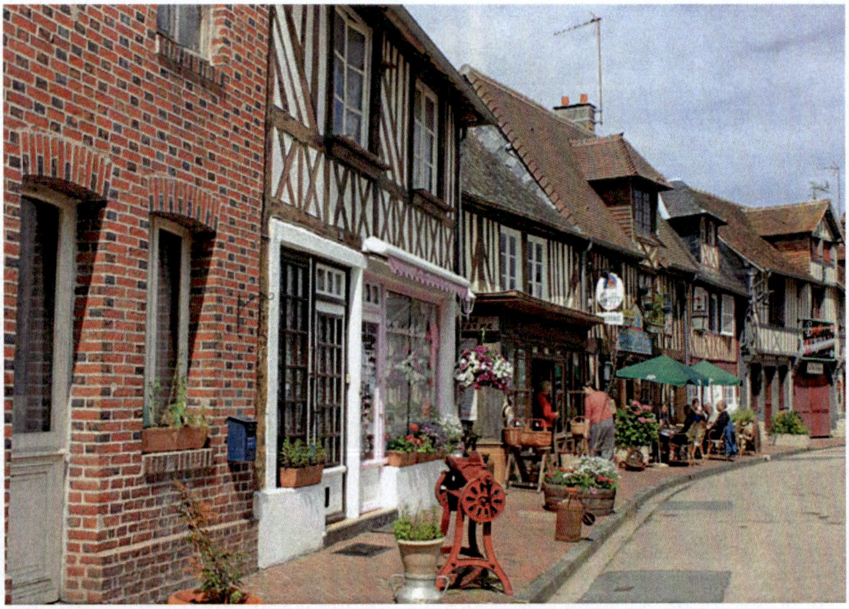

Fig. 14.1 In old cities such as in Beuvron-en-Auge, France, craftsmen and merchants often lived above their shop

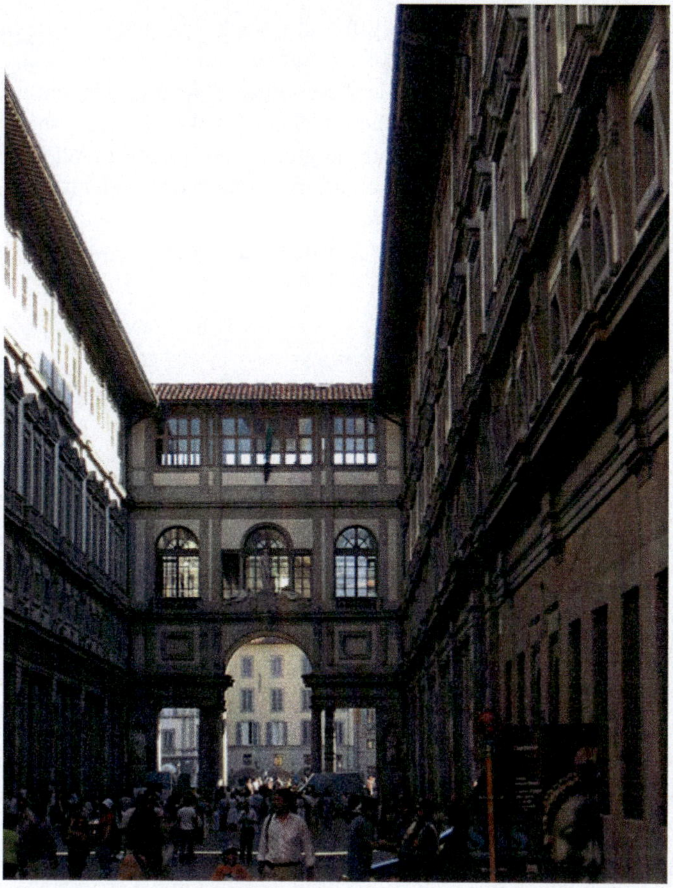

Fig. 14.2 The Uffizi Gallery built by the Medici family in Florence, Italy in 1580 is a prime example of one of the first buildings designed for administrative work

other social marketing and pyramid business schemes. Creative professions such as artists and writers also commonly worked from home at the time (Reynolds and Bibby n.d.).

Heavy private vehicle-induced traffic, the 1970 Clean Air Act, and the 1973 OPEC Oil Embargo were among the driving factors that put telecommuting at the forefront of the workforce in the 1970s. Mitigating the impacts of the gas crisis and air pollution were major sources of concern at the time, making remote work a temporary solution to these preoccupations. By the 1980s, large American companies such as American Express and General Electric started to implement and grow their remote work programs. In the 1990s, US Congress, federal agencies, and US President Clinton backed remote work by issuing a memorandum directing executive branch agencies to create more flexible family-friendly work arrangements (Reynolds and Bibby n.d.). Throughout the early 2000s, improved remote collaboration tools and software were being developed. While the number of remote workers was slowly climbing, the usage and demand for these technological tools saw a sharp increase in 2020 as millions of workers around the globe were forced to WFH during the COVID-19 pandemic. These global circumstances ended up accelerating a trend that had already been underway. Now that remote forms of work are more common than they ever have been, adapting homes accordingly provides a new set of challenges and opportunities.

14.2 Advantages and Challenges of Working from Home

Though people's feelings about WFH remain divided, it can provide a distinctive advantage in terms of economic and potentially environmental sustainability. One of the most obvious advantages of remote work is the reduced amount of time spent commuting. According to an analysis conducted by the Texas A&M Transportation Institute, the average American who drives to work spends approximately 54 h in traffic yearly. Studies have also found a link between commuting by car and increased stress, pollution, and respiratory illnesses (Cramer and Zaveri 2020). Therefore, eliminating or reducing the frequency of commuting can potentially provide psychological and physical benefits. Importantly, the time saved on commuting can help provide a better work/life balance by allowing more freedom for leisure, spending time with family, and physical activity.

In terms of environmental sustainability, remote work can reduce greenhouse gas emissions (GHG) and other forms of pollution although much research is left to be done on the magnitude of this impact. According to the Global Workplace Analytics (GWA), if every American worked remotely half of the time, 51 million metric tons of greenhouse emission from vehicle travel could be eliminated (Cramer and Zaveri 2020). In 2020, during the height of the COVID-19 pandemic, travel restrictions and lockdowns forcing people to remain at home led to a significant decline in concentrations of pollution in many densely populated cities like Los Angeles (Plumer and Popovich 2020). However, even though people work remotely, they may still use their vehicles for reasons other than work making the positive impact of remote working potentially less significant. Office commuters only make up approximately 18 percent of all traffic congestion (Cramer and Zaveri 2020). Nonetheless, satellite images pulled from studies conducted on travel behavior and pollution levels during the pandemic have legitimized the fact that temporarily reducing travel behavior patterns has positive effects on reducing air pollution. Therefore, working from home can become part of the solution if implemented at a large enough scale.

Financial savings are also among the favorable reasons for working from home. For employers, remote work means purchasing fewer office supplies and cutting office rental costs. For example, the US Patent and Trademark Office estimated that in 2015, it saved over $38 million US by having less office space. On the other hand, the GWA estimated that employees could save an average of $2,000 to $6,500 US annually by not having to spend as much on gasoline, public transit fares, and day care services for their children (Cramer and Zaveri 2020).

Although still highly debated among researchers, some studies have also found that working from home can help employees be more productive. A study led by Stanford professor Nicholas Bloom in 2014 examined the behaviors of remote workers at a Chinese travel agency and found a 13 percent higher efficiency rate than their office-based colleagues. Other benefits include higher job satisfaction and less employee sickness due to decreased instances of physical contact and sharing of objects (Cramer and Zaveri 2020). A higher number of career opportunities that are non-restrictive to one's geographic location is also beneficial as it encourages inclusivity and diversity (Courtney 2021).

WFH of course also poses its own unique challenges. Feelings of social isolation, boredom, and lack of physical exercise are some of the flagrant reasons why some people dislike working from home (Fig. 14.3). Importantly, working from home can cause financial burdens as the purchase of office supplies, the cost of internet fees, and office equipment can mean that remote work is simply inaccessible or financially unsustainable. A multitude of distractions at home can also affect how productive employees are. Televisions and cell phones can become an easy source of distraction. For parents with younger children, working from home can be difficult as the separation between family life and work becomes highly confounded. Therefore, although working from home presents many opportunities and benefits, it is far from being a desirable or feasible option for everyone. Even so, design can help play an important role in making working from home more enjoyable and desirable.

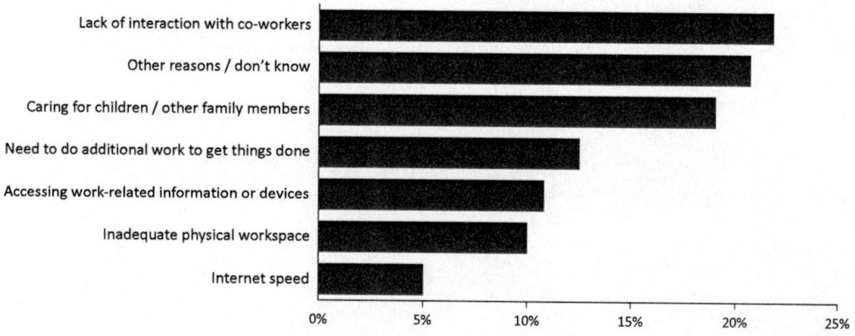

Fig. 14.3 Working from home can pose challenges for some. Feelings of social isolation, boredom, and lack of physical exercise are some of the flagrant reasons why some people dislike working from home

14.3 Types of Remote Work and Needed Accommodations

For some, working from a kitchen table can be the easiest option when a place is sought. However, for those whose job requires to work from home regularly, such a configuration is not optimal. Investing in a workspace setting that will fulfill needs such as comfort and isolate forms of distractions within the home is a key. When conceptualizing a home suitable for remote work, the use of space and its ability to create a productive and efficient environment is also an integral part of quality design (Fig. 14.4). Apart from identifying where the workspace will be located, furniture components are also a fundamental part of fostering creativity, organizational practices, reducing injury, and strain (Wrike n.d.-a).

The common WFH job types involve office work. In its most basic form, this type of work requires a desk, storage space for filing paperwork, a chair, and access to multiple outlets for electronic devices. Other types of remote work could involve product fabrication. In this case, the worker may need machinery, non-slip flooring, access to ample lighting, and protective equipment. Fabrication can also take on lighter forms requiring fewer intensive resources, machinery, and tools such as making clothing items or food. Essentially, each type of career will require its own set of personal configurations and equipment in the home for optimal comfort, safety, and productivity.

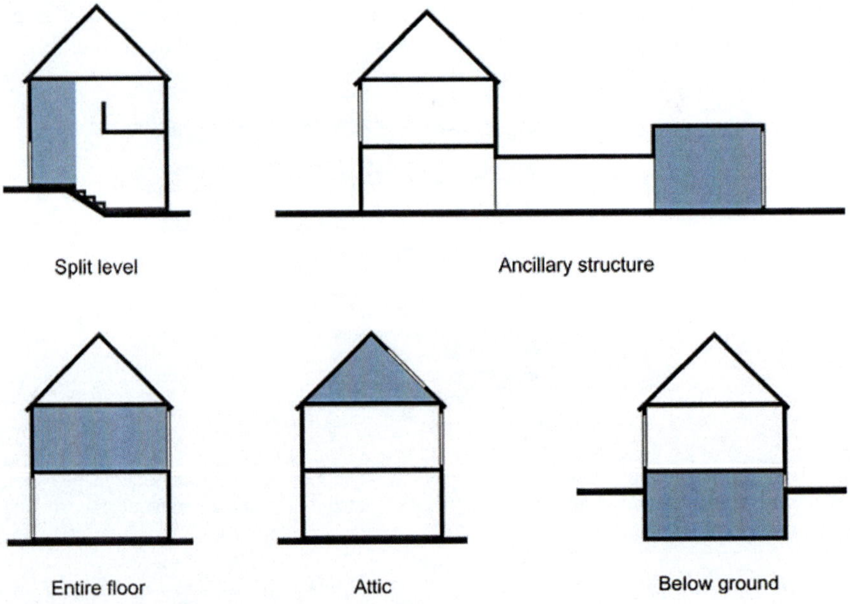

Fig. 14.4 When designing a home suitable for remote work, various locations, and their ability to create productive and efficient environments need to be considered

No matter what type of remote work one is involved in, there are distinct design principles to consider when designing a workspace at home. These include division of space, lighting, workroom flow, storage capacity, temperature regulation, and decor. Of course, variations of these principles are expected during the design phase to tailor to the worker's specific needs.

Locating the workspace in a room separated by a door helps to create a clear delineation between home and work when the space permits (Wrike n.d.-a). This can provide psychological benefits for the homeowner as this design principle also isolates distractions within the home and can make it easier to stay on task. Similarly, once the workday is over, the homeowner can step away from the workroom and tuck it away in the back of their mind to resume personal activities just as they would in a regular on-site job.

Lighting is equally important as it can influence the overall comfort of the workspace. Fluorescent lighting can cause drowsiness while natural light promotes alertness and productivity (Wrike n.d.-a). Natural light also helps to reduce eyestrain and headaches during the day. Depending on the type of work, an optimal workspace will have at least one window to allow ample natural light to enter the room (Roberts 2021). In the northern hemisphere, the placement of the workspace should be on the south side of the house as it gets more sun exposure throughout the day. For those living in the southern hemisphere, the workspace should be placed on the north side of the home (Vigliarolo 2020). According to Roberts (2021), if possible, the window should provide a view.

The flow of the workroom should also be considered throughout the design and decoration phase. Furniture and equipment need to serve the homeowner in a purposeful way (Roberts 2021). In other words, understanding the workflow that is required for certain types of work should be prioritized before the purchase of furniture and equipment. Ideally, the equipment and furniture should accommodate the use of space without overwhelming it. Hence, selecting the right size and placement of furniture and equipment will help optimize the efficient use of space and help the homeowner to not feel overwhelmed or constrained in the workspace if floorspace is limited.

Storage capacity and ample shelving plays an essential role in creating an efficient workspace (Fig. 14.5). As seen in Chap. 12, clutter can be linked to increased instances of stress. Therefore, having an organized workspace will alleviate stress and reduce the time spent searching for items. Additionally, working from home can quickly become overwhelming if personal and professional material is mixed (Wrike n.d.-a). Therefore, keeping personal and work documents and equipment separate is a key organizational practice that should be facilitated by storage furniture. For those working with confidential documents, a lockable door that separates the workspace from the home can also provide added safety beyond the benefit of dividing work activities and the home.

The ability to regulate the temperature of the workspace will be a determining factor in the level of comfortability for a homeworker. Studies show that an overly warm room tends to cause drowsiness while excessively cold rooms can be a source of

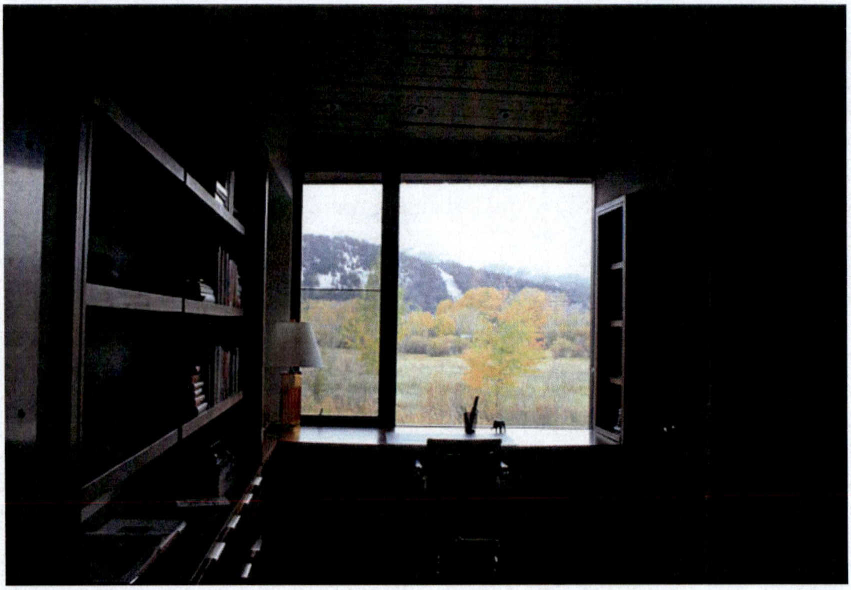

Fig. 14.5 Storage capacity and ample shelving plays an essential role in creating an efficient workspace

distraction and can even lead to more typing errors in office-like settings (Wrike n.d.-a). Each workspace should be designed to include a thermostat device for temperature regulation. This is not exclusive to office-like settings as it is equally important for those doing indoor manual work as well.

In terms of decor, the color of the walls and accessorizing the room are crucial parts of creating an environment that is creative, comfortable, and familiar within one's home. Colors like green and blue are known to positively influence mood and productivity levels while bright colors such as red or yellow can become sources of distraction (Wrike n.d.-a). Therefore, the choice of color becomes important in creating a work setting that is both personalized and encourages a productive environment. Additionally, decorating the workspace with personal items, quotes, and pictures can increase motivation while adding low-maintenance plants can potentially help reduce stress and purify the air (Wrike n.d.-a). It is also encouraged to bring a sense of playfulness to the room by adding color, lounge furniture such as bean bags, or even an indoor swing. Playful features in the room can positively influence creativity and reduce boredom while working, making the workroom more enjoyable.

14.4 Selecting the Location of the Workspace

Once more, the type of employment and their accompanying activities will influence the preferred placement of the workspace in the home. For instance, manual workers may want to have their workshops in the basement or the garage. These areas may offer more sound and vibration insulation compared to other rooms in the home, making them good locations to minimize disturbances to other members of the home while working. Safety is of utmost priority when considering manual types of work. Having a space separated by a door as well as being located further away from the main living area can be desirable for reducing the risk of injury and preventing dust and other particles from contaminating the rest of the dwelling. This is especially important to consider when young children are members of the family. If the fabricated product is large and heavy, one may consider adding a door to connect the basement to the outdoor for ease of access and ventilation. A basement workshop can also help reduce the need to use an air conditioning system during the summer, lowering both energy consumption and utility bills.

The transformation of the garage into a functional workshop also provides adequate access to the exterior as well as provides the necessary space for proper ventilation and any large equipment. This design strategy, that conserves space, is less resource intensive and costly to choose since the structure is already included in the home. Alternatively, a separate workshop can be built in the yard if there is no basement, garage, or enough space to add an extension to the side of the dwelling.

For office work, there is much more flexibility in selecting the location of the workspace since activities are generally far less disruptive and hazardous than fabrication and manual work. For office workers who meet with clients on a regular basis, one may consider adding an extension to the side of the home to create a consulting office that is directly accessible for clients from outside as well as accessible for the homeowner from the home's interior.

In smaller dwellings, a quality office area does not have to come at the expense of a small space. Considering the office design ideas discussed in Chap. 12, residual spaces can be used for a unique office setting. Places like under staircases can create an intimate work area that is both convenient and makes for efficient use of space. Large, shallow closets and cabinets can also be converted into an office by adding adequate task lighting, shelving units, and a desk. This unique workspace can easily be hidden away with a sliding door at the end of the workday or designed without a door to incorporate it seamlessly into the rest of the home. A home with high ceilings could opt for a mezzanine or loft style office configuration on the ground floor. Because this design focuses on an elevated component, floor space within the home can be entirely conserved. The mezzanine's design will give the homeowner the feeling of an open floor concept when looking down at the living area below while enjoying more privacy (Jerde et al. 2020).

If more office space is desired, the basement or attic can also be turned into workrooms that offer quiet and privacy (Elmkies 2020). The basement configuration can also accommodate client meetings adequately. The addition of a separate side door leading into the basement's office is recommended for this type of work to ensure that the homeowner's privacy is considered. Again, the basement configuration also provides temperature-regulating benefits. However, outdoor seasonal environmental conditions may be important to consider when selecting the location of the workspace as this will affect the frequency of regulating temperature. For example, a basement workspace may need more heating during the winter while an attic workspace arrangement may require an air conditioning system in the summer since it may tend to get warmer than in other parts of the home. As a result, seasonal changes are to be considered when selecting the location of a workspace as this will influence heating and energy consumption.

14.4.1 Eco-Consciousness in Workspace's Design

To make any workspace more sustainable, various methods can be employed. When it comes to the purchase of equipment, opting for energy efficient devices will help reduce the cost of utility bills and reduce the home's carbon footprint. For workspaces with large windows, curtains, and drapes can be hung over them for decorative purposes and to winterize the space. Heat loss through windows accounts for up to 30 percent of residential heating energy used during colder months. Though simple, thermal insulated curtains can reduce heating loss up to 10–25 percent during winter as well as maintain indoor temperature control during the summer (Delgado 2020).

Office furniture should be complementary to other rooms and style of the home rather than settling on plain items. The key to a desirable office setting is its aesthetic appeal (Roberts 2021). While aesthetics is important, there is potential for office furniture to be sustainable as well. Just as with the fashion industry, cheap, mass-produced furniture sourced in unsustainable and unethical ways damages the environment. Known as *fast furniture*, this consumption phenomenon is damaging because these mass-produced items tend to get discarded by consumers quickly and travel extensive distances to get to markets. Therefore, the phenomenon of fast furniture is equally as damaging as fast fashion and feeds into the self-perpetuating cycle of overconsumption (Vigliarolo 2020).

As mentioned throughout Chaps. 9, 10, 11, and 12, various product certifications such as Forest Stewardship Council (FSC) and B Corp can help customers make better-informed decisions about how the products they purchase were produced and sourced from. Again, choosing to purchase second-hand furniture can help promote a circular economy and reduce excessive consumption. While getting rid of printers and discontinuing the use of desktop computers may not be feasible depending on the type of work a homeowner does, designing homes and office spaces that are conscious about material consumption, energy usage, and energy loss, are some of the best ways to create an eco-friendlier home environment.

14.5 Designing a Live-Work Home

Russell and Rame Hruska designed a building that houses their architecture office and family home on a vacant infill lot in Houston, Texas. Their firm, Intexure Architects, moved from a downtown warehouse to an urban neighborhood (Meinhold 2011). The design combines a studio on the ground floor with living space on the second. With the ability to substitute travel time with telecommuting and incorporate "green" construction principles, they show genuine commitment to sustainable living in an automobile-dependent city like Houston (Fig. 14.6).

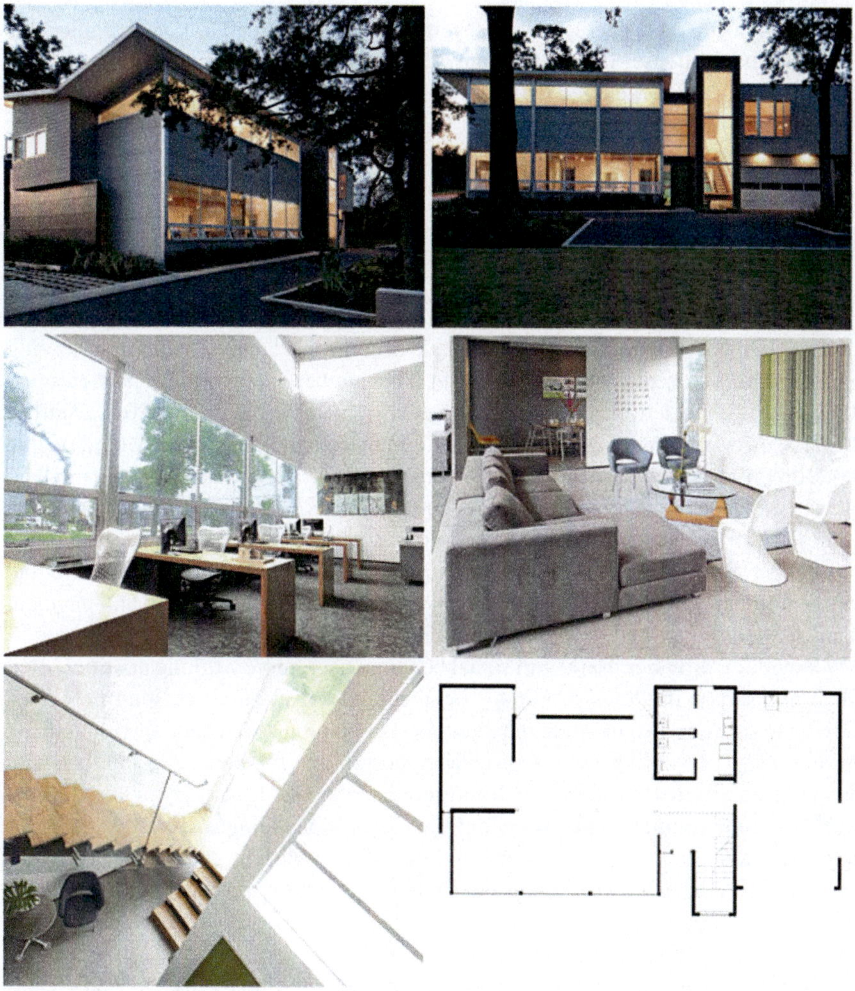

Fig. 14.6 Russell and Rame Hruska designed a building that houses their architecture office and family home on a vacant infill lot in Houston, Texas

The two-story home that measures 204 square meters (2,200 square feet) is minimally decorated and uses locally purchased materials such as steel columns, wood flooring, and ceiling coverings. The south-oriented studio takes advantage of natural light while using passive solar design strategies to gain heat in the winter and avoid the summer sun. The office's two-story tall ceiling with wall-to-wall windows creates an open space while providing a work area for up to 12 employees. Spaces like the second-floor materials library were created as multipurpose areas. During the day, the table could accommodate a meeting overlooking the office space below, while in the evening, it could be used as a dining table for the family. The second-floor kitchen offers flexibility as one can prepare a meal without interrupting activities in the working spaces.

The architects also incorporated several green technologies. For example, this LEED-certified house includes photovoltaic panels and a rainwater-collection system. The façade is cladded with recycled materials that were also sourced within 805 km (500 miles) when possible.

14.6 Working in Common

First created in 1995 by a group of hackers composed of computer engineers in Berlin, Germany, common workspaces have become extremely popular for working professionals in recent years (Mente 2021). The number of coworking spaces around the world is expected to reach approximately 41,975 by the end of 2024 (Statista 2021). Coworking involves the sharing of collaborative office spaces where people work on projects independently or as a team (Fig. 14.7). Generally, the people who use shared office spaces work for different companies and themselves. In other words, common workspaces are meant for people to work away from home or the office setting. Coworking spaces offer the supplies and equipment one would expect to find in a traditional office setting (DropDesk n.d.). Many offer membership and payment plans as well as pay-per-visit options making it flexible for users.

Coworking provides users with social benefits that remote working does not offer. Most noticeably, it allows people to socialize and expand their existing networks. These interactions also increase the likelihood of gaining social support from others. Additionally, coworking spaces provide opportunities for people to interact with people from different professional backgrounds, which is less likely in a traditional workspace environment. The intermingling of people encourages diversity and inclusivity, creative thinking, cooperation, and knowledge sharing. No matter how well-designed a home workspace may be, the social benefits provided by coworking are sometimes unmatched. Therefore, some like to alternate between shared office spaces, working at the office, and working from home.

As of late, it is important to consider the immediate impact that the COVID-19 pandemic has had on the use of shared offices. They saw large declines in use due to health and safety measures. Although shared workspaces have been allowed to resume their activities, the lasting concerns about health and safety may make some

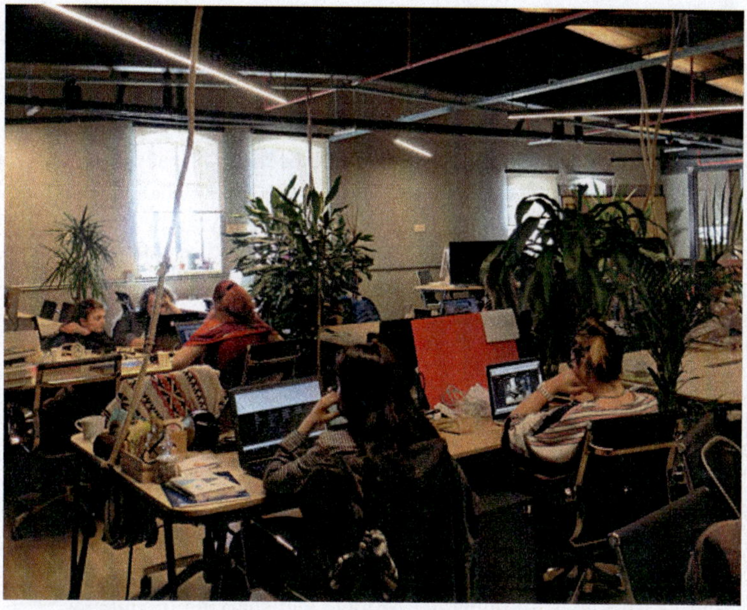

Fig. 14.7 Coworking has been becoming increasingly popular and involves the sharing of collaborative office spaces where people work on projects independently or as a team such as this space in Istanbul, Turkey

previous users reluctant to use the service again. Similarly, new potential users may be slower in adopting the service considering the circumstances. Nonetheless, there is no doubt that as time progresses, the use of coworking spaces will resume and keep steadily growing as people recognize the potential benefits they offer.

Importantly, coworking spaces are eco-friendly alternatives to traditional office settings. As people share resources like computers, printers, desks, and other office supplies, less equipment is required per coworker. Arguably, working in common can help reduce consumption by sharing and pooling together costly resources and supplies to be more sustainable.

14.7 A Live–Work Home

In response to the increasing number of people choosing to work from home, the Live-Work Home was designed to offer occupants a flexible solution for their home office. The goal of the design was to provide homebuyers a variety of choices for the location, size, and features of their office to accommodate a range of different households.

The design for a narrow-front multi-story dwelling included a variety of customizable options for workspaces, tailored to the occupants needs. Each option was designed with attention to access, privacy, light, and sound, to ensure the efficiency and comfort of both the home and the office space. For individuals working remotely, smaller home offices on the main or second floor could be used. Depending on the number of employees and the type of work being conducted, a separate point of access could be included. When a larger office or more privacy was necessary, a designated floor of the home could be used for the workspace. In addition, the option for retail or commercial office space on the ground floor with the owners' residence on upper floors could be accommodated. Regardless of the type of office, each option was designed to include sufficient access to natural light. Once the interior plan was selected, buyers would be offered a variety of fittings to customize the home office, including built-in modular storage options and worktops.

In the design process, buyers were provided with visualizations of example units to demonstrate some of the different options available for the interior plan and the design of the home office (Fig. 14.8). The basement option was ideal for occupants who desired more separation between the office and the rest of the home with a secluded room at the rear. For the first-floor option, the office would be located in the middle of the home, with the living area at the front of the dwelling and the kitchen at the rear. A large window along the longitudinal wall would provide the office with natural light. Partitions with integrated storage and shelving modules were used to enclose the office.

The option for a second-floor office offered a similarly sized space with more privacy, being in a less public zone of the house. Finally, an office space could occupy the attic floor, maximizing the degree of privacy from the rest of the home. The attic option would provide users with ample space and natural light. Overall, the Live-Work Home facilitated working from home by offering a range of different choices for office spaces, each with different features. It demonstrates how designers can accommodate the needs of different households in planning home office spaces.

14.8 Final Thoughts

Even though the world has gotten accustomed to working remotely, mixed feelings about its advantages and challenges persist. The COVID-19 pandemic has accelerated a paradigm shift that had already been underway for quite some time. Most remarkably, modern society has collectively made the return to a variation of the pre-industrial lifestyle where living and work activities are coupled under the same roof. Contemporary workers demand more flexible working options from their employers and desire more free time to achieve a better work/life balance. As hybrid forms of work offer an acceptable compromise for many, there is little doubt that it will become the preferred model for many employers. The future of remote work has yet to reach a stage of maturity. The sustained development of technology and collaborative software aimed at increasing the fluidity of remote work will continue to

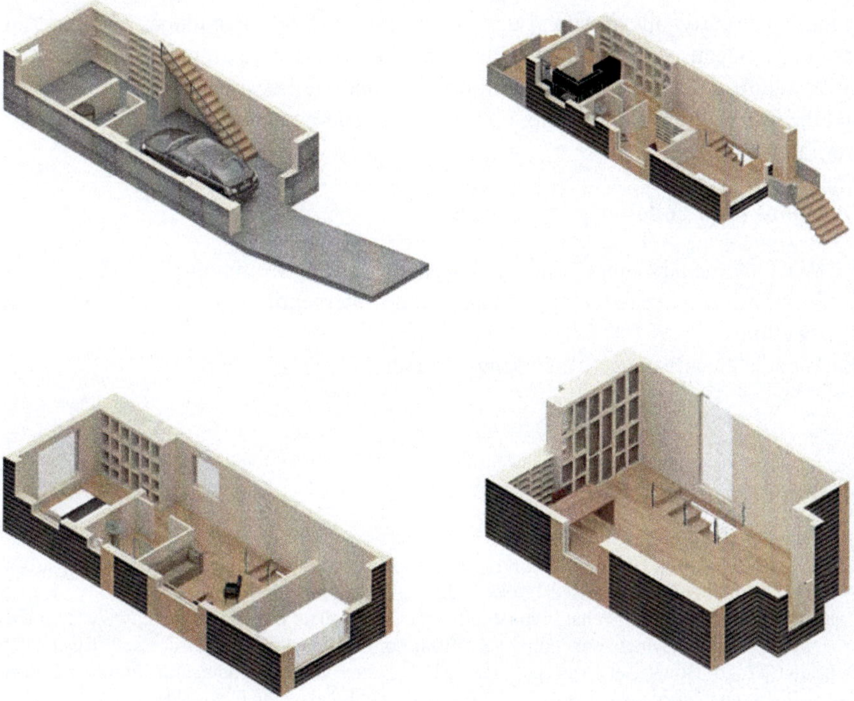

Fig. 14.8 Axonometric of the Live-Work Home's levels. Basement (top left), first floor (top right), second floor (bottom left) and attic (bottom right)

support the societal shift in favor of discarding the rigid work model of the industrial revolution.

However, despite the many benefits offered by remote work, it is still not accessible for all. While some governments and employers provide subsidies and equipment to support their workforce, not all workers are fortunate enough to have this assistance. Many marginalized, indigenous, and low-income communities struggle to gain adequate access to internet services and equipment for personal use, let alone for employment purposes. In the developing world, similar challenges are observed as a lack of infrastructure and resources remains a key impedance to human and economic development. As a larger portion of the global workforce moves to non-traditional ways of working, there is hope that more governments and employers will provide the necessary resources to allow workers to benefit from more flexible work opportunities. Beyond the provision of physical and financial resources, increased forms of social support will be required to ensure the mental wellbeing of the remote workforce, seeing as a lack of social interaction can be difficult to handle. With options like coworking spaces, the challenges, and barriers to working from home can be partly alleviated through the sharing and pooling of human and material resources. Because coworking can lower these barriers, more users may be expected in the future. The combination of an efficient and well-designed home workspace and the availability of coworking options can encourage more sustainable forms of work to take shape.

Questions for a Follow-Up Discussion

1. What are the advantages and challenges of working from home?
2. What are the needed accommodation and interventions for remote work in a dwelling?
3. What is the advantage of working in common?

References

Courtney E (2021) The benefits of working from home. https://www.flexjobs.com/blog/post/benefits-of-remote-work/. Accessed 16 Oct 2022

Cramer M, Zaveri M (2020) What if you don't want to go back to the office? https://www.nytimes.com/2020/05/05/business/pandemic-work-from-home-coronavirus.html. Accessed 16 Oct 2022

Delgado C (2020) How to set up an eco-friendly home office you won't hate. https://www.architecturaldigest.com/story/eco-friendly-home-office-ideas. Accessed 16 Oct 2022

DropDesk (n.d.) What is coworking? Everything you need to know about coworking spaces https://drop-desk.com/what-is-coworking. Accessed 16 Oct 2022

Elmkies R (2020) 11 ideas for working remotely when you don't have a home office. https://www.bobvila.com/slideshow/11-ideas-for-working-remotely-when-you-don-t-have-a-home-office-48561. Accessed 16 Oct 2022

Jerde K, Minton M, Mather L (2020) 65 Home office ideas that will inspire productivity https://www.architecturaldigest.com/gallery/home-offices-slideshow. Accessed 16 Oct 2022

Meinhold B (2011) Intexure's architecture office doubles as beautiful eco home, Last updated 14 Jan 2011. Accessed on 6 May 2011. http://inhabitat.com/houston-architecture-office-doubles-as-beautiful-eco-home/. Accessed 16 Oct 2022

Mente J (2021) The history of coworking: How flexible office space became a force in the working world. https://cobaltworkspace.com/the-history-of-coworking/. Accessed 16 Oct 2022

Plumer B, Popovich N (2020) Traffic and pollution plummet as U.S. cities shut down for coronavirus. https://www.nytimes.com/interactive/2020/03/22/climate/coronavirus-usa-traffic.html. Accessed 16 Oct 2022

Reynolds BW, Bibby A (n.d.) The complete history of working from home. https://www.flexjobs.com/blog/post/complete-history-of-working-from-home/#:~:text=More%20employees%20working%20remotely%20than,they%20grew%20to%20over%2059%25. Accessed 16 Oct 2022

Roberts G (2021) 10 tips for designing your home office. https://www.hgtv.com/design/rooms/other-rooms/10-tips-for-designing-your-home-office. Accessed 16 Oct 2022

Statista (2021) Number of coworking spaces worldwide from 2018 to 2020 with a forecast to 2024. https://www.statista.com/statistics/554273/number-of-coworking-spaces-worldwide/. Accessed 16 Oct 2022

Vigliarolo B (2020) 5 ways to make your home office eco-friendly. https://www.techrepublic.com/article/5-ways-to-make-your-home-office-eco-friendly/. Accessed 16 Oct 2022

Wrike (n.d.-a) Setting up a home office. https://www.wrike.com/remote-work-guide/home-office-setup-equipment/#importance-of-an-effective-wfh-office-setup. Accessed 16 Oct 2022
Wrike (n.d.-b) Remote work statistics. https://www.wrike.com/remote-work-guide/remote-work-statistics/. Accessed 16 Oct 2022

Illustrations Credits

Figures not listed below are in the public domain or have been conceived, drawn, or photographed by the author and/or members of his research and design teams. Their names are listed in the Acknowledgements and the Projects' Teams pages. Every effort has been made to list all contributors and sources. In case of omission, the author and the publisher will include appropriate acknowledgment or correction in any subsequent edition of this book.

Chapter 1

Figure 1.4: After Hodges, T. (2010). Public Transportations Role in Responding to Climate Change. The Federal Transit Administration, U.S. Department of Transportation
Figure 1.5: After Statistics Canada (2006) CYB Overview 2006: Population and Demography

Chapter 2

Figure 2.8: After Dimond, J, (1976), Residential Density and Housing Form, Journal of Architectural Education, Vol. 3, February.

Chapter 4

Heijman ONE
Photography: MoodBuilders

Chapter 7

The Catskills House

A. Friedman, *Fundamentals of Innovative Sustainable Homes Design and Construction*, The Urban Book Series, https://doi.org/10.1007/978-3-031-35368-0

Photography: Amanda Kirkpatrick

Chapter 8

TK-33
Photography: Jan Ove Christensen & Peter Jørgensen

Chapter 10

BS House
Photography: Alessandra Bello

Chapter 11

Tenhachi House
Photography: Akihide Mishima

Chapter 13

Villa Deys
Photography: René de Wit, Rien van Rijthoven, Femke Bijlsme

Chapter 14

Russell and Rame Hruska Live-Work Home
Photography: Don Glentzer, Keven Alvarado

Project Teams

I would like to thank the firms whose projects are featured in the book. Case studies that were produced under my direction were designed either in my private practice or as part of my university research. I have attempted to recall all the firms and my team members. If I have omitted someone, my sincere apology and I will do my best to correct it in future editions.

Chapter 1: Sustainable Dwellings for Changing Times

Skaftkarr
Sitra
Itämerenkatu 11-13, PO Box 160, 00181 Helsinki
www.sitra.fi

Komoka, Ontario, Canada
Avi Friedman, Architect
Team: Josie White, Nyd Garavito-Bruhn

Chapter 2: Denser Living

2 McIntyre Drive
MGS Architects
10-22 Manton Lane, Melbourne, VIC 3000, Australia
www.mgsarchitects.com.au

APT1
Avi Friedman & Charles Grégoire

Chapter 3: Quality Affordable Dwellings

Bloembollenhof

A. Friedman, *Fundamentals of Innovative Sustainable Homes Design and Construction*, The Urban Book Series, https://doi.org/10.1007/978-3-031-35368-0

S333 Architecture + Urbanism
70 Cowcross Street, London, UK, EC1M 6EJ
www.s333.org

The Grow Home
Avi Friedman, Witold Rybczynski & Susan Ross

Chapter 4: Comfortable Small Interiors

Heijman ONE
MoodBuilders
Klokgenbouw 144, 5617 AB Eindhoven nr 144-21, the Netherlands
www.moodbuilders.nl

The Next Home
Avi Friedman, Principal Designer
Team: Jasmin S. Fréchette, Cyrus M. Bilimoria, David Krawitz, Doug Raphael
Consultants: R. Kevin Lee, Julia Bourke, Richard Gingras, Vince Cammalleri

Chapter 5: Attractive and Energy Efficient Facades

Västra Hamnen
Klas Tham Architect & Planner SAR/MSA EF
Karl Gerhards väg 21, SE13335 Saltsjöbaden, Sweden
klas.tham@spray.se

The Affordable Prefab Home
Avi Friedman & Vince Cammalleri

Chapter 6: Innovative Construction Practices

Floating Houses in IJBurg
Marlies Rohmer Architecture + Urban Planning
P.O. Box 2935, 1000CX Amsterdam, the Netherlands
www.rohmer.nl
Photography: Avi Friedman

The Pod Home
Avi Friedman & Charles Grégoire

Chapter 7: Utilities Systems for Sustainability

The Catskills House

J_spy Architecture |
White Lake, NY, United States
jspyarchitecture.com
Photography: Amanda Kirkpatrick

Domus ex Machina
Avi Friedman, Principal Designer
Team: Charles Grégoire, Isabella Rubial, & Zhong Cai

Chapter 8: Green and Healthy Materials

TK-33
Tegnestuen LOKAL
tegnestuenlokal.dk
Photography: Jan Ove Christensen & Peter Jørgensen

The Green Grow Home
Avi Friedman & Vince Cammalleri
Team: Jim Nicell, Francois Dufaux, Joanne Green, Susan Fisher, Aud Koht, Kevin Lee, Aryan Lirange, Denis Palin, Mark Somers, Nicola Bullock, and Michelle Takoff

Chapter 9: Energy Efficient Dwellings

Hammarby Sjöstad
White
Östgötagatan 100, Box 4700, 11692 Stockholm, Sweden
www.white.se
Photography: Avi Friedman & David Auerbach

Energy Efficient Home
Avi Friedman & Vince Cammalleri

Chapter 10: Home Automation

BS House
Vicolo Bressa 3/2, 31044 Montebelluna (TV), Italy
Reisarchitettura | reisarchitettura.it
Photography: Alessandra Bello

The Smart Home
Avi Friedman, Principal Designer

Team: Charles Grégoire, Isabella Rubial, Zhong Cai

Chapter 11: Cooking and Dining at Home

Tenhachi House
Tenhachi Architect & Interior Design |
Kanagawa, Japan
ten-hachi.com
Photography: Akihide Mishima

The Sustainable Kitchen
Avi Friedman, Principal Designer
Team: Charles Grégoire, Isabella Rubial, Zhong Cai

Chapter 12: Storing Stuff and Furnishing a Home

3500 Millimeter House
AGo Architects
South Jakarta, Indonesia
agoarchitecture.com
Photography: Kafin Noe'man

The Max Storage Home
Avi Friedman, Principal Designer
Team: Charles Grégoire, Isabella Rubial, Zhong Cai

Chapter 13: Getting Old at Home

Villa Deys
Paul de Ruiner
Valschermkade 36D, 1059 CD Amsterdam, The Netherlands
www.paulderuiter.nl
Photography: René de Wit, Rien van Rijthoven, Femke Bijlsme

Adaptable Home
Avi Friedman, Principal Designer
Team: Jasmin S. Fréchette, Cyrus M. Bilimoria, David Krawitz, Monica Slanik, Shawn Lapointe, Hor Hooi Ping (Agnes)

Chapter 14: Working from Home

Russell and Rame Hruska Live-Work Home

Intexure Architects
1815 Southmore Boulevard, Houston Tx
www.intexure.com
Photography: Don Glentzer, Keven Alvarado

Live-Work Home
Avi Friedman, Principal Designer
Team: Charles Grégoire, Isabella Rubial, Zhong Cai

Bibliography

Abdellatef Y, Khan MA, Khan A, Alam MI, Kavgic M (2020) Mechanical, thermal, and moisture buffering properties of novel insulating hemp-lime composite building materials. MDPI. https://www.mdpi.com/1996-1944/13/21/5000. Accessed 11 Oct 2022

ADA.gov (n.d.) Introduction to the ADA. https://www.ada.gov/ada_intro.htm. Accessed 16 Oct 2022

Adhitya S (2017) A rhythmic approach to urban design. In: Musical cities: listening to urban design and planning (DGO-Digital original). UCL Press, pp 26–33 https://doi.org/10.2307/j.ctv550cz9.10. Accessed 5 Oct 2022

Aging in Place (2018a) Important updates and modifications for the bedroom. https://aginginplace.org/aging-in-place-important-updates-and-modifications-for-the-bedroom/. Accessed 15 Oct 2022

Aging in Place (2018b) Kitchen of the future: remodeling for comfortable aging in place. https://aginginplace.org/kitchen-of-the-future-remodeling-for-comfortable-aging-in-place/. Accessed 15 Oct 2022

Aging in Place (2018c) What to do when you redo bathroom. https://aginginplace.org/what-to-do-when-you-redo-your-bathroom/. Accessed 15 Oct 2022

AHAM (2021) AHAM history. Association of Home Appliance Manufacturers. https://www.aham.org/AHAM/AuxAHAMHistory. Accessed 13 Oct 2022

Ahmed MMS, Radwan A, Serageldin AA, Abdeen A, Abo-Zahhad EM, Nagano K (2021) The thermal potential of a new multifunctional sliding window. Sol Energy. https://www.sciencedirect.com/science/article/abs/pii/S0038092X21007015. Accessed 13 Oct 2022

Allan K, Phillips AR (2021) Comparative cradle-to-grave life cycle assessment of low and mid-rise mass timber buildings with equivalent structural steel alternatives. MDPI. https://www.mdpi.com/2071-1050/13/6/3401. Accessed 11 Oct 2022

Ander GD (2016) Daylighting. National Institute of Building Sciences. https://www.wbdg.org/resources/daylighting. Accessed 5 Oct 2022

Apis Cor (n.d.) The world's largest 3D printed building. https://www.apis-cor.com/dubai-project. Accessed 23 Sept 2022

Aquacell (2019) Inside salesforce tower's water recycling system. Aquacell. https://aquacell.com.au/v2/wp-content/uploads/2019/09/Saleforce-Tower.pdf. Accessed 13 Oct 2022

ArchDaily (2016a) How to correctly design and build a kitchen. https://www.archdaily.com/789894/how-to-correctly-design-and-build-a-kitchen. Accessed 16 Oct 2022

ArchDaily (2016b) Tenhachi house/Tenhachi architect & interior design. https://www.archdaily.com/789221/tenhachi-house-8-tenhachi-architect-and-interior-design. Accessed 16 Oct 2022

ArchDaily (2019) 3500 millimetre house/Ago architects https://www.archdaily.com/909456/3500-millimetre-house-ago-architects. Accessed 15 Oct 2022

A. Friedman, *Fundamentals of Innovative Sustainable Homes Design and Construction*, The Urban Book Series, https://doi.org/10.1007/978-3-031-35368-0

ArchDaily. (2011) Floating houses in IJburg/Architectenbureau Marlies Rohmer. Last updated 20 Mar 2011. http://www.archdaily.com/120238/floating-houses-in-ijburg-architectenbureau-marlies-rohmer/. Accessed 13 Oct 2022

Arendt RG (1996) Conservation design. ACRD. https://www.acrd.bc.ca/cms/wpattachments/wpID188atID760.pdf. Accessed 28 Sept 2022

Arnold RD, Wade JP (2015) A definition of systems thinking: a systems approach. Proc Comput Sci. https://www.sciencedirect.com/science/article/pii/S1877050915002860. Accessed 11 Oct 2022

Asdrubali F, Desideri U (2018) Building envelope. In: Handbook of energy efficiency in buildings. https://www.sciencedirect.com/science/article/pii/B9780128128176000395?via%3Dihub. Accessed 13 Oct 2022

B Corp (n.d.) About B Lab. https://www.bcorporation.net/en-us/movement/about-b-lab. Accessed 15 Oct 2022

Baker B, SCAN Foundation (2018) Green retirement communities are sprouting. https://www.nextavenue.org/green-retirement-communities-are-sprouting/. Accessed 15 Oct 2022

Balke EC, Healy WM, Ullah T (2016) An assessment of efficient water heating options for an all-electric single-family residence in a mixed-humid climate. Energ Build. https://www.sciencedirect.com/science/article/pii/S0378778816309136. Accessed 13 Oct 2022

Balling RJ (2016) Architecture/design: a car-free, polycentric city, with multi-level skybridges and inter-building atria. CTBUH J 3:28–33. http://www.jstor.org/stable/44154343. Accessed 27 Sept 2022

Bathsystem (2019) Prefabricated bathroom pods. Bathsystems. https://www.bathsystem.com/en/products/prefabricated-bathroom-pods/. Accessed 13 Oct 2022

Becker J (n.d.) 21 surprising statistics that reveal how much stuff we actually own. https://www.becomingminimalist.com/clutter-stats/. Accessed 15 Oct 2022

Behroozy A, Keegel TG (2014) Wet-work exposure: a main risk factor for occupational hand dermatitis. In: Safety and health at work. https://pubmed.ncbi.nlm.nih.gov/25516808/. Accessed 13 Oct 2022

Belniak S, Leśniak A, Plebankiewicz E, Zima K (2013) The influence of the building shape on the costs of its construction. J Financ Manage Prop Constr. https://www.emerald.com/insight/content/doi/10.1108/13664381311305096/full/html. Accessed 28 Sept 2022

Bertram N, Blanco JL, Mischke J, Andersson N (2019) Scaling modular construction. In: Global infrastructure initiative by McKinsey & Company. https://www.mckinsey.com/~/media/mckinsey/business%20functions/operations/our%20insights/voices%20on%20infrastructure%20scaling%20modular%20construction/gii-voices-sept-2019.pdf. Accessed 13 Oct 2022

Bharne V (2011) Humanizing high-rise urbanism: design strategies and planning tools. CTBUH J 4:18–23. http://www.jstor.org/stable/24192542. Accessed 27 Sept 2022

Boston University C of C (2018) WiFi and the future of connection | Center for Mobile Communication Studies. https://sites.bu.edu/cmcs/2018/09/14/wifi-and-the-future-of-connection/. Accessed 13 Oct 2022

Bradford A (2022) 8 best smart blinds. https://www.familyhandyman.com/list/best-smart-blinds/#:~:text=Smart%20blinds%20are%20window%20coverings,honeycomb%2C%20roller%20and%20light%20filtering.text=Blinds%20can%20be%20hardwired%20or,solar%2C%20battery%20or%20electrical%20cord. Accessed 13 Oct 2022

Britnell G (2015) CIPP rehab of water mains running through backyards. https://ferpalinfrastructure.com/wp-content/uploads/2015/11/mstt-journal-2015.pdf. Accessed 13 Oct 2022

Brown H, Proust K, Newell B, Spickett J, Capon T, Bartholomew L (2018) Cool communities-urban density, trees, and health. MDPI. https://www.mdpi.com/1660-4601/15/7/1547. Accessed 28 Sept 2022

Bunbury A (2015) Designed to last: long live sustainable housing. ReNew Technol Sustain Fut 132:64–67. https://www.jstor.org/stable/renetechsustfutu.132.64. Accessed 27 Sept 2022

CAA (2022) Drains and foundations: prevention or repair? CAA Quebec. https://www.caaquebec.com/en/at-home/guides/drains-and-foundations-prevention-or-repair/. Accessed 13 Oct 2022

Campbell C (2013) Visualizing residual spaces in a new light. https://www.sjsu.edu/urbanplanning/docs/honors-reports/URBP%20298%20%20Honors%20Report%20-%20Campbell.pdf. Accessed 15 Oct 2022

Canada Green Building Council (n.d.) LEED: the international mark of excellence. https://www.cagbc.org/our-work/certification/leed/. Accessed 28 Sept 2022

Canada Mortgage and Housing Corporation (CMHC) (2000) Flex housing: the professional guide. Publish by Canada Mortgage and Housing Corporation, Ottawa

Canada Mortgage and Housing Corporation (CMHC) (2021) Understanding Canadian households self-reported sense of social inclusion: the social inclusion index (SII). Research Insight June 2021. https://assets.cmhc-schl.gc.ca/sites/cmhc/professional/housing-markets-data-and-research/housing-research/surveys/canadian-housing-survey/2021/canadian-households-self-reported-sense-social-inclusion-index-69774-en.pdf?rev=f51ebc5b-ef8e-4f66-b988-1e9809841a3a. Accessed 23 Sept 2022

Canada NR (2014) Medium-density fibreboard. https://www.nrcan.gc.ca/our-natural-resources/forests-forestry/forest-industry-trade/forest-products-applications/taxonomy-wood-products/medium-density-fibreboard/15849. Accessed 13 Oct 2022

Candido C (2011) Adaptive comfort: passive design for active occupants. Environ Des Guide 69:1–5. http://www.jstor.org/stable/26151893. Accessed 5 Oct 2022

Cantrell J (2022) Natural capital. World Bank. https://www.worldbank.org/en/topic/natural-capital#1. Accessed 11 Oct 2022

Center for Transit-Oriented Development (2014) Creating connected communities—HUD USER. HUDuser. https://www.huduser.gov/publications/pdf/Creating_Cnnted_Comm.pdf. Accessed 28 Sept 2022

Chen J (2021) What is a zero-lot-line house? Investopedia. https://www.investopedia.com/terms/z/zero-lot-line-house.asp. Accessed 28 Sept 2022

Chow WK, Wood A, Li J (2013) Reduction in evacuation time for tall buildings through the use of skybridges. J Archit Plan Res 30(2):146–166. http://www.jstor.org/stable/43031086. Accessed 27 Sept 2022

Christian K (2022) Sustainability and ethics certifications: what do they actually mean? https://www.thegoodtrade.com/features/sustainable-certifications-and-standards. Accessed 15 Oct 2022

Clarke D (2017) Green roofs. Sanctuary Mod Green Homes 38:82–84. https://www.jstor.org/stable/90000611. Accessed 27 Sept 2022

Clifton C (2020) Methods of venting plumbing fixtures and traps in the 2018 IPC. https://wabo.memberclicks.net/assets/pdfs/Plumbing_Venting_Brochure_2018.pdf. Accessed 13 Oct 2022

Cohn D, Passel J (2018) A record 64 million Americans live in multigenerational households. https://www.pewresearch.org/fact-tank/2018/04/05/a-record-64-million-americans-live-in-multigenerational-households/#:~:text=Record%2064%20million%20Americans%20live%20in%20multigenerational%20households%20%7C%20Pew%20Research%20Center. Accessed 15 Oct 2022

Colville C (2022) Sustainable storage solutions. https://www.countryandtownhouse.com/interiors/sustainable-storage-inspiration/. Accessed 15 Oct 2022

Cooke P (2021) Future shift for 'big things': from starchitecture via agritecture to parkitecture. J Open Innov Technol Mark Complex 7(236):119. https://doi.org/10.3390/joitmc7040236. Accessed 23 Sept 2022

Coolen H, Meesters J (2012). Private and public green spaces: meaningful but different settings. J Hous Built Environ 27(1):49–67. http://www.jstor.org/stable/41487473. Accessed 27 Sept 2022

Courtney E (2021) The benefits of working from home. https://www.flexjobs.com/blog/post/benefits-of-remote-work/. Accessed 16 Oct 2022

Cox W (2022) Demographia international housing affordability. Demographia. http://demographia.com/dhi.pdf. Accessed 13 Oct 2022

Cramer M, Zaveri M (2020) What if you don't want to go back to the office? https://www.nytimes.com/2020/05/05/business/pandemic-work-from-home-coronavirus.html. Accessed 16 Oct 2022

Cutland Consulting Ltd, Eco Design Consultants Ltd (2016) The challenge of shape and form. NHBC foundation. https://www.nhbcfoundation.org/wp-content/uploads/2016/10/NF-72-NHBC-Foundation_Shape-and-Form.pdf. Accessed 13 Oct 2022

Dahms K, Ducharme A (2022) Housing affordability monitor—2021q4—NBC. National Bank of Canada. https://www.nbc.ca/content/dam/bnc/en/rates-and-analysis/economic-analysis/housing-affordability.pdf. Accessed 28 Sept 2022

Danish Architecture Centre (2014) Porvoo: energy efficient residential area. Last updated 21 Jan 2014. https://www.slideshare.net/SitraEnergia/porvoo-skaftkrrsyksy2011eng. Accessed 29 Sept 2022

Delgado C (2020) How to set up an eco-friendly home office you won't hate. https://www.architecturaldigest.com/story/eco-friendly-home-office-ideas. Accessed 16 Oct 2022

Delgado JA, Nearing MA, Rice CW (2013) Conservation practices for climate change adaptation. Adv Agron. https://www.sciencedirect.com/science/article/pii/B9780124076853000025?via%3Dihub. Accessed 13 Oct 2022

Delile H, Keenan-Jones D, Blichert-Toft J, Goiran J-P, Arnaud-Godet F, Albarède F (2017) Rome's urban history inferred from Pb-contaminated waters trapped in its ancient harbor basins. Proc Natl Acad Sci 114:10059–10064. https://doi.org/10.1073/pnas.1706334114. Accessed 13 Oct 2022

Deng L (2010) The value of regulating building facades: a case study of masonry ordinances in Chicago suburbs. J Architect Plan Res 27(4):271–286. http://www.jstor.org/stable/43030912. Accessed 5 Oct 2022

Derananian AR, Donlon MK (2019) Panelized wall systems—joint detailing for success. IIBEC. https://iibec.org/wp-content/uploads/2019-cts-derananian-donlon.pdf. Accessed 13 Oct 2022

Development Services Department (2019) Design manual for small and narrow lots. Fayetteville, AR. https://www.fayetteville-ar.gov/DocumentCenter/View/15360/Small-Lot-Infill-Standards-Manual?bidId=. Accessed 28 Sept 2022

Dhar TK, Hossain M, Rahaman KR (2013) How does flexible design promote resource efficiency for housing? A study of Khulna, Bangladesh. Smart Sustain Built Environ 2(2):140–157. https://doi.org/10.1108/SASBE-10-2012-0051. Accessed 5 Oct 2022

DOE (2021) Microhydropower systems. https://www.energy.gov/energysaver/microhydropower-systems. Accessed 13 Oct 2022

Dollarhide M (2021) Mass customization. Investopedia. https://www.investopedia.com/terms/m/masscustomization.asp. Accessed 13 Oct 2022

Donn M, Thomas G (2003) Designing comfortable homes. CDN. https://cdn.ymaws.com/concretenz.org.nz/resource/resmgr/docs/ccanz/ccanz_tm37.pdf. Accessed 28 Sept 2022

Donn M, Thomas G (2010) Designing comfortable homes. https://cdn.ymaws.com/concretenz.org.nz/resource/resmgr/docs/ccanz/ccanz_tm37.pdf. Accessed 13 Oct 2022

Douay N (2018) Urban planning in the digital age. Wiley. Vol 6. Ch. 1. https://www.wiley.com/en-us/Urban+Planning+in+the+Digital+Age-p-9781119539476. Accessed 27 Sept 2022

DropDesk (n.d.) What is coworking? Everything you need to know about coworking spaces. https://drop-desk.com/what-is-coworking. Accessed 16 Oct 2022

Du FL, Li T (2013) Selected, peer reviewed papers from the 2013 international conference on applied mechanics and materials (ICAMM 2013), November 23–24, 2013, Zhuhai, China. (2014). Study on architectural design for indoor space of small area house. Appl Mech Mater 477–478, 1128–1131. https://doi.org/10.4028/www.scientific.net/AMM.477-478.1128. Accessed 5 Oct 2022

Easterseals (n.d.) 16 ways to make your kitchen more accessible. https://blog.easterseals.com/16-ways-to-make-your-kitchen-more-accessible/. Accessed 16 Oct 2022

Easton D, Easton T (2014) Modern rammed earth construction techniques. In: Modern earth buildings. https://www.sciencedirect.com/science/article/pii/B9780857090263500149. Accessed 11 Oct 2022

Eckley SN (2019) Carbon sequestration. Encyclopædia Britannica. https://www.britannica.com/technology/carbon-sequestration. Accessed 13 Oct 2022

Editors of Encyclopaedia BT (2022) Prefabrication. Encyclopædia Britannica. https://www.britannica.com/technology/prefabrication. Accessed 13 Oct 2022

Edmonton Senior Council (n.d.) Making our houses lifelong homes. https://www.seniorscouncil.net/uploads/files/AccessibleHousing_LORES_Feb171.pdf. Accessed 15 Oct 2022

EESI (n.d.) Energy efficiency. EESI. https://www.eesi.org/topics/energy-efficiency/description. Accessed 13 Oct 2022

Ellen MacArthur Foundation (2022) New to circular economy overview. Ellen MacArthur Foundation. https://ellenmacarthurfoundation.org/topics/circular-economy-introduction/overview. Accessed 11 Oct 2022

Elmkies R (2020) 11 ideas for working remotely when you don't have a home office. https://www.bobvila.com/slideshow/11-ideas-for-working-remotely-when-you-don-t-have-a-home-office-48561. Accessed 16 Oct 2022

Enercare (2019) 8 ways your smart home can save energy. https://www.enercare.ca/blog/smarter-home/8-ways-your-smart-home-can-save-energy. Accessed 13 Oct 2022

Energy Project Z (2019) Positive energy homes. Zero energy project. https://zeroenergyproject.org/buy/positive-energy-homes/. Accessed 13 Oct 2022

Energy Star (n.d.-a). What is Energy Star. https://www.energystar.gov/about. Accessed 13 Oct 2022

Energy Star (n.d.-b). Smart home tips for saving energy. https://www.energystar.gov/products/Smart_home_tips. Accessed 13 Oct 2022

Environmental Protection Agency US (EPA) (2009) A guide to energy-efficient heating and cooling. Energy Star. https://www.energystar.gov/sites/default/files/asset/document/HeatingCoolingGuide%20FINAL_9-4-09_0.pdf. Accessed 13 Oct 2022

EPA SA (2022) Black and grey water management https://www.epa.sa.gov.au/environmental_info/water_quality/programs/grey_and_black_water_discharge. Accessed 13 Oct 2022

Esteves A, Esteves MJ, Mercado MV, Barea G, Gelardi D (2018) Building shape that promotes sustainable architecture evaluation of the indicative factors and its relation with the construction costs. Archit Res. http://article.sapub.org/10.5923.j.arch.20180804.01.html. Accessed 28 Sept 2022

Euroleague for Life Sciences Summer School (2014) Designing the nature of the Green Belt. Ila + Irub. Boku. https://boku.ac.at/fileadmin/data/H03000/H85000/H85200/2014_Summerschool/Summerschool_2014_projects_neuweb.pdf. Accessed 28 Sept 2022

European Environment Agency (EEA) (2020) Why is resource efficiency important? https://www.eea.europa.eu/themes/waste/resource-efficiency/why-is-resource-efficiency-important. Accessed 13 Oct 2022

Faehnle M, Söderman T, Schulman H, Lehvävirta S (2015) Scale-sensitive integration of ecosystem services in urban planning. GeoJournal 80(3):411–425. http://www.jstor.org/stable/24432627. Accessed 27 Sept 2022

Falk A (2021) Summary report on heat pumps in district heating systems. Celsius City. https://celsiuscity.eu/wp-content/uploads/2021/05/Heat_pumps_in_district_heating_systems_RISE_2021.pdf. Accessed 13 Oct 2022

Fan F (2020) IOPscience. IOP Conf Ser Earth Environ Sci. https://iopscience.iop.org/article/10.1088/1755-1315/474/7/072091. Accessed 28 Sept 2022

Ferris S (2021) 10 ways to design a kitchen for aging in place. https://www.houzz.com/magazine/10-ways-to-design-a-kitchen-for-aging-in-place-stsetivw-vs~56039514. Accessed 16 Oct 2022

Fialko M & Hampton J (2011) National association of city transportation officials. NACTO. https://nacto.org/docs/usdg/activating_alleys_for_a_lively_city_fialko.pdf. Accessed 28 Sept 2022

Fisher-Gewirtzman D (2017a) The impact of alternative interior configurations on the perceived density of micro apartments. J Architect Plan Res 34(4):336–358. http://www.jstor.org/stable/44987241. Accessed 5 Oct 2022

Fisher-Gewirtzman D (2017b) The impact of alternative interior configurations on the percieved density of micro apartments. J Archit Plann Res 34(4):336–358. https://www.researchgate.net/publication/328555089_The_impact_of_alternative_interior_configurations_on_the_perceived_density_of_micro_apartments. Accessed 28 Sept 2022

Foletta N, Field S (2014) Europe's new low car(bon) communities. http://www.gwl-terrein.nl/files/artikelen/low%20carbon%20communities.pdf. Accessed 5 Oct 2022

Friedman A (2001) The grow home. McGill-Queen's University Press, Montréal

Friedman A (2002) The adaptable house: designing for choice and change. McGraw-Hill, New York

Friedman A (2012) Fundamentals of sustainable dwellings. Island Press, Washington, DC

Friedman A (2013) Innovative houses; concepts for sustainable living. Laurence King Publishing, London

Friedman A (2014) Planning small and mid-sized towns; designing and retrofitting for sustainability. Routledge, London

Friedman A (2020) Prefab living. Thames & Hudson Ltd, London

Friedman A, Cammalleri V (1992) Industrialization of narrow-front rowhousing using panelized wall systems. McGill University, Montreal

Friedman A, Cammalleri V (1995) The environmental impact of building materials in the North American home building industry. Build Res Inform 23(3):162–166

Friedman A, Cammalleri V (1996) The impact of R-2000 building technology on Canadian Housing. Build Res Inform 24(1):5–13

Friedman A, Cammalleri V (1997) Selection criteria for energy-efficient windows. Build Res Inform 25(4):234–238, July/August

Friedman A, Krawitz D (2002) Peeking through the keyhole. McGill-Queen's University Press.

Frontczak M, Wargocki P (2011) Literature survey on how different factors influence human comfort in indoor environments. Build Environ 46(4):922–937. https://doi.org/10.1016/j.buildenv.2010.10.021. Accessed 5 Oct 2022

Fuller RJ, Crawford RH (2011) Impact of past and future residential housing development patterns on energy demand and related emissions. J Hous Built Environ 26(2):165–183. http://www.jstor.org/stable/41261677. Accessed 27 Sept 2022

Garcia AL, Achaiah AT, Bello J, Donovan T (2017) Ferrock final paper 4.24.17 John-bello—IronKast. Iron Kast. http://ironkast.com/wp-content/uploads/2017/11/USC-Ferrock-Final-Paper-4.24.17.pdf. Accessed 11 Oct 2022

Giaquinto G (2019) Design ideas for the social kitchen. https://blog.kitchenmagic.com/blog/bid/174536/design-ideas-for-the-social-kitchen. Accessed 16 Oct 2022

Gilmore JH, Joseph Pine B (2014) The four faces of mass customization. Harv Bus Rev. https://hbr.org/1997/01/the-four-faces-of-mass-customization. Accessed 13 Oct 2022

Glennon M (2018) What are the benefits of motorized blinds? https://www.somfysystems.com/en-us/blog/post/what-are-the-benefits-of-motorized-blinds. Accessed 13 Oct 2022

Global WG (2022) Difference between blackwater and greywater. Global Water. https://www.globalwatergroup.com.au/our-blog/difference-between-blackwater-and-greywater. Accessed 13 Oct 2022

Goldstein B, Gounaridis D, Newell JP (2020) The carbon footprint of household energy use in the United States. PNAS. https://www.pnas.org/content/117/32/19122. Accessed 13 Oct 2022

Gong J, Sumathy K (2016) Active solar water heating systems. Penn State. https://pennstate.pure.elsevier.com/en/publications/active-solar-water-heating-systems. Accessed 13 Oct 2022

Government of Canada (2016) Thinking about aging in place. https://www.canada.ca/en/employment-social-development/corporate/seniors/forum/aging.html. Accessed 15 Oct 2022

Green Building (n.d.) In APA—the engineered wood association. https://www.apawood.org/green-building. Accessed 5 Oct 2022

Green Storage (n.d.) What makes us green. https://www.greenstorage.ca/what-makes-us-green. Accessed 15 Oct 2022

Group M (ed) (2022) Central heating. The Master Group. https://www.master.ca/consumer/central-heating. Accessed 13 Oct 2022

Grundig (n.d.-a) How sustainable plywood is changing the way we design kitchens. https://www.grundig.com/ktchnmag/blog/how-sustainable-plywood-is-changing-the-way-we-design-kitchens/. Accessed 16 Oct 2022

Grundig (n.d.-b) How to organize your kitchen waste system. https://www.grundig.com/ktchnmag/blog/how-to-organise-your-kitchen-waste-system/. Accessed 16 Oct 2022

Grundig (n.d.-c) Sustainable homes that generate more electricity than they consume. https://www.grundig.com/ktchnmag/blog/sustainable-homes-that-generate-more-electricity-than-they-consume/. Accessed 16 Oct 2022

Grundig (n.d.-d) The complete guide to kitchen layouts. https://www.grundig.com/ktchnmag/blog/the-complete-guide-to-kitchen-layouts/. Accessed 16 Oct 2022

Halbe SB, Mark RK (2018a) Why small living and for whom. In: Interior design for small dwellings, Routledge, pp 10–30. https://doi.org/10.4324/9780429506604. Accessed 5 Oct 2022

Halbe SB, Mark RK (2018b) Elements and principles of design for small dwellings. In: Interior design for small dwellings. Routledge, pp 82–106. https://doi.org/10.4324/9780429506604. Accessed 5 Oct 2022

Halbe SB, Mark RK (2018c) Color and light. In: Interior design for small dwellings. Routledge, pp 107–138. https://doi.org/10.4324/9780429506604. Accessed 5 Oct 2022

Halbe SB, Mark RK (2018d) Finding storage space. In: Interior design for small dwellings. Routledge, pp 189–201. https://doi.org/10.4324/9780429506604. Accessed 5 Oct 2022

Hammoud M, Baker W, Scheeren O, Parakh J, Hean CK, Lochhead H, Murray P, Azad MK, Pasquarelli G, Wood A, Hellmuth W, Wimer R, Lubin J, Xu Y, Thapar C, Chang A, Fiala R, Hardwick T, Lumpkin J, Murray W, Halliday P, Parry E (2018) CTBUH 2018 conference special: polycentric cities—the future of vertical urbanism. CTBUH J 4:44–51. https://www.jstor.org/stable/26614165. Accessed 27 Sept 2022

Hampton A (2010) Thermal mass and insulation for temperate climates. Environ Des Guide 1–11. http://www.jstor.org/stable/26150778. Accessed 5 Oct 2022

Hankin A (2022) 5 reasons millennials aren't buying homes. Investopedia. https://www.investopedia.com/news/real-reasons-millennials-arent-buying-homes/. Accessed 13 Oct 2022

Hao JL, Cheng B, Lu W, Xu J, Wang J, Bu W, Guo Z (2020) Carbon emission reduction in prefabrication construction during materialization stage: a BIM-based life-cycle assessment approach. Sci Tot Environ. https://www.sciencedirect.com/science/article/pii/S0048969720313838?via%3Dihub. Accessed 13 Oct 2022

Harris A (2021) U.S. self-storage industry statistics. https://www.sparefoot.com/self-storage/news/1432-self-storage-industry-statistics/. Accessed 15 Oct 2022

Hayes A (2021) Consumerism. https://www.investopedia.com/terms/c/consumerism.asp. Accessed 15 Oct 2022

Hays J, Mason B (2020) 8 Transforming furniture solutions for small spaces. https://www.thespruce.com/transforming-furniture-for-small-spaces-4058276. Accessed 15 Oct 2022

Herzog T, Krippner R, Lang W (2017) Facade construction manual, 2nd edn, DETAIL. https://doi.org/10.11129/9783955533700. Accessed 5 Oct 2022

Hillel D (2005) Water harvesting. In: Encyclopedia of soils in the environment, vol 4. Elsevier/Academic Press, pp 264–270. https://pubs.giss.nasa.gov/abs/hi09000a.html. Accessed 13 Oct 2022

Hoang M, Vandal A (2017) Micro-MACRO living in the global high-rise. CTBUH J 3:32–37. https://www.jstor.org/stable/90020855. Accessed 28 Sept 2022

Hohenadel K (2015) These sleek pop-up rentals designed for broke millennials redefine prefab housing. Slate Magazine. http://www.slate.com/blogs/the_eye/2015/01/23/heijmans_one_are_movable_rentals_for_young_single_person_households.html. Accessed 5 Oct 2022

Honic M, Kovacic I, Rechberger H (2019) IOPscience. IOP Conf Ser Earth Environ Sci. https://iopscience.iop.org/article/10.1088/1755-1315/225/1/012073. Accessed 11 Oct 2022

Hopton ME, White D (2011) A simplified ecological footprint at a regional scale. J Environ Manage. https://www.sciencedirect.com/science/article/pii/S0301479711002556. Accessed 13 Oct 2022

Horrocks T (2013) Come together. Monument 20th anniversary issue, pp 58–63

Howe AS, Ishii I, Yoshida T (1999) Kit-of-parts: a review of object-oriented construction techniques. ISARC Proc. https://www.iaarc.org/publications/proceedings_of_the_16th_isarc/kitofpartsa_review_of_objectoriented_construction_techniques.html. Accessed 13 Oct 2022

Hoyt H (2020) Harvard Joint Center for Housing Studies. https://www.jchs.harvard.edu/sites/default/files/media/imp/harvard_jchs_gramlich_design_and_construction_strategies_multifamily_hoyt_2020_3.pdf. Accessed 13 Oct 2022

Hoyt H, Schuetz J (2020) Thoughtful design can create high-quality affordable multifamily housing. Joint Center for Housing Studies of Harvard University. https://www.jchs.harvard.edu/blog/affordable-housing-doesnt-have-to-look-cheap-inside-or-out. Accessed 23 Sept 2022

HQ Editorial Group (2020) 6 Tips on where to place your wireless router for the best signal/coverage. TP-Link. https://www.tp-link.com/us/blog/87/6-tips-on-where-to-place-your-wireless-router-for-the-best-signal-coverage/. Accessed 13 Oct 2022

Huang HB, Mapes J, Newell J, Wolch J, Seymour M, Reynolds K (2010) The forgotten and the future: reclaiming back alleys for a sustainable city. https://css.umich.edu/publication/forgotten-and-future-reclaiming-back-alleys-sustainable-city. Accessed 28 Sept 2022

Huebner GM, Shipworth D (2016) All about size? The potential of downsizing in reducing energy demand. Appl Energy. https://www.sciencedirect.com/science/article/pii/S0306261916302033. Accessed 13 Oct 2022

Hunter RB (1941) Methods of estimating loads in plumbing systems. https://www.govinfo.gov/content/pkg/GOVPUB-C13-a13c5a2020dd0a7b1df8b91b165745ca/pdf/GOVPUB-C13-a13c5a2020dd0a7b1df8b91b165745ca.pdf. Accessed 13 Oct 2022

Hydro Quebec (n.d.) What is the breakdown of household electricity use? Hydro Quebec. https://www.hydroquebec.com/residential/customer-space/electricity-use/electricity-consumption-by-use.html. Accessed 13 Oct 2022

IDEA (ed) (2019) District heating. International District Energy Association. https://www.districtenergy.org/topics/district-heating. Accessed 13 Oct 2022

IEA (2021) District heating—analysis. IEA. https://www.iea.org/reports/district-heating. Accessed 13 Oct 2022

Illinois University Library (2021) Sustainable product design: sustainable design principles. https://guides.library.illinois.edu/c.php?g=347670&p=2344606. Accessed 23 Sept 2022

International Institute for Sustainable Development (IISD) (2020) Green homes grant: £2 billion for energy efficiency in UK. https://www.iisd.org/sustainable-recovery/news/green-homes-grant-2-billion-for-energy-efficiency-in-uk/. Accessed 23 Sept 2022

International Telecommunication Network (ITU) (n.d.) Internet use. https://www.itu.int/itu-d/reports/statistics/2021/11/15/internet-use/. Accessed 13 Oct 2022

Javed AR, Fahad LG, Farhan AA, Abbas S, Srivastava G, Parizi RM, & Khan MS (2021) Automated cognitive health assessment in smart homes using machine learning. Sustainable Cities and Society, 65, 102572. https://doi.org/10.1016/j.scs.2020.102572. Accessed 13 Oct 2022

Jefferson R (2019) More seniors are embracing technology. But can they use it? UCSD researchers suggest asking them. https://www.forbes.com/sites/robinseatonjefferson/2019/06/28/more-seniors-are-embracing-technology-but-can-they-use-it-ucsd-researchers-suggest-asking-them/. Accessed 15 Oct 2022

Jelle BP (2016) Nano-based thermal insulation for energy-efficient buildings. In: Start-Up Creation. https://www.sciencedirect.com/science/article/pii/B978008100546000008X. Accessed 11 Oct 2022

Jerde K, Minton M, Mather L (2020) 65 home office ideas that will inspire productivity. https://www.architecturaldigest.com/gallery/home-offices-slideshow. Accessed 16 Oct 2022

Ji HW, Koo DD, Kang J-H (2020) Short- and long-term structural characterization of cured-in-place pipe liner with reinforced glass fiber material. Int J Environ Res Public Health 17:2073. https://doi.org/10.3390/ijerph17062073. Accessed 13 Oct 2022

John Desmond Limited (2016) A brief history of the kitchen. https://www.johndesmond.com/blog/design/a-brief-history-of-the-kitchen/#:~:text=The%20origins%20of%20the%20kitchen,were%20hanging%20above%20the%20fire. Accessed 16 Oct 2022

Johnson R, Menzie K (1992) Here are seven of the basic rules for getting your ducts in a row. Baltimore Sun. https://www.baltimoresun.com/news/bs-xpm-1992-07-04-1992186077-story.html. Accessed 13 Oct 2022

Jowitt SM, Mudd GM, Thompson JFH (2020) Future availability of non-renewable metal resources and the influence of environmental, social, and governance conflicts on metal production. Nature News. https://www.nature.com/articles/s43247-020-0011-0. Accessed 11 Oct 2022

Kantamneni S, Chandramouli DK, Chaitanya JSN, Sai AN (2021) A study on modular construction technique. http://www.ijmtst.com/volume7/issue07/48.IJMTST0707087.pdf. Accessed 13 Oct 2022

Kashef M (2009) Sense of community and residential space. Research Gate. https://www.researchgate.net/publication/41668003_Sense_of_Community_and_Residential_Space_Contextualizing_New_Urbanism_within_a_Broader_Theoretical_Framework. Accessed 13 Oct 2022

Kell S (n.d.) 13 clutter statistics that will shock you. https://www.suzykell.com/blog/clutter-stats. Accessed 15 Oct 2022

Khan M, Bilyk O, Ackman M (2021) Update on housing market imbalances and household indebtedness. Bank of Canada. https://www.bankofcanada.ca/2021/04/staff-analytical-note-2021-4/. Accessed 23 Sept 2022

Khatib JM (2016) Sustainability of construction materials, 2nd edn. Elsevier, Amsterdam

King A (2020) Mitigating criticality, part III: improving the stewardship of existing supplies. Crit Mater. https://www.sciencedirect.com/science/article/pii/B9780128187890000074. Accessed 11 Oct 2022

Klocker N, Gibson C, Borger E (2017) The environmental implications of multi-generational living: are larger households also greener households? https://ro.uow.edu.au/sspapers/3182. Accessed 15 Oct 2022

Kral H (2021) 5 main benefits of a smart sprinkler system. https://www.familyhandyman.com/list/benefits-smart-sprinkler-system/. Accessed 13 Oct 2022

Kramer M, Sobel L (2014) Smart growth and economic success: investing in infill development. United States Environmental Protection Agency. https://www.epa.gov/sites/default/files/2014-06/documents/developer-infill-paper-508b.pdf. Accessed 28 Sept 2022

Krangsås SG, Steemers K, Konstantinou T, Soutullo S, Liu M, Gieancola E, Prebreza B, Ashrafian T, Murauskaité L, Maas N (2021) Positive energy districts: identifying challenges and interdependencies. MDPI. https://www.mdpi.com/2071-1050/13/19/10551/htm. Accessed 13 Oct 2022

Kurvinen A, Saari A (2020) Urban housing density and infrastructure costs. MDPI. https://doi.org/10.3390/su12020497. Accessed 28 Sept 2022

Lamnatou C, Chemisana D (2021) Solar thermal systems for sustainable buildings and climate change mitigation: recycling, storage and avoided environmental impacts based on different electricity mixes. Sol Energy. https://www.sciencedirect.com/science/article/pii/S0038092X21009816?via%3Dihub. Accessed 13 Oct 2022

Larrichia F (2022) Consumer spending on smart home related devices worldwide from 2019 to 2025. https://www.statista.com/statistics/873607/worldwide-smart-home-annual-device-sales/. Accessed 13 Oct 2022

Larrimore J, Schuetz J (2017) Assessing the severity of rent burden on low-income families. The Fed. https://www.federalreserve.gov/econres/notes/feds-notes/assessing-the-severity-of-rent-burden-on-low-income-families-20171222.htm. Accessed 28 Sept 2022

Larsen MSS, Lindhard SM, Brunoe TD, Nielsen K, Larsen JK (2019) Mass customization in the house building industry: literature review and research directions. Frontiers. https://doi.org/10.3389/fbuil.2019.00115. Accessed 13 Oct 2022

Lasky J (2018) Two designers create a small but Luxe Retreat in the Catskills. Dwell, April 9, 2018. Accessed 9 Feb 2019. https://www.dwell.com/article/two-designers-create-a-small-but-luxe-retreat-in-the-catskills-953fcd77. Accessed 13 Oct 2022

Lasquety-Reyes J (2021) Number of smart homes forecast in the world until 2025. https://www.statista.com/forecasts/887613/number-of-smart-homes-in-the-smart-home-market-in-the-world. Accessed 13 Oct 2022

Latha H, Arunraj E (2014) Compressive strength of insulated concrete form blocks. Research Gate. https://www.researchgate.net/publication/261798817_COMPRESSIVE_STRENGTH_OF_INSULATED_CONCRETE_FORM_BLOCKS. Accessed 11 Oct 2022

Law Insider (2022) House sewer definition. Law Insider. https://www.lawinsider.com/dictionary/house-sewer. Accessed 13 Oct 2022

Lee H (2021) Are millennials leaving cities? Yes, but young adults are not. Joint Center for Housing Studies. https://www.jchs.harvard.edu/blog/are-millennials-leaving-cities-yes-young-adults-are-not. Accessed 13 Oct 2022

Lee S, Koschinsky J, Talen E (2018) Planning tools for walkable neighborhoods: zoning, land use, and urban form. J Archit Plan Res 35(1):69–88. http://www.jstor.org/stable/45215820. Accessed 27 Sept 2022

Lelieveld J, Klingmuller K, Pozzer A, Burnett RT, Haines A, Ramanathan V (2019) Effects of fossil fuel and total anthropogenic emission removal on public health and climate. Proc Nat Acad Sci USA. https://pubmed.ncbi.nlm.nih.gov/30910976/. Accessed 13 Oct 2022

Lessard G (2018) Principles and best practices. Larpent. https://www.larpent.ca/wp-content/uploads/2019/04/AccessoryDwellingUnitsPrinciplesAndBestPractices.pdf. Accessed 13 Oct 2022

Lewitin J (2021) Cork flooring review: pros and cons. https://www.thespruce.com/cork-flooring-pros-and-cons-1314688. Accessed 16 Oct 2022

Li J, Just RE (2018) Modeling household energy consumption and adoption of energy efficient technology. Energy Econ. https://www.sciencedirect.com/science/article/pii/S0140988318301440?via%3Dihub. Accessed 13 Oct 2022

Li KF (2013) Smart home technology for telemedicine and emergency management. J Ambient Intell Humaniz Comput 4(5):535–546. https://doi.org/10.1007/s12652-012-0129-8. Accessed 13 Oct 2022

Lindsay M, Williams K, Dair C (2010) Is there room for privacy in the compact city? Built Environ (1978-) 36(1):28–46. http://www.jstor.org/stable/23289982. Accessed 27 Sept 2022

Line C (2019) What is a plumbing stack? Clean Line Plumb. https://cleanline.ca/what-is-a-plumbing-stack/. Accessed 13 Oct 2022

Litman T (2022) Understanding smart growth public service cost savings. Victoria Transport Policy Institute. https://www.vtpi.org/sgcp.pdf. Accessed 28 Sept 2022

Livingston A (2022) What is the tiny house movement—plans, resources, pros & cons. https://www.moneycrashers.com/living-tiny-house-movement-plans/. Accessed 15 Oct 2022

Longarini N, Crespi P, Zucca M, Giordano N (2014) (PDF) the use of fly ash in high strength concrete mix design. Research Gate. https://www.researchgate.net/publication/282605450_The_use_of_fly_ash_in_high_strength_concrete_mix_design. Accessed 11 Oct 2022

Macieira M, Mendonca P (2016) (PDF) building rehabilitation with dry and wet systems. Research Gate. https://www.researchgate.net/publication/305787506_Building_Rehabilitation_with_Dry_and_Wet_Systems_-_Embodied_Water_Comparison. Accessed 13 Oct 2022

Magay AA, Bulgakova EA, Zabelina SA (2018) Organizing vertical layout environments: a forward-looking development strategy for high-rise building projects. E3S web of conferences. https://www.e3s-conferences.org/articles/e3sconf/abs/2018/08/e3sconf_hrc2018_01016/e3sconf_hrc2018_01016.html. Accessed 28 Sept 2022

Marckmann B, Gram-Hanssen K, Christensen TH (2012) Sustainable living and co-housing: evidence from a case study of eco-villages. Built Environ (1978-) 38(3):413–429. http://www.jstor.org/stable/23290271. Accessed 27 Sept 2022

Martin's Tree Service Inc. (2020) The difference between coniferous and deciduous trees. Martin's Tree Service Inc. https://www.martinstree.com/blog/difference-coniferous-and-deciduous-trees/. Accessed 13 Oct 2022

Martínez DM, Ebenhack BW, Wagner TP (2019) Residential and commercial sector energy efficiency. In: Energy efficiency: concepts and calculations. https://www.sciencedirect.com/science/article/pii/B9780128121115000081?via%3Dihub. Accessed 13 Oct 2022

Martín-Gómez C, Zuazua-Ros A, Del Valle de Lersundi K, Saiz-Ezquerre BS, Ibáñez-Puy M (2021) Integration development of a ventilated active thermoelectric envelope (VATE): constructive optimization and thermal performance. Energy Build 231:110593. https://doi.org/10.1016/j.enbuild.2020.110593. Accessed 13 Oct 2022

McNeill Lalonde & Associates (2021) Wood vs. concrete: how structural materials are chosen. MLA Canada. https://mlacanada.com/newsfeed/wood-vs-concrete-how-structural-materials-are-chosen#:~:text=Upfront%20cost%20and%20long%2Dterm%20value&text=As%20a%20natural%20resource%20that,more%20expensive%20to%20build%20with. Accessed 28 Sept 2022

Meadows D (1999) Leverage points places to intervene in a system. Sustainability Institute. http://www.donellameadows.org/wp-content/userfiles/Leverage_Points.pdf. Accessed 23 Sept 2022

Medical Guardian (2018) The most common places you're most likely to fall around the home. https://www.medicalguardian.com/medical-alert-blog/senior-safety-/the-most-common-places-youre-most-likely-to-fall-around-the-home. Accessed 15 Oct 2022

Meinhold B (2011) Intexure's architecture office doubles as beautiful eco home. Last updated 14 Jan 2011. Accessed 6 May 2011. http://inhabitat.com/houston-architecture-office-doubles-as-beautiful-eco-home/. Accessed 16 Oct 2022

Mendelsohn H (2021) 27 clever room divider ideas to get the most out of any space. https://www.housebeautiful.com/room-decorating/g27348734/room-divider-ideas/?slide=15. Accessed 15 Oct 2022

Mente J (2021) The history of coworking: how flexible office space became a force in the working world. https://cobaltworkspace.com/the-history-of-coworking/. Accessed 16 Oct 2022

Moe K (2012) Insulating North America. Constr Hist 27:87–106. http://www.jstor.org/stable/44215886. Accessed 5 Oct 2022

Mohammadi-Hamidi S, Beygi Heidarlou H, Fürst C, Nazmfar H (2022) Urban infill development: a strategy for saving peri-urban areas in developing countries (the case study of Ardabil, Iran). MDPI. https://www.mdpi.com/2073-445X/11/4/454/htm. Accessed Sept 2022

Monark (2017) The heart of your home: a history of the kitchen. https://monarkhome.com/history-of-the-kitchen/. Accessed 16 Oct 2022

Morel J-C, Charef R, Hamard E, Fabbri A, Beckett C, Bui Q-B (2021) Earth as construction material in the circular economy context: practitioner perspectives on barriers to overcome. The Royal Society Publishing. https://doi.org/10.1098/rstb.2020.0182. Accessed 11 Oct 2022

Moreno M, Lampard C, Williams N, Lago N, Emmett S, Goikoetxea M, Barandiaran MJ (2015) Eco-paints from bio-based fatty acid derivative latexes. Prog Org Coat. https://www.sciencedirect.com/science/article/pii/S0300944015000107. Accessed 11 Oct 2022

Moreno VL, Martín Ruiz ML, Malagón Hernández J, Valero Duboy MÁ, & Lindén M (2017) The role of smart homes in intelligent homecare and healthcare environments. In: Dobre C, Mavromoustakis C, Garcia N, Goleva R, Mastorakis G (eds) Ambient assisted living and enhanced living environments. Butterworth-Heinemann, pp 345–394. https://doi.org/10.1016/B978-0-12-805195-5.00014-4. Accessed 13 Oct 2022

Murphy J, Harshaw J (2011) Understanding chilled beam systems. 38. https://www.trane.com/content/dam/Trane/Commercial/global/products-systems/education-training/engineers-newsletters/airside-design/adm_apn034en_1209.pdf. Accessed 13 Oct 2022

Murray A, Skene K, Haynes K (2015) The circular economy: an interdisciplinary. JSTOR. https://www.jstor.org/stable/44164300. Accessed 11 Oct 2022

Murray C, Schuetz J (2019) Is California's apartment market broken? Brookings. https://www.brookings.edu/wp-content/uploads/2019/07/20190711_metro_Is-California-Apartment-Market-Broken-Schuetz-Murray.pdf. Accessed 28 Sept 2022

Nägeli W, Tajeri N, O'Donovan JR (2016) Small interventions: new ways of living in post-war modernism. Birkhäuser, part of Walther de Gruyter GmbH. Retrieved May 28, 2022, from https://mcgill.on.worldcat.org/oclc/1061074982. Accessed 5 Oct 2022

National Geographic (2011) Xeriscaping. http://www.nationalgeographic.org/encyclopedia/xeriscaping/. Accessed 13 Oct 2022

National Institute on Aging (NIH) (2017) Aging in place: growing older at home. https://www.nia.nih.gov/health/aging-place-growing-older-home#support. Accessed Oct 15 2022

National Post Staff (2022) A standard house costs almost twice as much in Canada as it does in the U.S. National Post. https://nationalpost.com/news/a-standard-house-costs-almost-twice-as-much-in-canada-as-it-does-in-the-u-s. Accessed 5 Oct 2022

Natural Resources Canada (NRCAN) (2003) Stand-alone wind energy systems—NRCAN. Wind Energy Systems. https://www.nrcan.gc.ca/sites/www.nrcan.gc.ca/files/canmetenergy/files/pubs/WindEnergy_buyersguide_ENG.pdf. Accessed 13 Oct 2022

Natural Resources Canada (NRCAN) (2019) Government of Canada. Natural Resources Canada. https://www.nrcan.gc.ca/energy/products/categories/heating/heat-pumps/air-source/ductless/21316. Accessed 13 Oct 2022

Natural Resources Canada (NRCAN) (2021) Government of Canada. Natural Resources Canada. https://www.nrcan.gc.ca/energy-efficiency/products/product-information/heating-equipment-for-residential-use/13740. Accessed 13 Oct 2022

NC (n.d.) What is your carbon footprint? The Nature Conservancy. https://www.nature.org/en-us/get-involved/how-to-help/carbon-footprint-calculator/#:~:text=A%20carbon%20footprint%20is%20the,is%20closer%20to%204%20tons. Accessed 13 Oct 2022

Ncube A, Matsika R, Mangori L, Ulgiati S (2021) Moving towards resource efficiency and circular economy in the brick manufacturing sector in Zimbabwe. Science Direct. https://doi.org/10.1016/j.jclepro.2020.125238. Accessed 13 Oct 2022

Nelson A (2018a) Once we were small: traditional and contemporary homes. In: Small is necessary: shared living on a shared planet. Pluto Press, pp 21–43. https://doi.org/10.2307/j.ctt1zk0mpz.8. Accessed 5 Oct 2022

Nelson A (2018b) Apartment living in cities. In: Small is necessary: shared living on a shared planet. Pluto Press, pp 44–71. https://doi.org/10.2307/j.ctt1zk0mpz.9. Accessed 5 Oct 2022

New PT (NPT) (2019) NPT—new polyureathane technologies—sealing and bonding In: Adhesives & sealants for wood flooring and parquetry. https://www.nptsrl.com/wp-content/uploads/2020/06/CATALOGO-PARQUET-NPT-2019-ENG-mod.pdf. Accessed 11 Oct 2022

Ng L, Poppendieck D, Dols WS, Dougherty BP, Emmerich SJ (2018) Balancing energy and IAQ: NIST net-zero energy residential test facility. ASHRAE J. https://www.ncbi.nlm.nih.gov/pmc/articles/PMC6508590/. Accessed 13 Oct 2022

Nystul J (2022) 13 brilliant things you can do with a glass jar. https://www.onegoodthingbyjillee.com/13-ways-to-use-glass-jars/. Accessed 16 Oct 2022

Obrinsky M, Walter C (2016) Energy efficiency in multifamily rental homes: an analysis of residential energy consumption data. J Sustain Real Estate 8(1):2–19. https://www.jstor.org/stable/24876479. Accessed 27 Sept 2022

O'Ceallaigh C, Conway M, Mehra S, Harte AM (2021) Numerical investigation of reinforcement of timber elements in compression perpendicular to the grain using densified wood dowels. Constr Build Mater. https://www.sciencedirect.com/science/article/pii/S0950061821007509. Accessed 11 Oct 2022

Office of Mobile Design Corp Eof (2021) Nomad live & work-electric-truck. OMD. http://www.designmobile.com/#/nomad-live-work-electric-truck/. Accessed 13 Oct 2022

Ohio University (2020) 3D printed buildings guide: required materials, tips, and resources. https://onlinemasters.ohio.edu/blog/3d-printed-buildings-guide/. Accessed 23 Sept 2022

Oka M (2011) Toward designing an environment to promote physical activity. Landsc J 30(2):280–298. http://www.jstor.org/stable/43324379. Accessed 27 Sept 2022

Oreskovic NM, Roth P, Charles SL, Tsigaridi D, Shepherd K, Nelson KP, Bar M (2014) Attributes of form in the built environment that influence perceived walkability. J Archit Plann Res 31(3):218–232. http://www.jstor.org/stable/44114605. Accessed 5 Oct 2022

Organisation for Economic Co-operation and Development (OECD) (2019) OECD work on green growth. https://issuu.com/oecd.publishing/docs/gg_brochure_2019_web. Accessed 23 Sept 2022

Paay J, Kjeldskov J, Skov MB (2015) Connecting in the kitchen: an empirical study of physical interactions while cooking together at home. Proceedings of the 18th ACM conference on computer supported cooperative work & social computing, pp 276–287. https://doi.org/10.1145/2675133.2675194. Accessed 16 Oct 2022

Pal RK, Goyal P, Sehgal S (2021) Effect of cellulose fibre based insulation on thermal performance of buildings. Mater Today Proc. https://www.sciencedirect.com/science/article/pii/S2214785321019027?via%3Dihub. Accessed 11 Oct 2022

People's Architecture Office Eof (2013) Courtyard house plugin. People's Architecture Office. http://www.peoples-architecture.com/pao/en/project-detail/3. Accessed 13 Oct 2022

People's Architecture Office Eof (2018) Shangwei urban village plugin houses. People's Architecture Office. http://www.peoples-architecture.com/pao/en/project-detail/27. Accessed 13 Oct 2022

People's Architecture Office Eof (2022) Our vision. Plugin House. https://www.paopluginhouse.com/our-vision. Accessed 13 Oct 2022

Plumer B, Popovich N (2020) Traffic and pollution plummet as U.S. cities shut down for coronavirus. https://www.nytimes.com/interactive/2020/03/22/climate/coronavirus-usa-traffic.html. Accessed 16 Oct 2022

Polat Z (2021) The identity and hierarchy of urban parks: planning to design and management. https://www.researchgate.net/publication/348716388_THE_IDENTITY_AND_HIERARCHY_OF_URBAN_PARKS_PLANNING_TO_DESIGN_AND_MANAGEMENT. Accessed 28 Sept 2022

Pradhan S, Al-Ghamdi SG, Mackey HR (2019) Greywater recycling in buildings using living walls and green roofs: a review of the applicability and challenges. Sci Total Environ 652:330–344. https://doi.org/10.1016/j.scitotenv.2018.10.226. Accessed 13 Oct 2022

Price M (n.d.) Aging in place? Here's 8 kitchen tools to make cooking easy. https://www.rehabmart.com/post/aging-in-place-heres-8-kitchen-tools-to-make-cooking-easy. Accessed 16 Oct 2022

Raman S (2010) Designing a liveable compact city: physical forms of city and social life in urban neighbourhoods. Built Environ (1978-), 36(1), 63–80. http://www.jstor.org/stable/23289984. Accessed 27 Sept 2022

Ramenah H, Tanougast C (2016) Reliably model of microwind power energy output under real conditions in France suburban area. Renew Energ. https://www.sciencedirect.com/science/article/pii/S0960148115304377. Accessed 13 Oct 2022

Rardin N (2015) Forced air heating system guide. HVAC solutions. https://www.hvac.com/resources/what-are-forced-air-systems-here-are-3-things-you-should-know/. Accessed 13 Oct 2022

Ray E (2014) Living on air. The Age, Friday, 13 Sept 2014, p 18

Reynolds WB, Bibby A (n.d.) The complete history of working from home. https://www.flexjobs.com/blog/post/complete-history-of-working-from-home/#:~:text=More%20employees%20working%20remotely%20than,they%20grew%20to%20over%2059%25. Accessed 16 Oct 2022

Richardson S (2022) Multigenerational homes: a buyer's guide. https://www.rockethomes.com/blog/home-buying/multigenerational-homes. Accessed 15 Oct 2022

Richman RC, Cianfrone C, Pressnail KD (2010) More sustainable masonry façades: preheating ventilation air using a dynamic buffer zone. J Build Phys 34(1):27–41. https://mcgill.on.worldcat.org/oclc/646281834. Accessed 5 Oct 2022

Roberts G (2021) 10 tips for designing your home office. https://www.hgtv.com/design/rooms/other-rooms/10-tips-for-designing-your-home-office. Accessed 16 Oct 2022

Roberts T (2020) Designing a sustainable kitchen. https://www.buildwithrise.com/stories/sustainable-kitchen. Accessed 16 Oct 2022

Rosso F, Pisello AL, Pigliautile I, Cavalaglio G, Coccia V (2020) Natural, bio-based, colored linoleum: design, preparation, characteristics and preliminary life cycle assessment. J Clean Prod. https://www.sciencedirect.com/science/article/abs/pii/S0959652620322496. Accessed 11 Oct 2022

Rubeiz C (2004) Benefits of PEX cross linked polyethylene for plumbing pipe. Green Building Solutions. https://www.greenbuildingsolutions.org/blog/flexing-pex-plumbing-possibilities-cross-linked-polyethylene-pipes/. Accessed 13 Oct 2022

Rukikaire K, Collins T (2020) Building sector emissions hit record high, but low-carbon pandemic recovery can help transform sector—UN report. UN Environment. https://www.unep.org/news-and-stories/press-release/building-sector-emissions-hit-record-high-low-carbon-pandemic. Accessed 13 Oct 2022

Saieh N (2009) Villa Deys/Paul de Ruiter Architects. https://www.archdaily.com/17627/villa-deys-paul-de-ruiter. Accessed 15 Oct 2022

Sarbu I, Sebarchievici C (2017) Solar heating and cooling systems || Solar water and space-heating systems. https://www.sciencedirect.com/science/article/pii/B9780128116623000050?via%3Dihub. Accessed 13 Oct 2022

Saroglou T, Meir IA, Theodosiou T (2020) Façades: improving the energy efficiency of a Mediterranean high-rise envelope. CTBUH J 2:20–27. https://www.jstor.org/stable/27034277. Accessed 5 Oct 2022

Saxton M (2019) Who is buying tiny homes? https://www.buildwithrise.com/stories/who-is-actually-buying-tiny-homes. Accessed 15 Oct 2022

Schmidt R III, Austin S (2016) Adaptable architecture theory and practice, 1st edn. Routledge. https://doi.org/10.4324/9781315722931. Accessed 15 Oct 2022

Seyam S (2018) Types of HVAC systems. In: Kandelousi MS (ed) HVAC system. In Tech. http://www.intechopen.com/books/hvac-system/types-of-hvac-systems. Accessed 13 Oct 2022

Shuhaiber A, Mashal I (2019) Understanding users' acceptance of smart homes. Technol Soc 58:101110. https://doi.org/10.1016/j.techsoc.2019.01.003. Accessed 13 Oct 2022

Simpson A (2021) The effect of moisture on the thermal property of a reed thatch roof during the UK heating season. Energy Build. https://www.sciencedirect.com/science/article/abs/pii/S0378778821010616. Accessed 11 Oct 2022

Sims A, Ulloa G (2020) 20 genius storage ideas for small spaces. https://www.architecturaldigest.com/story/storage-ideas-for-small-spaces. Accessed 15 Oct 2022

Snow M, Prasad D (2011) Building-integrated photovoltaics (BIPV). Environ Des Guide 1–15. http://www.jstor.org/stable/26151891. Accessed 5 Oct 2022

Soo VK, Peeters J, Paraskevas D, Compston P, Doolan M, Dufleu JR (2017) Sustainable aluminium recycling of end-of-life products: a joining techniques perspective. J Clean Prod. https://www.sciencedirect.com/science/article/abs/pii/S0959652617332298. Accessed 11 Oct 2022

Sotayo A, Bradley D, Bather M, Sareh P, Oudjene M, El-Houjeyri I, Harte A M, Mehra S, O'Ceallaigh C, Haller P, Namari S, Makradi A, Belouettar S, Bouhala L, Deneufbourg F, Guan Z (2019) Review of state of the art of dowel laminated timber members and densified wood materials as sustainable engineered wood products for construction and building applications. In: Developments in the built environment. https://www.sciencedirect.com/science/article/pii/S2666165919300043. Accessed 11 Oct 2022

Southworth M, Ben-Joseph E (2017) Reconsidering the Cul-de-Sac. Real Estate Wharton Business School. https://realestate.wharton.upenn.edu/wp-content/uploads/2017/03/513.pdf. Accessed 28 Sept 2022

SpaceWise (2021) Upgrade your kitchen with these 17 sustainable ideas. https://www.extraspace.com/blog/home-organization/diy-projects/how-to-design-an-eco-friendly-kitchen/. Accessed 16 Oct 2022

SpaceWise (2022) 17 creative under stairs storage ideas you need to try. https://www.extraspace.com/blog/home-organization/diy-projects/creative-under-stairs-storage-ideas/. Accessed 15 Oct 2022

Stahel WR (2016) The circular economy. Nature. https://pubmed.ncbi.nlm.nih.gov/27008952/. Accessed 30 Jan 2022

Stanley J (2021) 41 genius small kitchen storage and organization ideas. https://www.familyhandyman.com/list/40-kitchen-organizing-ideas-that-will-save-your-sanity/. Accessed 16 Oct 2022

States K, Mcquaid J (2022) Hand & portable power tools. Office of Environmental Health and Safety. https://ehs.princeton.edu/workplace-construction/workplace-safety/hand-portable-power-tools. Accessed 13 Oct 2022

Statista (2021) Number of coworking spaces worldwide from 2018 to 2020 with a forecast to 2024. https://www.statista.com/statistics/554273/number-of-coworking-spaces-worldwide/. Accessed 16 Oct 2022

Statistica (2021a) Average number of people per family in Canada from 2000 to 2019. https://www.statista.com/statistics/478948/average-family-size-in-canada/. Accessed 23 Sept 2022

Statistica (2021b) Affordability of single-family detached homes in Canada 2nd quarter 2021b, by market. https://www.statista.com/statistics/720848/affordability-of-single-family-detached-homes-by-market-canada/. Accessed 23 Sept 2022

Statistica (2021c) Global housing market—statistics & facts. https://www.statista.com/topics/5466/global-housing-market/#dossierKeyfigures. Accessed 23 Sept 2022

Steiner F, Simmons M, Gallagher M, Ranganathan J, Robertson C (2013) The ecological imperative for environmental design and planning. Front Ecol Environ 11(7):355–361. http://www.jstor.org/stable/43187631. Accessed 27 Sept 2022

Stephan A, Crawford RH (2016) The relationship between house size and life cycle energy demand: implications for energy efficiency regulations for buildings. Energy. https://www.sciencedirect.com/science/article/pii/S0360544216314724?via%3Dihub. Accessed 13 Oct 2022

Sterley M, Serrano E, Källander B (2021) Building and construction: timber engineering and wood-based products. In: Adhesive bonding, 2nd edn. https://www.sciencedirect.com/science/article/pii/B9780128199541000216. Accessed 11 Oct 2022

Stinson L (2018) Prefab 'plug-in houses' help revitalize dilapidated buildings. Curbed. https://archive.curbed.com/2018/12/18/18145769/prefab-plugin-houses-help-revitalize-dilapidated-buildings. Accessed 13 Oct 2022

Suzuki N, Nakayama Y, Nakaoka H, Takaguchi K, Tsumura K, Hanazato M, Hayashi T, Mori C (2021) Risk factors for the onset of sick building syndrome: a cross-sectional survey of housing and health in Japan. Build Environ. https://www.sciencedirect.com/science/article/pii/S0360132321003802. Accessed 11 Oct 2022

Tabbitt S (2021) The technology that is driving the future of smart mobility. https://www.smartcitiesworld.net/special-reports/special-reports/the-technology-that-is-driving-the-future-of-smart-mobility. Accessed 13 Oct 2022

Tang J, Galbraith N, Truong J (2019) Living alone in Canada. Statistics Canada. https://www150.statcan.gc.ca/n1/pub/75-006-x/2019001/article/00003-eng.htm. Accessed 5 Oct 2022

Tang W, Khavarian M, Yousefi A, Cui H (2020) Properties of self-compacting concrete with recycled concrete aggregates. In: Self-compacting concrete: materials, properties and applications. https://www.sciencedirect.com/science/article/pii/B978012817369500009X. Accessed 11 Oct 2022

TechPrefab (2021) Innovative construction: panelized vs modular. Tech Prefab. https://www.techprefab.com/post/panelized-vs-modular. Accessed 13 Oct 2022

Tetteh N, Amponsah O (2020) Sustainable adoption of smart homes from the Sub-Saharan African perspective. Sustain Cities Soc 63:102434. https://doi.org/10.1016/j.scs.2020.102434

The Federal Highway Administration (2006) Neo-traditional neighborhood design. The Federal Highway Administration. https://safety.fhwa.dot.gov/PED_BIKE/univcourse/pdf/swless16.pdf. Accessed 28 Sept 2022

The Good Trade (n.d.) 9 sustainable furniture brands for a stylish home in 2022. https://www.thegoodtrade.com/features/eco-friendly-furniture-brands-for-a-stylish-and-conscious-home. Accessed 15 Oct 2022

The Tiny Life (n.d.) What is the tiny house movement? https://thetinylife.com/what-is-the-tiny-house-movement/. Accessed 15 Oct 2022

Thongchareon N, Khongtong S, Srivaro S, Wisadsatorn S, Chub-Uppakarn T, Chaowana P (2021) Development of structural insulated panels made from wood-composite boards and natural rubber foam. Polymers. https://pubmed.ncbi.nlm.nih.gov/34372099/. Accessed 11 Oct 2022

Tighe D (2020) Share of consumers that shopped online for the first time since the coronavirus pandemic was declared in 2020, by country. https://www.statista.com/statistics/1192388/first-time-online-shoppers-since-covid-19/. Accessed 13 Oct 2022

Tiseo I (2021a) Most expensive prices of tap water in selected cities worldwide in 2021a. https://www.statista.com/statistics/478870/leading-cities-by-highest-freshwater-prices/. Accessed 13 Oct 2022

Tiseo I (2021b) Plastic waste in the UK—statistics & facts. https://www.statista.com/topics/4918/plastic-waste-in-the-united-kingdom-uk/#topicHeader__wrapper. Accessed 15 Oct 2022

Triforce (2015) A little bit of Joist history…. TRIFORCE® open joist. https://www.openjoisttriforce.com/a-little-bit-of-joist-history/. Accessed 13 Oct 2022

Triforce (2020) A prefab home manufacturer's quest for the ultimate floor joist leads it to TRIFORCE®. Phoenix Build. https://phoenixbuilding.ca/wp-content/uploads/2020/06/triforce-case-study-maisons-bonneville.pdf. Accessed 13 Oct 2022

Trubka R, Newman P, Bilsborough D (2010) The costs of urban sprawl—predicting transport greenhouse gases from urban form parameters. Environ Des Guide 1–16. http://www.jstor.org/stable/26150802. Accessed 27 Sept 2022

Turner L (2017) A roof over your head: choosing the right roofing materials. ReNew Technol Sustain Fut 138:52–60. https://www.jstor.org/stable/renetechsustfutu.138.52. Accessed 27 Sept 2022

U.S. Green Building Council (USGBC) (n.d.) Value of LEED. https://www.usgbc.org/leed/Why-leed. Accessed 23 Sept 2022

Union of Concerned Scientists (2016) What's driving deforestation? Union of Concerned Scientists. https://www.ucsusa.org/resources/whats-driving-deforestation. Accessed 11 Oct 2022

United Nations Climate Change (2021) COP26 explained. https://ukcop26.org/uk-presidency/what-isa-cop/. Accessed 28 Sept 2022

United Nations Habitat (2020) World cities report 2020 the value of sustainable urbanization. https://unhabitat.org/World%20Cities%20Report%202020. Accessed 23 Sept 2022.

United Nations Habitat Innovation (2021) Current projects, initiatives and partnerships. https://unhabitat.org/sites/default/files/2021/01/innovation_unit_brochure_final_1.pdf. Accessed 23 Sept 2022

United States Environmental Protection Agency (EPA) (2016) WaterSense labeled controllers. https://www.epa.gov/watersense/watersense-labeled-controllers. Accessed 13 Oct 2022

United States Environmental Protection Agency (EPA) (2017) Shower better. https://www.epa.gov/watersense/shower-better. Accessed 13 Oct 2022

University of Alberta Office of Sustainability (2021) What is sustainability? https://www.mcgill.ca/sustainability/files/sustainability/what-is-sustainability.pdf. Accessed 23 Sept 2022

Urban Cultivator (n.d.) Eat better food at home. https://www.urbancultivator.net/kitchen-cultivator/. Accessed 16 Oct 2022

US Department of Energy (2010) Guide to home water efficiency. https://www.energy.gov/sites/prod/files/guide_to_home_water_efficiency.pdf. Accessed 13 Oct 2022

US Department of Energy (2022) Water-efficient technology opportunity: rainwater harvesting systems. Energy.gov. https://www.energy.gov/eere/femp/water-efficient-technology-opportunity-rainwater-harvesting-systems. Accessed 13 Oct 2022

Vachon M (2018) The ever-shrinking condo. Can J Urban Res 27(2):37–50. https://www.jstor.org/stable/26542035. Accessed 5 Oct 2022

Velenturf APM, Archer SA, Gomes HI, Christgen B, Lag-Brotons AJ (2019) Circular economy and the matter of integrated resources. Sci Tot Environ. https://www.sciencedirect.com/science/article/pii/S0048969719330220. Accessed 11 Oct 2022

Verizon (2022) What is Wi-Fi? | Definition, meaning & explanation | Verizon Fios. https://www.verizon.com/info/definitions/wifi/. Accessed 13 Oct 2022

Vermont Journal of Environmental Law (2014) Vol. 14, no. 4, Summer 2013 of Vermont Journal. Jstor.org. https://www.jstor.org/stable/vermjenvilaw.14.issue-4. Accessed 11 Oct 2022

Vertical PR (2019) Common plumbing challenges in multi-story buildings. Vert Pipe Repair. https://www.verticalcipp.com/news/2019/7/17/common-plumbing-challenges-in-multi-story-buildings. Accessed 17 Apr 2022

Vigliarolo B (2020) 5 ways to make your home office eco-friendly. https://www.techrepublic.com/article/5-ways-to-make-your-home-office-eco-friendly/. Accessed 16 Oct 2022

von See A (2021) Smart home security market value worldwide 2018–2025. https://www.statista.com/statistics/1056057/worldwide-smart-home-security-market-value/. Accessed 13 Oct 2022

Wang L (2018) Danish home champions wood over concrete for lower carbon emissions. Inhabitat. https://inhabitat.com/danish-home-champions-wood-over-concrete-for-lower-carbon-emissions/. Accessed 11 Oct 2022

Ward A, Wilson A (2009) Design for adaptation: living in a climate-changing world. Building Green. https://www.buildinggreen.com/feature/design-adaptation-living-climate-changing-world. Accessed 23 Sept 2022

Webb M (2022) Biomimetic building facades demonstrate potential to reduce energy consumption for different building typologies in different climate zones. Clean Technol Environ Pol 24(2):493–518. https://doi.org/10.1007/s10098-021-02183-z. Accessed 5 Oct 2022

Webster M (2022) Wet work definition & meaning. Merriam-Webster. https://www.merriam-webster.com/dictionary/wet%20work. Accessed 13 Oct 2022

Weinreb AR, Rofè Y (2013) Mapping feeling: an approach to the study of emotional response to the built environment and landscape. J Archit Plann Res 30(2):127–145. http://www.jstor.org/stable/43031085. Accessed 5 Oct 2022

Wells NM, Evans GW, Yang Y (2010) Environments and health: planning decisions as public-health decisions. J Archit Plann Res 27(2):124–143. http://www.jstor.org/stable/43030900. Accessed 27 Sept 2022

White (2014) Hammarby Sjostad. http://www.white.se/en/project/15-hammarby-sjostad. Accessed 13 Oct 2022

Whitelaw F (2014) Bamboo build. Jstor.org. JSTOR201. https://www.jstor.org/stable/sanctuary.28.72. Accessed 11 Oct 2022

Wong MS, Hassell R, Yeo A (2016) Garden city, megacity: rethinking cities for the age of global warming. CTBUH J 4:46–51. http://www.jstor.org/stable/90006409. Accessed 27 Sept 2022

Wood S (2015) The look and feel of a place: character, community, and the compact city. J Archit Plan Res 32(1):23–39. http://www.jstor.org/stable/44113096. Accessed 27 Sept 2022

World Finance (2021) Solving the global housing crisis. World Finance. https://www.worldfinance.com/infrastructure-investment/solving-the-global-housing-crisis. Accessed 28 Sept 2022

World Health Organization (WHO) (2021) Aging and health. https://www.who.int/news-room/fact-sheets/detail/ageing-and-health. Accessed 15 Oct 2022

Wright K (2010) The relationship between house density and built-form energy use. Environ Des Guide 1–8. http://www.jstor.org/stable/26150782. Accessed 27 Sept 2022

Wrike (n.d.-a) Setting up a home office. https://www.wrike.com/remote-work-guide/home-office-setup-equipment/#importance-of-an-effective-wfh-office-setup. Accessed 16 Oct 2022

Wrike (n.d.-b) Remote work statistics. https://www.wrike.com/remote-work-guide/remote-work-statistics/. Accessed 16 Oct 2022

Wu W, Skye HM (2021) Residential net-zero energy buildings: review and perspective. Renew Sustain Energ Rev. https://www.sciencedirect.com/science/article/pii/S1364032121001532?via%3Dihub. Accessed 13 Oct 2022

Yascaribay C (2019) Average kitchen sizes: what are the standard sizes for kitchens? https://marble.com/articles/average-kitchen-sizes. Accessed 16 Oct 2022

Ye YY, Xu P, Mao JC (2014) Selected, peer reviewed papers from the 2014 2nd international conference on renewable energy and environmental technology (REET 2014), August 19–20, 2014, Dalian, China. (2014). Energy performance of external versus internal shading devices in residential buildings. Appl Mech Mater 672–674, 546–549. https://doi.org/10.4028/www.scientific.net/AMM.672-674.546. Accessed 5 Oct 2022

Yuan C, Shan R, Adelia AS, Tablada A, Lau SK, Lau SSY (2019) Effects of vertical farming on natural ventilation of residential buildings. Energ Build 185:316–325. https://doi.org/10.1016/j.enbuild.2018.12.028. Accessed 5 Oct 2022

Yuan X, Deng F, Liu Z (2021) 14th international symposium on computational intelligence and design (ISCID) Hangzhou, China 2021 Dec 11–2021 Dec 12 (2021). 2021 14th international symposium on computational intelligence and design (ISCID). In Design of Chinese smart kitchen based on users' behaviour, essay, IEEE, pp 301–305. https://doi.org/10.1109/ISCID52796.2021.00076. Accessed 5 Oct 2022

Zairul M (2021) The recent trends on prefabricated buildings with circular economy (CE) approach. Clean Eng Technol. https://www.sciencedirect.com/science/article/pii/S2666790821001993. Accessed 13 Oct 2022

Zemella G, Faraguna A (2014) Evolutionary optimisation of façade design: a new approach for the design of building envelopes. Springer. https://doi.org/10.1007/978-1-4471-5652-9. Accessed 5 Oct 2022

Zhang J, Cui Y, Lu H, Lin Z (2021) Prediction of outflow parameters for a shower head. Water Res 202:117436. https://doi.org/10.1016/j.watres.2021.117436. Accessed 13 Oct 2022

Zhou Y (2019) Adaptable prefabricated interiors in urban dwellings. Post-professional Master of Architecture: urban design and housing McGill University, Peter Guo-hua Fu School of Architecture.